IN LETTERS OF FIRE

*Thirteenth Regiment
West Virginia Infantry Volunteers*

*Drawing a Line in the Wild – Lord, the glory of it
1863*

HEIDI C. EADS JONES, M.A.
EDITOR AND COMPILER

ISBN: 978-1-970730-90-6 (hardcover)
        978-1-970730-89-0 (paperback)

Contact the author by mailing correspondence to:

Heidi C. Eads Jones
heidi.eads.jones@gmail.com
10375 Paradise Boulevard
Treasure Island, FL 33706

Published by

Fideli Publishing, Inc.
119 W. Morgan St.
Martinsville, IN 46151

www.FideliPublishing.com

Some Opening Words and Acknowledgments ...

"Well 'Comp,' you and I quit business at Lynchburg, but the boys got into it in the valley under Sheridan. I have seen but very little said in the papers about the 13th Regiment. But it is often said, and it has been my experience, that the bravest and those who did the most, say the least. This may be the case with the 13th West Virginia."

— Fred Ohlinger,
Private, Company D 13th West Virginia Volunteer Infantry Regiment to the "Editor of the Pomeroy, Ohio *Leader*" (dated "Haines City, Florida, March 9, 1913"), "Fred and 'Comp' Get in Touch After Almost 50 Years," Pomeroy, Ohio, *Leader*, No. 342, pp. 127-28, courtesy of Terry Lowery.

How much can be told 158 years later about a regiment of men that 'say the least'? Were they among the bravest as supposed by Fred Ohlinger? Here follows the story of a regiment with tremendous grit and resilience, who in this, the third year of the Civil War were "tried in no common crucible." They looked to themselves and time and again dashed into the enemy or stood their ground in line of battle when others were driven back. They stepped-up, sacrificed, loved freedom and the "Union." It is good to remember that freedom, self-determination and courage in the face of destruction and death are not ordinary. I maintain with General George Crook, who commanded them, that the story of this regiment and indirectly the story of the Army of West Virginia needs to be recalled and told, that it may "be engraved upon the hearts of the American people in letters of fire."

Many people generously shared their time, talent, research, expertise, enthusiasm and encouragement for this project and without them it would never have come to be. Finding out about the "Bloody 13th," who they were and what they did, became pretext for adventure, travel through beautiful country and meeting marvelous people and I would not have missed a moment of it.

I am humbled and supremely grateful for the opportunity and for all of the help received over a period of some 20 years in pressing forward this project. I would like to thank *all* of my family (many now 'beyond the veil'), most especially my husband and son, Bruce and Donald, for going with me to hunt up old camps and battlegrounds in East and West Virginia — rollicking trips in the heat and dust of Virginia summers armed with maps (old and new); James E. Taylor's drawings; and various verbal descriptions of battles and fields and for, as a general rule, affording me, 'lo, these many years,' time and opportunity to collect research and to write. Librarians and archivists from Kansas to the eastern seaboard made available important materials and my gratitude goes out to them, each and all.

Special thanks to Terry Lowery with the Charleston Archives and History Library for all of his help; to Mike Musick at the National Archives in Washington D.C.; Michael Greenburg at the Bethesda Medical History Library; Darl Stephenson; Larry Strayer; Dr. Richard Summers at Carlisle Barracks Army History Library; Richard Hunt; Elijah Myers; and to Robin Surfacee for her expertise and candor in guiding the publication of this volume.

Table of Contents

In Letters of Fire

Major General George Crook addressing a re-union of the Army of West Virginia in 1884 spoke the following:

> It is good that we should revive the feelings of devotion to our great country, its laws, its integrity, its grandeur, which strengthened us in the darkest hour of our terrific struggle, and made us face unflinchingly and without murmur,—cold and hunger, sickness, danger and death, that the Union—One and Indivisible, should not perish from the face of the Earth. [...] You men of West Va., [...] I know your sacrifices, I know of the homes, the wives and children whom the call of your country compelled you to leave. Few soldiers in other armies had the same experiences to endure, or the same sacrifices to make. The Mountains of West Virginia furnished as a class the purest patriots in the army. They fought for love of freedom, for love of country.
>
> The Mountaineer is always free.
>
> The time will come when your simple story of love of country and devotion to principle shall be engraved upon the hearts of the American people in letters of fire [...].

Maj. Gen. George Crook, U.S. Army, Address at the re-union of the Army of West Virginia, at Cumberland, Maryland, September 1884, pp. 1-2.

Heidi C. Eads-Jones, M.A.
Martinsville, Indiana, in the 157th year
since the regiment was mustered in.

Abbreviations

A.A.G.	Assistant Adjutant General
A.C. or C.	Army Corps
A.G.O.	Adjutant General's Office
A.N.V. or A.N. Va.	Army of Northern Virginia
A.O.S.	Assistant Orderly Sergeant
Adj. or Adj. Gen.	Adjutant General
anon.	anonymous
Arch.	Archives
Art.	Artillery
Asst. Adj. Gen.	Assistant Adjutant General
b.	born
Brig. Gen.	Brigadier General
C.H.	Court House
CM	Carded Medical Files, NARA
C.S.A.	Confederate States of America (the Confederacy)
ca.	circa or about
Capt.	Captain
Cav.	Cavalry
cmdg.	commanding
Co.	company or county
Col.	Colonel
Coll.	Collection
comm.	commissioned officer
comp.	compiler
Corpl.	Corporal
cos.	companies
D. W.Va. and Dept. of W. Va.	Department of West Virginia
Dist.	District
Div.	Division
ed.	editor
frag.	fragment
G.O.	General Order
Gen.	General field officer or General Orders
Gov.	Governor

HdQtrs. or Hdqtrs.	Head Quarters or Headquarters
i.e.	that is
HPL	Hayes Presidential Library
IHS	Indiana Historical Society, Indianapolis
illeg.	illegible
Inf. or Inftry.	Infantry
infra.	below
Lieut. Col.	Lieutenant Colonel
Maj. Gen.	Major General
Pub. Lib.	Public Library
M.E.	Methodist Episcopal Church
manu.	manuscript
micro.	microfilm
Misc.	Miscellaneous
Mtn. Dept.	Mountain Department
n.d.	no date
n.p or n.pp.	no page number(s)
NARA	National Archives and Records Administration, Washington D.C.
No.	Number
O.C.	o'clock
O.R.	*Official Records of the War of the Rebellion*
OHC	Ohio Historical Center, Columbus
orig. manu. or orig. pub.	original manuscript or publication
O.S.	Orderly Sergeant
O.V.I.	Ohio Volunteer Infantry
p. or pp.	page or pages
Prov. Mar.	Provost Marshal
pseud.	pseudonym, nom de plum or pen name
Pt.	Part
Pt. Pleasant	Point Pleasant
Q.M.	Quarter Master
Reg. or Regt.	Regiment
Rev.	Reverend
RG	Record Group
Sec.	Secretary
Ser.	Series
Sergt.	Sergeant
S.O.	Special Order
Spec.	Special (as in Special Orders)

Supp. O.R.	Supplement to the *Official Records of the War of the Rebellion*
supra.	above
trans.	transcription
U.S. or U.S.A	United States of America (the Union)
unpub.	unpublished
Uncat.	Uncatalogued
V.V.I.	Virginia Volunteer Infantry
Va.	Virginia
Vet.	Veteran
Vol.	Volume
Vols.	Volunteers
W.Va.	West Virginia
WVSA	West Virginia State Archives, Charleston
WVU	West Virginia University, Morgantown

IN LETTERS OF FIRE

Thirteenth Regiment
West Virginia Infantry Volunteers

Drawing a Line in the Wild – Lord, the glory of it
1863

HEIDI C. EADS JONES, M.A.
EDITOR AND COMPILER

In Letters of Fire

... "the gamest set of fellows" and "the purest patriots in the army"...

In the new year, as in the old, the 13th West Virginia Infantry Regiment continued to garrison, scout, fortify and engage in counter-insurgency operations to preserve the peace and support the Union cause in Western Virginia.

> "During the winter of 1862-63 the regiments [of the Kanawha
> Division] had been separated and stationed at different posts
> from Charleston, its Headquarters, to the outposts beyond
> Gauley Bridge on the several roads leading out from that point,
> and during the Winter the most important of these posts were
> fortified by the regiments occupying them."
>
> — R.B. Wilson, 842 Lincoln Avenue, Toledo, O[hio], "Kanawha
> Division: Its Campaigns. [Part] I-continued," *The National
> Tribune* (Washington, D.C.), February 18, 1897, p. 2.

The year 1863 opened with the armies of the North and South arrayed in opposition. In Tennessee, the Battle of Murfreesborough was being fought. On the hills outside Fredericksburg, Virginia, the Army of the Potomac, commanded by Major General Ambrose Burnside, was held at bay by General R.E. Lee. Vicksburg, Mississippi, would soon be the target for renewed assaults by General Ulysses Grant. Northerners strongly criticized the government at Washington, the generals and the Army for their management of the war and particularly for the terrible defeat at Fredericksburg (December 15, 1862).

At the end of January, Major General Burnside was replaced by General Joseph Hooker as commander of the Army of the Potomac. Burnside was placed in command of the Department of the Ohio to which the 13th Virginia/West Virginia Infantry Volunteers belonged. With the new year, the Emancipation Proclamation went into effect. Pro-Union West Virginians and Union military authorities at Western Virginia outposts were heartened when they learned that on January 5 the last rail of the B&O Railroad was laid completing the reconstruction of the line linking Baltimore and Wheeling. On January 6, trains began running.[1]

By February 1863, members of the third Wheeling Convention had finished drafting the new State constitution. Next to the national conflict, the new State and her amended State constitution (amended to provide for the manumission of slaves held within her borders) were discussed everywhere in Western Virginia. Feelings ran high. Officers from the Provisional State government traveled to the western counties of the future new State to give speeches. The date for referendum on the amended constitution was set for Thursday, March 26, 1863.[2] On February 1, President Lincoln issued orders for the draft of 500,000 men to serve three years or "during the war."

January found General Jacob D. Cox's force in the Kanawha Valley numbering just under 8,000 present for duty.[3] According to Cox, every man was needed and could not be spared for use in other fields. In the new year, as in the old, the army of the Kanawha, comprised largely of soldiers recruited from West Virginia and Ohio, longed to go on an all-out offensive to once and for all drive the Confederates from the counties west of the Allegheny Mountains. At the departmental level, however (i.e., the War Department at Washington and the Department of the Ohio, at Cincinnati, to which military department of Western Virginia belonged), authorities wished only to 'hold' West Virginia with a minimum of men and to syphon off what was considered 'surplus' soldiery (anything above 6,000 men) to other larger theaters of war.

During the first week of January, Brigadier General Eliakim P. Scammon, at Fayetteville, commanding the 2nd Kanawha Division, contemplated a joint expedition with Brigadier General George Crook to cut the Virginia and Tennessee Railroad.[4] While this was an excellent plan and indeed, was successfully undertaken to advantage the following year, Major Generals Henry W. Halleck and Horatio G. Wright, for the time being, would curtail such proactive operations, as they both wanted to withdraw some of Cox's force for use in other fields. Some troops had already been withdrawn. On December 28, 1862, four veteran regiments: the 30th, 37th, 47th Ohio and 4th Virginia Infantry, under command of Brigadier General Charles Ewing, were ordered from the Kanawha Valley. On January 1, 1863, they were en route for Cincinnati, from there to join General Ulysses Grant's army on the Mississippi River, where they were would be attached to the 15th Army Corps.[5] Cox remonstrated against such detachments from his command countering that nearly or quite as great a force as was already in West Virginia now was needed. Troops should be withdrawn only if there was a pressing exigency.[6] Wright conceded to Cox in this instance but believed the Kanawha Valley could, during the winter months of 1863, be held with less men (6,000 to be sufficient for this purpose), as the danger of raids was considered to be remote in that season.

On January 21, 1863, however, Cox's concerns for the Kanawha Valley were over-ridden when Brigadier General George Crook was ordered to move with four Ohio regiments (the 11th, 36th, 89th, and 92nd Ohio Infantry Regiments) from Western Virginia and to report to Wright at Cincinnati. From there, they were to be sent to re-enforce Major General William Rosecrans' Army of the Cumberland, in Tennessee. In Crook's absence, the command of the Kanawha Division devolved upon General E.P. Scammon.

Confederate troops in the area, perhaps catching wind that the Kanawha Valley would be further depleted of troops, shifted about, seemingly to gather intelligence. This caused a slight delay in the drawing down of Kanawha troops. On the same day that Cox received orders to transfer Crook, Scammon reported that the enemy were sending flags of truce to different points in his front, apparently with the intention of finding the positions and the strength of his forces[7] and that further, he had received reports that the enemy with three regiments of infantry and one regiment of cavalry were at Raleigh. As a result, some slight delay occurred before transferring troops from West Virginia to Tennessee that military authorities might determine the enemy's intentions at Raleigh. By January 22, however, Cox's Aide de Camp Captain S.L. Christie wrote to Brig. Gen. Scammon confirming that the withdrawal of forces would proceed as directed and that:

> [t]he Fifth [West] Virginia will remain at Ceredo; the Thirteenth [West] Virginia, half at Point Pleasant and half at Coalsmouth; at Charleston and Camp Piatt, each half a regiment. You will command the whole Kanawha District.[8]

On January 7, Colonel Rutherford B. Hayes, of the 23rd Ohio Infantry Volunteers[9] was placed in command of the 1st Brigade of the Kanawha Division (also known as 3rd Division Department of West Virginia). The brigade consisted of the 23rd and 89th Ohio Volunteer Infantry and two cavalry units, commanded by Captains George W. Gilmore and [William B.?] Harrison. Hayes moved from Camp Reynolds to Camp White, located across the river from Charleston. After this change of base, his brigade was augmented by the 5th and 13th West Virginia.[10]

The Union army in Western Virginia could now look forward to being locked into a holding pattern occupying the territories which would become the new State of West Virginia. There was no hope entertained of initiating an offensive. The regiments of the Kanawha Division after the transfer of forces to Rosecrans numbered 6,000 strong.[11] They continued to be scattered widely at different outposts. Home Guards provided some military presence in the areas between outposts but the situation offered a tempting opportunity to Confederates to attack in detail. This was particularly so as secret Confederate mail routes kept rebel partisans inside and outside Federal lines apprised of Federal strength, location and movements.[12] A letter from "Grotius," with the 5th Virginia Infantry at Ceredo (dated "Jan. 11, '63") gives some sense of the status of military affairs in this section. He began, "I propose to let them ['friends of the Union'] hear from this very little thought of but very important Department of our Army," and he continued:

> Though there is nothing particularly interesting or important going on at present, it will, perhaps, be gratifying to know even this much. That we are in no immediate danger of attack is also gratifying, at least to us. Since the 5th has been here, their time has been principally employed in scouting the hills of Big Sandy, picketing and foraging. Several prsioners have been captured and sent to Wheeling. On Friday night last, twelve prisoners were shipped for Headquarters on board the Ohio No. 3, in charge of our Adjutant. They belonged to Gens. Jenkins and Floyd's army. [...] We are continually picking up stragglers and men who have ventured home on furlough. The 5th is recruiting slowly [...].[13]

The problem of the previous year had not been resolved (i.e., the lack of mounted forces in the Kanawha Valley to cope with lightning-strike-type assaults made by mounted rebel partisans behind Federal lines), and there was concern that the lack of mounted troops in the Kanawha Valley would permit a replay of the failures and losses of the previous year once active campaigning resumed in Spring. There was talk at this time and, indeed, a positive movement afoot to mount the 13th Virginia (West Virginia) Regiment. The 13th soldiers heartily welcomed the prospect of being mounted. On horseback, they would enjoy a decided jump in advantage in dealing with partisan fighters. Riding was preferable to walking, and the cavalryman enjoyed a prestige that the foot soldier did not. The movement to mount the 13th, however, came to naught. Although they would not be mounted themselves, their lot in life did improve when Captain John S. Witcher's company (Company G 3rd Virginia Cavalry) was attached to the 13th and 5th Virginia Regiments to both support their operations and be supported when necessary by infantry.

The Confederates, for their part, were far from deterred from the idea of giving up West Virginia to the "Pierpont government." General Robert E. Lee still had not given up on regaining the Kanawha Valley with its valuable resources and wanted

Union forces that occupied the Valley driven out. Confederates remained determined to maintain their position along the New River Narrows to hinder any Federal movements against the Virginia & Tennessee Railroad. Still, no great movements were undertaken after General W.W. Loring's forces were driven from the Kanawha Valley the previous autumn. Confederate partisan operations continued, however, in the form of quick lightning raids through Union lines and beyond to break up and capture detachments of Federal military and squads of Home Guards, to disrupt county government, terrorize Union citizens and to obtain horses and other supplies. Partisan warfare in Western Virginia continued to be conducted in disregard of the "usages of civilized warfare" and private citizens, who realized the limitations of the army in this regard, clamored to be granted immunity from the law (civil and criminal laws) when dealing with rebel sympathizers actively furthering the rebel cause as spies and informers.

Kidnap and heinous murders of pro-Union citizens continued. The break up of the "restored government" continued to be a pressing concern for Confederate government and military. Major General Sam Jones, in command of the Confederate Department of Western Virginia, for one, was well aware that Federal forces posted in West Virginia were not concentrated in sufficient numbers to advance south on offensive operations with any force which Jones could not drive back.[14] R.E. Lee, for another, wrote to General John D. Imboden on January 20, conveying his approval of Imboden's idea "of capturing the municipal officers of the Peirpoint government." "It would relieve the people in the northwest very much to catch them all," continued Lee, "and render the office of sheriff as dangerous a position as possible."[15]

Cabell County, where detachments of the 13th would spend considerable time, was subject to rebel incursions to such an extent, that court could not be held there without securing a guard of United States soldiers. On January 3 (or Saturday the 10th), half a dozen rebel cavalrymen rode into Barboursville, the Cabell County seat and carried off the commissioners books (i.e.,the county tax books) from the Cabell County Courthouse. The unknown raiders carried off the strike with impunity as the men of the available force there, though put under arms, were so drunk they didn't care what happened and didn't bother to interrupt the raiders.[16] January 5, Cabell County justices decided to transfer the court to Guyandotte, where protection could be more readily obtained. The strongly secessionist town of Guyandotte had been heavily garrisoned by Union forces since the massacre of Union citizens and soldiers there on November 10, 1861. The Court would remain at Guyandotte until the end of the war.[17]

On January 22, Cox wrote to Scammon, at Fayette Court-House, to confer as to what course to follow in the Kanawha Valley now that their forces had been reduced. Both generals agreed that "the diminution of numbers must be compensated for by extreme activity," the cavalry at Scammon's disposal was potentially to be of "great service." Cox made further recommendations. The imminent reduction of force in the lower valley would "make it needful to keep watch of the country between the Kanawha and Guyandotte Rivers, as there will be efforts made to establish lines of contraband communication there."[18] To keep Scammon's men busy with "plenty to do," Cox also suggested fortification projects, such as

> lining the crest of Cotton Hill with abatis, covering the paths and roads leading to Gauley, by both upper and lower routes, with some defenses in the nature of breastworks, and doing the like at the principal points of the crest of the Loup Creek Hills, where paths pass over the hills, so that if it should become necessary to concentrate at Gauley, that position could be held while provisions should last, even if the enemy reached the river below.[19]

R.B. Wilson, a member of Kanawha Division during its years of service, wrote to the *National Tribune* in 1897 describing the campaign of 1863 in the Kanawha Valley in terms which suggested that it was a time of preparation for the extreme hardships to come during the campaign of 1864:

> While the campaign of 1863 was not distinguished by any great events, it had been a season of additional special training, as events proved, for an important campaign that was to follow, which in some respects was the most unique and remarkable one of the war. The division had become thoroughly acclimated to the mountains. The hard marching of previous campaigns had accustomed its soldiers to mountain climbing under all conditions of weather, at all seasons of the year, and their powers of endurance had been tested in every way. The nature of the warfare carried on in these mountains required movements of the utmost celerity and secrecy, if anything was to be accomplished by them. Many were made at night, when icy streams had to be forded and guides followed through mountain by-paths. Sometimes they lost their way and marched many miles out of their course. It required also that the soldiers should be equipped in the lightest possible marching order. Knapsacks had long since been discarded as a useless burden, ordinary tents were an impossibility, and shelter tents too burdensome. The open sky was their one tent and the bare earth their common couch. A battered tin cup for making coffee and a forked stick for broiling their salt pork was their cooking outfit, which was never delayed by accident to wagontrain. A rubber blanket,

worn as a shoulder sash in good weather and as a capote in bad, was their bedding at night. Thus equipped, they sped away over the mountain roads with a long swinging stride, at a pace over distances unknown to infantry movements of the large armies.[20]

State Agent Colonel Jacob Hornbrook and Colonel J.P.R.B. Smith, cmdg. 106th Mason County Militia, would continue to render important humanitarian services to West Virginia soldiers during 1863, assisting the Governor under the official title of "Aids to the Military Department of the State of West Virginia."[21] Hornbrook especially showed many a kind attention to the wants of sick and wounded soldiers in hospitals and to those in the field. He visited soldiers when they were paid and "with the utmost fidelity and promptness" carried home their money to their families and friends.[22]

The new year opened with problems relating to the sutler of the 13th West Virginia, Lieutenant Samuel Keenan of the 8th Virginia Infantry. Officer John H. Oley, 8th Virginia Infantry (later the 7th West Virginia Cavalry Regiment), wrote to H.J. Samuels, Adjutant General of Virginia, from Wheeling on January 10, 1863, regarding Lieutenant Keenan, who had apparently absented without leave from the 8th Virginia to work as sutler for the 13th. Oley recommended that Keenan (Kenan) be dismissed from the U.S. Service:

> I called once or twice this P.M. to see what you had done about Kenans case—but did not see you—I spoke to the Governor this evening + he told me to leave a note for you about it—
>
> Kenan has been absent from the Regiment without leave since Aug. 12, 1862 + ought to be dismissed at once, as we can not legally fill his place until that is done
>
> He is now Sutler for the 13th Va Regiment—I can have him tried by a Court Martial but it will be much shorter to have dismissed at once—which I wish you would please have done as soon as possible[23]

Still undersized, recruiting for the 13th continued but went along so very slowly that some thought that the skeleton regiments "scattered about" should be consolidated. Private citizens, who had provided food, lodging and other services to support recuits and recruitment for the 13th, applied for reimbursement with the provisional government at Wheeling with more or less success.[24]

Muster rolls indicate how spread out the regiment was at this time. 13th Regiment Field and Staff Muster Rolls for 1863 preserve that for the months of January and February about half of the companies were stationed at Point Pleasant, Mason

County. Company A was stationed at Winfield, Putnam County, a river town situated on the Kanawha. Company G was at "Colesmouth, V[irgini]a."[25] Companies B and D were stationed at Hurricane Bridge (Putnam County). Company B was stationed at Hurricane Bridge until at least April 10. Company H was stationed some sixty miles farther south and west as the crow flies from Winfield.[26]

Morning reports for the 13th indicate that Company A was stationed at Winfield for the entire month of January. The company had present and absent: three commissioned officers (captain, 1st and 2nd lieutenants) and 82 enlisted men. Company A had a drummer but no fifer, no bugler, farrier, blacksmith, artificer and no horses. Terms of "Extra Duty, Special Duty and Detached Service" were detailed as follows: from January 1 to 11, one non-commissioned officer had extra duty; January 1 to 5, one private had extra duty; January 6 to 11, two privates had extra duty; and January 19 to 31, two privates had extra duty on nine occasions. No privates had extra duty on January 25, 28, 29 and 30.

Extra duty assignments for Company A soldiers seemed to predominately involve getting fuel for fires. For the week of January 1 through 5, Thomas Moore and Robert Gray had extra duty digging coal. January 6 through 8, Thomas Moore, Robert Gray and William Gray were digging coal. On January 9, only William Gray had coal-digging duty. January 10 through 12, Robert Gray and William Gray were again at coal digging. January 19 through 24, Richard George and John Green had extra duty chopping wood. The coal diggers and wood-choppers were off duty Sunday, the 25th, but on Monday and Tuesday, January 26 and 27, Richard George and John Green were again detailed for wood chopping. "Remarks" accompanying January morning reports for Co. A note in an entry for January 31 that Richard George and John Green were again "wood chopping," and that they had both been at it for "10 days." Two, three, four and five privates had "daily duty" and groups of one to twelve soldiers had detached duty during the month. Daily duty could include anything related to camp maintenance and hospital. This included cooking, cleaning, building shelters, guard duty, hauling or digging trenches, serving as nurses in regimental hospitals, etc. Returning from detached service were Corporal Andrew J. Davis, James Light and Jim Spradling, who returned to camp from Charleston on January 22, and Lieutenant Slack with ten men, who returned from a scout at 12 o'clock on January 14.[27]

During the month of January 1863, Company A reported a considerable number of men "present sick" in camp and "absent sick," hospitalized away from camp or sick at home. Men recorded as present sick were hospitalized at the regimental field hospital at Winfield. A high of eighteen men were reported present sick and a minimum of ten were absent sick over the course of January. One to six enlisted men were absent sick at any one time with that number declining toward the end of the month.

Entries in regimental reports record that on Friday, January 23, Private George W. Ramsey reported back to camp from Regimental Hospital Point Pleasant; January 28, Sergeant George W. King returned from his home at Charleston, Kanawha County (the record does not positively indicate that he returned from sick leave); January 20, Sergeant Amos C. Brown "returned from home sick;" and Friday the 30th, "A[mos C]. Brown, Nat[h]en Snodgrass, S[amuel] Snodgrass, R[alph] Pauley and G[eorge] W. Ramsy were all ordered to R[egimental] Hospital, Point Pleasant." January 31, Amos Brown, Nathan Snodgrass, Samuel Snodgrass and Ralph Pauley returned to Company A from Regimental Hospital Point Pleasant. Co. A was apparently well-behaved during the month as no one was recorded in arrest or confinement.[28]

January morning reports for Company B provide another piece in the mosaic of army life for soldiers of the 13th. Co. B was still stationed at Camp Brown in Putnam County. The strength of the company held steady for the month at three commissioned officers and ninety-two enlisted men present and absent. January 4 and 5, there was one drummer reported present in camp to drum out the rhythm of camp life. One to five enlisted men were absent with leave; one man being absent for the first two weeks of January. Captain Milton Stewart was detailed on special duty January 14 and 15. Four to seven privates had extra duty during the month. The number of enlisted men on extra duty increased as the month progressed. Three and four privates had daily duty January 1 through 16. Three groups of enlisted men were ordered on detached service during the month: a squad of three men; of five; and the last squad was comprised of six men. Four to eight men were present sick and one musician was present sick for most of the month. Four to fourteen men were absent sick over the course of the month with fewer men absent sick at the beginning and end of the month. One non-commissioned officer was reported in arrest or confinement for the period of January 10 through 20. January 23, Co. B "[r]eceived from Lieut. Col. J.R. Hall 1000 rounds of cartridges."[29]

January reports for Company C indicate that the company was stationed at Point Pleasant for the month. The company had one drummer present January 1 through 31 and one fifer present January 22 through 31. At the beginning of the month, the number of men present and absent was reported as three commissioned officers and eighty enlisted men. One enlisted man had a three week leave of absence beginning January 1. Private James Parsons deserted from camp at Point Pleasant on January 29. He remained absent for the remainder of the month. The company gained one enlisted man, farmer Marcus L. Jones, who was enrolled in the regiment by Captain Van D. McDaniel on January 22, 1863. Jones eventually became one of the regimental musicians. Regimental mathematics may not have been too strong an area, for despite this gain, at the end of the month, the number of men present and absent was still reported as three commissioned and eighty enlisted men.

A military "tightening of the reigns" seems to have been necessary as on January 5 Sergeant John Plant, Sergeant Elijah M. Roseberry and Corporal Andrew B. Shobe were reduced to the ranks for disobeying an order (by order of Colonel W.R. Brown). To fill the vacancies, Corporal William B. McCauley and Private James S. Kelly were promoted Sergeants. Privates William G. Davis and John Patton were promoted to Corporals, all of these promotions ordered on January 9, 1863.

Five privates had extra duty each day of the month. One commissioned officer was detailed for detached duty for the entire month. Company C had eleven to eighteen men present sick including Captain Van D. McDaniel, who was sick from January 1 through 20. Andrew J. Sheppard was reported sick in Post Hospital at Point Pleasant beginning January 1. James L. Woodyard was detailed for nurse in hospital at Point Pleasant January 21. None were absent sick, and no one was present or absent in arrest or confinement.[30]

January morning reports for Company D show that the company was stationed at Camp Brown at Winfield, (West) Virginia. On duty for the month was one drummer and one fifer. No one was absent with leave. Two enlisted men, William Burchfield and Nathaniel Burchfield, were reported deserted at Winfield on January 10,[31] leaving Company D at the end of the month with an aggregate of three commissioned officers and eighty enlisted men present and absent. One to six privates had extra duty every day over the course of the month, and one to four privates had daily duty. For the period of January 1 through 14, two groups from Company D under command of one commissioned officer were detailed for detached service. January 3, Captain Simon Williams and twenty men went on a recruiting scout, and four men and one corporal went to "Tazes [Teays] Valley," south of Winfield, "to attend a funeral." January 12, Lieutenant George Snowden and ten men went on a scout to Hurricane Bridge. Two to nine men were present sick during the month with higher numbers of present sick reported for the last half of the month. Three to five enlisted men were absent sick. No one was present or absent in arrest or confinement.[32]

Company F reports show that the company was stationed at Point Pleasant for the month. One drummer and one fifer were present each day and no one was absent with leave. One enlisted man was reported absent without leave January 18 through 25. Private Henry W. Hays died in 13th Regimental Hospital at Point Pleasant of measles on January 21. At the end of the month, Company F tallied three commissioned officers and seventy-seven enlisted men. One or two non-commissioned officers had extra duty January 1 through 31. Two to four privates had extra duty each day and four privates had daily duty each day. Two enlisted men were reported on detached duty January 13 through 26, and one enlisted man had detached service January 27 through 31. Compared with other 13th Companies, Company F enjoyed fairly good health at this time. None were absent sick and only five to nine privates were present sick over the course of the month. No one was reported present or absent in arrest or confinement.[33]

January morning reports for Company H indicate that the company spent the month stationed at Point Pleasant. A drummer and fifer were on duty each day. One to five enlisted men had leave of absence during the month. One enlisted man was reported absent without leave January 1 through 7, and again, another (or the same) enlisted man was so reported January 29 through 31. January 5, Benjamin F. Kimberling enlisted and joined Co. H on January 8.[34] On January 1, 2nd Lieutenant John B. Bumgardner's resignation was accepted, to date from December 31, 1862, by order of Major General H.G.Wright. January 17, Private William Perdue was promoted to 2nd Lieutenant to rank from December 31, 1862.

There was a loss to the company of two enlisted men, William Reynolds, discharged for disability on January 8, and Isaac West, who died at Post Hospital Point Pleasant, January 24, 1863, "of disease &c." At the end of the month, Company H tallied present and absent three commissioned officers and seventy-six enlisted men. One to three privates had daily duty each day of the month. A whopping eight to twenty-seven privates were reported present sick for the month; none were absent sick; neither was anyone reported present or absent in arrest or confinement.[35]

Morning reports for Company I begin on January 28.[36] These record that the company was stationed at Point Pleasant, January 28 through 31. On January 28, the company, still very small in size, had present a 2nd lieutenant and thirty-six privates. Three enlisted men were reported absent with leave January 28 and 29, and no one was reported absent without leave. One enlisted man, John Young, died of "disease convulsions &c." in Regimental Hospital Point Pleasant on January 25, 1863.[37] The last day(s) of January, James Mathena (Matheny) and Thomas James were each demanded at the "insistence of their fathers" and "discharged for being minors."[38] At the end of the month, Company I had present and absent one commissioned officer and forty enlisted men. On January 24, Sergeant Peter Darnel was detailed for extra duty as drill sergeant by Colonel William R. Brown commanding the regiment. Darnel was "to drill this detachment of the regiment." January 25, John Faidley was detailed on daily duty. No one of Company I had special duty or detached service and none were present or absent in arrest or confinement. January 26, Private John R. Comer was sent to hospital by Captain William E. Feazel (cmdg. Co. I), where he remained through the end of January and perhaps longer. March 3, 1863, Comer died in Regimental Hospital Point Pleasant.[39]

Women of the area selflessly donated foods and other articles in short supply to ease the lot of men in regimental hospitals. A card of thanks was written on behalf of the 13th Regiment by W.W. Harper, Sergeant Major, Co. K, for publication in the local newspaper, the *Weekly Register*. Harper was himself serving unofficially as regimental chaplain for the 13th. Harper gratefully recognized the charitable work and donations made by local ladies' aid societies. Harper wrote for publication in the *Register*:

> We the members of this Regiment, feel it our duty to return our grateful acknowledgements to the ladies that belong to the Aid Societies of Pomeroy and West Columbia, for the kind and patriotic regards shown towards our sick in contributing so bountifully of those items that are so necessary to impart aid and comfort to the aching heads and hearts of our sick and dying soldiers.[40]

Colonel William R. Brown's General Orders Nos. 4, 5, 6 and 7 suggest that another tightening of discipline was necessary 'for the morale' of the regiment. On January 4, 1863, Brown issued General Order No. 4, clarifying who had authority to make arrests and further:

> II Certain acts of insubordination having lately occurred among the men composing this command resulting in gross outrage upon peaceable citizens in this place, it becomes the imperative duty of the commanding Officer for the sake of peace and quiet in this community and for the sake of both military and civil Law which have been insulted and violated and also for the fair reputation which the Regiment has now by hitherto uniform good conduct to see that all persons engaged in such acts do not go unpunished. Insubordination defiance of the rul[e]s and regulations for the government of the Armies of the U[nited] S[tates] if not nipped in the bud and suffered to go *unpunished* must inevitably result in total demoralization and we shall become a disgrace to ourselves and a reproach to our friends.
>
> It is therefore ordered that in future all arrests upon whatever charges shall be made only by the proper authorities and in a proper manner and that all may know who the proper authorities are it is hereby declared that the commanding officer is the only person to judge of the necessity for such arrests and will issue his orders accordingly.
>
> In future all persons whatever officers or men in this command will be held strictly accountable for all illegal + unwarrantable acts committed by them if found guilty will be promptly and severely punished.[41]

On January 5, Colonel Brown fired off General Order No. 5, setting the rhythm of the day and ordering reductions in rank in relation to an insurrectionary incident involving some of the non-commissioned officers of Company C:

> II. Hereafter untill further orders Reveille will be at 6 o'cl[oc]k A.M. First call for Guard mounting, 10 minutes to 7 and 2ⁿᵈ call at 7 o'clock A.M. at which time precisely the Detail from each company will assemble on the Regimental Parade.
>
> II II. Sick call will be at 8 o'cl[oc]k A.M. at which time the Orderly Sergeant of each com[pany] will report all the sick of his company to the Surgeons office or Quarters.
>
> II II II. Sergeant John Plants Co. C Sergeant Elijah M. Rosebery Co. C Sergeant James R. Walkup Co. F Corporal Andrew B. Shobe Co. C having wilfully committed the highest offence known to military law (viz), the violation of the 8ᵗʰ, 9ᵗʰ and 21ˢᵗ Articles of war in this that they did on the night of Jan. 3ʳᵈ 1863 after taps engage in a mutiny or sedition encouraging and abetting the same instead of endeavoring to suppress it and did not give information thereof to the commanding officer that they did on the same night and at the same time + place, when ordered to desist and return to their quarters by the officer of the Guard refused to obey and broke through the Guard absenting themselves from camp without leave from their commanding officer. Therefore it is ordered that they be reduced to the Ranks, and they are hereby reduced. John Plants Elijah M. Rosebery James R. Walkup Andrew B. Shobe and Morrow Crook [William M. Crook] will hereafter do duty as privates in their respective companies.
>
> The commanding officer deeply regrets that so many of his non-commissioned officers have so far forgotten their duty as to render it necessary to punish and hopes this way be the last case of disobedience to use no harsher term which shall come to his notice.[42]

January 8, General Order No. 6 was issued from headquarters at Point Pleasant regarding provision of proper documentation for soldiers absent with authority from the regiment.

> II. In compliance with request contained in Circular from Adjutant Gen[era]l's [office] dated at Wheeling Jan. 1, 1863 and for reasons recited therein as follows The general negligence of company commanders failing to furnish absentees from their com-

panies with proper descriptive lists for any exigency that may arise results in great hardship to the soldiers and individual suffering confusion and trouble in procuring them when needed and great delay in obtaining discharges to sick wounded and discharged soldiers.

It is therefore ordered that in every case of absenteeism from their companies whether caused by sickness disability detail or furlough the soldiers of this Regiment be furnished with proper Descriptive Lists and accounts of pay and clothing by their company officer all failures to do so will be reported to the Col[o]n[el] Com[man]d'g, who will hold such comp[any] Commanders responsible for a direct violation of orders.[43]

G.O. No. 6 may well have been issued to protect men absent by authority from their respective commands. A "Circular" (dated "Wheeling Jan. 15, 1863") issued by Joseph Darr, Jr., Major and Provost Marshal General of the State, receiving wide publication in Virginia and Ohio counties indicated as much. Darr quoted top military authority for citizen arrests (by mayors, sheriffs, police and private citizens) of soldiers absent from their commands. Citing General Orders No. 3, Headquarters Department of the East, New York City, January 12, 1863, issued by command of Major General John E. Wool and General Orders No. 92, War Department, Adjutant General's Office, Washington, July 31, 1862, Darr specifically directed "earnest appeal" to "loyal citizens of Virginia and of counties in the State of Ohio bordering on the Ohio river opposite Virginia, to do everything in their power to apprehend and hand over to the nearest authorities soldiers who are deserters from their regiments or corps."[44] On January 9, 1863, General Order No. 7 was issued from 13th West Virginia regimental headquarters regarding promotions made to fill the reductions in rank ordered in Gen. Order No. 5:

> To fill the vacancy caused by the reduction of the Non-commissioned Officers designated in General Orders No. 5 part Third on recommendation of their company commanders the following nominations are announced in company C viz Corporal William B. McCauly to Sergeant Private James S. Kelly to Sergeant Private James [William] G. Davis to Corporal + Private John B. Patten to Corporal they will be obeyed and respected accordingly.[45]

Captain John V. Young reporting on his company's movements (Co. G) sent the following to Col. Brown. Notice in Young's account how alive with enemy operations was the country around Mud River and Hurricane Creek.

Hurricane Bridge Putnam Co W. Va. 22 J[anuary] [18]63
Wm. Brown
Col Comdg 13th Regt Va. Vol., Infty

Dear sir

I have the honor to report that on the 13th of this month at 4 o clock p.m. the sentinel on the fortification reported that a number of cavalry was approaching from the direction of Winfield I sent two of our cavalry to asertain who they was, I also sent Corp Cash accross the fields, he returned in a few minutes and reported rebels. I rallied my men to the fortification, from here I could see the whole rebel force stretched from Jack Conner to the turnpike below us. Their line of skirmishers advanced three times almost in range of our guns, the last time they advanced in quick time as though they was determined to make the attack I at this ordered the Flag hoisted and it was don[e] in a hurry, At the sight of the Flag the whole line halted paused for a few minits then commenced falling back throwing down fences in evry direction. Here they formed in line of battle in three different squads.

We remained in our position until dark about this time we learned that our cavalry men was captured we could see squads seting on there Horses in evry direction at dark I took my me[n] and took possession of a brushey hill to the south of our Quarters where we could over look the neighborhood here we remained until 10 o clock. From this place we could see that they had built fires in all directions but southward. And here I would say that I have learned since that Co. C. was left behind. It was raining & coald and my men becom verry much chilled, and at 10 oclock I started for Coals mouth we traveled about five miles in the direction of Mud River then taking the divide between Mud & the Hurricane Creek we marched some 12 or 15 miles through an almost trackless wilderness & on our rout we come verry near runing in a squad of Guerrillas numbering about 80 or one hundred, we arrived at Coals Mouth next morning at 8 oclock reported to Lt Col Cromly [Comly] received orders to collect my things to gather and remain at Coals Mouth until further orders which I did

I beg leave to say in this report that I had but 90 men with me Lt. Elkins & 25 men was at Coals Mouth

By order, Sergt Morris was detailed on scout with 8 men.

respectfully
Your obt servt
J V Young
Capt Comdg Post[46]

Capt. Young, cmdg. Co. G at Coalsmouth, continued to be a law unto himself and on January 6, Lieutenant Robert Brook, Young's first officer, wrote to Lieutenant Colonel James. R. Hall, cmdg. Post at Winfield, complaining of Young's treatment of him:

> I am sorry to be under the necesity of informing you that all hope of a plesent connection with Capt Young is gone from me and I am driven to the necessity of seeking Redress.
>
> You know how he done in relation to the <u>Blankets</u> for the sick men causeing me to pay for them. By the request of the Dr I had to B[e]g a Bed. Cup for his the Capts nephew sence he was <u>shot</u> Butr mor Reciently. When the teamster was bringing me a load of <u>coal</u> the Capt wout not allow him to do so and yesterday we got a boat load of coal and when some was brought up I sent <u>Joe</u> to bring some to the Room and Sargent P. Elkin ordered him not to take any Saying that I shout B[e]g my Coal. Thus they seek to do me all the injury they can in a low un manly way. I have born so many ronges that I am chafed in my feelings and in a lawfull way am now determined to have Redress or fall in the attempt. I did believe that the Reckles and disobedient Course pursuit by the Capt would have led to his arrest. But it is not so and his daughter is now hear Riding the <u>Gray Mare</u>, that Capt. Williams took from J. Hill and gave to Mrs. Young. Hill is a quiet citizen and although I do not know what Report was made to you as to the mare I regarded it as a Cool disrespect of Gen Crooks order and I am told that the Capt. has a Colt taken from Nelson that he put in the care of Jos Gary to be wintered for him (the Capt.) So much as 6 or 7 Horses has acording to information been stoped in the hands of Capt Young and a buggy also all this not being investigated Brings the impression that it will perhaps be difficult to do any thing with him But By the Truth I will now stand or fall.

He seems to doubt the extent of your authority and enquired of Col. Hage [Hayes?] yesterday if you ordered him in when he was out on a scout under hyer authority if he was competed to obey your order now some man has the authority and I propose to find the man

I know, I am disgraced by long forbearance Don't deem me harsh, But allow what I feel after every thing being done that he could do to deprive me of my place and pay and after much Misrepresentation and unjust treatment still is doing those little low things. With much impatience I will wait for Relief[47]

Sunday, January 25, J.V. Young wrote to his wife, Paulina, from Winfield referencing the escalation of troop movements in the Kanawha Valley, his willingness to sacrifice all for the Union and an incident between himself and General Crook, who ordered Young to return horses taken from secessionists, as described in Brook's letter above. Young wrote as follows:

The 8th Regiment, Ohio is here and we expect two more Regiments in a few days. There is a great move among the troops in the valley but what is the cause I don't know. I fear that we will be ordered away. I would dislike to go very much but if the government needs me I am willing to go. God knows that I am willing to make any & every sacrifice for our country and our holy religion.[48] And I hope that we will be amply paid for all our sacrifices that we have made for the sake of our country. The time is coming when we and our children will be applauded for our loyalty to the Union.

In regard to General Crook's order to me to return those horses, making me enemies, it has made me many friends. The loyal people of West Virginia are after him with a sharp stick. His order to me was sent to Washington and he is now ordered away from this Valley. I suppose on account of his orders to me. My friends at the Point wrote to general Crooks that Captain Young did the right in taking those horses and if that was the course he allowed to pursue he had better leave the valley and let someone else take the command that would not protect rebels. I tell you he is in a hot place and if he gets out he may be thankful. I will come tomorrow and tell you all about it.[49]

On January 1, new year's day, William W. Harper, Sergeant Major, Company F, wrote two letters from "Headq'rs 13th Reg't Va. Vols., Point Pleasant" to the editors of the Wheeling *Intelligencer.*[50] His letters are alive with enthusiastiasm about President Lincoln's signing of the bill admitting West Virginia to the Union. He writes in his dramatic way of his elation at being "disenthralled from all connection with a haughty rotten Eastern aristocracy."

> How could we but feel joy and gladness thrill through our inmost souls when the glad tidings was received that the iron grasp of Eastern tyrants and rebels was about to be stricken off, and that we were about to realize to its fullest extent that for which we have long been struggling.[51]

It should be remembered that the issue of West Virginia's admittance to the Union as a new State was of enormous importance to her Union soldiers. Though very much in midst of war, the prospect of becoming a "New State" separate from the 'Old Dominion' brought with it a new and powerful optimism and the hope that natural resources (human and otherwise) west of the Alleghenies would at last be developed. It was hoped that admittance to the Union would in future result in increased property values and a more prosperous economy brought about by a host of new of measures.[52] In addition, it was hoped that with separate Statehood internal improvements would be made to better the lives of the people and attract new settlers, as had occurred in States adjacent to West Virginia to the north. These improvements included establishment of a statewide free school system, construction of better roads, railroads canal systems, local insane asylums and penitentiaries. When it was argued that the Restored Government at Wheeling was unconstitutional and illegitimate and the admission of West Virginia to the Union without the consent of the lawful State government at Richmond, the counter-argument offered was that it had been unconstitutional for Virginia to secede from the Union without the consent of her Western counties.

F.H. Pierpont, engineer of the Restored Government of Virginia and her first Governor, wrote as follows to the Wheeling *Intelligencer* shortly after the war, when arguments were in circulation, in Ohio and elsewhere, that the State of West Virginia should be reunited with Virginia. His remarks explain just how important the New State issue was to Union soldiers and to many people living west of the Alleghenies in early 1863, when the issue of new statehood for Western Virginia—not generally favored in Washington—might easily have died:

> [...] the people residing in what is now Virginia, always refused equal political rights to the people who lived in what is now West Virginia; they divided the state into arbitrary geographical divisions, Tidewater, Piedmont, Valley, and Trans Allegheny; West Virginia is Trans-Allegheny except for two or three counties; it

embraces what Washington called West Augusta; it had no representation in the colonial legislature; western Virginians paid taxation for the extensive system of railroads and canals in eastern Virginia; 'These acts made the iron which went into the hearts of the West Virginians;' at the break of the war Eastern Virginia went into secession and invited the armies of the south into her territory; men of the south said the war was transferred to the Potomac and the Ohio; the Knights of the Golden Circle, an organization of democratic secessionists who considered democracy and secession as synonymous, were more numerous in southwestern Pennsylvania, southern Ohio, Indiana and Illinois, than they were in West Virginia; they loved the south because it was democratic; they hated the north because it was Republican; the action of Unionists of West Virginia checked Maryland, Kentucky and Missouri, in their mad cause of rebellion; [... I] talked with Mr. Lincoln four days before his death when he remarked

"I believe I never told you the circumstances under which I signed the bill for the formation of W.Va. into a new State. My cabinet was divided on the subject — Mr. Blair and Mr. Bates were opposed to the bill, Mr. Wells took no interest in it — Mssrs. Seward, Chase, and Staunton were strongly in favor of it. I was undecided, halting over the matter, and the day before the time expired I received your telegram — in which you stated that you greatly desired that I should sign the bill for the formation of West Virginia: That the Union Soldiers in the field had their hearts set on it; that the Union people of the State had also their hearts set on it, and should the bill fail you feared general demoralization of the soldiers and the people. On reading the telegram, I said to myself, There is no Constitutional question in this case — it is purely political. Here is the nation struggling for its existence. It needs all the help it can get. This seems to me to be one step. It strengthens the Union cause. It weakens the enemy. I will sign it."

There were nearly 20,000 soldiers in the field as loyal Virginians, a large majority of them had gone to the field before the requirements were formed; some had gone in squads with only a 2nd Lieutenant to command them; the formation of the State was an achievement of earnest men; it was a step to strengthen the Union and weaken the rebellion; had the men of West Virginia joined the southern side of the fight the war would have gone into

Pennsylvania, Ohio, Indiana, and Illinois, and Congress would not be disturbed this winter by a petition from Ohio to pay for a few houses lost in the Morgan raid; the loss would have been so appalling that it would have bankrupted the nation to attempt to pay it; West Virginians defended their home, their flag, and their country, and so served as a breakwater to southwestern Pennsylvania and Ohio; they restored their State government, carried terror to the Knights of the Golden Circle; they gave backbone to the North and West [...].[53]

The Civil War was not the cause but the occasion for the West's breaking away from Eastern Virginia. It is interesting to note that the equal protection amendments to the U.S. Constitution passed after the war by the U.S. Congress to protect Black Americans were exactly what West Virginans had been demanding from their State government at Richmond in the decades before the war.

Recruiting and the month of January wears on

By mid-January efforts to recruit another company for the 13th Regiment yielded a sufficient number of volunteers to require the services of a mustering officer. On Friday, January 16, Col. William R. Brown wrote to H.J. Samuels, State Adjutant General, requesting that a mustering officer be sent to Point Pleasant to muster in "a detachment of (45) forty five new recruits" raised in Kanawha County by 2nd Lieutenant William E. Feazel. Brown further requested that Lieut. Feazel, who had for some time sought to command a company in the 13th, "be commissioned 1st Lieut. so then he can be mustered with [the] detachment."[54]

13th soldier, Private Mark E. Robison of Company E, wrote home on January 16, 1863, about the anticipated arrival of the paymaster; the hard rains; sickness in Co. E; the rebels; and becoming mounted infantry. He wrote:

> Pt Pleasant Va
> Jan 16, 1863
>
> My Father + Mother
>
> I take the present opportunity to write you a few lines to let you know that I am still in good health + I hope that when thease come to hand they may find you all well and doi[n]g well we have not been paid off yet but we look for the paymaster on in a few days our quartermaster got a letter from wheeling stating that the paymaster was there and would be here by the 18[th] of the month he will be here in a few days at foillett [Fayette]. It is raining very hard here and has bean raining for two days we expect big water for the river is now rising and it is very mudy + slopy

There is twenty eight of the boyes of this company in the hospital sick most of them with the measles some of them is very sick but not dangerous Alax and Bat Clark [Alexander Clark and Squire Clark, both Co. E] both has the measles

You must keep a sharp look out for the rebels for it would not surprise me much if they came back in the spring but they will have to fight more than they did when they came before if we cannot starve them out this winter nor drown them out we have a bat time of it next sumer I expect to hear the bull dogs bark but I think that we are cuting thier railroads + I think that they will be compeled to [saunter ?] out from their strong holds and then we will whip them all to peaces

There is talk of making us mounted infantry of us we will have to tend to horses but we will not have to take long marches + ca[rr]y our own napsack + I think that we would be sure to stay in the vally as it will be done upurpose to get some regt to keep down Bushwhackers + Bur[s or g]er

I should like to come home once more but cannot do so now but I think that we will get furlough when we ar paid off

I must bring my letter to a close by giving my respects to all inquiring friend and my love to you all I remain your son

Truly
Mark E. Robison

To Mr John Robison[55]

Miscellaneous correspondence provides evidence of concerns regarding reimbursements and death in the regiment. On January 17, Lieutenant Colonel James R. Hall wrote to Adj. Gen. Samuels regarding rations furnished to 13th Regiment recruits by Hutchinson McDaniel, who conducted a hotel and restaurant in Point Pleasant. Hall wrote as follows: "Enclosed please find duplicate account of Rations furnished Recruits by H. McDaniel of Point Pleasant Virginia — Please send Draft payable to H. McDaniel."[56] On Monday, January 19, John Bodkin, Private, Co. E, died at Point Pleasant, of "fever." His corpse was "Sent home."[57] January 20, the 13th Regiment drew clothing for the second time. They would next draw clothing on April 30.[58] January 23, Brigadier General George Crook, commanding the Kanawha Division, issued Special Orders No. 17 providing that "The 2nd Regt. Va. Cav. 5th + 13th Regts. Va. Inf. and Simmonds Battery of Light Artillery are hereby ordered to report to Brig. Genl. Scammon for duty."[59] On January 23, Capt. A.F. McCown wrote to Adj. Gen. Samuels as follows: "Enclosed please find duplicate account for transportation of recruits for my Company, the steam boat has brought suit against me personally for the account — You will please get the account approved by the Governor and enclose me a check on Point Pleasant for the amount."[60]

There were losses due to deaths and desertions. On the 23rd, W.J. Nelson, Hospital Steward at General Hospital Point Pleasant, informed 13th Regiment Adjutant William I. Mathews per letter, that "privates Isaac West of Co. 'H,' William Wells and George Slater of Co. 'E,' died at this Hospital last night."[61] Isaac West died of "pleuritic."[62] William H. Wells died of consumption, and George Slater died of erysipelas. The last two named were both buried at Point Pleasant. Saturday, January 24, Captain John V. Young wrote to his wife Paulina, from Winfield, saying that five of his men deserted while he was away at Charleston. "Two Rucers are gone,"[63] wrote Young. On January 25, Private John Young, Company I, died at Post Hospital Point Pleasant, of convulsions. Private Young was supposed to have entered the service about January 1, 1863, "perhaps sooner or later,"[64] but he had apparently not been in the service long enough to have been entered upon company rolls and there was consequently no evidence whatsoever of his enlistment, muster in or service.[65]

About the fourth week of January, John Deriah Carter (John D. Carter), Captain cmdg. Co. E at Point Pleasant, wrote to "a member of the [State] Legislature" regarding the slow pace of recruitment not only of men for the 13th Regiment but for other skeleton regiments in West Virginia whose recruitment seemed to have reached a standstill. He described the slow business of recruiting and the burden this placed upon the Federal Treasury as follows:

> Our regiment gets along very slowly in the recruiting business. I am afraid we will not do much good as a regiment unless we can be filled up in some way. I think that the Skeleton Regiments, now scattered about, with three or four hundred men in them, should be consolidated, and all unnecessary officers mustered out of the service at once. Let those that remain remain because they are fit for the various offices they propose to hold. [...] At the present status of the Virginia Regiments, it takes more money to pay the officers of the various regiments than to pay all the non-commissioned officers and privates. That is not right, and it ought to be stopped and that speedily. [...] Will you not therefore, urge upon the proper authorities the immediate necessity of doing something to fill up these parts of regiments.[66]

Wednesday, January 28, J.V. Young wrote again to his wife from Winfield. "The paymaster has gone to Ceredo to pay the 5th Regt. Va. Vols. and will return this week to pay us, and I think in all probability he will be here between now and Saturday. Nothing new here. The troops are all gone, Gen. Crooks with them."[67] On January 28, Young's command, Co. G, was ordered to Coalsmouth, Kanawha County, to quiet the country there.[68] The mouth of the river and country along the Coal for some ten miles below Charleston was well-adapted for the predatory activity of partisans. The mountains encroached upon the mouth of the river furnishing hiding places and protection for bands to lay in ambush and prey upon squads of Federal soldiers, some merchant on his route, sutler with commissary goods or upon steamboats.

Steamboats had heretofore often stopped upon being hailed from shore to pick up passengers. In such rugged countryside, miscreants could easily make good their escape into the adjacent mountains and hide in comparative security. Because of this state of things, many scouting forays were made into the country along Coal.[69] Young's command would spend the next months posted in this countryside scouting and cleaning out such predators. Meanwhile, he waited for signs of the paymaster. Thursday, January 29, he wrote home from Winfield reporting that the elusive paymaster had still not come.[70]

Payment of unpaid bills incurred by the 13th seems to have been the business of the start of the new year of 1863. January 30, George W. Tippett, proprietor of the Point Pleasant newspaper, the *Weekly Register*, wrote to Adjutant General Samuels regarding an unpaid claim for the printing of posters advertising recruitment of the 13th Regiment and enlistment blanks for the same:

> I have an account against the Commonwealth of Virginia, for printing 'Volunteer Enlistment' blanks, which were ordered to be done by Lieut. Col. James R. Hall of 13th Va., and Lieut. Wm. E. Feazel Recruiting officer, also some Volunteer Posters for Lieut. Feazel — amounting in all to about $31.00. Now I write to you for information as to how I shall proceed to get the above amount due me. Col. Hall told me that he was authorized to certify my account but since the killing of Lewis Wetzel by his father and his armed assault upon myself, he has obstinately refused to do so, I have written him twice but have received no answer in return. Lieut. Feazel is here. He says he will certify the account if you authorize him to do so.[71]

On March 23, 1863, Tippett, again wrote to Wheeling (to Samuels) regarding the payment of this bill, saying:

Yours of the 1st of March has been received and in reply would say your action in respect to my accont [account] is approved by me. I am glad you did it for me, it is true the accont [account] was made about that time (September.) I hope you will try and process the amount of said bill for me as soon as possible, for it is hard to make collections at this time and the $29.50 would come in 'good play' at present. Any time you conveniently throw work into my hands, would like to have you do so cant you persuade some of your city Merchants to Advertise down this way. Anything that you may do in that way will not be forgotton at the proper time by your friend + well wisher Geo W Tippett publisher Register[72]

The *Weekly Register* was a "doubtful enterprise" from the outset. "[T]he proprietor," wrote Tippett, referring to himself, did not make "one dollar for his time" that first year (publication began end of February/beginning March 1862).[73] Printing materials of all kinds, especially paper were very high-priced at this place and time. As the war and straightened times continued, he advertised to buy people's rags from which to make paper on which to print his four-page weekly. Tippett's publishing business was slow, publication of his weekly was a struggle, and Tippet had a large family to support.

Tippett needed to collect on such jobs done on credit as the one he complained of in his letter to Samuels, as he operated too close to the margin and the life of his paper ($1.00 for a year price of subscription) very likely depended on collection of monies owed on print jobs. This particular job, incidentally, seems to have been ordered by Lieut. Col. James R. Hall but for reasons of bad blood resulting from the shooting of Tippett's editor, Lewis Wetzel, in October of the previous year by Hall's father, Tippet could get no money by appeal to Hall the son.

Despite such difficulties, the *Register* continued to play a vigorous role in West Virginia's bid for statehood. During the weeks preceding March 26, 1863, when the people in the Western counties would vote to ratify or reject the anti slavery amendment to the new State constitution, it was the handful of pro-Union newspapers in Western Virginia that played a prominent part in publicizing and supporting the referendum. Besides Tippett's *Weekly Register* and the Wheeling *Daily Intelligencer*, other papers supporting the New State movement were the Wellsburg *Herald*, Fairmont *National*, the Tyler County *Plain Dealer* and the Ritchie *Press*. These newspapers had all given strong support to the Union cause from the start of hostilities and now, with particular passion and devotion, pressed the importance of the upcoming referendum on the New State Constitution.

With time, the *Register* not only persisted but thrived. By October 20, 1864, it could boast that it had the "largest circulation of any paper published in this section of West Virginia and the only one in this Congressional District that is published regularly." ("To Advertisers," Point Pleasant *Weekly Register*, October 20, 1864, [p. 3].)

Saturday, January 31, Demarcus Ward, Private, Company H, wrote to his father from regimental hospital at Point Pleasant:

Point Pleasant
Jan the 31ˢᵗ 1863

Dear Father I take the opportunity of dropping you a few lines to inform that I am in tolerable health I have had the measles and have been in the hospital for some time but I am getting well again and think I will be able for duty in a few days I hope this letter will find you all well and enjoying yourselves finely I think you might write oftener than you do we are expecting to be paid off to day or Monday but I do not know whether we will or not but I hope it will be so there is no news of importance astir here it is all quiet at Point Pleasant as well as on the Potomac and it is the same in Wayne County there is no talk of us leaving hear yet and I supose our Regiment will remain here till Spring and the probability is that we will then be called upon to act in the uper part of the Kanawha Vally since I commenced writing the Pay Master has passed up and has gone up to pay the 8ᵗʰ Va which is at Buckhanon he told our Col that he would be back here Wednesday or Thursday but we have lost all confidence in him so I close my letter and ask you to write to me soon and often give my love to all the Children and Mother and all the friends and Remember your affectionate Son

Demarcus Ward to
William R Ward

The Union for Ever
May The Star spangled
Banner still wave
Oer the Land of the free
And the home of the brave[74]

February

Divided by companies at various posts

As during the month of January, 13th West Virginia headquarters continued to be at Point Pleasant in February. Companies A, B, D and H were transferred to Hurricane Bridge during the course of the month. Young's Company G continued to be stationed at Coals Mouth.[75] On February 2, Col. Brown wrote to Governor F.H. Pierpont recommending J.E. Barret, assistant surgeon of the 23rd Regiment Ohio Volunteer Infantry for the position of surgeon in the 13th. Brown urged that Barret be commissioned immediately, as "the Regiment now number 960 men and is very much in need of an additional medical officer."[76] As in the old year, Col. Brown had occasion to be reminded that he was again late with his reports. General Scammon wrote to Brown regarding his January reports:

> In obedience to General Order No. 2, Head Quarters Dist. W.Va. You will immediately make and forward to these Head Qrs a return of your command. Hereafter, until otherwise ordered, tri-monthly returns will be made on the 7th 17th and 27th days of each month – 'Monthly Reports' on the last day of each month.[77]

Saturday February 7, Col. Brown was sent word that he would be receiving ordnance stores from Allegheny Arsenal in Pennsylvania:

> The Stores designated in the Invoice herewith enclosed were this day turned to the Quartermaster's Department for transportation to your address. On their delivery to you be pleased to forward duplicate receipts therefor to me, at this Arsenal.[78]

At the end of the first week of February, Point Pleasant attorney, W.H. Tomlinson, complained bitterly to Major General J.D. Cox at Marietta, Ohio, that the Point

Pleasant Courthouse had been authorized for use as a hospital by 13th Regimental Surgeon Dr. Samuel G. Shaw, and that highly contagious diseases were being treated there. This, continued Tomlinson, was creating a hazardous situation for the community. On February 20, General Cox informed Colonel Brown of the complaints made against "your Surgeon" for using "the Court House at Pt. Pleasant for hospital purposes," and he ordered Brown to have the building vacated and "thoroughly cleaned" before turning it back to the "charge of the proper civil authorities."[79]

Morning reports show that for the month of February 1863, it was business as usual for the 13th at their various posts, except that the paymaster — whose visit had long been despaired of, had at last visited the regiment and the soldiers received their first pay, this being for the period of their enlistment to December 31, 1862. An interesting article appeared in the Point Pleasant newspaper just prior to the paymaster's visit. In it, the writer levelled complaints at the Provisional State government for still being in arrears to the regiment, and citing as good leadership in this regard, Indiana Governor Oliver P. Morton. Governor Pierpont, stated the writer, should intercede on behalf of the men as had Governor Morton.

> With mingled feelings of shame and indignation we again recur to the fact, that our own 13th Regiment, *have not been paid a cent of their monthly wages.* They were to have had a month's pay in advance![80]

The writer proposed that it was not the government's inability to pay but some "incompetency or rascality" at work here. He continued,

> Here is a quiet, sober, law abiding Regiment, composed of our best citizens, doomed to inglorious guard duty, for five long months, who never saw but are continually taunted by *hearing of,* the paymaster. It's an unmitigated outrage upon the abused, deceived men and a burning shame upon the authorities, we care not who they are.— We remember their promised bounty was by red tapeism long withheld and now their monthly pay is still in arrears. Couldn't Governor Pierpont interpose in behalf of *our* neglected volunteers as Governor Morton does.[81]

The weather around Point Pleasant was disagreeable during the third week of February but this did not curb the uptick in spirits now that the 13th soldiers had been payed. The local newspaper chronicled with the greatest pleasure upon the "startling fact" that "a veritable Paymaster has at last been seen" and congratulated the officers and privates of the 13th upon finally receiving their pay, cautioning them against "improvident use of 'their pile' because if the past is any indication of the future it may be another 6 months before they next see the U.S. Paymaster."[82] 13th soldier Mark E. Robison wrote another letter home on Sunday, February 15. His remarks describe in humorous detail how "the boys" went about enjoying 'their pile.'

Pt Pleasant Va

February 15, 1862 [1863]

Dear Father + Mother

I take my pen in hand to let you know that I am still well and I hope that when theas shall come to hand they may find you all well and doing well I should like to come home to bring my money but can not tell whether I will get to come or not if you get a chance you had better come down and get it I have got fifty dollars and ninty cents I do not like to have that much in camp

Joel is not much better he is going to try to get a sick furlough to come home but cannot tell when he will get to come

we have a high times in camp since we was paid off some of the boys got tight last night but we all right this morning and I think will stay so they say that pay day don't come often and when it does come they mean to make the best of it there was no quarreling and fighting with them we also have great trading with us buying + selling old watches + some of the boys have spent a large part of thier money already

I got a letter from Brother Moses dated the fourth of February he was well He says the weather was very cold and the snow was then fifteen inches deep he was at Buckhanan Upsher county Va

I have nothing more at preasent to write about remember me to all my friends and my love to all of you I remain your son Truly Mark E Robison

To Mr John Robison Sr[83]

Local Southern sympathizers at Point Pleasant had made themselves so brazenly conspicuous—right under the noses of 13th companies C, E and F and officers at regimental headquarters—that in mid-February, the editor of the *Register* warned "the Rebel Congress" that met "daily at a certain Rebel Store House in Point Pleasant" to "dry it up" or they would "be attended to."[84] Concerns were also voiced in the paper regarding the accumulation of mail which had been allowed to collect at Point Pleasant and how this stoppage of the mail could effect soldiers and injure prospects for the new State government. Mail facilities had been curtailed and Point Pleasant was no longer the point at which the mail route ended "thus depriving the people in the Valley and thousands of volunteers of all opportunities of hearing from their friends or transmitting money." "Cart-loads of mail" had accumulated at Point Pleasant and the "Government Wharfboat" there had become the storehouse of mails down the Kanawha, "with none to overhaul or forward."[85]

As the question of the viability of the "New State" hung in the balance both at home and in Washington and a vote upon the issue lay before citizens of Western Virginia, 13th Regiment officer, Sergeant Major William W. Harper (W.W.H.) put pen to paper to give voice to his views on the subject in no uncertain terms. In a letter to the Wheeling *Intelligencer* dated "Point Pleasant, Va., Feb. 7, '63" he wrote the following:

> If you want to forge with your own hands the chains by which you and your children will be forever fettered to a haughty eastern aristocracy. If you want to cast your and your children's destinies in with a negro confederacy, ruled by despots and tyrants. Then vote against a new State.
>
> If you want to make West Virginia, the bloody battle ground on which contending armies shall meet in deadly strife to decide this fierce contest, and which would spread desolation and ruin all around. Then vote against a new State.
>
> If you want to have our own beautiful Ohio, made crimson by the blood shed by constant murder, perpetrated by antagonistic forces, stationed on either shore. Then vote against the new State.
>
> If you want to see your cities, towns and villages, all along the border, becoming depopulated and crumbling to ruins. Then vote against a new State.
>
> If you want to drive the sturdy laborer, the ingenious mechanic, the scientific farmer and in fact, all the inestimable blessings which the onward march of civilization and the progress of the age, will necessarily impart. Then vote against the New State.
>
> If you want to reduce the population of the country to one third of its present number. If you want to convert millions of broad acres of rich productive soil into a barren wilderness, overgrown with briars, thorns and thistles and infested with owls; serpents; and lizards. Then vote against a new State.
>
> If you want still to keep locked up in the eternal fastness of our mountains those exhaustless stores of mineral wealth for ages yet to come. If you want to stultify and stagnate all enterprise. If you want to blot out the last glimmering ray of hope that flits along the horizon of the future and which has cheered and animated the

spirit of all loyal people of West Virginia, almost for ages past. Then vote against the new State.

If you are opposed to colleges, academies and free schools and all other institutions that cultivate [learning]. Then vote against the New State.

If you want all the great blessings of [illeg.] knowledge which alone refines and exalts the soul and opens up the only avenues through which true bliss imparted to man's immortal being, confined exclusively to the aristocracy for privileged classes. Then vote against the new State.

If you want the world to know that you are at heart a rebel and deeply in sympathy with treason. If you want to do a kind act for the Southern Confederacy, if you want to please such men as Jeff Davis, Rhett, Yancey, Cobb, Keit, Wise, Letcher, Toombs, Pryor, Wigfall, &c.,[86] if you want to fill all rebeldom with shouts of joy over the defeat of this great and noble project, and one vote alone will do it. Then sir, you had better cast your vote against the new State.

If you want to worse than throw away the vast sums of money already expended by the loyal people of West Virginia, in their honest struggles to shake off the yoke, that Eastern despotism has placed upon them. Then vote against a new State.

If you want to thwart and render null and void all our efforts hitherto to build up for ourselves and our children, a government at once, grand and glorious and which will secure to all the rights of freemen. Then vote against the new State.

If you want to heap upon the people of West Virginia a burthen of taxation to help pay the expenses of this Southern rebellion, that will grind you and them down for twenty generations to come, and perhaps not one single dollar [go] in the benefit of those from whom it has been extorted. Then vote against a new State.

If you want to establish a government under which no man can excercise the right of suffrage, unless he is worth $5,000 or has a few negroes, and where all the power is in the hands of an aristocracy. Then vote against a new State.

If you want to establish permanently and forever the system of human slavery. If you want to set up a slave market where human beings are sold like horses and mules and where fathers will sell their own offspring, and under the cover of which is perpetrated crimes so heinous and revolting, that even a devil might shudder to commit them, and which ignores and puts to blush, every sentiment and principle of the christian religion. Then we say vote against the new State.[87]

It is likely that some process was already afoot, to determine whether Capt. Young's company (mustered in as Co. G 11th Regiment Virginia Infantry Volunteers) would remain with the 13th under command of Col. Brown or be returned to the 11th Virginia, to which it properly belonged but to which Young did not wish to be returned. Officers of the 13th also favored that the present arrangement be preserved and on Sunday, February 15, Brown and seven commissioned officers of the regiment wrote to Adj. Gen. Samuels, making earnest appeal that Young's company be permanently attached to the 13th:

> We the undersigned commissioned officers of the 13th Regiment V.V.I. feeling a deep + intense interest in the cause of our country + the good of this Regiment and for other reasons herein after mentioned do earnestly petition + insist that Co. G under command of Capt. J.V. Young be permanently attached to this Regiment.
>
> First Capt. Young who alone raised the company + has command[ed] it since it's organization is exceedingly anxious to be connected with said Regiment And furthermore he asserts in most positive terms that if the company is taken away and attached to any other Regiment that it will become completely demoralized. They were enlisted in the first place for cavalry but were forced into infantry at the point of the bayonet and now that the Regiment will probably be mounted they claim that it is but just and right that they be permitted to remain with the Regiment. This company was raised in this region of country and know it thoroughly from the great Kanawha to the Kentucky line + from the Ohio to Tazewell county and would therefore be of invaluable service as scouts + spies. This company has been doing a great deal of heard [hard] service all through this country and are willing still to labor and toil on for the good of the country And if the Regiment is to remain in this region there is no one company here perhaps that would be so effective for the above and other considerations not named we urge the necessity of complying with our requests and we will ever pray
>
> Names
>
> W.R. Brown Col 13th Va Vol I

J.V. Young
Jas. R. Hall Lt Col 13th Va
S. Comstock R[egimental] Q[uartermaster] the 13th Va
A.F. McCown Capt. Co. F 13th Va Vol
V.D. McDaniel Capt Co C 13th Va. Vol
J.D. Carter Capt Co E 13th Va Vols
T.W. Hampton Capt Co H 13th V. V. I.
O.W. Griswold Lt [Co H 13th V. V. I.][88]

On Thursday, February 19, Gov. Pierpont wrote to Col. Brown, at Point Pleasant informing him that the 13th Regiment "will be mounted — No doubt about it. I cant say what day — but soon enough for the Spring Campaign." Pierpont added: "I wish you would detail some of your men to accompany the Commissioners of the revenue in the dangerous parts of the Co[unty] — of Mason."[89] February 21, Provost Marshal Joseph Darr wrote to the "Comd'g Officer at Pt. Pleasant" requesting that

> you will make some effort to arrest deserters from the U.S. forces who may be in the Counties near your Post, bordering on the Ohio River, in the State of Ohio. The reward for the apprehension of deserters to be paid by the Ass't Quarters Masters U.S.A., is stated in General Orders, War Dep't, July 31st 1862. U.S. Depty Marshals, Constables, and Police officers may thereby be induced to assist in this matter. Forward to Cincinnati, to Comd'g Officer there, under guard, all deserters belonging to Regiments serving South of you, and to me all deserters from Regiments serving North of Parkersburg, and all deserters from Regiments serving in Kanawha Valley forward to their respective Regiments, giving the date and place of arrest, a descriptive list of the soldier as far as you may be able, and a statement of the cost of his arrest and delivery to you.[90]

February 26, Colonel Brown issued Special Orders No. 3 providing that:

> Lieut Col. Jas. R. Hall Com'd'g Detachment 13th VVI. Hurricane Bridge, Putnam Co. Va. is directed to hand over to Lieut. Saml. S. Mathers his commission as 2nd Lieut 13th Va Vol. Infty and place him on duty with Co A until some [special?] action in regard to his case. He will for the present be born on the returns of Co A.[91]

Also on the 26th, Colonel Carr B. White, cmdg. District of Kanawha, issued General Order No. 3 providing that "Sutlers of regiments in this command are hereby forbidden to sell goods to citizens without first getting permission from the commandant of the post where they may be and then only to persons known to be loyal to the U.S. Govt."[92] Once again, Col. Brown was reminded about his regimental returns. On Friday, February 27, 1863, the Adjutant General's Office at Washington wrote to Brown informing him that his "Regimental Returns except for the month of October 1862 have not been received at this Office." Brown was directed to Articles of War 19,458 and General Orders No. 169 of 1862 providing that each month's returns as to the number of men present and absent, killed wounded and missing from the command, be reported in triplicate. "[O]ne copy for the Adj. Gen. at Wash. D.C.; one for the brigade or other immediate commander and one to be retained."[93]

February morning reports for February 1 through 11 show that Company A was stationed at Winfield as they had been during the previous month but that on February 12 they transferred to Camp Hurricane Bridge (at Hurricane Bridge) pursuant to orders received on February 11. 2nd Lieutenant Theophilus Maher, aged about fifty years, resigned on February 1,[94] and a vacancy existed in the company February 9 through 22, when with the appointment of Samuel Mathers on February 23, a full complement of three commissioned officers was again reported present. The number of enlisted men held steady for the month at eighty-two. One drummer was reported present and on duty in camp for the month. February 1 through 10, from one to three privates had extra duty. On February 1, Perry Gatewood was detailed for "extra duty at Regimental Hospital Point Pleasant," and February 2 through 10, Richard George and John Green were again on extra duty chopping wood. Two privates, Elijah F. Newell and John T. Newell, were assigned to "daily duty in Regimental Hospital at Winfield." Elijah F. Newell returned to the company on February 13 after serving as hospital nurse at Regimental Hospital at Winfield for "12 days." John P. Newell returned from duty as hospital nurse on February 14 after having served "13 days." Beginning February 9, and continuing to the end of the month, various detachments were made from Co. A numbering from just one man to groups of from three to eight men. Between February 25 and including the 28th, seven commissioned officers and one enlisted man and then four commissioned officers and one enlisted man were on detached service. February 10, Lieutenant Slack with five men were ordered on a scout to Mud River.

Company A continued to be plagued by sickness with up to twenty-five reported sick on February 15. On February 23, ten men must have been sent to hospital away from camp as ten men were reported and remained "absent sick" for the remainder of the month. Then followed, to the end of the month, a gradual improvement in health as the numbers of men sick declined until just four men were reported present sick at the last of February. February 16, Samuel Snodgrass and Nathan Snodgrass returned from hospital at Winfield. Sergeant Thomas Moore returned from Winfield on the 17th. February 19, Sergeant Amos Brown and Ralph (also called Ref and Rief) Pauly were ordered to "Regimental Hospital" at Point Pleasant. On February 25, 1st

Sergeant George W. King, John Edens, Richard George, W. Ramey (or Ramsey), Samuel Flake, Samuel Snodgrass, and Nathan Snodgrass were absent with leave from the Surgeon at "Post Hospital Hurricane Bridge." No one from Company A was reported in arrest or confinement.[95]

Morning reports submitted for Company B for the month of February preserve that February 1 through 12, the company was still stationed at Winfield as it had been the preceding month but February 13 through 28 the company had been moved and was stationed at Hurricane Bridge. Private Jesse Hart's jaunty letter to his father written from camp at Hurricane Bridge on Tuesday, February 24, suggests that army life was not all scouting and danger:

> herican brig Putnam county Virginia
> February the 24, 1863 my Dear father
>
> it is with the gratest of pleasure that I take my pen in hand to inform you that I am well at presant and I hope that when these few lines comes to hand that will find you all well and injoying mutch love I received you letar the 23 of this month and I was glad to hear from you I am sorow to say that I dident get to come home before we left winfield I don't know when I can come home now and I have sent you thirty dolars by James A. Rayburn and he left it at his fathrs at the Rayburns and if you hant got it yet go and get it and get you good cloths and plenty to eat and will come home as soon as I caan don't bea uneasy a bout me for reports goes fast a perpes to make people uneasy and don't you beleave it eny thing till you know it is so they are plenty of rebles out heir and we don't know how soon we will bea at tcked but we hant a fied [afraid] of them for they must remember that this is the blody 13 Regment and can capture more chickens than you ever saw I started a ltar the other day the mail is stoped up Knawah and we cant get any letars with out some of our boys comes to the point and direct your letars to point pleasant in care of col Brown her I directed one to show you how write soon J[96]

Company B tallied at the start of the month, a total of three commissioned officers and ninety-two enlisted men and this number held steady for the month. No additions or alterations to Company B were noted in morning reports, although Private Vincent D. Rice had been discharged at Point Pleasant on February 8 by order of Major General H.G. Wright. One to six enlisted men had leave of absence for the month, and two enlisted men, Fisher Barnett and James Barnett, were reported absent without leave on February 28. On the morning of the 28th, the company's orderly sergeant with four men started in pursuit of the absentees. The company had no extra staff: no musician, farriers, blacksmiths and no "Artificiers" (a military mechanic; one who can 'make things') present.

On February 8, the company received one thousand rounds of cartridges from Lieutenant Colonel J.R. Hall. As to delegation of duty, no one had special duty nor daily duty. Seven privates had extra duty February 1 through 11, and one to three privates had extra duty February 12 through 28. Considerable detached duty was ordered suggesting that heavy scouting duty was performed. Large detachments consisting of one to two commissioned officers and from fourteen to twenty-three enlisted men were detached on February the 10th, 12th, 13th, 14th, 15th, 16th, 17th, 21st and 27th. Smaller detachments, numbering in size from three to eight enlisted men, often under command of a commissioned officer were on detached duty all other days of the month.

The company had from one to twenty-two men present sick. The number of men absent sick hovered around six to eight men until February 12, when the number of absent sick was reported as twenty-five. There was some improvement in these numbers at the end of the month, when on February 28 the number of men absent sick dropped to sixteen. Two privates were reported present with the company in arrest or confinement, February 5 through 7.[97]

Morning reports for Company C show that the company continued to be stationed at Point Pleasant for the month of February. Co. C had one drummer and one fifer present each day of the month and no one absent with leave. Enlisted men reported absent without leave were: one absent February 1 through 9; two absent February 10 through 22; and one absent February 23 through 28. The company lost two enlisted men, who died of disease at U.S. General Regimental Hospital Point Pleasant. One of these was James R. Dicken, formerly a resident of Point Pleasant, who died of measles that had gone into pneumonia, on Sunday, February 8. His family, also residents of Point Pleasant, reported that it was the fever that had killed him. The other death was that of Private Adam Dunlap, aged about 44 years, who died February 19 "at 9-1/2 oclock p m."[98] of measles complicated by "quinsy," an inflammation of the throat brought on by measles.[99] One enlisted man was reported transferred on February 19. February 5, Ira Williams was reported sick in hospital at Point Peasant and continued so for the entire month. February 11, Alexander Clonch likewise reported sick to Point Pleasant General army hospital and remained there. On February 23, Samuel S. Mathers was promoted to 2nd Lieutenant of Co. A 13th Regiment, and was dropped from Co. C rolls.

At the end of the month the strength of Co. C, present and absent, was reported as three commissioned officers and seventy-seven enlisted men. As to the delegation of duty, none had special or daily duty. Five privates were detailed for extra duty each day of the month. One commissioned officer had detached duty each day. Ten to nineteen soldiers were reported present sick including Captain Van D. McDaniel, sick from February 5 through 12. No one was reported present or absent in arrest or confinement.[100]

Morning reports submitted for Company D indicate that it was stationed at Winfield, Putnam County, February 1 through 11 and then at Mount Vernon, Hurricane Bridge, for the period February 12 through 28. Co. D had a drummer and fifer present for the month. Leaves of absence were granted. February 19 through 28, three men at a time were given leave of absence, including one commissioned officer absent with leave from February 19 to 25. Co. D had no loss or gain in strength and finished the month with three commissioned officers and eighty enlisted men present and absent. 1st Lieutenant James W. Hanna was detailed for special duty on February 19. One private had extra duty every day of the month, and two to three privates had daily duty each day of the month. Detached duty was ordered as follows: six enlisted men February 11 to 13; four enlisted men on February 14; and two enlisted men on February 26 or 28, were detailed to go in pursuit of Harry F. Sherman, a deserter from Company D. The company had comparatively good health during the month with only one to two men present sick in camp February 1 through 11 and one to two present sick February 16 to 28. Three to eight men were absent sick over the course of the month. A comparable number of men were absent sick during the last half of the month. No one was present or absent in arrest or confinement.[101]

February 10, William W. Whittaker, Private, Company E, was discharged for disability. On Wednesday, February 25, two soldiers of Company E died at Point Pleasant Hospital. Eighteen-year-old Private John Pauley (born September 12, 1844) died of measles and erysipelas. His body was sent home to his parents, Peyton and Mary A. Pauley. The other death was that of Private Owen B. Cark, who also died of erysipelas.[102]

Company F morning reports show that the company was stationed at Point Pleasant February 1 through 28. One drummer and one fifer were present each day; no one was absent with leave; and one enlisted men was reported absent without leave February 26 through 28. "Remarks" following reports record that on February 14 Co. F received their pay from date of enlistment to December 31, 1862, from one Major Doddridge. The company gained one enlisted man, Private Gustavus A. Andrews, a shoemaker, who was enlisted by Capt. A.F. McCown at Point Pleasant, on February 24.

At the end of the month, Company F tallied an aggregate of three commissioned and seventy-eight enlisted men. No one was delegated for special duty; two non-commissioned officers had extra duty February 1 through 28; and two to seven privates had extra duty each day. Among those specified for extra duty were Private

Peter Yeager detailed as "nurse in Regimental Hospital" on February 9. February 26, Private Joseph Pounds was detailed, by order of Col. Brown, for extra duty in the "Post Commissary's Office, Point Pleasant." Four privates had daily duty each day. One enlisted man had detached duty February 1 through 17. On February 17, Captain McCown and Lieutenant Timothy Russell started for Hurricane Bridge with forty men as escort for Lieutenant Colonel J.R. Hall. This detachment returned from Hurricane Bridge on February 24. February 25 through 28, one enlisted man had detached duty. Company F reported six to eleven men present sick during the month. There was improvement even over these low numbers towards the end of the month. Captain McCown was himself among the sick February 5 through 16. No one was absent sick nor was anyone reported present or absent in arrest or confinement.[103] Private David Burrows wrote home to his wife, Lovina, on February 27, asking did she "get the money I sent home?" Further, he described the "hard trip" to Hurricane, "hard talk" about conscription [i.e., conscription was a "rich man's war, poor man's fight" and "what do people there think about it?"] and that the 13th Regiment was going to be "mounted sure and I don't care."[104]

Some explanation as to what raised objections and "hard talk" from soldiers regarding the new conscription law being discussed in Congress at Washington is of interest here. Unlike the Confederacy, which had a rigorous conscription policy in place, the North lacked an adequate system whereby men lost to death, injury and disease in the field might be replaced. With the prospect of Spring campaigning before them, certain legislators recognized the necessity of a conscription system (a draft) whereby the ranks of the armed forces could be restored. Despite opposition from anti-administration Democrats, on March 3, 1863, a bill calling out all able-bodied men (citizens and non-citizens) between the ages of twenty and forty-five went into effect. Those liable for service were divided into two classes. As the first pool of men was exhausted the second would be drawn upon. The first class embraced all men (married and single) between twenty and thirty-five years of age and all unmarried men between thirty-five and forty-five years of age. The second class consisted of all married men between the ages of thirty-five and forty-five.

Certain exemptions were permitted: those suffering from mental and physical debilities; convicted felons; Federal and State officers such as judges and governors; and those supporting dependent parents or orphaned children. It was the features favoring the affluent—a system of substitution or payment of a commutation fee—which prompted the 'hard talk' referred to by Burrows. A man of means had two options whereby he might avoid conscription. The law permitted a man called up by draft to hire a substitute to serve in his place. Alternately, he might buy his way out of duty providing he pay the requisite three hundred dollar commutation fee. These legal loopholes to conscription—for those who could afford to hire another or pay the communtation fee—engendered considerable bitterness and caused the war to be known as 'rich man's war, poor man's fight.'

Morning reports submitted by Company H indicate that the company was stationed at Point Pleasant February 1 through 10. February 10, Colonel Brown issued Special Orders No. 2 from regimental headquarters at Point Pleasant commanding that:

> II Lieut O.W. Griswold com'd'g Co. H. will take his company and report for duty to Lieut. Col. Jas R. Hall com'd'g Detachment 13th Va at Winfield, Va. He will without delay make all necessary preparations and be ready for the morning boat up the Kanawha.
>
> II II Lieut. L. Harpold Co. C. will return Lieut. Griswold from duty as officer of the Day.[105]

On February 11, the company was marching for their new post at Hurricane Bridge. They remained at Hurricane Bridge through the end of the month. Company H had present a full complement of commissioned officers, four sergeants, and six corporals. One drummer was present for the month, and a fifer was present February 1 through 11. One to five men (2nd lieutenant and enlisted men) were absent with leave for a total of almost three weeks of the month. Private Doliver Workman, who had deserted the company on December 13, 1862, returned about February 28. The charge of desertion standing against him was removed as he was absent "[b]y authority of Sec. of War F. Ainesworth Capt. and Asst. Surgeon U.S.A."[106] Another enlisted man, Andrew J. Allen, deserted February 12 or 13.[107] "Remarks" following reports indicate that Charles M. Lawrence enlisted in the regiment on February 24. At the end of the month, Company H tallied present and absent three commissioned officers and seventy-five enlisted men.

Three privates had extra duty February 1 through 11, and one private had daily duty February 22 through 28. Two enlisted men were detailed for detached duty February 12 through 26. Sickness pervaded and diminished the ranks of Company H this month. Company H had one to four present sick in camp February 2 through 28, and seventeen to twenty-six absent sick. No one was present or absent in confinement.[108]

On Saturday, February 7, Taylor W. Hampton, commanding Co. H, wrote to Adjutant General H.J. Samuels concerning difficulties with his appointment as Captain:

> After being introduced to you at the Virginia House[109] in October last, I enlisted as private in the 13th Regt. V.V.I. then and now quartered at this town, with an express promise by Col. Brown, yourself and Feazel if I would fill or recruit Co. H to the mini-

mum number, that I should be commissioned as Capt. of said Co., my commission to date from 5th November and to rank from 1st October, and to receive pay from date of commission. It was very difficult to recruit at that time, but notwithstanding this I filled the company after taking command of it, and have taken the entire responsibility as commander of said company, making all needful requisitions and returns for said company, up to the present time. I was not mustered by a mustering officer only a private until the 6th of January and was then informed by him that I would receive pay, only from date of Muster. Can I not be mustered back to date of commission, if his statement be true. Please answer by turn of mail.

> Yours Respectfully
> T.W. Hampton.[110]

Tall, darkly handsome Taylor W. Hampton, a lawyer by occupation, had been a resident of Gallia County, Ohio, at the time of his enlistment in the 13th on September 20, 1862. The difficulty referred to by Hampton in the letter above was not resolved. He was promoted to Captain to rank from November 5, 1862,[111] but his muster-in date of January 6, 1863, remained on the books and he resigned March 10, 1863. Perhaps he had a better opportunity for rank in his home State. Hampton's pension card indicates that he subsequently served as Lieutenant Colonel of the 16th Ohio Infantry and as Lieut. Col. of the 141st Ohio Infantry Regiment. In any event, William I. Mathews was appointed captain to fill Hampton's place. He assumed command of the company on April 27, 1863.[112]

February morning reports submitted for Company I record that the company was still stationed at Point Pleasant and that it remained there through the 28th. Still in the process of being recruited and ineligible for lack of numbers to select a captain, the company was commanded by W.E. Feazel, 2nd Lieutenant and Recruiting Officer. On Sunday, February 1, Col. Brown wrote to Lieut. Feazel directing him to "have Posters for the Recruiting service printed to the amount not to exceed five (5) dollars per month."[113] Also on the 1st, Brown issued Special Orders No. 1 directing that Sergeant Peter Darnel of Company F be "hereby detailed to drill Lieut W.E. Feazel's recruits and will report accordingly."[114] About the third week of February, Lieut. Feazel had the Point Pleasant newspaper run a notice advertising that recruits were still needed to fill the 13th and that the regiment was to be mounted. Feazel's notice was as follows:

> This Regiment, is by late order from the Secretary of War to be mounted.— The regiment is not yet quite full, and as it is now fully determined that it shall be mounted in time for the Spring

campaign, we know of no better opportunity for those who have any desire to go voluntarily into the service of their country than to come forward and immediately enlist in the Regiment. Congress has now passed a conscription law for the purpose of reaching all those who are able for the service but have refused hitherto from going into it as volunteers. All, therefore, who wish to escape the odium which will necessarily result from being forced into the ranks to defend their country and liberties from treason and rebellion, can do so by coming forward and enlisting in the 13th Va. Vols. A better set of officers from the Colonel down we venture to say, is not found in the service, kind, patient, and brave. Come forward then and do your duty for your country. We must have more men, and if they cannot be had by volunteering they must be had by conscripting. This government must be saved, if it takes the last man and the last dollar to do it. Permit then the noble instincts of patriotism to prompt you to do your duty, and come out from your hiding places and help us to save our country. Lieut. Wm. E. Feazel, Recruiting Officer.[115]

Morning reports for Company I indicate that Lieutenant Feazel was present with the company February 1 through 9. He then took leave of absence from February 10 through 22 to go to Wheeling. He returned on February 23 and was present with the company through the 28th. On February 1, thirty-five privates were noted as present in camp. On the 28th, thirty-four privates were reported present. February 7, Sanford Williams, Hiram G. Snyder, Joseph M. Cobb, Allen S. Cobb and Michael Bailey were reported absent with leave to Charleston. February 16, these men all returned from Charleston with the exception of Michael Bailey. On the 24th, Rezen P. Tulley was reported absent with leave. One enlisted man was absent without leave February 21, and four enlisted men were reported absent without leave February 22. The men were Ezra A. Hansbury (who had just enlisted on February 7, 1863), Samuel Hunter, Michael Bailey and Charles Brown. Brown returned February 23, which left still three men absent without leave for the period February 23 through 28. At the end of the month, Company I tallied present and absent one commissioned officer and forty-two enlisted men.

No soldier in Co. I had special, extra or detached duty. One private had daily duty February 3 through 27. For most of the month, the number of men present sick in camp hovered around fourteen or fifteen. On February 2, Alexander Craig, Martin Mallory, John S. Newhouse, James Truslow, James H. Good, Nathaniel M. Jackson, Charles Brown, John M. Cobb, John W. Casa, William Burns and John W. Riley were all sent to Post Hospital at Point Pleasant. February 12, Alexander Craig was detailed to "return to Hospital for duty." Soldiers reported as "returned from Hospital convalescent" were: Henderson Moore, who returned to camp February 14; Robert Samples, William Summerfield and James Truslow, all returned on February 25; John Riley and James H. Good returned February 26; Henderson Moore and John S. Moorehouse returned February 27; John M. Cobb, William W. Burns and Nathaniel M. Jackson, all returned from hospital on February 28. At the end of February just four men were reported present sick. No one was absent sick at home or in hospital and none were present or absent in confinement.[116]

Butler's Company

On Monday, February 9, Col. Brown wrote to State Adjutant H.J. Samuels from headquarters requesting that Joseph M. Butler be given a recruiting commission. Brown explained, "Butler has a company now in the process of formation and it needs all the help it can get."[117] While nothing remains in extant records to tell us what happened to Butler's company, an interesting story (dating probably to late 1862 early 1863) is related by Austin Butler about his father, Joseph Martin Butler. Before the war, J.M. Butler was a sawyer by trade and by inclination an ardent supporter of Abraham Lincoln. Austin was about 11 years old at the time these events took place. He recalled that about the time Lincoln was elected U.S. President, his father moved his family from Point Pleasant to Glenwood. They named their new place "Lincoln Hill" to commemorate the move and Lincoln's victory. Then, the war began. His father received a 2nd lieutenancy to recruit for the 13th Regiment (and perhaps other regiments) and in fact, that he joined 13th Regiment.[118]

> [T]hen the Bushwhackers and rebel sympathizers began to make it hot for us, (there being a lot of them). The rebels offered a thousand dollars for Grandfather's head, dead or alive. I couldn't tell you half that happened. [...] Sam and I [Sam and Austin, brothers, Butler's sons] slept upstairs on Lincoln Hill. Father was home and one morning right early, there was a column of soldiers passing by. [The brothers Sam and Austin] went to the window and commenced howling for Abe Lincoln. The soldiers shouted for Jeff Davis and come to find out they were rebel soldiers, Gen. Jenkins command. Jenkins stopped at the gate and asked father if he would give him a drink of water. Father took a pitcher of water out to him. He asked father if he knew Gen. Jenkins. Father told him yes. Well, he says, I'm Gen. Jenkins. Father told him he wouldn't have known him in uniform. They [Jenkins' men] had robbed a store at Ravenswood of Government uniforms and the head of the marchers had on Union uniforms and that is what fooled us. [...] father didn't hang around

there very long. He skipped back to Point Pleasant to his regiment. But in a day or two mother and us children were on our way to Alton [Illinois].[119]

New alarms

Tuesday, February 10, detachments of the 13th were on the move in response to new guerrilla operations and alarms. 13th Headquarters was moved from Point Pleasant to Hurricane Bridge, a distance of 48 miles.[120] "Companies A, B, D and H" were ordered to Hurricane Bridge in Putnam Co. "and erected an earth fortification."[121] Capt. J.V. Young, cmdg. at Coalsmouth, received word in a notation from James L. Botsford, Captain and Assistant Adjutant General, District of Kanawha Headquarters, dated February 10, that Lieut. Col. J.R. Hall had been ordered "to advance to Hurricane Bridge and scout in the immediate neighborhood."[122] During the month of February, the 13th would skirmish at Mud River.[123] Also on the 10th, Brig. Gen. E.P. Scammon wrote to Col. Brown, commanding post at Point Pleasant, regarding "powder for blasting purpose" en route for Charleston but stopped at Point Pleasant. Scammon wrote: "Mssrs. Morison + Oaks had permission from Genl Cox to bring powder for blasting purpose. It has been stopped at Pt. Pleasant. You will please see that it is sent up here. By command of Brig. Genl. Scammon."[124] Saturday, February 21, J.V. Young wrote to his wife, Paulina, from Coals Mouth regarding movements of the enemy in the Kanawha Valley and the scarcity of supplies:

> I expected to come home yesterday but the excitement in the country prevented me from coming. Floyd is bearing down on Guyan River [Guyandotte River] with twelve hundred Cavalry and you see that it behooves us to be on the alert. Brooks has gone to Point Pleasant and Lieut. Cunningham is out with a Scout, therefore you see that I can't leave my post. [...] I have money here and if you think you can keep it I will bring it home. I have paid out already $250 dollars, and owe more yet, but I think I can leave you four hundred for your own use. I can't get eggs or butter here without paying two prices. The soldiers buy all that comes in. [...] If you want hay and corn you had better buy it down in the Valley.[125]

A Confederate Mail-Route From Point Pleasant Carrying Secret Correspondence Fuels the Rebel Cause and the Case of Calvary Gibson

Correspondent "J.M.R." presumably a resident of Union Ridge, Mason County, wrote to the Point Pleasant *Register* from Union Ridge, on Wednesday, February 11, describing the perils to the loyal community when covert Confederate mail-routes were not aggressively uncovered and broken up. Rebels, of course, could not trust pro-Union postmasters with delivery of their mail and the secret mail routes they established for themselves quickly became conduits for intelligence which fueled guerrilla operations. The operation of one of these mail routes together with the brutal murder of a Home Guard volunteer required the involvement of the 13th Infantry Regiment and Company G of the 3rd Loyal Virginia Cavalry Regiment. The combined efforts of these forces led to the dispersal of a secessionist nest of partisans, who also seem to have been behind other robberies and abductions committed in Mason and adjoining counties. The transfer of 13th Companies (A, B, D and H) from their post at Winfield to Hurricane Bridge was accomplished within days of the incident related by J.M.R.

The rebel line of communication described by J.M.R. was a regular weekly mail route. It originated in Point Pleasant and ran through Union Ridge to different parts of the country. The mail route was enabled by private citizens—women and men—who traveled from twenty to thirty miles in one night to carry a package of letters to its next destination. The information provided to the enemy by this secret line of communication fueled the rebel cause in ways that proved particularly disastrous for Union citizens and soldiers alike. Correspondent J.M.R. offered the case in point. "Last Sunday night," February 9,

> a gang of horse thieves that infest our countryside made a run on the Big Hurricane Creek, Putnam county. Their plan was to surprise and capture the Home Guards, but their plans partially failed, finding they were discovered, they r[a]n into the house of Calvary Gibson, shot seven holes through him, bayonetted him several times and [...] they took the butts of their guns broke his skull in until the hammer of the lock was buried in his brains up to the barrel, all this was being done while he was begging for his life, and to make the crime of a deeper dye, caught the wife of the murdered man, and choked her until she was insensible [then] they threw a small child of the family into the fire. [...] They passed on, captured the Captain and one of the members of the Home Guards, both said to be murdered afterwards, they also left the country destitute of horses.[126]

"What can we do?" asked J.M.R. to throttle such conduits of information "kept up in our midst and under our noses," supplying partisans with names, times, locations and other valuable information with which to stage an attack. Pressing for suspension of the laws in this regard, J.M.R. continued:

If we report a man, he is arrested taken to military headquarters and in a few days is back with a fresh supply of information for the rebel leaders in this area. We are not allowed to shoot them down or we will be arrested for murder by civil authority. [...] Are we to be driven from this new State, we have helped to make, the name is sweet to us for we know what difficulties we have labored under to make it. We are now deprived of our horses for farming purposes, while the rebels are sporting upon the finest stock the country can produce. All then, we ask of the State Government is to give us the authority, and we will protect our homes and our property, but without this we must be driven from our homes, and that is just what the rebels want, for they are aware of the election that is to decide our future destiny, they know if they succeed in driving us away from the ballot-box and the homes we love so dearly, they will have no opposition, and can operate without molestation. — These are the difficulties we have to contend with and unless we have the co-operation of the State Government, our part of the New State will be left to the mercy of the rebels.[127]

A letter written by someone at 13th Regiment Headquarters at Point Pleasant, dated Tuesday, February 17, and reprinted on the front page of the Wheeling *Daily Intelligencer* also referenced the attack on the Gibson family. This article provided another perspective and additional details, illuminating the troubles with which regiments who garrisoned this countryside had to cope. The writer emphatically reiterated the oft-cited need for more mounted troops to effectively deal with these predators. The un-named correspondent wrote to the *Intelligencer* as follows:

Editors Intelligencer:

You will please excuse me for troubling you with a communication at this time and from this, Regiment. We have but three companies here now, the rest having gone to Hurricane Bridge, Cabell county, to give us protection as best they can under the circumstances, to the Union people in that section of country, which has of late, and in fact, ever since the outbreak of this infernal rebellion, been haunted by raids of guerrilla bands of rebels and murderers, and who have lately been making sudden dashes into the neighborhood, and after having accomplished their fiendish purposes, as suddenly disappear again. Mr. Morris, the Deputy Sheriff of Putnam county, was captured a few days ago, [on January 30, 1863, by a party of Rebels guerrillas] with all his officers, and carried off Southward, since when he has not been heard of. [Morris was out at Hurricane Bridge at the time, on official business attending to a sale. Fortunately, he carried but little money. Another man, A.L. Curry, was pursued by the same rebel party, but succeeded in escaping.] James B. Edwards, Commissioner of the lower district of this county, was captured on the 10th inst. [February 10th] and all his books and papers taken from him. They attempted to make him take the oath of allegiance to the C.S.A., but he refused. They then turned him loose and permitted him to go home without further molestation. About the same time, and perhaps by the same gang, a Mr. Gibson, of Putnam county, living in Tasey's Valley [Teays Valley], who was one of the Home Guards, was murdered in his own house, right in the presence of his wife and children, at the hour of midnight. They shot him through with seven balls, and then, as though that was not sufficient [...] they then broke his skull with the but of a gun, and a little daughter [...] who attempted to plead for his life was knocked into the fire by one of these [...] but happily escaped being much burned by the interference of her heart-broken mother [...].

Night before last a Government train was fired into by these marauding bands while going from Gauley to Fayetteville but were repulsed twice, and finally gave it up. Now the question naturally presents itself: What is the best remedy for such a state of things? Well, sir, in the first place, these guerrilla bands are all mounted and we are all on foot [...]. Let the Government station at several important points in this region of the country, two or three regiments of mounted infantry [...] The truth is, unless there is something like this done for this part of West Virginia and that speedily too, we will hear of more blood being shed and more robberies being committed.[128]

Clearly, concluded the writer, among their purposes was intimidation of the people into voting against the new State and to keep them from the poles so as to diminish the vote on the new State constitution as much as possible. The Gallispolis *Journal*, observing this deplorable state of affairs from across the river in Ohio, published that the country had been so thoroughly stripped "of everything in the way of supplies" already, plunder could not be the aim but rather to thoroughly discourage voter turnout as much as practicable:

A very small vote at the coming election in favor of the new State, will furnish the traitors South, and their friends North, with an argument against the 'usurpations' of the President, and 'unconstitutionality' of the act admitting her as a State. To affect this, the poorer classes, who are generally in favor of the measure reside at a distance from the places of voting, must be driven to the woods for safety. The wealthy slaveholders, nearly all favor secession, or the old State, and of course will either vote against the Constitution or not vote at all. Thus the whole vote must of necessity be greatly diminished, and the result taken by Jeff. Davis and his crew, as evidence of the strength of their cause in West Virginia. Active measures are being taken to protect Union men in the exercise of this great and, to them, all important right.[129]

During the last week of February, continued the *Journal*, the Deputy Sheriff of Putnam county was taken off by some guerrillas "to prevent his active co-operation with pro-Union citizens on the day of election."[130] In response to this act, Colonel Brown of the 13th Regiment had five noted secessionists—prominent Putnam County citizens—arrested by Captain Milton Stewart.[131] These were lodged in jail at Charleston held as hostages for the safe return of Sheriff Morris. The five citizens were Thomas Fife, W. Love, L.L. Bronaugh, Robert T. Harvey, and Alexander W. Handley. Fife was almost immediately paroled and returned to his farm near Buffalo. The fate of the others, however, hinged upon the fate of Morris.[132] Col. Brown's action in this case was uniformly approved on either side of the Ohio River. The editor of Point Pleasant *Register* pointedly remarked that "[a] little more wholesome hanging and a little less swearing, is the only thing that will put an end to these raids [...]. Whenever any of those rebel thieves, cutthroats or guerrillas are caught, they should be immediately shot."[133] The Gallipolis *Journal* reporting on the current "troublesome" nature of Western Virginia's guerrillas also approved Brown's holding citizen secessionists accountable for the actions of guerrillas, and added "[u]ntil after the election, 'untiring vigilance' should be the motto of the Union men. In their case, it is truly 'the price of liberty.' "[134]

Other terrorist acts of intimidation were perpetrated in the area at this time. Union man Isaac Vance living on the Spruce Fork of Little Coal River, Boone County, Virginia, was brutally murdered by a gang of rebel guerrillas. "They shot him dead in the road near his home, stripped him of all clothing, then the culprits disappeared without a trace."[135] In the week prior to February 12, "brave Southrons" also robbed John Winkley, who was out "buying furs for Mssrs. Leonard and Gates," a mercantile business in Point Pleasant. Winkley was robbed about "3 miles south of Winfield at the house of Joseph Forth" of about $50.00 and "every article (except his clothing), plus his horse." The writer concluded by making urgent appeal for more mounted regiments "indispensable to prevent such outrages."[136]

The Keaton gang

Another problem area in the bailiwick of the 13th and 5th West Virginia Regiments was the "Keaton settlement," rallying point for the "Keaton gang" led by "Lieut[enant] Keaton, a desperado," who had "been annoying the citizens of this part of Virginia for twelve months past"[137] and was now wanted for the murder of Home Guard militiaman Calvary Gibson (see February 9th, above). Geographic references contained in the newspapers reporting the stories associated with this settlement are of so general a nature that location of the place can be only loosely indicated as lying somewhere in south-central Lincoln County in vicinity of the Guyandotte and Mud Rivers. Mud and Guyandotte Rivers form a part of the Mississippi River watershed. As the partisan element operating in this section of Virginia used the arteries of the Guyandotte and Mud as well as settlements in this rugged section of country to cloak and launch their operations into the more northwestern areas of State, it is proper to acquaint the reader with a few facts concerning the courses of these waterways.

Mud River is a tributary of the Guyandotte River and the Guyandotte a tributary of the Ohio River. The Guyandotte arises from the confluence of two streams—the Winding Gulf and Stonecoal Creek—in Raleigh County. The Guyandotte flows, meandering considerably providing a ready conduit through Wyoming, Mingo, Logan, Lincoln and Cabell counties. Finally, it meets the Ohio from the south at Huntington. Mud River arises in Boone County and meets the Guyandotte at Barboursville. A quick look at a map reveals that both rivers provided arteries to the northwest, where Union sentiment and support for the New State Movement was most concentrated and supported by the presence of Federal forces. These routes were thus ideal for use by partisans to the south to interfere with the people of the rogue Northwestern counties and thwart their efforts to establish a new and separate State.

The 13th Regiment's involvement in movements undertaken to ferret out and dislodge Keaton's partisans, who were believed to be behind the recent wave of terrorist incidents in the Teays Valley area and below, was referenced by a soldier belonging to the 5th Virginia, whose undated letter was published in the Ironton *Register* and reprinted with fuller context in the Gallipolis *Journal*. As the 13th Regiment's records, such as morning reports and etc., are fragmentary and altogether missing for some companies, it is difficult to pinpoint when the 13th was at the Keaton settlement, where they reportedly burned a house in retaliation for being fired upon. What notations there are suggest that detachments made from Companies A, B and D, (all companies stationed at Hurricane Bridge) were probably the scouting party referred to by "One of the Fifth." The scouting party seems to have been a coordindated operation comprised of detachments and officers from Companies A, B and D.[138] Twenty-three men were detailed from Company A; two commissioned officers and twenty-two enlisted men were detailed from Company B; and Sergeant Hezekiah Scott and nineteen or twenty men were detailed from Company D to scout "on the Mud River" on March 12 and 13 (or 11th through 13th). No incidents pertaining to this scout were recorded in surviving regimental records. "One of the Fifth" was more forthcoming.

At this time, the 5th Virginia was stationed at Camp Pierpont, Ceredo. Their posting there was to ensure that the vote on the new gradual emancipation of slavery clause of the new State constitution might proceed unmolested. They had been scouting "the valley of the Little Kanawha and up the eastern side of the Big Sandy river."[139] They had been "continually picking up stragglers and men who ha[d] ventured home on furlough."[140] Other regiments stationed in the region were the 39th Kentucky Infantry "at Peach Orchard, on Big Sandy river, near its head and the 40th Ohio and 84th Indiana Infantries at the Forks of Big Sandy."[141] There were also some home guard militia but the men of the 5th and 13th Virginia Regiments and Lieutenant John S. Witcher's cavalry company (Co. G 3rd Va. Cav.) seem to have been the men of the hour on this occasion.

In his letter to the Ironton *Register*, "One of the Fifth" referenced an attack on the 13th—apparently the first troops to be sent into the area. After the attack on the 13th at the Keaton settlement, a two-day scout was undertaken by Witcher's cavalry company and a detachment of the 5th Virginia (as seems to be indicated by the use of first person in the following account) along the waters of the Guyandotte and Mud Rivers. Wrote "One of the Fifth," Witcher's company

> left this camp [Camp Ceredo, probably near Wayne Court House] at one o'clock P.M. on the 18th [of March], and marched that day to Poors Hill, a distance of twenty-five miles through a drenching rain. They camped overnight in a barn and next morning resumed their march, and brought up in the *Keaton* settlement, a distance of 40 miles from camp. Here is the place that a few days ago a company of the 13th Virginia was fired upon from the bushes in retaliation, for which a house or two soon disappeared. Here we found that a horseman had lately passed; we followed the trail up a path and a short turn in the way brought us (six in number, not including Lieutenant Witcher) upon a log house, at which were hitched four cavalry horses. Witcher ordered a charge, whereupon four men came hastily out of the house, armed and equipped and after a slight resistance and an exchange of half a dozen shots, they were captured. This proved to be a very important capture, as among the squad was the Leader of the gang, Lieut. Keaton, a desperado, that

has been annoying the citizens of this part of Virginia for twelve months past, and has eluded vigilance, heretofore of the military authorities. This same Keaton was engaged in the murder of Mr. Gibson, a citizen of Virginia; who lived on the waters of Hurricane. The facts of this case as related by the Prosecuting attorney, were that Keaton and his gang came to Gibson's house in the night and shot three balls into the room where they supposed he was, missing Gibson's wife's head by 3 inches. Gibson, however, was in another room and the gang went around to the other side of the house, and entered Gibson's room. As he was putting on his pants, Gibson cried out 'For God's sake don't murder me,' and they shot him in the left breast. He fell and they shot him four more times as he lay writhing in his own blood. Gibson's little girl ran about screaming to her Mother that her Father had been killed and in so doing got in Lieut. Keatons (of the C.S.A.) way and he threw her into the fire. The mother sprang to the child exclaiming 'Don't burn up my child, after killing my husband' to which Keaton said 'Shut up you damned union bitch, or I'll kill you too.' This happened about a month ago, and is fresh in the memory of many, and this is the kind of men and acts that (the rebel) Gov. Letcher in his late (inter-cepted) message recommends the State and C.S.A. employ, as their most useful troops. After the prisoners were properly secured, we shaped our course toward the Poor settlement. They approached the house of Doc Bledsoe and arrested him. This same Bledsoe is the one Keaton swore 'he could hold West Virginia with.' They pushed on and next came to the house of one Johnson, inside of which were three more of Keaton's command. They had ordered dinner and were about to eat it when they were apprehended.[142] After dinner, the Union cavalrymen pushed on with their prison-ers and horses, for about another three miles. Their advance came upon 3 more of Keaton's men, one of whom was Marion Adkins.[143] He had been one of the murderers of Lawrence Nixon, a son of William Nixon, Sheriff of Wayne county, Va. A few shots were exchanged and Adkins was severely wounded. One of the other two captured was the noted guerrilla by the name of Bias. These men had stolen from a number of merchants in the area Adkins from a drygoods store in Barboursville. The gang had robbed Cox's store boat about a month previous of about $800 worth of goods. On Keaton's person were notes and accounts amounting to about

$4,000, showing the sale of about 100 horses (all of which he had stolen). His accounts also show that he was dealing largely in the mercantile business.[144] On the different leaves of his passbook was found charged here and there, a pair of drawers, &c., with various articles such as pocket knives, combs, &c. Besides the above, Lieut. Keaton had a C.S.A. mail with letters to various parties at different points inside our lines, addressed to persons in Parkersburg, Point Pleasant, &c. — It is of great importance and no doubt will be made use of by the authorities. We returned to camp after an absence of two days traveling through drenching rains and swollen stream[s], bringing back nine of the most desperate characters in the Northwest Virginia, besides their guns, revolvers and 9 splendid horses, well equipped, one horse is valued at $250. Witcher's Company has already struck terror in the hearts of rebels in Cabell in adjoining counties. As the men are acquainted with every bridle path through this region.[145]

The weeks before the referendum on the New State issue, slated to be decided on March 26, were critical ones. The question of the previous summer, "Will we be driven from the Kanawha Valley?" became: "[Will] we will hold this country or leave it. There is no half way. Which shall rule here the friends of the U. S. government, or friends of the rebellion?"[146] On February 27, 1863, then candidate A.I. Boreman, (first elected Governor of the State of West Virginia after expiration of Gov. F.H. Pierpont's term of office) wrote in confidence to Governor Pierpoint from Parkersburg with instructions that the letter be read and then destroyed. It wasn't destroyed and is a powerful letter, rich in detail and alive with emotion. The letter reveals that not only were partisans employing terrorist tactics to intimidate the Western population to vote against the new State constitution on May 26, or keep them from the poles entirely, but that pro-Southern politicians—and according to Boreman—self-serving Union men seeking power and position from which to profit and weather the storm were using the New State proposition as a pretext to put forward their own program. The latter program, according to Boreman, was cleverly designed with a variety of arguments to undermine and demoralize West Virginia military volunteers (a largely illiterate group) and the general population at large. Boreman, expressed concerns that he was "more fearful for the future of West Virginia than ever before," inasmuch as these "treasonable speeches" were being made at many points in Western Virginia. While Unionists in West Virginia were not able to safely travel inland beyond sight of the Ohio River, speakers such as "Sherrard Clemens, Davis and others" went with perfect safety into the heart of guerrilla country to make their speeches, unmolested by rebels and brigands and were getting up "a state of feeling" more effectively than the rebels could themselves do left to their own devices.

These pro-Southern politicians, of course, devoted a certain amount of their speech-making to the "usurpations of the government" and the "New State proposition," as was to be expected. It was evident, however, from speeches Boreman had opportunity to hear in and around Wheeling and Harrison County that a larger agenda was being served up and that the speakers' chief purpose was to

> decry the general government — bring it into disrepute — depreciate the currency — discourage the soldiers and induce them to believe that the government is gone — that they are fighting to free the negro, and nothing else — that the constitution is being trampled in the dust by the government and its officers — and to produce the impression on the public mind generally that the rebellion cannot be suppressed — and that we may as well acknowledge the rebel government, and become a part thereof.[147]

The citizens of the West Virginia counties were according to Boreman, teetering on the edge of a knife, and he feared the effects of these speeches on West Virginia soldiery. If these conditions are allowed to persist, he wrote, "we will soon have no soldiers in West Virginia." These speakers, continued Boreman,

> go before the people in the guise of pretended loyalty, and claim that they are their best friends, and are only warning them against usurpations and efforts to take away their liberties — and thus, insidiously, ingratiate themselves into the affection and friendly feelings of the masses. And when they find themselves successful in this, then they will take the last and fatal step, and we are gone![148]

For the present, continued Boreman, public sentiment was against them but it "may not be so in future." It may do to permit such exercise of freedom of speech in loyal States, where civil strife, "murder, rapine, devestation and ruin have not yet been commenced by actual acts" but West Virginia was differently situated. Boreman outlining the incalculable value the service regiments such as the 13th were giving continued:

> Nearly the whole territory is <u>battleground</u>. After you get a short distance below the 'Pan Handle' it is <u>not safe</u> for a loyal man to go into the interior out of sight of the Ohio River. We are within the lines of the army, whose presence is absolutely necessary to the protection of our persons from being carried off and imprisoned — our property from being carried off before our eyes, and in fact to

save our lives. Threatened as we are every day that the rebel forces will be upon us and take the country before the vote on the 'New State' proposition. Would Generals Rosecrans or Hooker permit this thing within the lines of their armies? And are we to permit this for fear of the cry of violating the liberty or speech and of the press? I tell you, in the honesty of my heart, that if this thing is permitted we will soon have no soldiers in West Virginia. There is not one of these speeches made, but it is heard by more or less of the soldiers — and they are generally illiterate men — and with what feelings and opinions do they leave after hearing such speeches? Evidently, in many cases, with the opinion that the government is gone — not worth fighting for – can't sustain itself — that the money with which they are paid is a cheat and a swindle — the government is so in debt it can never redeem it — that the rebel government is just about as good as ours — and that it matters little to them under which they live — and a thousand other such impressions, which all go to the utter demoralization of the soldiery in our midst. And the effect upon the people generally is just as bad. Having the same impressions above mentioned, they the people, will refuse to volunteer — they will resist a draft or conscription, because these speakers teach them that that is right. They too will take up the cry of the indebtedness and bankruptcy of the country and the utter worthlessness of the currency, and thus will West Virginia be left in an utter helpless and hopeless condition. The government too timid or tame in its policy to protect us, and we too weak or too cowardly to protect ourselves. The question will soon be upon us — the crisis is fast approaching — [...] whether we will hold this country or leave it. There is no half way. Which shall rule here the friends of the U. S. government, or friends of the rebellion?[149]

Boreman's concerns were also felt at army headquarters. Soon after assuming command of the Department of the Ohio on March 25, 1863, Major General Ambrose E. Burnside took positive action against such talk. He too had become anxious about the harm done by disloyal politicians in his department and its demoralizing effects on the army. Burnside reported:

Letters were being sent into the army for the purpose of creating discontent among the soldiers, newspapers were full of treasonable expressions, and large public meetings were held, at which our

Government authorities and our gallant soldiers in the field were openly and loudly denounced for their efforts to suppress the rebellion. Our military prisons were full of persons arrested for uttering disloyal sentiments and committing disloyal acts.[150]

Burnside had the leaders of this movement arrested, tried, convicted and banished. For the time being, this had a quieting effect in the military district.[151]

During the months of February and March, efforts were made to fill the appointment of Major for the 13th Regiment. Members of the Convention at Wheeling petitioned Gov. Pierpont to appoint Lieutenant William E. Feazel.[152] By February 23, however, Col. W.R. Brown, cmdg. the 13th, learned from Pierpont that the 13th was "certainly to be mounted," and Brown requested that not Feazel but regimental Quarter Master Stephen Comstock be considered to fill the position for Major.[153] On March 2, Lieut. Col. James R. Hall wrote to Adjutant General H.J. Samuels thwarting Lieutenant Feazel's bid for the position of Major, perhaps Feazel was not the most popular of fellows. Hall wrote:

> Col. William R. Brown informs me that it is probable that Lieut. Wm. E. Feazel will be appointed as Major of the Regiment — I think it would not be satisfactory to the officers and would request that if the appointment be not already made that it be withheld for the present — I know that Capt. Johnson of Co. A has friends who expect him to be the Major — but we have not the number of men to entitle us to a Major — I think that when we are mounted we can easily recruit to 1000 men.[154]

On March 6, Hall wrote to Feazel, from 13th Head Quarters as follows:

> Sir
>
> Your letter to me in regard to the appointment of Major of this Regiment is before me and I would say that some of the Line Officers have been speaking to me and I know that they have been expecting to be allowed the privilege of expressing a wish in regard to the appointment and there are those whose rank + Services should not be entirely ignored. When even the Officers have a fair expression if that choice falls upon you then I will be perfectly satisfied.
>
> Very Respectfully
> James R. Hall
> Lieut. Col. 13th Va. Infantry.[155]

Meanwhile, during the second half of February, E.C. Carson of Meigs County, Ohio, was recommended for some appointment in the regiment. This recommendation was made perhaps with a thought to have him fill the position of Major. Political maneuvering for the rank of Major became a most strenuous thing. Carson, it was asserted, had served in the 4th Virginia from the time of its organization as captain and lieutenant "enjoying the entire confidence of those under his command and respect of his superiors."[156] While serving in the mountains around Gauley during winter (the winter then still in progress, of 1862-63), Carson's health had "became so impaired as to cause him to resign; since which time he has been under medical treatment." He had, by this time, become "so far recovered as to think himself once more able to enter the service" and was in hopes of obtaining some appointment in the 13th West Virginia yet in the process of organization.[157]

A number of Carson's personal acquaintances endorsed Carson for appointment. These included John A. Plants, formerly Captain of the Mason County 106th Militia, now recently reduced to the ranks of Co. C 13th Regiment. Less tarnished contemporaries also stepped up to recommend Carson. These included Col. William R. Brown, commanding the 13th; J.J. White, Sheriff of Meigs County, Ohio; A.B. Donnally, Meigs County Clerk; W.H. Lasley [Laidley], Probate Judge; S.S. Paine, Meigs County Recorder; George M. McQuigg, accountant to the department of State Revenue; Daniel Polsley, Lieut. Governor of the Restored Governement of Western Virginia and 2nd Lieut. William E. Feazel, detached on recruiting service for the 13th Regiment. Feazel wrote from Point Pleasant, on March 2, appending the following to Polsley's endorsement of Carson, the whole to be forwarded to Gov. Pierpoint:

> Whereas by order of Brig Genl Millroy I was detached from my company to recruit and having recruited part of a company and it being necessary to fill it up in hast[e], I have this day agreed that if E.C. Carson will fill it [illegible word] that he is to have command of the company with the following conditions to wit, that all arrangements made by me with regard to Lieutenants and Sergeants is to be carried out in full unless satisfactorily changed with my consent and the consent of the parties interested in the appointment.
>
> Wm. E. Feazel 2nd Lt. 13th Va. Vol. Infty. Rect. Officer[158]

Lieut. Col. Hall whole-heartedly recommended Carson, writing on March 2 from 13th Headquarters Point Pleasant:

> His Excellency F.H. Peirpoint Gov. of Va.
>
> Sir:
>
> I have known Mr. Carson for some time knew him when in the 4th Va Regt and have no hesitancy in recommending him for any position believing that he will be a usefull man in the service — He has made an arrangement by which he is to have command of the Company now commanded by Lieut Wm. E. Feazel of this Regiment.[159]

Carson was never appointed to serve in any capacity with the 13th.

On March 9, Col. Brown wrote to Adjutant Samuels with regard to the promotion of another officer, provided the 13th was mounted.

> As the 13th Regt. Va. Vol. Infty is about to be changed into Cavalry I would respectfully recommend that you Commission Francis W. Sisson Lieutenant and Regimental Commissary for the Regt. provided the change is made. He has been connected with the Commissary and Quartermaster office ever since the Regiment was commenced and is entitled to the position not only by promotion but by the ability with which he has filled the office which he now holds being that of Quartermaster Sergeant. And I also consider him fully qualified for the position which I now ask for him.[160]

'taint all honey an' pie, mebbe, but Lordy, the glory of it!

To shift to another topic, Leonard Oliver, Private, Company D 13th W.V.V.I., told the following humerous story 'about Jim Frye,' which occurred while the company was stationed at Winfield:

> We were camped at Winfield, W.Va., and we had a fellow by the name of Jim Frye in our regiment,[161] who was a shiftless, good-natured, witty and —- lazy. Partly because of his imperturbable good-nature and partly because he was too lazy to parry the attacks made upon him, Frye became the butt of all raillery and fun abroad in camp.
>
> Well, one morning some of the boys were lounging in a store near the suburbs of the town, discussing the probability of the truth of certain rumors afloat to the effect that rebel spies had been seen about the outskirts of the camp, and that various depredations had been committed, in which small stock and poultry had

suffered to an alarming extent. During the discussion, Frye, who sat astride of the coal box, whittling and spitting at the shavings he made, kept a remarkable silence, all the while eying the floor meditatively, as if he had never considered the fact that pork meant ham, sausages, 'fat, salt and unctuous,' and fowl meant savory stews, and gravy, and various other luxuries not always purchasable in camp—though seldom wanting.

One of Jim's good qualities consisted of being a hunter. When he felt like it he would sally out with his gun, and seldom returned to camp without having bagged some game. The air had become thick with stories of the rebels and their misdoings in the country, and the people in the neighborhood were in a continual state of alarm, and the troops were kept wide awake, and on the alert.

It was at this time that Jim announced his intention to his comrades in the store, to go our for a day's hunt. 'You look out,' called one of the boys, 'or you'll get nabbed and hauled into camp here for a reb!' 'Dunno as I'd object to that seein' as the tramp back allers sets hard on me anyways,' answered Jim, as he sauntered off, and, as he went, a person of fine observation might have noted Jim's eyes light up as if the friendly admonition had suggested a brilliant idea to him.

It was a bright day in February, and patches of snow gleamed and sparkled in the sunlight and there upon the hills; and air was bracing and almost chilly, but the warmth of the sunshine bespoke soft ground and mud later in the day. Soft fleecy clouds, lovely in their white repose, floated in the blue heavens, and rested lovingly against the great silent hills.

Jim had a great deal of what we fellows dubbed 'poetry of nature,' and he was not alltogether blind as a bat to the beauty around him. Indeed, the day was so serene and delightful, the forest so quiet and restful, and he found the air so exhilarating that he wandered on many miles further that he usually did, in search of game. Once down to work, however, he had no end of luck in filling his game bag. True, the quails kept provokingly shy of him, but woe unto the unsuspecting chicken that came within range of his gun. No matter if the rabbits did go s[c]urrying across the fields. Jim solaced himself with a ten-pound gobbler that strayed up to him. The day had worn pretty well along when the vivid question arose in his mind how to get his spoils into camp, for, as he had averred, the 'home stretch' bore hard on him, and his indolent nature recoiled from the exertion.

At this juncture, an idea occurred to him, and he forthwith proceeded to put it into execution. The contents of the game bag he secured in a manner intended to disarm suspiscion, and defy inspection. This done, he set out for the farmhouse nearest at hand. The worthy farmer, and his boys were engaged in unloading a cart in the yard, and they eyed Jim's approach suspiciously, a fact which Jim noted as being propitious to the furtherance of his scheme.

In accordance with his request, Jim was taken into the house and regaled with 'a cold bit,' after doing justice to which, he casually remarked that he was a rebel soldier, and supplemented his words with the startling announcement that the Confederate troops were within six hours' march of that locality. He also dwelt long and significantly upon the harrowing fact that the rebels were preparing to scatter ruin and desolation through the country and lay waste the farms, burn dwellings and make prisoners of the farmers themselves.

Jim was not slow to discover that his words had not fallen unheeded. Fear and consternation were depicted upon the faces of those around him, mysterious glances were exchanged between members of the family, and faint whispers betokened suppressed excitement. Nothing loth, Jim seated himself before the fire and awaited results, which, as he fondly hoped, would complete his scheme. He fully expected the farmer and his sons would make a prisoner of him and take him to camp, and as hasty preparations of some kind began in other parts of the house, he felt certain of success.

There was flitting here and there, and hurrying back and forth through the chambers overhead, and excited consultations were held by the family. He found it hard to repress a chuckle as he waited in momentary expectancy of the desired arrest. But the hours grew apace, and not a finger did the patriotic farmer raise toward making him a prisoner. The bustling and hurrying about ceased, and the house became suddenly and strangely quiet. It was unaccountable, and Jim concluded to investigate matters a little. He peeped into several rooms and finally discovered that the premises were deserted, and it dawned upon his mind that the whole family had given him the slip, and, somewhat crest-fallen, he shouldered his gun and weighty game bag, and set out for the next house to try his joke again.

Upon arriving at the house he found its only occupant was a purring cat stretched on the carpet before the fire, while the disordered condition of things told him that his story about the rebs had preceded him. Jim began to think that his plot was no good, and by the time he had gone into several houses along the road he was sure of it, and, tired of stalking from house to house, he set off for camp across the muddy fields, and reached there before 'drill.'

It was about 4 o'clock in the afternoon that the country folk began to pour into the town of Winfield. By 6 o'clock the town was a stirring mass of anxious looking men, white faced women, and crying children.

Our Colonel was nonplussed. He had made several attempts to find the true cause for the existing state of alarm, but having failed, he took extra precautions and doubled the pickets, all of which had a tendency to augment the excitement. All that could be extracted from any of the coolest headed of the citizens was, that a suspicious looking character had been skulking about through the country, and that he had stopped at the farmhouses and warned the people of the dangerous proximity of the rebels. All the stories differed, but one fact was noticeable, and that was to the effect that the description of the suspicious person was about the same in every instance. The mention of a pair of new blue over-alls conjured in the Colonel's mind the image of 'Lazy Jim Frye.'

Shortly before nightfall, unlucky Jim put in his appearance. He looked most 'aesthetically weary,' and his new blue over-alls were splattered with mud by his long and tiresome tramp; moreover, it needed no second glance at his habiliments and accoutrements to make sure that they were identical with those worn by the often described individual who had been the cause of the present alarm. A new light dawned upon the Colonel's mind. He ordered Jim to be put under arrest and brought before him. The farmer who had furnished Jim the 'cold bite' identified him as the self-avowed rebel who had frightened him and his family by his story about the rebels, and numerous others said that he was the same man whom they had seen in the woods. Jim, seeing that he was in for it, confessed the truth, and told the whole story.

At 9 o'clock the detachment of cavalry sent out to reconnoiter, returned and reported the country quiet for miles around. The citizens, being assured there was no danger, soon wended their way to their respective homes, and by midnight order and quiet was obtained.

And Jim! Well Jim's trouble had just fairly begun. Colonel Brown was too vexed over the affair to allow the offense to pass unpunished, but bless you, you could *never* guess the manner of punishment! It was this: Every day for ten consecutive days, at dress parade, Jim was marched out, accompanied by fife and drum, and after being assisted to mount to his shoulder a hod full of bricks, he was required to carry it up and down before the line of men six times. Jim was an overly modest chap at the best, and to be so made the cynosure of all eyes was too much for him, and being born chronically tired, too, he was fearfully cut up about it.

Even at this late day I can see poor old Jim's abashed countenance, red and streaming with perspiration as he carried his heavy load up and down, keeping step to the inspiring strains of the fife and beats of the drum, and I can almost hear the banter of his comrades and the laughter with which they assailed his ears. 'There's nothing like serving yer country, old feller!' a rollicking friend would call out. 'Well, 'taint all honey an' pie, mebbe, but Lordy, the glory of it!' Jim would reply, and so it went, day after day, until his time was out.[162]

March 1863

New State meetings

While States to the north were reeling from the news that the first Conscription Act in the nation's history had been passed by Congress on March 3, 1863,[163] in West Virginia, Union supporters rallied together during the month of March with renewed vigor in support of the new State and against "copperheads" and in favor of the draft which would fall upon Union and Secession citizens alike. On Monday, March 2, a large and enthusiastic New State meeting was held at the Mason County Court-house to drum up support for the 'immediate abolition of slavery amendment' to be voted upon March 26. The Mason County assembly came to resolutions. These were, that all fair and honorable means would be used by the loyal people of the county to secure full attendance at the polls on March 26 for a united vote in favor of the amended constitution.[164] The assembly also approved the act of the General Assembly at the late session disfranchising disloyal persons.

At the same time, the soldiers held their own meeting. In the evening of March 3, "[t]he officers and privates of the 13th Virginia Regiment, held a meeting [...] at the [Point Pleasant] Courthouse for the purpose of expressing their indignation concerning the course of the infamous copperheads throughout the North."[165] It was their fervent wish that copperhead mails (from Indiana, etc.) be stopped at the local post office. The soldiers' resolutions were embodied in an article (reprinted below) entitled "How the Officers and Men of the 13th Virginia Reg't Feel About the Butternuts of the Free States." It was published in the *Register*.

> Below we give the preamble and resolutions adopted at the soldier's meeting, held at the Courthouse on Monday the 3d inst. They are to the point.
>
> Whereas, Our hearts are filled with sensations of the most painful character whilst looking over the proceedings of those

cliques of rebel sympathizers who are poisoning the political atmo-
sphere all over the loyal States, paral[y]zing the arm of the administration, by uttering sentiments of hostility to the war which is
now being waged by the Government expressly for the purpose
of crushing down treason, and saving our flag and country from
being disgraced by the Catalines of the South, of whom even we
may be proud, when compared with these contemptible sneaks and
dishonorable peacemongers of the North.

And whereas, It is our firm and settled conviction that there
is but one way to honorably settle this difficulty, and to wipe out
now and forever, the last symptoms of treason and subdue the last
abettor of rebellion and to restore a permanent peace and a perfect
Union to this country and that is by an untiring and determined
prosecution of the war, backed up and sustained by the whole force
of the loyal States. And that this will do it, there can be no doubt,
in the mind of an honest loyal man, in the country.

Resolved, That we, the officers and men of this Regiment hold
that all cliques of men in the Free States, and elsewhere, who are
constantly finding fault with the war policy, and yet too cowardly
to take up arms against armed treason, are trying to divide and
bewilder the Union element in the country, as base, contemptible
menials, fit only to crouch down to, and do the drudgery of the
traitors of the South.

Resolved, That the principles and spirit that lie at the foundation of these Men's actions, are meaner and more villainous
and treasonable, if possible, than that which gives inspiration and
energy to the leaders of the armed legions of the conspiracy against
our country.

Resolved, That in a time of civil war like the present, when the
very existence of the nation is in jeopardy, and when more than
590,000 bayonets are raised by the hands of treason, and pointed
directly at its heart, that every man who will refuse to sustain the
legal, constitutional authorities in their honest struggle to save its
life, is a perfect nullity, a ba[d] coward or a traitor, and deserves
[two illeg. words] contempt of all loyal men everywhere.

Resolved, That we the soldiers of the 13th Reg. V.V.I., are sound
to the core. That love of country as we love our existence, and that
for the Union of our country, the defence of her Constitution,
and the untarnished splendor of her flag, we have sworn to fight,

and so help us God, that flag shall be honored, that constitution shall be defended, and the unity of the country shall be maintained, the opinions of the rebels north or south to the contrary notwithstanding.

Resolved, That we as a Regiment, loathe and detest any man or set of men who may be making speeches, passing resolutions and throwing them in the way of the onward march of our armies, simply because the administration, from the inevitable force of circumstances, has been compelled to strike at the great cause of the rebellion, which we believe is human slavery; but we say perish all — save the Government.

Resolved, That it is our deep settled conviction, that no compromise, no pacific measures whatever, presented by us to appease the wrath of traitors or to stay the flow of blood, could be honorable to the nation, or affect in the least a permanent or a lasting peace.

Resolved, That in the army alone, is found the only effectual remedy for the overthrow of rebellion and the crushing out of treason, and therefore, we insist that if these men are too cowardly or too treasonable to take up arms in defence of the liberties purchased by the blood of our revolutionary fathers, then we ask them to let us alone and we will carry the emblem of our liberties triumphantly through the storm, and guide our noble ship safely o'er the breakes, and the priceless boon of peace, shall again perch upon our banners, with a united country to live in and freedom's banner o'er us.

Resolved, That the ascending flames of burning cities, the charred and crumbling walls of stately edifices, the silent halls of Legislation, the forsaken alters of God's Temples of worship, the blighted withered heaps of a nation once great, glorious and prosperous and with a brighter future than ever dawned upon any other nation upon earth, the wailings of the widow, the cries of the orphan all over the land, is the only sure, certain and unavoidable results of the policy that these peace men are pointing out.

Resolved, That we most heartily agree with, and will sustain our Postmaster General in forbidding the transmission of all disloyal papers through the mail [apparently a copperhead paper from Shelbyville, Indiana, had found its way to Mason County[166]] and that we have no confidence whatever, in the pretended loyalty of such papers as the Cincinnati Enquirer, the N.Y. Herald, the N.Y. World, the Dayton Empire, the Chicago Times and the Wheeling Press, and we look upon their course as covered ever with the shape of treason, "the poison of Aspasia under their lips and the way of truth they have not proven." And therefore, it is our opinion that they ought not to be permitted to pass through the mails.

Resolved, That these resolutions be published in the Weekly Register and all loyal papers throughout the country, be requested to copy them.[167]

The posture of the 13th Regiment as to the Copperhead movement in the Northern States is unmistakable; their loathing for the movement is manifestly clear from the text and representative of the sentiments held by many Union soldiers and militia volunteers from this section of the reorganized State. Unlike Ohio, Pennsylvania and Indiana, West Virginia had no "peace party," and West Virginia Unionists were extremely critical of the fact that their neighbors did not suppress such movements within their borders. Loyal West Virginians having selected their side of the question with eyes open and having endured many difficulties to hold their position would hear nothing of the peace proposed by Copperheads. They wanted no compromise and the only acceptable terms of peace were that those in rebellion would lay down their arms and submit to the Government of the United States. Only then, would the people of West Virginia agree to making peace with the Southern States.

Loyal West Virginians had already risked and endured much. What would their position have been had Copperhead peace been accepted and adopted? If not forced back into the "old Dominion" of Virginia they would have lived sharing a border with a hostile country, a prospect almost as intolerable for them to contemplate as what their plight would be if the South actually won the war by force of arms. Should the plan espoused by the Copperheads be accepted and implemented all the expenditure of blood and treasure would have been for naught. Hopes for the future development of West Virginia's resources to enrich the State (foremost among these being proper education and opportunity for her young people); greater say in their State government; and an economic role in the nation's commerce and enterprise long hoped for but essentially thwarted by the State legislature at Richmond—all would be sidelined.

About the same time as the 13th soldiers meeting at the Mason County Court House, "[t]he following was picked up in the street," and handed-in to the Point Pleasant *Weekly Register* with the request that it be published for the "gratification of the boys."[168]

The Gallant 13th Virginia Volunteers

Hurrah for the gallant Virginia boys,
A noble Spartan band,
Who are for Liberty and right,
And the Union heart and hand.

Chorus.
Hurrah for West Virginia,
Hurrah for her, oh then
Hurrah for the 13th lads
Composed of fighting men.

Remember the cheerful homes you leave
And the friends both true and kind,
Remember the girls, the true-hearted girls
The girls you left behind.

Chorus.

When through the far off dusky skies,
Beams forth the evening star
We'll think of the boys the gallant boys
The boys that gone to war.

Chorus.

Let cowards wear the garb of peace
And watch their goods and chattels
But we'll sing the praise of the soldier boys,
Who fight our Union battles.

Chorus.

Three loud cheers for the 13th lads
The center of our joy.
Three cheers for the glorious stars and stripes
And six for the West Virginia boys.

Chorus.

And when amid the battles roar,
Where cannon smoke doth make you blind
Strike one full blow for your flag and homes
And the girls you left behind.

Chorus.

We love the boys of the Buckeye State,
And hail to old Kentuck,
But dearer to each fair one's heart,
Is West Virginia's pluck.

Chorus.

May each of the gallant 13th lads
Receive his share of honor,
As our lovely flag whose glories fall
Like living light upon her.

Chorus.

When around our warm fire sides we sit,
We'll think of those afar
And breathe a blessing on the name
Of the boys that's gone to war.

Chorus.

And when this cursed war is o'er
You at home we hope to find
To cheer the hearts of Union girls
The girls you've left behind.

Chorus.[169]

13th Regiment soldiers met again on Tuesday evening, March 10, for discussion of questions relating to formation of the New State. 1st Lieutenant Timothy Russell[170] was called to the chair and W.W. Harper (Sergeant Major 13th Regiment) appointed Secretary. Addresses were made by Harper and E.M. FitzGerald[171] and a meeting of soldiers and citizens was announced for Saturday evening, March 14, at the Point Pleasant Court House.[172] State Auditor Samuel Crane was also present at the March 10th meeting, and he spoke to the group. His points were apparently delivered "with great force, to the delight of the large and appreciative audience."[173] Crane sought to change the minds of those indifferent to or opposed to Western Virginia becoming a new State. The gist of his remarks were published in the Point Pleasant *Register*:

> Where does the opposition come from? from rebels and their sympathizers. Be careful of your company, New State and Union *twin sisters*. Delegates remain same, regardless of population. Separation indispens[a]ble. Glorious future in store, if separated *now*.

What the fathers prayed for, we *are permitted* to see. It's advent will be heralded with enthusiasm by all. Congressional dictation ridiculed –'the raw head and the bloody bones.' Soldiers fight with the hope that West Virginia will be independent. He himself [Crane] a slaveholder and slaves remained with him, though *permitted to do as they please.* Slavery and rebellion doomed and the *South herself struck the blow.* West Virginians always ready to repel John Brown raids. Dorr case cited as answer to Constitutional quibbles. Ordinance of Secession a nullity and voted down. The negro humbug exposed.[174] Sued for debt of Virginia impossible. Boards in *name of board P.W.* may be sued. Sends to Dixie per balloon the Copperheads. Public debt discussed. Expense of old Government and new $200,000 on hand, $2000,000 more due the State. Sword alone will determine. The Union must be preserved. *Work* for the New State. Urged all to subscribe for the REGISTER.[175]

The 13th is attached to Colonel Rutherford B. Hayes' Brigade

In mid-March a reorganization of the military machine occurred in West Virginia with a view to consolidate control and promote efficiency. Certainly the idea was a welcome one for Unionists in general but particularly so at this precarious juncture in West Virginia's bid to become a new State. The result, however, was just the opposite of what was intended with the result being more of the the same: fragmentation of her military forces into isolated and unsupported detachments. For starters, the department was attached to the Middle Military Department, Major General Robert C. Schenck commanding. He had been in command since September 1862 with his headquarters at Baltimore, Maryland. Schenck was far too distant from West Virginia to exercise efficient command and to make matters worse, the dynamic General Jacob D. Cox was transferred out of the District of Kanawha and West Virginia permanently. His command there was then divided between four different officers. Each of these operated "pretty much on his own" with "little supervision exercised" from above.[176] Brigadier General Eliakim P. Scammon commanded on the Kanawha. He commanded Third Division to which the 13th Virginia (West Virginia) Regiment belonged.

Colonel Rutherford B. Hayes, commanding 1st Brigade, Third Division, District of the Kanawha, received orders to garrison Charleston. The 5th and 13th Volunteer Infantry Regiments and Lieutenant John S. Witcher's Company G 3rd Virginia Cavalry, all comprised mainly of Virginia and Ohio volunteers, had for some time been working in co-operation to maintain Union control in the Kanawha Valley. They were attached to Hayes' Brigade on March 17.[177] Colonel Hayes took the 23rd Ohio

Volunteers with him (his regiment) to Charleston. He set up headquarters at Camp White. Camp White was located near Charleston, opposite the mouth of Elk River across the Kanawha River, on its left bank, just below (or down river from) Charleston. The other regiments of his brigade (now including the 5th and 13th Virginia Infantry and Witcher's cavalry company) were strung out along a "line," like beads on a necklace, from "Gauley to the Kentucky line."

Hayes wrote in his journal for Sunday, March 15: "My brigade is 23rd Ohio, 5th Va. Col. Ziegler, 13th Va. Col. Brown. [...] 13th at Coalsmouth and Hurricane Bridge."[178] With headquarters centrally located at Camp White, Hayes intended to "go in both directions often"[179] to check on his command. Hayes wrote to his uncle, Sardis Birchard, from Camp White on March 22:

> We seem intended for a permanent garrison here. We shall probably be visited by the Rebels while here. Our force is small but will perhaps do. My command is Twenty-third Ohio, Fifth and Thirteenth Virginia, three companies of cavalry, and a fine battery. I have some of the best, and I suspect some of about the poorest troops in service. They are scattered from Gauley to the mouth of Sandy on the Kentucky line. They are well posted to keep down bushwhacking and the like, but would be of small account against an invading force. We have here three weak, but very good regiments, Twenty-third, Twelfth, and Thirty-fourth Ohio, some, a small amount, of good cavalry and regiments of indifferent infantry.[180]

We remain in the dark as to what "regiments of indifferent infantry" Hayes is referring to, but there is the sense that he had a rather low estimation and little confidence in the local West Virginia soldiery. He would come to think differently of them. On March 17, the same day that the 13th was transferred to Hayes' Brigade, Hayes wrote to Brown reminding him of the necessity of timeliness in submitting his reports:

> Pursuant to Special Order No. 36 dated Hd. Qtrs. Dist. Kanawha, Charleston, Va., March 17, 1863 a copy of which is enclosed herewith you will report to these Hd Qtrs. Tri Monthly Reports must be transmitted promptly of the 6th, 16th and 26th of each month and a copy of the Monthly Report on the last of each month.[181]

The 13th continued to serve under Colonel Hayes for another 18 months — a period in which the regiment experienced "considerable hard service."[182] This was an understatement.

Shortly before March 12, 1863, Stephen Comstock, a millwright now serving as regimental quartermaster for the 13th Regiment, sold his share in Eagle Mills leaving H.J. Benedict as sole proprietor. The "high price of coal &c.," the fuel that powered mechanical mill machinery, had proven such a hardship that Comstock had already given public notice the previous month that "the miller would grind for customers only on Saturdays."[183] Likely also, was that Comstock's duties as quartermaster occupied so much of his time that he could no longer attend to his milling business.

The ongoing problem of finding a qualified medical officer for the 13th Regiment was again referenced in a letter dated "March 26, 1863," from W.W. Holmes, Medical Director for the District of Western Virginia to the Governor of Virginia:

> On the 10th February Reports of Medical Board, and papers recommending Drs. O. Nellis and J.P. Wilson for positions in the Med. Staff of Va Regiments, was forwarded to you. March 3rd the subject was again alluded to in a communication to you from this office. Dr O. Nellis has received his commission, but nothing has been heard of Dr. J.P. Wilson's recommendation. The 13th Va Regt (to which Dr Wilson was recommended) is in need of his services.
>
> Hoping that the matter will receive early attention I have the honor to be
>
> Very Respectfully
> Your obdt Servant
> W.W. Holmes,
> Medical Director
> District Western Va[184]

The upcoming referendum on the State constitution would have been cause for extraordinary vigilance on the part of the 13th soldiers and her officers. They were tasked with the responsibility of ensuring that no major disruption of the vote be permitted to derail the political process. The tedium of garrison duty and the seemingly endless need to scout and scour the countryside continued during the month of March and culminated at the end of the month in the fights at Hurricane Bridge and Point Pleasant between detachments of the 13th Regiment and Albert G. Jenkins' command.

Regimental records for March 1863

Morning reports for Company A indicate that the company was stationed "at Hurricane Bridge, Putnam County, Virginia," for the entire month of March. Present and absent, the company numbered three commissioned officers and about eighty-three enlisted men. On March 1, three young farmers, Samuel Jones, Philip Wintz and William Hurel enlisted at Hurricane Bridge and joined Co. A. Hurel would become fifer of the company. On March 28, Henry Sands, enlisted man, died of wounds received in action at Hurricane Bridge. Details for detached duty were ongoing throughout the month. March 12 to 13, a group of twenty-three enlisted men were detailed for detached duty. March 24, Sergeant Miletus Grinstead and nineteen or twenty men were ordered on a scout along Mud River under command of Captain Milton Stewart of Co. B. March 26, the scouting detachment to Mud River returned to camp. All other days, from one to six men had detached duty. March 27, Lieutenant Greenbury Slack and five men (John Thomas, William Means, [illeg. name], Ashar Ramsy, and Lewis Humphrey) were detailed "to guard a prisoner to Charleston named Elijah Hemming."

A small number of enlisted men and commissioned officers (up to a total of six in all) were permitted leave of absence. On March 25, Richard George returned from home and reported for duty. March 1 through 13, ten to fourteen men were reported "absent sick." The health of the company improved toward the end of the month when the number of absent sick dropped to eight or nine. Over the course of the month, from one to nine men were reported "present sick." No one was in arrest or confinement.[185]

Morning reports submitted for Company B for March record that the company was stationed at Hurricane Bridge for the entire month. Except for March 26, one drummer was reported present each day. One to five enlisted men were absent with leave during the month, and two enlisted men were absent without leave March 1 through 5. Co. B finished the month with an aggregate of three commissioned officers and ninety-one enlisted men present and absent. One new recruit, Adam W. Roberts, a carpenter by trade, enlisted at Hurricane Bridge and was duly sworn in on March 14. Two enlisted men, Jesse Hart and Ultimus Young, died in action of wounds received in the attack made on their detachment at Hurricane Bridge on March 28. James A. Rayburn, Corporal, was wounded. No one had "special" or "daily" duty. Two to four privates had "extra duty" each day of the month. Co. B was again this month detailed for a good deal of detached service. Beside the large detachment detailed from Co. B for March 12 to 13 (consisting of two commissioned officers and twenty-two enlisted men); from three to fourteen men had detached duty each day. Over the course of the month, the company had two to six men present sick and ten to sixteen men absent sick. March 6 through 9, two privates were in arrest or confinement.[186]

Morning reports for Company C indicate that the company was stationed at Point Pleasant from March 1 through 28. On the evening of March 28, after the attack on Hurricane Bridge, the company marched from camp at Point Pleasant to Hurricane Bridge. They arrived there at 5 o'clock a.m. on the morning of the 29th and remained (at Hurricane Bridge, their new station) March 29 through 31. The company had one drummer and one fifer present each day in camp and on the march. No one was absent with leave. One enlisted man was reported absent without leave March 1 and March 4 through 22. The company gained three new recruits; all three were enrolled by Captain Van D. McDaniel at Point Pleasant. On March 9, John Patterson enlisted; March 26, Andrew J. Long joined; and on March 26 or 27, Elisha Hoschar enlisted. Private William Sheline was reported "deserted" from the company at Point Pleasant on March 23. At the end of the month, present and absent, the company reported three commissioned and seventy-nine enlisted men.

The record would suggest that duty was light. Five privates had extra duty each day of the month. One commissioned officer (Capt. Van D. McDaniel on recruiting service) had detached duty each day and one enlisted man had detached duty March 2 through 31. Over the course of the month, six to ten privates were recorded present sick in hospital at Point Pleasant. None were sick at home. Eight men of Co. C remained at Point Pleasant (probably in hospital) when the company marched to Hurricane Bridge. These eight men joined in the defense of Point Pleasant when on March 30 Confederates under command of General Albert G. Jenkins attacked the town. These men suffered no casualties in that fight.[187]

Morning reports for Company D indicate that for the month of March 1863 the company was stationed at Hurricane Bridge. Co. D had a drummer present for the entire month and one fifer present March 1 through 11 and 14 through 22. There was considerable coming and going from the company. March 6, George Smallcomb and John Sealey had leave of absence to go to their homes in Mason County. March 7, William Swain and John Bishop returned from their homes in Jackson County. March 8, James Robinson returned to the company at Hurricane Bridge, and Corporal Arthur Darnel, Privates A.J. Gibbs and Junius R. Gibbs had leave to go to their homes in Mason County. Between March 9 and 24, three to eight enlisted men were granted leave of absence. Among these were Henry C. Williamson and Newman Swain, who had leave to go to their homes in Jackson County, Virginia, on March 12. March 15, George Smallcomb, John Sealey, Corporal Darnel and A.J. Gibbs returned from Mason County. March 29, William Burchfield and Nathaniel Burchfield returned to camp at Hurricane Bridge. One enlisted man was reported absent without leave on March 8; two enlisted men were absent without leave on March 15; and one enlisted man was absent without leave on March 21. On March 20, Harry F. Sherman returned to the company—he had been absent without leave. He was restored to duty by order of Col. Brown. On March 12, eighteen-year-old farmer, Marshall Statton was enlisted in the United States Service at Hurricane Bridge, by Captain Simon Williams. He joined Co. D.[188] Two more enlisted men returned from desertion on March 29. The company finished the month with a total of three commissioned officers and eighty-one enlisted men.

2nd Lieutenant George Snowden (Co. D) had special duty on March 31. Each day, one private had extra duty and two privates had daily duty. Detached service was delegated as follows: March 2, seven enlisted men were detailed to scout with Lieutenant George Snowden on Mud River. This scouting party appears to have been out for about a week. The men returned to camp on March 7. March 11 through 13, Sergeant Hezekiah Scott and twenty enlisted men were detailed to scout with Capt. James W. Johnson on Mud River. Again twenty enlisted men were detailed for detachment March 24 and 25. Co. D had over the course of the month, two to five men present sick, including Captain Simon Williams sick March 9 through 16. Eight to thirteen men were reported absent sick with more men sick men during the last half of the month. On March 28, Henry Hoffman was wounded in the head at the "skirmish at Hurricane Bridge."[189]

Morning reports for Company F indicate that the company was stationed at Point Pleasant March 1 to 28. On the 29th, the company was sent to reinforce the detachment, which had been attacked at Hurricane Bridge, and the company remained there March 29 and 30. Drummer and fifer were present with the company throughout the month and except for Private John Twaddle, who deserted at Point Pleasant on March 2, the entire company was on duty, no one being absent with or without leave. At the end of the month, the company tallied three commissioned officers and seventy-seven enlisted men. As to delegation of duty, one to five non-commissioned officers had extra duty March 1 through 28 and one to three privates had extra duty each day of the month. Among those so detailed was Private Thomas Williamson ordered by Colonel Brown for extra duty as "Blacksmith in the Regimental Quartermaster's Department" on March 23. On March 28, Corporal Rezin Bumgarner and Private Peter Yeager were relieved from extra duty and returned to duty in the company. Four to five privates had daily duty each day. March 19 through 22, ten enlisted men were detailed for detached duty. March 29 through 31, one commissioned officer and fourteen enlisted men had detached duty. Over the course of the month, Company F had comparatively good health with just four to eight men present sick March 1 through 28 and four enlisted men absent sick March 29 to 31. There were more sick men reported overall toward the end of the month. No one was present or absent in arrest or confinement.[190]

The ejecting of citizens actively supporting the Confederate cause was among duties performed by the 13th companies at this time. Once apprehended, these civilians were put beyond Federal lines. Evidence of one such case is preserved at the Regional History Library at West Virginia University, at Morgantown. On March 6, Col. Brown instructed Capt. J.V. Young, cmdg. at Coalsmouth:

I herewith return Mr [Matthew M.] Cole to your charge and desire that he be sent down the Kanawha river, the first opportunity with the positive understanding that if he is ever seen again in this country or in any manner connecting himself with rebels he will be shot and nothing but his life will pay the penalty for degressing [digressing] from this order,

[H]e can have an opportunity of sending for his clothes, or if you desire you can send a man or two with him to get them and let him stay a day longer — this done send him down the river at once + let him go

By order of Col W R Brown
E J Bridgeman Actg Adjt[191]

On March 9, Young wrote from Barboursville to Emma, his daughter, regarding tensions in his company (Co. G) and his obvious dislike for Brigadier General George Crook. In a moment of uncharacteristic equainimity, Young also gave expression to his desire to restore Virginia and her people—all of her people including it would seem, secessionists.

Col. Brown has just arrived and I have turned over the command to him again, which relieves me very much. [...]

General Crooks has put B[r]ook[s] and his company under command of Company K. Lieut. McCoy was here last night and says Brooks is very mad and says that he will report General Crooks to the Governor and if he don't do something for him he will disband his company. This is what he ought to do and do it quick, for he is a disgrace to any people and a perfect tyrant. I suppose he is done making raids on helpless women and children. I hate treason as much as Brooks but when Secessionists submit to the laws they should have protection and shall have as long as I have a sword or a gun. I want Virginia restored with the citizens in it, not a waste territory with people and property destroyed. No, Emma, I have not been fighting to destroy but to restore Virginia to her [wanton?] greatness; and if tyrants are permitted to make raids on helpless families and destroy them and their property we have missed our aim. May the Lord restore Virginia and bring back her poor deluded children [...].[192]

Morning reports submitted for Company H show that the company was stationed at Hurricane Bridge for the entire month of March. A drummer was present each day. One to two enlisted men had leave of absence for almost the whole month, and two enlisted men were absent without leave March 6 through 31. One man returned from desertion, the other absentee, Private Benjamin F. Kimberling, a sixteen-year-old Mason County youth, remained absent for the month (or more).[193] The company gained a new recruit. James H. Murphy, a young farmer from Wayne County was enlisted in the regiment on March 2 at Hurricane Bridge by O.W. Griswold. George W. Fulwiler went missing in action at Hurricane Bridge on March 28. Leroy Newman was wounded in the thigh in the same engagement. No one was present or absent in confinement. At the end of the month, Company H tallied present and absent two commissioned officers and seventy-seven enlisted men. No record of special, extra or detached duty was entered on the roll. One private had daily duty. Sickness increased in Company H as the month wore on. Over the course of the month, the company had five to fifteen men present sick and five to eighteen men absent sick. More men were reported sick at the end of the month.[194]

Morning reports for Company I show that the company was stationed at Point Pleasant March 1 through 31. The company had present for duty 2nd Lieutenant William Feazel, present March 1 through 3 (he left camp on March 3) and thirty-three privates. On March 4, Sergeant Peter Darnel with twenty-four men accompanied Lieut. Feazel on a recruiting expedition up Elk River in Kanawha County. They remained absent recruiting through March 31. This left just "ten Privates present for duty" at Point Pleasant. One enlisted man was absent with leave on March 1, and three enlisted men were reported absent without leave March 1 through 31. The company gain a new recruit, who joined on March 3. At the end of the month, Company I tallied present and absent Lieut. Feazel and forty-three enlisted men. Company I had on average just six privates present sick. March 1, David K. Young was sent to and Rezen P. Tulley returned from regimental hospital Point Pleasant. On March 3, John R. Comer died in Point Pleasant Hospital.[195]

Thursday, March 5, Lieutenant Col. James Comly was ordered to inspect the 5th Virginia Infantry at Ceredo and "investigate the advisability of changing location of troops, and report in person to Maj Gen Wright at Cin[cinatti]. On return [Comly was also] to inspect 13th Va Vol. at Point Pleasant and Hurricane Bridge."[196] Comly inspected the 13th Regiment from March 19 through March 21. He entered the following in his journal:

> 13th at Pt. Pl[easant] under Lcol Hall [illeg. words] 110 for duty [illeg words] Remainder Rifled muskets, Accoutrements only a week old. 3 wagons + 2 amb. here 5 wagons + 2 amb H. B., Good. The men have only been paid once [illegible word] the officers not at all. Consequently the sutler was never [placed] [illegible words] and he has now no stock on hand. Some of the Comp [drew?] Co.

savings which are taken care of for use of the Co. The [sutler?] had never had any funds. There is a 6 pounder smooth bore which was left by Col. Lightburn, it has never been recapted for by the 13ᵗʰ and should be taken charge of by some one. [illeg. words] order. (Get Lt. Col. Hall's Map.) 13ᵗʰ Va at Hurricane under Col. Brown 4 companies all Rifled Muskets. Have new accoutrements not yet issued. Knapsacks complete Hav. + C[illeg. 2 letters]. Clothing[197]

Citzens and soldiers vote in the referendum on the new State Constitution

At mid-month, General Scammon, commanding at Charleston, anticipating a rebel raid to derail the upcoming State refendum, held his men on alert and even proposed making a raid of his own to stop the anticipated incursion.[198] Despite hard times and partisan efforts to intimidate pro-Union people, on March 12, the Point Pleasant *Register* reported that "We have cheering news from all parts of the State, the people are rallying *en masse* for the New State." This despite the difficulties publicizing the issues at stake in the upcoming vote as laid out in the following editorial from the Point Pleasant *Weekly Register*:

> Owing to the recent derangement in our mail facilities, and the few papers that have survived the hard times, the people in the Southern and Western portions of the State, have not been able to gain much information in regard to our present election, yet as far as they are not prevented by the rebels, they will vote for the new State, and many are coming out of their rebel neighborhoods so as to vote — fully realizing how much depends upon present separation. Never before have this people been called upon to decide so important a matter.[199]

Indeed, despite such challenges, a large vote was polled at the general election held Thursday, March 26. The Western Virginia electorate approved the modifications made by Congress to the revised State constitution regarding the abolishment of slavery within its borders in a vote of "20,622 to 440 against."[200] The modifications made and passed were called the Wiley Amendment. This amendment changed the slavery clause of the first State constitution from one providing for the gradual emancipation of slavery to one requiring immediate emancipation of slaves.

Colonel J.P.R.B. Smith, Captain John Bowyer (sheriff of Putnam County in 1862, now commander of local militia) and Nelson B. Coleman had all been appointed commissioners to take the vote of all Virginia troops in the United States service "south of the Little Kanawha Valley and the streams running into the same place."[201] By May 26, the date set for the referendum, all soldiers ("not less than 10,000"[202]) had already cast their votes. The soldiers voted in the field "wherever it was practicable."[203] Smith and Bowyer reported that of those soldiers polled up to Tuesday, May 24,[204] "twelve thousand votes had been cast by the soldiers for the amended Constitution, to eleven against."[205]

It would be fairly safe to say that all polling places expected to be raided and troops, however small in number, were concentrated at polling places. Rumors that rebel bands were postured to disrupt the election had some effect if not the desired effect. This was particularly the case in the interior counties and border counties with a strong pro-Southern population such as Cabell. In such counties, the vote had to be polled over a period of days. In Cabell County, "the vote was polled over the 26th, 27th and 28th."[206]

Lieutenant John S. Witcher was patrolling at Barboursville with his cavalry company "during the 3 days previous to the election but returned to Ceredo on the 26th" as an attack was anticipated at Ceredo and one "Lt. Cummins" with just "22 of Capt. Bragg's men" protected Barboursville.[207] The 5th Virginia was on duty at Ceredo. The 13th was on duty at Point Pleasant and at Hurricane Bridge almost certainly policing the area around polling places in these counties (Mason and Putnam).

On Tuesday, April 7, 1863, at the Convention held at Parkersburg the Executive Committee opened and tallied the March 26 votes as far as received. The result was 27,785 for and 571 against the amended State constitution. The trend in voting indicated a victory for the New State movement but certain counties—all in the interior of the State—had not been heard from at that time. These were: Braxton, Calhoun, Clay, Fayette, Greenbrier, Logan, McDowell, Mercer, Monroe, Pochahontas, Putnam, Raleigh, Webster and Wyoming. The Executive Committee adjourned April 15. It was expected that the entire vote would go for ratification of the Wiley Amendment. It would then be announced and certified as to the results and forwarded to President Lincoln, who would then issue a proclamation regarding the new State and 60 days later, the new State government might then officially go into operation. These results and news of certification were published in the Point Pleasant *Register* on April 16, but whether all missing tallies from the interior counties made it to the Executive Committee by that point in time seems doubtful.[208]

"… the gamest set of fellows, I ever met …"

— Brigadier General Albert G. Jenkins about the men of the 13th
West Virginia Infantry Regiment, as quoted by B.J. Redmond,
attorney-at-law with the firm Hoge & Redmond, Point Pleasant,
W. Va. to F[rancis] H. Pierpont, Governor of West Virginia, dated
"Point Pleasant, April 2, 1863," published in the Wheeling *Daily
Intelligencer*, April 9, 1863, p. 1.

"... cool as an iceberg we fell into line,
To defend the bright stars which on our banner doth shine,
And to hold our position at Hurricane Bridge,
In spite of the rebs on each neighboring ridge."

— M.V.B. Edens, Co. A 13th Virginia Volunteers, "Battle of
Hurricane Bridge!," Point Pleasant *Weekly Register*,
April 30, 1863, [p. 1].

It was perhaps no coincidence that the timing of Brig. Gen. Jenkins' raid and attack upon Hurricane Bridge and Point Pleasant coincided with the referendum on the Wiley Amendment to the new State constitution. In hindsight however, Jenkins' objectives seem to have had less to do with derailing that particular vote and much more to do with obtaining supplies and to specifically harass detachments of the 13th Infantry — an "untried" regiment of renegade Virginians, so-called, from his old bailiwick. Confederate army soldiers, originally from the Point Pleasant area, had reportedly "long [...] desired and threatened to take Point Pleasant" but until now no serious attempt had been made.[209]

Circumstances that came together to make the raid a go seem to have been the following. It must have been known that the number of troops guarding the Kanawha Valley had been diminished. Moreover, during the winter of 1862-1863 word got round behind Confederate lines that the U.S. army had stock-piled a large quantity of government stores at Point Pleasant and that a number of horses were corralled there as well. Lightburn's retreat from Charleston the previous autumn (after the battle of Charleston fought September 13, 1862) with its heavy train of army supplies had, of course, terminated at Point Pleasant but in point of fact, Point Pleasant was hardly an important depot for army supplies. There were other factors as well. Point Pleasant and other towns in Mason County had been the locus since the previous year for important pro-Union meetings, recruiting stations and the home of important members of the Wheeling Conventions that made possible the Restored Government (upon a basis loyal to the United States). It hardly needs mention but the confluence of the Ohio and Kanawha Rivers which formed 'The Point' where the town Point Pleasant lay was obviously part of significant throughway for a sizeable fleet of military steamers plying waters of the Ohio and Kanawha to transport troops and materials. To control the Point if even for just a short time would have sent a message which would have undermined confidence in the new State's ability to protect itself.

There had been no a *general* invasion of the Kanawha Valley since September 1862 and by all appearances it appeared to military authority outside West Virginia that the threat of a *general* invasion of the Kanawha Valley had passed. This played in the Confederate advantage. A minimum of soldiers were stationed at Point Pleasant. Just enough soldiers were posted here to effectively suppress secessionist activities and ensure the safety of Union citizens; to guard what army supplies were stored; to maintain lines of communication; and to service the army hospital which had been established there.[210] Jenkins had probably also learned from his sources of intelligence in the area that the 13th Regiment was divided in detachments, at Point Pleasant and at Hurricane Bridge. Indeed, the fact that Jenkins' advance went undetected until he demanded the surrender of the 13th at Hurricane and the fact that he seemed to know the location of both detachments of the 13th cannot be mere coincidence.

Major General Sam Jones, C.S.A., commanding the Department of Western Virginia, had authorized Jenkins' expedition in the hope that he would secure a delivery to Mason County of "a lot of beef cattle" from Ohio. The Confederate Department of Western Virginia was greatly in need of beef cattle. In addition, the expedition would also accomplish "the destruction of much valuable transportation, and perhaps more substantial results." This was also an opportune time for Jenkins to make a raid, for so far as Jones could learn, little opposition would be encountered by him on his way to Mason County, as there were reportedly no Federal troops between Charleston and the mouth of the Kanawha just "a number of boats, wagons and mules."[211]

Jenkins was at this time in command of a brigade of cavalry encamped at Dublin Depot on the Virginia and Tennessee Railroad about 200 miles distant from his target. Accordingly, he took a detachment of about five hundred and fifty of his men, consisting of portions of the 8th and 16th Virginia Cavalries over the mountains and down the Kanawha through enemy occupied territory to accomplish what he might.[212] While Jenkins' men occupied national forces in the lower Kanawha Valley, Colonel John McCausland would occupy Federal troops at Fayetteville with his command creating a diversion in Jenkins' favor.

On about May 20, Jenkins set out with his men for Point Pleasant. Officers who accompanied Jenkins on this raid were Captain Alexander (Alex) H. Samuels of Cabell County (brother of H.J. Samuels, State Adjutant General of the Restored West Virginia government), commanding Co. E 8th Virginia Cavalry Regiment; Dr. Charles W. Timms of Buffalo, Putnam County, Assistant Surgeon to the 8th Va. Cavalry; Colonel James M. Cohorn (or Corns) of Wayne County, cmdg. 8th Va. Cavalry; Colonel Milton J. Ferguson, of Wayne County, cmdg. 16th Virginia Cavalry; Captain William R. Gunn of Point Pleasant, Mason County, cmdg. Company D 8th Va. Cavalry; and Lieutenant Holderby (probably 1st Lieutenant George W. Holderby of Cabell County), cmdg. Co. E 8th Va. Cavalry). Among Jenkins' enlisted men were Henry and William Bryan and John Tafft of Mason County.[213]

Three officers of the 8th Virginia Cavalry, Capt. Alex Samuels; Dr. C.W. (Robert) Timms, and Lieutenant Holderby, captured when Point Pleasant was attacked on

March 30 related the following regarding the raid. Jenkins' troops had traveled north from Salem, Roanoke County, with about four or five hundred men, on foot and not as cavalry. They made their way "through the woods until they struck the Guyandotte River, where they built rafts upon which they came down to a point not far from Hurricane Bridge whence they marched across to the Kanawha, and after attacking the steamers came down that stream in a large flat boat to Point Pleasant."[214]

Statements made by contemporary witnesses, correspondents and later historians that Jenkins' movements went undetected until he arrived with his men at Hurricane Bridge and demanded the surrender of the 13th companies posted there are not entirely accurate. A few days before the attacks on Hurricane Bridge and Point Pleasant, there were rumors in circulation that a general advance on all Kanawha posts was in the chute. As a result, Union forces posted along the Kanawha exercised greater vigilance. Point Pleasant attorney B.J. Redmond for one, wrote to Governor Pierpont before the fact, indicating that the 13th Virginia was the target at Hurricane Bridge:

> [a]bout the 25th of March, we were told by captured rebels that Gen. Jenkins was advancing with his brigade on our force at Hurricane Bridge, and that he had declared that he intended to serve the 13th as he had the 11th only worse. As the 13th were all Virginians he intended to capture or kill them all.[215]

Another Point Pleasant correspondent wrote to the Wheeling *Intelligencer*:

> Jenkins was fully advised to the exact location of what few Union troops were in the vicinity of Point Pleasant, most of which had gone up to Hurricane bridge, a day or two before the raid. The rebels hitched their horses a short distance out of town and marched in on foot.[216]

Certainly Jenkins' troops were better advised as to the positions and numbers of the 13th companies, than they were of *his* whereabouts. While Jenkins' sudden appearance at Hurricane Bridge and Point Pleasant (the former perhaps a diversionary attack to draw troops from what was presumed a rich supply depot) may have produced some surprise, the troops there were not only in short order prepared to receive him ("ready for action in 15 minutes" when "cool as an iceberg they fell into line") but offered stiff resistance which surprised him.

At this time, the 13th was divided into three detachments posted as follows. Companies A, B, D and H were in the southern part of Teays Valley at Hurricane Bridge, on Hurricane Creek, Putnam County. The military post at Hurricane Bridge was under the command of Captain James W. Johnson of Co. A.[217] Companies C, E and F remained at Point Pleasant and Company G continued to be stationed at Coals

Mouth. On Saturday, May 28, at "[a]bout 6 o'clock"[218] in the morning, Jenkins command reached Hurricane Bridge just as "twilight was lighting up the eastern skies."[219] The Confederates "sent in"[220] and Captain Johnson's pickets brought in a flag of truce with a note from General Jenkins,[221] demanding an "unconditional surrender"[222] of the forces at Hurricane Bridge. "Major Nouning," C.S.A (likely James H. Nounann, Major, 16th Virginia Cavalry[223]), reportedly, had been the officer sent in under the flag of truce to deliver Jenkins' note. "Col. Brown who was in command of that post had been necessarily called off on business, leaving Capt. Johnson (seconded by Captains Stewart and Williams) in command, Capt. Johnson replied he would fight first. Jenkins' force consisted of 850 men, and Johnson's less than 200."[224] Jenkins note as transcribed in Capt. Johnson's report was as follows:

> Hurricane Bridge Va March 28' 1863
> Colonel Brown, Comdg 13th Regt U. S. Vols
> Hurricane Bridge,
>
> Colonel
>
> I have now an overwhelming force, so disposed as to completely surround you, and cut off your retreat. A humane desire to avert the loss of life, induces me to demand your Surrender. In the event of your Compliance, and the Surrender in good faith of all the forces under your command, they shall receive the treatment warranted by the usages of war, and both officers and men will be paroled. Twenty minutes will be allowed for the consideration of this note and to return a reply.
>
> I am Colonel very Respectfully
> Yr obt Servt
> A G Jenkins
> Brig Genl C. S. A.[225]

Capt. Johnson reported that

> Upon the receipt of the above note, I immediately sent in reply. That I should not surrender the forces under my Command, unless forced to do so by an exhibition of his boasted Strength. [The Confederate officer who had brought the note took his leave.] And I immediately set about making the best possible disposition of the limited forces under my command.
>
> In 15 minutes we were ready for action. All our available force, numbering about one hundred and fifty effective men, were drawn

up inside our fortifications. When the enemy appeared in force and opened a furious fire upon us. Simultaneously on three sides, from as many different hills, owing to the high elevation of which and the unfinished condition of our works, exposed our men to a most galling Cross fire, which they withstood and returned with firmness of Veterans.

The enemy's Sharp Shooters posted on the adjacent heights and armed with globe sighted Rifles, were constantly endeavoring to pick off Officers and men. After about five hours of brisk and animated firing from both sides, the enemy sullenly withdrew his forces, leaving a few of his wounded who fell into our hands, from whom we learn that the enemy's force engaged did not number less than 500 men.[226]

The 13th Regiment detachment at Hurricane Bridge numbered in all about 160 effective men.[227] It was in fact attacked at the appointed time by eleven companies of Virginia Confederate dismounted cavalry numbering from five hundred to seven hundred men.[228]

The Confederate attack was "replied to with vigor."[229] "A battle ensued, which lasted from six to eleven o'clock A.M."[230] Johnson's "men fought under cover of temporary fortifications."[231] After about five hours of continual musketry fire — the only arms in possession on either side — the attack was abandoned. Accounts vary as to why the Confederates withdrew at this point in time. One account claimed that "[t]he General with his forces being severely punished, withdrew," and that "[t]he loss on the rebel side, is though[t], to be v[e]ry heavy."[232] Another report stated that Jenkins having "rightly conclud[ed] that the capture of the place would involve a great loss of life"[233] withdrew his men leaving behind four men, who were severely wounded. These wounded were taken prisoner by the 13th.

Captain Johnson complimented his command, writing in his report:

> To both Officers and men I return my most sincere thanks for the bravery and gallantry displayed during the engagement, where so many heroic deeds were performed, it would be unjust to mention individual acts of gallantry. It is enough to say that all behaved in the most noble and gallant manner.[234]

Jenkins himself was said to have paid compliment to the men of the 13th, or so it was claimed by B.J. Redmond in his letter to Governor Pierpont. Redmond wrote:

> It is not our purpose to speak of the gallantry of our officers and men. Jenkins did enough of that before leaving the neighborhood

of Point Pleasant [after the battle there on March 30]. He said that he never would send another flag to the 13th; that they were the gamest set of fellows he ever met.[235]

Martin V.B. Edens, Corporal, Company A 13th West Virginia, wrote a poem entitled "Battle of Hurricane Bridge!" which was published on the front page of the Point Pleasant *Register* of the April 30, 1863, edition. The poem supplies additional details of the tactics and course of the battle found in no other sources.

BATTLE OF HURRICANE BRIDGE!

By M.V.B. Edens, Co. A 13th Va. Vol.

Between a high hill and a circuitous ridge,
Is the neat little town, known as Hurricane Bridge,
Or at least, 'twas a neat little village before—
The rebellion arose in the bright days of yore.

But rebellion has banished those pleasures so sweet,
And the wail of despair is now heard in the street,
Its many fine buildings in ashes are laid
The wealth of its inmates forever decayed.

The 13th Virginia is camped near this town,
Who in a late battle have won great renown.
The regiment is formed of a brave set of boys,
Who accomplish their work without making much noise.
But Albert G. Jenkins not aware of that fact,
Came early one morning to make an attack,
With a yard of white cotton unfurled to the breeze,
His staff came into camp, while Jenkins stayed behind the trees.

'Twas Samuels[236] who brought in the banner of truce,
Says Johnson, to surrender I this morning refuse.
In the time which is given to my works, I'll repair,
And give you a battle on principles fair.
When cool as an iceberg we fell into line,
To defend the bright stars which on our banner doth shine,
And to hold our position at Hurricane Bridge,
In spite of the rebs on each neighboring ridge.

For Jenkins marched forth with his menacing host
To capture or drive us away from our post,

He threw his men round on the top of each hill
To make us surrender or fly at his will.

He thought with his men on the left and the right,
He would capture our squad without having to fight,
And with Yankee clothing he his men would equip,
And thus would secure successful his trip.

But Albert was much disappointed to find
To surrender that morning we were not inclined
Then he told all his men to prepare for the blow
And soon they'd conquer the insolent foe.

From each hill top they fired on the 13th in vain,
For each at his post did boldly remain.
For the space of four hours he fought very hard
But his incessant firing we not much did regard.

Like Jeff. Davis' scrip that's more plenty than good,
They sent in their lead from each neighboring wood.
Like the insects that sing in the long days of June,
The shells from the rebs almost whistled a tune.

But it was not sufficient to induce us to dance,
[illegible word] like Maso[n] in England or Slidell in France,[237]
Who after they made known their treacherous plan,
Have to end their hard labor just where they began.
With stomach's quite empty and feet without shoes,
Each felt like a bankrupt with a spell of the blues,
Like a dishonored guest ere the bankquet is o'er
Arrives at the feast and is kept from the door,

They knew in our camp were crackers, coffee and meat,
And they longed to get in to get something to eat.
Like Cruso's man Friday, who on the mountain did stand,
And beheld in the distance, his own native land.

With obstacles in front which he could not surmount
Was not able to settle the lengthy account,
And fell 'ere he crossed o'er the dark rolling wave
And found in the deep sea a watery grave.

So Jenkins after giving us a hard shower of hail
And finding with muskets he could not prevail

Sent out his command to retreat from the field,
For the 13th Virginia was too stubborn to yield.

Being tired and hungry, they left in disgust,
Aware of the fact that Albert won't do to trust;
He promised to give them both food and attire,
In short everything that a reb could desire.

Except a good thrashing they left as they came
With nothing to brighten Jeff. Davis' fame;
And Jenkins found out that one blunder he made
In seeking for sunshine he found nought but shade.[238]

Casualties sustained by the 13th Regiment at Hurricane Bridge on March 28 were variously reported. Captain Johnson commanding the detachment at Hurricane Bridge submitted "[o]ur loss was 3 killed and four wounded, one of whom has since died."[239] *Field History of the 13th Regiment* preserves that "[t]he 13th lost four killed and three wounded."[240] Captain A.F. McCown reported in morning reports submitted for Co. F that a total of "four men" were killed and "<u>five</u> were wounded."[241] A 23rd Ohio report in Lieutenant Colonel James Comly's papers stated that the 13th lost "3 killed, 6 wounded, one mortally."[242] Attorney B.J. Redmond wrote to Governor Pierpont that three soldiers were wounded slightly.[243] My own research yields the following information as to casualties:

> Company A had two men wounded and lost Private Henry Sands, killed in action at Hurricane Bridge. Inventory and final statements for Sands were forwarded. He had "no effects."[244]
>
> Privates Jesse Hart and Ultimus Young, both Company B, died in action of gun shot wounds.[245] Jesse Hart was a farmer from Dutch Flats, Mason County. Ultimas Young was from near Point Pleasant.[246] Young was buried in Greer Cemetery, Ambrosia, West Virginia.[247] James A. Rayburn, Corporal, Co. B, sustained a head wound.[248] He recovered sufficiently to return to active service.
>
> Company D had one man wounded.[249] Company H lost George W. Fulwiler, who went missing in action and 1st Corporal Leroy Newman was severely wounded in the thigh; the gunshot breaking the femur "about the middle."[250] Acting Adjutant Emory J. Bridgeman of Company F was mortally wounded and died in Regimental Hospital at Hurricane Bridge, on April 2, 1863.[251] 1st Lieutenant and Adjutant Bridgeman had risen from the rank and file and was

"a most valuable young officer."[252] He was born December 24, 1841, and prior to enlistment had been employed as a clerk in the Syracuse Coal and Salt Company Store, at Syracuse, Ohio (from roughly June 1858 to December 1858 and from September 1859 to October 1860). Before the war, Bridgeman had resided with his mother in Meigs County, Ohio. He assisted her on their farm. His wages from the Company Store were used to pay the cost for a short term of school at Duff's Commercial College at Pittsburgh, Pennsylvania. Emory was "a young man of excellent business qualifications." The Bridgemans were a fairly well to do family. They had farm property situated on "hill land" near Syracuse consisting of "a minimum of 99 acres" and also owned land in the town. The family homestead consisted of a brick house and two barns. Before joining the 13th Virginia Volunteers, Emory was a member of Company F 63rd Regiment Ohio Volunteer Infantry. His younger brother, Austin A. Bridgeman, served in the same unit. Emory was discharged from the 63rd Ohio on June 26, 1862, on a "Surgeons Certificate of disability." He was enlisted in the 13th Regiment at Point Pleasant by Lieutenant Theophilus Maher, on October 7, 1862. He was mustered in on October 9. Statements made by the War Department in a letter to the Bureau of Pensions in regard to N. Bridgeman's application for pension (mother's pension) indicate, that Emory was probably first detached for service as clerk to Colonel Brown on December 31, 1862, and that he was commissioned 1st Lieutenant and Adjutant of the 13th Regiment on April 1, 1863, one day before his death. His commission having terminated with his death, it was returned.[253]

Immediately after the fight at Hurricane Bridge, the 13th force stationed there "commenced enlarging and strengthening the[ir] fortification."[254] Within hours reports of the attack on the post reached Point Pleasant where sister Companies C, E and F were stationed under command of Colonel William R. Brown. Straight away, all the available force at Point Pleasant, except Captain Carter's Company E (namely, Companies C and F, 13th W. Virginia), were ordered by Brig. Gen. E.P. Scammon (cmdg. Kanawha Division) to march for Hurricane under command of Col. Brown and reinforce the position at Hurricane. They moved immediately, and on the evening of March 28 Companies C and F marched from camp at Point Pleasant for Hurricane Bridge under command of Colonels Brown and Hall. The distance from Point Pleasant was "46 miles."[255] They arrived at Hurricane Bridge between four and five o'clock on the morning of the 29th.[256] The Confederates, for their part, left

Hurricane Bridge immediately after withdrawing from the engagement there "by a wholly different route, from that taken by the reinforcements under Lieutenant-Colonel Hall"[257] and so avoided being discovered as they proceeded towards Point Pleasant. In the ensueng days, the force at Hurricane Bridge did heavy duty scouting and guarding. "We have been on one big scout since we have been up here. [... W]e are in Dixie now. I would have wrote sooner but I was on guard & scouting & I had not time to write," wrote Private David Burrows with Co. F from Hurricane Bridge.[258]

A.G. Jenkins' Troops Make for Point Pleasant

The next day, March 29, Jenkins' men passed through Winfield, following the Kanawha River towards Point Pleasant. At 12 o'clock noon, two government steamers traveling down river from Charleston were attacked by Jenkins' men in the vicinity of Hall's Landing, Putnam County, located "4 or 5 miles below" Winfield[259] and thirty miles below Charleston.[260] The U.S. Steamer *Victor No. 2* (Captain Frederick "Fred" Ford, of Gallipolis, Ohio, cmdg.) was in advance[261] and the *General Meigs* (Captain Summers, cmdg.) in rear. The *Victor No. 2* was unarmed and there were no muskets on board. She had left Charleston and had just touched at Winfield on her way down the Kanawha to deliver mail and inquire as to the possibility of "any danger below" but upon receiving word that "it was quite possibly safe" to proceed, the *Victor* continued on.[262] On board the *Victor* were passengers from Charleston. Passengers included a Paymaster to the Federal military, Major B.R. Cowan, who carried in his possession considerable government funds. There was also Quartermasters Capt. E.P. Fitch and E.C. Rieumbaugh, two families including women and children (reportedly the families of Fitch and Rieumbaugh), numbering in all 14 people and a number of others, who had been up to Charleston to visit friends and relatives in the army.

The eyewitness accounts of Captain Ford and Major Cowan suggest that the Confederate "rascals" had "made extensive preparations to capture the boats" and were "deployed along the river for a distance of two miles"[263] on either side in ambush. As the *Victor* was descending the river and had reached nearly "opposite Hall's house and landing" (the house being located near the landing[264]), "four hundred"[265] of Jenkins' men "issued from Rob Hall's house and four other houses from behind "fences trees and logs" and fired a veritable "streaming blaze" directly into the steamer.[266] Paymaster Cowan related that the attacking party "opened a heavy fire, which, fortunately, at first did little or no damage,"[267] but this initial attack was signal for the main body of Confederates heretofore concealed behind a clump of trees to emerge, and they in their turn poured "[v]olley after volley [...] into the little craft, but the pilot had the presence of mind to keep her directly in the channel, while the engineer put on all the steam he had and for 5 minutes she ran through a complete hail of leaden bullets."[268]

The passengers screened themselves as best they could behind bulkheads and other shelter.[269] The direction of the fire may have suggested that Jenkins' purpose was not to destroy but to hijack the boats, presumably, to hasten their own arrival at Point Pleasant.[270] Captain Ford went up on the hurricane deck, where he was necessarily much exposed, to stand by the pilot and encourage him as he stood to his post. It was said that thirty-six bullets passed through the pilot house, and six balls struck the wheel but the pilot did not leave the wheel.[271]

For one entire mile, officers and crew stood to their posts as the boat steamed through the torrent of bullets completely exposed. All casualties on the *Victor* were private citizens. One man and one horse were killed outright with the first attack and one man was mortally wounded. Several horses aboard the transport were also wounded.[272] The second steamer, the *General Meigs*, some distance behind the *Victor*, was also fired upon "but her officers being advised of what had happened to the other boat, put her passengers down in the hold"[273] and consequently, she passed though the blockade of Confederates without loss of life. The boats were "literally riddled with bullets."[274] Having passed the danger, the *Victor* continued on her way down river followed in rear by the *Meigs*, both boats making for Point Pleasant and Gallipolis to notify the citizens of their danger.

Captain Ford first stopped at Point Pleasant. Ford's officers notified Captain J.D. Carter (Co. E 13th Va. Regiment), commanding the small group of Union forces at the Point Pleasant military post that there was a sizeable Confederate force in the Kanawha Valley; that the *Victor* had been attacked; and that the Confederates could quite possibly be headed for Point Pleasant. This taken together with the news that the 13th detachment at Hurricane Bridge had been attacked on the 28th probably raised considerable alarm in Point Pleasant. Capt. Carter "made the best preparation that he could, with his small force to receive them [the enemy], occupying the court-house."[275] John Hall, then out on a $30,000 bond and awaiting trial for the murder of Lewis Wetzel,[276] upon receipt of Ford's news "sent so far as possible, Union women and children into nearby Ohio towns for safety."[277]

One eyewitness, a "Mr. Brammer," who seems to have been on hand in Gallipolis when the boats arrived there, related that the *Victor* traveled the "remaining twenty-six miles to Gallipolis, in record time and reached the French settlement" that evening "before dark"[278] on the 29th. Upon her arrival at Gallipolis (the French settlement), "crowds gathered to observe the result of the attack and to hear particulars. One man counted 167 bullet holes in the lar board side of the boat."[279] At Gallipolis, the passengers of the two boats were unloaded and consequently were spared the experience of "the Point Pleasant affair which occurred next day."[280]

Anxious that the government wharf-boat lying at Point Pleasant (containing $86,000 worth of government property) not fall into the raiders' hands, Captain

H.H. Boggess at Gallipolis ordered the *Victor No. 2* back to Point Pleasant to guard it. The wharf-boat was tied up at the Point Pleasant wharf on the west side of town on the Ohio River. The *Victor*, Captain Ford and crew lay alongside the wharf-boat all that night ready at any moment to take her out of danger.

The Attack on Point Pleasant

"On the whole the rebels were defeated. Captain Carter of the
13th and his men fought like tigers."

— [R.L. Stewart, ed.?], Gallipolis *Journal*, Thursday,
April 2, 1863, p. 2.

Who actually commanded the Confederate attack at Point Pleasant was a matter of some conjecture, at least among Union citizens. The weight of opinion, however, seems to indicate that Brigadier General Albert G. Jenkins and Alexander Samuels (the latter called variously, Lieutenant, Lieutenant Colonel, Captain and Major), both of Cabell County, were in command at Point Pleasant. One eyewitness account reported that the "rebels" who attacked Point Pleasant were commanded by "Jenkins and Ferguson [Colonel Milton J. Ferguson commanding 16th Virginia Cavalry Regiment] respectively of Cabell and Wayne counties."[281] Another account stated that Jenkins in person, "together with Fitz Hugh" [Lieutenant Colonel Henry Fitzhugh, Jr., 8th Virginia Cavalry Regiment][282] conducted the raid on the town; and yet another published source stated that Lieutenant Alexander Samuels had soul command.[283] Another person noted with some sarcasm that "Jenkins and Fitzhugh were not even with the raiding party but were waiting behind a rock outside the town for the entire fight."[284] Captain J.D. Carter, who was probably in the best position to judge who led attacking force, stated that the enemy was under command of the so-called "invincible A.G. Jenkins" and of Alexander Samuels.[285]

As the reader may have already noticed, there is a looseness, variety, summary and open-ended-ness to accounts and interpretations of what happened before the 'ball opened' at the Point, so much so in fact, that one can see how under the circumstances *all* might convey the gist of what the raiders did without one version contradicting another or offering more than an additional minor detail or two. Col. R.B. Hayes writing home about the incursion from headquarters at Charleston some ten days after the fact, when in that interval something more specific might have been learned wrote blandly that after Gen. Jenkins had been repulsed by Capt. Johnson at Hurricane Bridge, he "then crossed the Kanawha twenty miles from the mouth or less and attacked Point Pleasant at the mouth."[286] Another source stated that having failed to capture the government steamers, Jenkins' men proceeded down river "taking horses, cattle, &c."[287] This latter statement cannot be entirely confirmed but it is plausible. Another source (more in the know) stated that the Confederates after leaving Hurricane Bridge followed the Kanawha to Frazier's Bottom and then went over land taking a short-cut until they struck the river again at Point Pleasant. They arrived at the latter place in force (federal military authorities estimated the Confederate force to be something less than five hundred men) the next morning (Monday, March 30), at about ten o'clock. Fighting erupted sometime thereafter at close to 11 o'clock a.m.[288]

Not unexpectedly, as is often the case in the historical record of all things 13th Virginia, accounts also vary widely as to what happened at Point Pleasant during the fight. Local attorney B.J. Redmond wrote that after firing upon the boats up river on the 29th the rebels

> procur[ed] flat boats in the night, they floated down and concealed themselves in the immediate vicinity of Point Pleasant, capturing our scouts and pickets, and were nearly into town before being discovered. But we deny that it was a complete surprise.[289] Capt. Carter having previously got his ammunition into the Court House, and other preparations made; drew his men into the Court House about the right time, and commenced the fight in fine spirits.[290]

Another person from Point Pleasant stated that Jenkins' men had "hitched their horses a short distance out of town and marched in on foot."[291]

It will be recalled that Captain John D. Carter's Company E had been left to hold Point Pleasant while Companies C and F (13th W. V.V.I.) were sent from the Point Pleasant military post to reinforce Hurricane Bridge following the attack there on March 28. This left at "the Point" on the 30th just Company E numbering only "60 men."[292] They were camped "between Main and Viand streets, 2 blocks from the court house"[293] on a commons above the courthouse just north of 8th Street.

Jenkins' raiders overwhelmed the defenders at Point Pleasant with a force estimated in the accounts of the time as just under five hundred to as many as eight hundred men. Firing began in earnest "after some fighting in the streets,"[294] when Carter's Co. E "took possession of some brick buildings."[295] The buildings appropriated by Carter included the Mason County Courthouse. Indeed, in his letter to the Gallipolis *Dispatch* defending his conduct during the defense of the town, Carter indicated that street-fighting erupted as Jenkins' troops entered the town, and that his small company resisted Jenkins' men in the streets and continued to fight them retreating as they fell back to the county courthouse.[296] "The few Union soldiers who were stationed in the town" were compelled to fight "the greatly superior force of rebels" as best they could from their refuge in the courthouse and this they did "for five hours."[297] The "unequal contest" was kept up "till near 3 P.M."[298] or "3-1/2 o'clock

P.M."[299] Meanwhile, citizens of Point Pleasant fled across the Ohio River and spread the alarm. "[A]fter 4 or 5 hours hard fighting" on the part of Company E, Jenkins' men "were driven [and] retreated" from the town.[300]

Carter defended the town with Company E and some few other soldiers—probably convalescents in the army hospital there. Eight men of Company C and eleven men of Company F (all 13th Virginia Regiment) were left behind convalescent at Point Pleasant, when their companies were transferred on the 29th to Hurricane Bridge. These engaged in the defense of Point Pleasant on March 30.[301] Despite criticism aired in the local presses, Colonel R.B. Hayes, Carter's superior officer, praised Carter and his company in a letter to his wife writing that they "defended themselves gallantly.[302] Another eye-witness stated that Carter and his men repulsed all the attacks made upon them and "pepper[ed] it to the rebels when ever they appeared in sight."[303]

From their position at the court-house, however, for all its central location, they "could not keep the whole town clear of rebels,[304] and a number of Jenkins' men scattered through the town in search of supplies to plunder. The Cincinnati *Daily Commercial* reported that "a portion" of the Confederate "command fired several houses, and ransacked a number of dwellings and stores" and "also burned 7,000 bushels of Government corn."[305] Mason County Home Guards (the 106th Virginia Militia Regiment) commanded by Col. J.P.R.B. Smith (i.e., those who "could find a gun to use") also fought in conjunction with Carter to protect the town.[306] The Point Pleasant *Register* reported that,

> Our town militia were mostly without arms or more of the filthy vagabonds would have bit the dust. Long have these gallant, chivalrous friends of our townsmen threatened to take Point Pleasant but this is their first attempt at capturing it, and as they received so thorough a thrashing, we doubt not they will hereafter 'let us alone.' Hurrah for Capt. C. and Company E and those of the Point Pleasant militia that could find a gun to use.[307]

Meanwhile, at the water's edge almost simultaneously with the attack on the town "at 10 A.M.," reported Captain Fred Ford aboard the *Victor No. 2* on the Ohio River side of Point Pleasant, while the *Victor* yet lay alongside the government wharf-boat Jenkins' men appeared "on the bank."[308] The Confederates at once made attempts to capture the wharf-boat and the *Victor*. Captain Ford ordered his crew to cut the lines of the wharf-boat so that he might tow it away to safety in this way saving a large amount of government stores from capture and destruction. Detachment of the wharf boat from its moorings was done "under heavy fire from the enemy."[309] The men of the *Victor* freed the wharf-boat "midst a shower of balls, and towed her safely to Gallipolis."[310] Upon delivery of the wharf-boat at Gallipolis, Ford and his crew "reported [themselves] ready immediately to return to P[oin]t Pleasant with re-enforcements, which after some detention was done."[311]

Published narratives and remarks concerning the course of events as witnessed from the Virginia side of the river reveal the human side of the drama unfolding in Point Pleasant during lulls in the battle and the degree to which town residents were involved in the contest. The *Kanawha Republican* published an incident related to them by a "friend from Point Pleasant," which occurred during the fight. A young girl residing near the courthouse took position in an upper window and at the top of her voice cheered Captain Carter and his men, "hurrahing for the Union and the Union boys." Supposing 'the boys' holding the courthouse would be glad for some refreshment and taking advantage of a slight cessation in the firing, she filled a basket with victuals and bore it off to them. Continued the *Republican*:

> Our friend could not give us the name of this noble little girl, but we intend to have it that it may be known to our readers. The patriotic conduct of this girl is in fine contrast with the woman who bore the flag of truce from Jenkins to Capt. Carter, demanding his surrender, telling him that Gen Jenkins would parole him and his men and treat them kindly; but if he declined, Floyd would soon be there with a large army, and would capture them certainly, and would show them no favors. On her third imploring visit under the flag of truce to the Court House, Capt. Carter it is said, told her not to come again—that if she did, his men would fire on her. And the fight went on.[312]

Attorney B.J. Redmond's account suggests that some excited informant prematurely reported the defeat of Carter's men, a point which in the ensuing days would become an exceedingly sensitive issue:

> Jenkins' force here [at Point Pleasant] consisted of about 700 men. Carter had only 60 men in the Court House. The fighting was severe for sometime until Jenkins sent in a flag demanding a surrender of the place, to which the gallant commandant replied that he entered the service to fight and not to surrender, and to take him if he could;[313] after which Gen. Jenkins sent some rebel ladies with a flag to beg Captain Carter to surrender, and to say to him if he would do so private property would be respected, otherwise he would burn "the d—n town." Failing again in obtaining a surrender he then meanly commenced threatening Union ladies and their property, and begging them to plead with Carter for a surrender.

In this too he failed, for some of the ladies instead of asking for a surrender went to the Court House, amidst a shower of bullets, and told Capt. Carter they didn't want him to surrender if their property was burned. A short time after this a retreat of the rebels was ordered, and their advance was at least two miles above town before the Ohio troops were across the river, although there was still a good many in town engaged in plunder when they did cross.

Capt. Carter hadn't sufficient force to march out and attack them, but they were completely whipped before the reinforcements arrived. It is not true then that the town was taken and retaken, and the man who dispatched you that statement knew that our boys were still waving the stars and stripes out of the window when the Ohio troops left the Ohio side of the river for Point Pleasant.

The truth is sir, there never was a more gallant fight made, than was made by our little band of Virginia troops. Lieut. Hawkins and others are now suffering from the effects of most painful wounds. We hold in utter contempt the man who could have the heart to do injustice to such noble fellows, and it gives us pleasure to say that the Ohio soldiers and the officers that came here showed much anxiety to get here long before they could get permission to do so, and when they did get here, showed an energy and courage in the pursuit of the retreating foe, which would do credit to any officer or soldier. But there were others higher in authority, who showed much less sympathy for us and courage to defend us. We begged of the authorities at Gallipolis the morning before the fight, and offered the most ample security for a few guns to defend ourselves, but were told the red tape would not allow.[314]

When the Confederates entered Point Pleasant, many citizens, among these "defenceless women and children,"[315] crossed the river in skiffs. "[S]howers of bullets were sent after them, but strange to say not a single person was struck with a bullet."[316] It was the shooting death of Major Andrew Waggoner at the hands of Confederate soldiers out scouting for horses to steal, however, which out of all the outrageous behavior engaged in by the raiders most incensed people in the area.

Major Waggoner was a "beloved citizen"[317] of Point Pleasant and "one of the few staunch Union men in that quarter and deserved a better fate."[318] He was the son of Colonel Andrew Waggoner, veteran of the French and Indian War and American Revolution. He had moved to Mason County to settle on his father's military land grant.[319] Waggoner was eighty-two or eighty-four years old at the time of his death and a revered veteran of the War of 1812. Known as the hero of Craney Island, he had been awarded a handsome sword by the Virginia Legislature for his heroism and repulse of the British in the defense of Norfolk at the battle of Craney Island. Waggoner had for many years also served as Clerk of the Mason County Circuit Court but had retired to live with his son, Colonel Charles B. Waggoner, at his residence just a short distance from Point Pleasant.

The Major, upon hearing that the rebels were approaching, got on his horse and went to town. As he was returning to his home, just about a quarter of a mile out of town, he met a squad of Jenkins' men on the road (possibly pickets), one of whom caught Waggoner's horse by the head. Jenkins' men had been given general orders to capture all available horses and the soldier was simply following orders although he did not know with whom he was dealing. Waggoner immediately drew his cane and began to beat the soldier with it when the soldier (or soldier's comerade) shot Waggoner dead with his carbine and then robbed him of his horse.[320]

Meanwhile, back in Point Pleasant town, Captain Carter's men were at last relieved by re-enforcements coming by boat from Gallipolis and vicinity. Reinforcements came in the shape of two companies of militia—the Gallia County Guards and Trumbull Guards. These were accompanied by a number of citizens from Gallipolis as well as patients from the Gallipolis General Army Hospital. These latter "turned out to the number of 100, armed and ready for fight, though many were really very ill."[321]

The reinforcements of militia, citizens and convalescents came armed "with musketry and a rifled cannon."[322] They arrived opposite Point Pleasant "about the same time" as the steamboat *Victor* hove into sight of the Point. The *Victor* had been armed with two or three pieces of artillery[323]—guns from the *General Meigs*. Militia, citizens and hospital patients were picked up and conveyed across the river aboard the *Victor* (and perhaps also by the *General Meigs*). The two militia companies were "among the first to cross the river,"[324] and working in concert, they aided in driving the rebels from the town.[325] The Trumball Guards received especial praise on both sides of the river. The Gallipolis *Journal* published that the Guards "conducted themselves like veterans" at Point Pleasant. "Their prompt action and bravery [... was] worthy of the highest praise."[326] Captain Carter, in his own inimitable style, also praised the Trumball Guards by name and the Gallipolis citizens as well, writing

> I award great credit to the Trumbell Guards and citizens, who did cross to our assistance notwithstanding the enemy *had retreated*, for they crossed as soon as they could. The steamer was detained for a reason that I will not now explain, but which I will, if any are curious to know, at a future time. And I will take this occasion to thank the soldiers and citizens of Gallipolis who did come to our relief. I thank you gentlemen, soldiers and citizens in the name of

Company E, 13th Va. V.I. , and will say, also, if ever you get into trouble we will not be slow to your rescue. *We know* that your will was good to come to our assistance sooner; but we know you had no means to cross the river, unless you had swam it.[327]

The Fog of War

The fact that Capt. Fitch claimed for himself and his crew the lion's share of credit for driving the rebels out of Point Pleasant with his hastily constructed "impromptu gunboats from Gallipolis"[328] resulted in considerable rancor, understandably so if one accepts that at the time Fitch's steamer finally hove up on the Ohio side (about 3 o'clock p.m. according to Poffenbarger below) the crisis had not only already passed but as a first counter measure, Fitch very nearly destroyed the town and Carter's small command when the artillerymen aimed the guns at the courthouse at the town center: the plan apparently being to shell the raiders out of the town center before crossing the government steamer to the West Virginia side to unload the miltiamen. Carter himself referred in what one might call an almost humorous vein to the "devilish poor sense of judgement" shown by the artillerymen in their "first attempt." The crisis had passed, wrote Carter, when help hove into sight in the shape of the *Victor No. 2*:

> It is true the [hea]vy ordnance came up on the Ohio side of the river; and it is also true (if citizens can be believe[d]) that the first act w[hich] men in command of the artillery [made] was to range their course for the Court-house and was only prevented from firing upon us by the earnest entreaty of some of the citizens of Point Pleasant who happened to be on that side of the river.— While I admire the courage of the artillerymen, I think they exhibited devilish poor judgement, in their first attempt.[329]

Carter continued with his account indicating that by the time the Gallia County reinforcements "arrived on the Virginia side of the river there were only thirty of the enemy in sight."[330] After re-sighting the floating battery of howitzers, "a few shots"[331] were fired toward the fleeing rebels. These shells "fell short" of their mark although "the noise may have scared the Rebs some," causing some "consternation" among them.[332]

Considerable hard feeling arose on either side of the Ohio River as both Point Pleasant and Gallipolis argued over who should have credit and who should bear blame and for what. Actions and failures to act spawned a body of correspondence rife with indignation and a compulsion "to set the record straight" on both sides of the Ohio at Gallipolis and Point Pleasant. The drama which played out in the press in support and against one side or the other and what was done or not done and when is revealing of what happened in terms of effective (and ineffective) martial might and of straight up human foibles. For a number of reasons, from the onset of the attack on the town at about 11 a.m. through the aftermath of the fight, the defense of Point Pleasant was left pretty much to those at Point Pleasant (i.e., Captain J.D. Carter's detachment; local militia; convalescent soldiers in hospital; and those of her citizens who had arms.) There was the unwillingness to send guns to the Mason County militia prior to the attack (probably for fear that the guns would be captured and turned on Gallipolis) and the inexplicable tardiness on the part of Gallipolis authorities in getting reinforcements over the river. This tardiness, when it became known that a large Confederate force was upon the Kanawha, cannot easily be accounted for.

Denizens of Gallipolis, whose direct experience with raids was practically nil as compared to the people of Point Pleasant, generally supposed that the Confederates would not attempt to capture Point Pleasant and by extension would *certainly not* hazard an attack on Gallipolis. This may have contributed to their lack of preparedness and promptness in responding to the attack on their neighbors across the Ohio. Another possible contributing factor to the less than prompt reinforcement of Point Pleasant was that once the battle had commenced, boats on the Ohio River were notified by telegraph to stay away from the Virginia side. Telegraph lines between Clarksburg and Charleston had also been interrupted so that military headquarters at Charleston was cut off from the "outside world and from that part of its department where Jenkins was operating."[333] Some reinforcements were ordered forward by General J.D. Cox, then presumably still at Marietta. He had summoned Colonel John L. Ziegler's regiment, the 5th Virginia Infantry, then at Ceredo and had sent a boat after them but these did not arrive at all.

This same lack of purpose characterized the conduct of the pursuit. Little was done to prevent the raiders from again crossing the Kanawha and making their escape southwards. Col. R.B. Hayes, cmdg. 1st Brigade at Charleston, advised that Capt. Fitch at Gallipolis be sent word "to run his steamboats up Kanawha" to intercept the rebels and prevent them from re-crossing but this like the ordering forward of the 5th Virginia by Cox, was not done either because it was too late or because in the aftermath of the fight the directive was overlooked. Hayes also ordered Lieutenant Colonel Comly "to rig a steam boat so as to protect men and go down the river to prevent Jenkins from recrossing the river." March 31, Comly set out as ordered from Coal's Mouth at daylight (about 7:30 a.m.) and headed down river. At 8:30 a.m., however, Hayes received a dispatch from Comly, then at Red House, saying that Jenkins was already "supposed to have recrossed the river five miles above Point Pleasant."[334] See also John Hall to Adj. Gen. H.J. Samuels (dated "Pt. Pleasant, April 6, 1863") in which Hall complained of Scammon's handling of the pursuit in which both the 23rd

Ohio and 13th detachment under Colonels Hall and Brown at Hurricane Bridge were to have co-operated in preventing Jenkins' men from escaping across the Kanawha to the south.[335] The upshot of this lack of concentrated and directed efforts with a central head was that there was a chase after the raiders in the ensuing days but always the pursuers found themselves out of striking range.

Questions in the aftermath of the raid

The delay occasioned by Gallispolis military authorities in coming to the aid of Point Pleasant sparked sharp criticsm at the Point. E.M. FitgGerald, editor of the Point Pleasant *Register*, in most direct language, laid blame for loss of life and property and Jenkins' escape with the bulk of his troops squarely upon Chief Quarter Master Captain at Gallipolis, E.P. Fitch. FitzGerald stated that when Captain Ford arrived at Gallipolis with the wharf-boat in tow and with the news that Point Pleasant had been attacked, Fitch, who had control of the government boats delayed for more than two hours (until about noon) before permitting any boats to return and succor the town. Ford stood ready to conduct the *Victor* with its battery of howitzers back to Point Pleasant to protect the bails of hay there (government forage) and relieve the town. Fitch, however, declined to permit any boat, including Ford's to go up the Kanawha after Jenkins and, blasted FitzGerald, it was at this point that the needless delay occurred in which no action was taken. "[F]inally," wrote FitzGerald, "after the enemy had abandoned all hope of taking the town, the boat was permitted to go to Point Pleasant. [...] Had the boat and re-enforcements arrived promptly, [Point Pleasant would] not have suffered in men and property."[336]

The Gallipolis *Journal* (James Harper, proprietor and editor and R.L. Stewart, assistant editor) publishing in defense of Gallipolis, maintained that there was no delay. When news of the "notorious Jenkins" attacking Point Pleasant reached Gallipolis at "12 M[idday]," this information put the Ohio town "instantly in commotion, and in a very short time a large force of armed men left for [Point Pleasant]."[337] Captain Ford having arrived at Gallipolis with the wharf-boat in tow and having given warning of the attack at Point Pleasant was re-enforced by local Gallipolis militia — the Trumble Guards (Lieut. Gilmer, cmdg.) and the Gallia Guards (commanded by Captain James Harper, also editor of the Gallipolis *Journal?*), together with a number of Gallipolis citizens. These then left immediately for Point Pleasant aboard the *Victor* with its floating battery of howitzers.

Nor was the question of delay the extent of the issue between the two towns as evidenced by E.M. FitzGerald's remarks below. Gallipolis claimed the lion's share of credit in repulsing the attack, claiming that it was the arrival of reinforcements from Gallia County, Ohio, that compelled the Confederates to retreat and which saved the town of Point Pleasant. Point Pleasant, for her part, claimed that the arrival of the reinforcements from Ohio being 'tardy,' the battle was won before ever the Ohioans arrived upon the scene, but that 'thank you very much,' their service was greatly appreciated in 'hunting out' and capturing Confederates still about the town and bringing them back to the boat to be conveyed to Gallipolis and imprisoned there.

Captain J.D. Carter fired off a missive of his own to the editors of the Wheeling *Intelligencer* to correct statements made by Quarter Master E.P. Fitch in his dispatch to the *Intelligencer* (pub. March 30, 1863). Among these statements was a claim that Point Pleasant was "taken" and so, the insinuation went, it was arrival of Fitch's steamer bearing artillery and militia which saved the beleaguered Point. "Point Pleasant was not taken and retaken," responded Carter. "This is," continued the Captain, "and I will call it by as soft a name as I can, [a] mistake. Point Pleasant was not taken, and consequently could not be retaken." Carter then stated briefly that he was attacked "at 11 o'clock, A.M." by the enemy "with 500 to 800." Carter's force amounted "to 60 men. We fought them until 3:30 P.M., when the enemy retreated." Then, continued Carter,

> [a]bout 4 o'clock the Trumbull Guards, [commanded by] Lieut. Gilmer, arrived and heartily engaged the rear of the retreating foe. They would have arrived in time to have taken part in the general engagement, had it not been that the man who controlled the steamer would not ferry them across the river; and anyone who understands army matters, knows who controls Government steamers when a Quartermaster is present. The stars and stripes waved over Point Pleasant all that day and still does. I, so far as I am individually concerned, care but little for such misrepresentations, but on account of the handfull of brave men that were with me, I deem it my duty as commander of this post to disabuse the public mind, and I think I have a better chance to know the facts than your informant Fitch from the fact that I was on the <u>Virginia</u> side of the river during the fight.[338]

Also refuting claims that Point Pleasant had been 'captured' during the late battle at the Point was E.M. FitzGerald, editor of the Point Pleasant *Register*. In a stinging letter to the editors of the Gallipolis *Journal,* he expounded that not only did the burning of the corn-cribs and government stables, beyond the town limits; the robbing of two stores (owned by pro-Union citizens); nor the necessity of cutting loose the wharf-boat, *not* constitute capture of the town, but the fight was already decided and the Confederates already retreating and "disgracefully routed" previous to "the tardy arrival of the *Victor* with reenforcements." The men of the Gallia County Militia (i.e., the Trumball Guards) in "hunting out the skulking thieves" with "alacrity" was greatly appreciated and for taking them "with a greater number otherwise captured, to the boat," the Guards deserved "great praise." "[T]hough," continued FitzGerald, "their earlier arrival would have been more opportune, yet all here entirely exculpate *them* from censure."[339]

The back and forth in the local press notwithstanding, the soldiers of the 13th Virginia (West) gained in reputation and stature in wake of the attacks on Hurricane Bridge and Point Pleasant in a way that was beyond dispute. "Carter's men successfully defend[ed] themselves against five times their number," reported the Gallipolis *Journal* (April 2, 1863, p. 2). Capt. Carter quipped in his letter to the *Dispatch* that there were "enough soldiers in Point Pleasant to *whip* the rebels which they did before help arrived."[340] Col. R.B. Hayes, Carter's superior officer, wrote to his wife praising the soldiers of Co. E. They had "defended themselves gallantly" and had repulsed all the Confederates attacks made upon them.[341] Civilian Brammer informed the *Ironton Register* that Co. E had "pepper[ed] it to the rebels when ever they appeared in sight."[342] As previously noted, from their position at the court-house, however, for all its central location, Carter's men "could not keep the whole town clear of rebels" and a number of the raiders scattered through the town in search of supplies to plunder.[343] The Cincinnati *Daily Commercial* reported that "a portion [of Jenkins'] "command fired several houses, and ransacked a number of dwellings and stores," and "also burned 7,000 bushels of Government corn."[344]

As to the feuding accounts, it would seem from this distance that while Carter's men held the courthouse and excercised control of the town center from their vantage point, they could not clear the town nor prevent mischief on the periphery nor in places out of their line of sight and range. Likely, it was the combined efforts of artillery, army and militia — the artillery firing upon the enemy from the boat and forces on land supporting — that soon cleared the raiders out of the town. Had reinforcements from Gallipolis arrived sooner, it seems fairly safe to surmise that loss of private and public property would have been less.

Finding the town could not be taken the Confederates began retreating at "3-1/2 o'clock P.M.[345]." By about 5 o'clock p.m., the Confederates had fallen back across the Kanawha. That night, they camped at the headwaters of Ohio Eighteen in south Mason County. Next day they were in Tazewell County, Virginia, and beyond the reach of Union forces.[346] Twenty-four prisoners were left behind in Federal hands.

Casualties and Other Losses taken at the battle of Point Pleasant, March 30, 1863

"[C]orn-cribs and government stables" located "beyond Point Pleasant town limits"[347] which had been without a force to hold them were plundered and burned.[348] Private property was vandalized.[349] Two stores owned by Unionists were partially robbed during the fight.[350] Capt. Carter driving home the exaggeration of the recent Gallipolis reports of the affair, stated in his letter to the Point Pleasant *Register* that a quantity of corn "belonging to the government" was burned but no "buildings" proper were put to the torch. Jenkins' men "burned the corn pens, a shed that was used as a bakery and a secesh stable. This was the extent, and *all* of their burning."[351]

The nature and extent of the damage done to Point Pleasant was widely and variously reported. Eyewitness John Hall, Esq., wrote that the Confederates had succeeded in stealing a box of ammunition, "a good [d]eal of clothing," and "about 200 horses from Union men."[352] As the Confederates "hastily skedaddled" from the scene of the fight, reported the Gallipolis *Journal*, "they stole at the same time about 25 mules belonging to the United States."[353]

A group of thirteen of Jenkins' men were captured on their retreat from Point Pleasant. These, (according to the account of one gentleman who provided eyewitness information to the Wheeling *Intelligencer*) had "before leaving the town supplied themselves abundantly with clothing and hats and shoes from the stores of Union men, and retiring a short distance took off their old clothes and put on the new."[354] It was while engaged in changing their clothes that they were captured. These hapless thirteen prisoners were taken to Gallipolis, confined, and refused parole.[355] On Saturday morning, April 4, they were conveyed to Wheeling, where they were committed to the Atheneum. On the lighter side, another casualty of the fight was the loss of Joseph Bromley's commission papers (commissioned as 2nd Lieutenant, Company F on October 8, 1862), which "was destroyed by Rebels in the fight at Point Pleasant."[356]

Little was gained by attacking Point Pleasant except the theft and destruction of a comparatively small amount of government stores and private property and the capture of a few citizens and soldiers, most of whom were subsequently released.[357] Confederates had overestimated the quantity and kind of government stores held at Point Pleasant. Someone had taken the precaution to stow government stores of significant value aboard the wharf-boat that was towed away to safety at Gallipolis. Not having acquired the stores they had expected to find and with the 13th Virginia and reinforcements from Gallipolis making it hot for them, after hours of fighting Jenkins' men withdrew "with severe loss."[358]

General Jenkins paid a steep price in effort and casualties in return for the little that was accomplished by the raid. There were few good horses and very little forage of any description to seize in the countryside through which they had passed. They were unable to dislodge the 13th at Hurricane Bridge; failed to capture the *Victor* or the *General Meigs* at Hall's Landing; and at Point Pleasant little was gained except the destruction of a comparatively small amount of government and private property and the capture of a few citizens and soldiers, most of whom were subsequently released.[359] In the final analysis, failing to capture the boats, the purpose of the raid on Point Pleasant may have been, as some supposed, simply to cause mayhem and get "something to eat and wear."[360]

The capture of Captain Alexander Samuels, Lieutenant Holderby and Dr. C.W. (Robert) Timms, Assistant Surgeon all belonging to Co. E 8ᵗʰ Virginia Cavalry, C.S.A.

Among the Confederate soldiers captured were officers, Captain Alexander (Alex) Samuels, Company E 8ᵗʰ Virginia Cavalry and Lieutenant Holderby (probably 1ˢᵗ Lieutenant George W. Holderby, Co. E 8ᵗʰ Virginia Cavalry), both inhabitants of Cabell County and Dr. C.W. (Robert) Timms, Assistant Surgeon to the 8ᵗʰ Virginia Cavalry (a Putnam County resident). Upon being questioned, their responses indicated that the capture of the two steamers had been an integral part of Jenkins' plan. According to "One Who Knows The Facts" (correspondent to the Point Pleasant *Weekly Register*), a "Rebel Captain, taken prisoner at Point Pleasant" disclosed what may indeed have been the crux of the plan. Jenkins had a spy at Gallipolis, who kept him informed of the departure and arrival of boats. He had been in hopes on taking the boat *Market Boy*, heavily loaded and just up from Cincinnati and then the two U.S. steamers, the *Victor* and *Meigs*. Jenkins intended to load his men upon the latter two boats to "come in them to Point Pleasant, attack and destroy the town then go to Gallipolis and fire the town, as Jenkins said that he was determined to burn the d-d town and [get] satisfaction for [the] past then to proceed to Guyandotte, burn the boats and escape to the mountains."[361]

Failure to take the boats seems indeed, to have reduced the attack to a low level raid for supplies and horses. Nor were the Confederates interested in taking prisoners. Men taken prisoner by them were nearly all immediately released on parole. For a few of the young men—back in their home neighborhoods—the raid it seems, became something of a lark and an occasion to visit 'the girl left behind.' A column in the Wheeling *Intelligencer* carried that the officers questioned divulged that had Jenkins captured the *Meigs* and the *Victor*, he intended to board them and "go down to Gallipolis to capture and parole the soldiers stationed there and to destroy the immense amount of Government stores at that point."[362] The plan, of course was frustrated by the unprecedented heroism of the crews—pilots and officers, who did not abandon their posts despite being showered with bullets and who successfully piloted their crafts through to safety.

In a more humorous vein, Captain Samuels was said to have stated that when he arrived at Point Pleasant with his men, he was "overcome by fatigue or something else, laid down and went to sleep and was only aroused by the rough handling of a Federal soldier who took him prisoner." Lieutenant Holderby called upon a young lady acquaintance soon after his arrival at the Point and "was so much engaged by her society that he was perfectly unconscious of the fact that the rebels had left the place and that Federal reinforcements had arrived from Gallipolis." Consequently, he was captured. "The rest of the rebel prisoners, some twenty in number were picked up in various parts of the town, having straggled behind the main body."[363]

The repulse of Jenkins' men at Point Pleasant became a feather in the cap of the Union people of that locale. It was cherished as a fitting comeuppance for the rebels of Point Pleasant, who reportedly, had "hailed the arrival of Jenkins with joy [...] fed the rebel soldiers, and directed them to the houses and stores of Union men, where they might plunder and destroy."[364] R.M. Stimson, owner-editor of the *Marietta Register*, wrote in his account that the prestige of the Union men in that neighborhood grew "more glorious," after the facts of the Point Pleasant engagement became more widely known. "Jenkins received such a 'drubbing' as he never before had," wrote Stimson.[365] Later that summer, the *Gallipolis Journal* in covering news of another strike by Jenkins' cavalry, this time at Greencastle and Chambersburg, Pennsylvania, remarked upon the lack of resistance offered by "the citizens of that country," who were largely of the Dunkard, Mennonite, United Brethren and Seven Day Baptist faith and thus pacifists, opposed to fighting and war. Jenkins had as a result of this posture a free hand and drove off "a large number of fine horses, cattle etc." The *Journal* made so bold as to compare the reception received by Jenkins at Point Pleasant in March with that offered him in Pennsylvania and advised Pennsylvania Governor A.G. Curtin:

> Jenkins' force in the late attack upon Pt. Pleasant was esti-
> mated at 650—and fully equal to the population of the town. Yet
> he was driven out of it by a very small military force, not a man of
> which had ever previously been under fire. With the same force he
> attacked two companies of the 13th Va. at Hurricane Bridge, and
> after a bombastic demand for surrender of the post, was forced to
> retreat with the loss of several men.
>
> We advise Gov. Curtin to procure the services of the Trum-
> bull Guards and the 13th Va., when Jenkins makes the next raid
> into Pennsylvania. The drubbing they gave him last March on the
> Kanawha has not been forgott[e]n. We predict the like results will
> follow any similar attacks.[366]

The 13th Regiment "gained great credit" for its part in the battles at Hurricane Bridge and Point Pleasant despite the fact that the part played by the regiment was misreported by citizens of Ohio, who were first to give their accounts to newspapers. A signal compliment came in the form of a petition signed by the "Citizens of Point

Pleasant" requesting that Co. E remain on duty protecting their town.[367] General Scammon in overall command congratulated the men and officers of the 13th in Special Order No. 51 issued Saturday, April 4. Scammon commended

> the troops of the Kanawha Division on the brilliant defense of Point Pleasant made by sixty men of the 13th Virginia. The loss of the Enemy exceeds the number of our troops engaged; and thanks are especially due to Captain <u>J. D. Carter</u> who so successfully conducted the defense against vastly superior numbers, as well as to the officers and men under his Command.[368]

Colonel R.B. Hayes, not expecting much from his Virginia troops, was heartened by the outcome of the engagements. In his diary Hayes entered: "Monday's fight at Point Pleasant was a fine affair; twenty Rebels killed and fifty taken prisoners, of whom twenty were wounded."[369] In a letter home to Lucy, Hayes' optimism stands in vivid contrast to his earlier remarks. The Kanawha army was strong enough, he believed, to hold its line against any threat the Confederates could bring to bear against it and "[h]e was especially pleased that his two West Virginia regiments, showing the results of good training, had developed into crack regiments."[370] Hayes' optimism was short-lived. By August 5, he wrote home that the Confederates "will run us out in a month or two, I suspect, unless we are strengthened, or they weakened. General Scammon is prepared to destroy salt and salt-works if he does have to leave."[371]

The number of casualties on the Union side in the engagement at Point Pleasant "were remarkably small considering the duration of the fight."[372] *Field History* of the 13th Regiment recorded the loss as "two killed and three wounded, of the latter 1st Lt W N Hawkins was seriously wounded in the shoulder."[373] Captain Carter reported "[o]ur loss was 1 man belonging to the 4th Va. Vols. who happened to be present and volunteered to help us Lt. W[illiam] N. Hawkins badly wounded."[374] The man belonging to the 4th Virginia, who was killed, was "Dr. Pritchett." He volunteered to fight with Company E and was "shot and fell dead in the Court House."[375] Allen Vickers, another casualty although not a regularly enlisted soldier was assisting the Home Guards when he was shot in the hand.[376] The number of Union men taken prisoner was not firmly determined as the following references indicate. Redmond recorded that thirteen men "principally from the hospital [were] taken prisoners and paroled."[377] Col. Hayes stated that "eight recruits for the 13th Va were captured by Jenkins during his late raid and immediately released on their parole."[378] Poffenbarger indicated in her post-war notes that "Carter lost 6 taken prisoner."[379] Capt. Fitch at Gallipolis dispatched to Kanawha Division Headquarters at Charleston that three Union army officers had been taken prisoner.[380] Pro-South diarist Henrietta

Barr living up river at Ravenswood, West Virginia, recorded that Jenkins had captured "about 40 men and a large quantity of commissary stores."[381] On April 18, with Special Order No. 5, Col. Brown without reference to specific men directed only that:

> The non-Commissioned Officers and Privates belonging to the 13th Regt Va Vol Inft, who were taken prisoners by Genl Jenkins during his late Raid in the Kanawha Valley men paroled and released by him in Violation of the Cartel, and are therefore not bound by the Parole. They will therefore return to duty immediately as if no parole had been given.[382]

Confederate losses tallied from official and unofficial Union sources suggest that the Confederates lost at Point Pleasant, all told, something over seventy men. *Field History of the 13th Regiment*, submitted to Gov. F.H. Pierpont by Col. Brown, reported that the Confederates sustained at Point Pleasant "a loss of 76 men in killed wounded and prisoners, including six officers among whom was Major Alexander Samuels Adjutant General on Genl Jenkens staff."[383] Muster Roll for Company E reported the total number of Confederate casualties as "twenty killed and fifty-two captured, twenty-five of whom were wounded, six of whom died."[384] B.J. Redmond's numbers are somewhat higher than Brown's and Carter's. In his letter to Pierpont, Redmond stated that the Confederates lost a total of 78 men, i.e., "rebel loss [is] said to be 25 killed, 26 wounded, 27 prisoners, among which are Lieut. Col. Samuels, 2 Captains, 2 Lieutenants and one Surgeon."[385] The surgeon seems to have been Dr. C.W. (Robert) Timms, Assistant Surgeon to the 8th Virginia Cavalry and one of the lieutenants was probably 1st Lieutenant George W. Holderby, Co. E 8th Virginia Cavalry.[386] Col. R.B. Hayes noted in his diary that "Monday's fight at Point Pleasant was a fine affair; twenty Rebels killed and fifty taken prisoners, of whom twenty were wounded."[387] Among the Confederates killed was Albert Neale of Ben Lomond, Mason County, and among the wounded left behind, was Edward Guthrie of Mason County.[388]

Local newspapers had their bit to add. The Gallipolis *Journal* published that in their haste to leave town the Confederates left their dead behind. They took "several" of their "wounded with them."[389] Capt. Fred Ford was quoted in the same issue of the *Journal* saying that "21 rebels" had been "found dead in and near the Point."[390] Mr. Brammer conveyed in his account to the *Ironton Register* that the nineteen Confederate dead who had been picked up "on the streets of Point Pleasant were buried on Tuesday last [probably March 31]."[391] The Point Pleasant *Register* reported in their edition of April 6 that thus far Confederate casualties were "20 killed, 25 wounded and 27 prisoners" but "dead rebels still are being brought in and scores of wounded were carried to Dixie."[392]

"the worst disappointed men you ever heard of"

— John Hall to General H.J. Samuels, manu. dated "Point Pleasant, April 6, 1863," Adjutant General Papers Miscellaneous 1861-65, West Virginia State Archives, courtesy of Terry Lowery.

Keen to repay 'hostilities,' the men of the 13th anticipated immediate orders to go in pursuit of Jenkins but in that they were disappointed by division headquarters. According to John Hall, who wrote to Wheeling about this, Gen. E.P. Scammon, commanding Kanawha Division, "forbid" Colonels Brown and Hall to go in pursuit of Jenkins until after they had been reinforced by Lieut. Colonel Comly with his detachment of the 23rd Ohio Volunteers. Consequently, the 13th was forced to wait for the 23rd to arrive and Jenkins was allowed to escape "with but little injury after leaving [Point Pleasant]." The 13th, to quote John Hall, became "the worst disappointed men you ever heard of."[393] Hall explained, writing that Scammon had sent Comly down the Kanawha from Charleston "on the day after the fite [the fight at Hurricane Bridge or Point Pleasant"] with about 500 of the 23d Ohio." Comly and his force "came along on the Boats" just as "Genl Jenkins had just got the last of his force over the River."[394] The rebels fired on Comly and his force as they approached Point Pleasant in the boats "but the boats run through came to this place [Point Pleasant] and on to Gallipolis then he was ordered back to Colesmouth and when there directed to send three of his Companies to Cols. Brown and Hall."[395] Col. Brown and Lieut. Col. Hall were directed to take five companies of men[396] and the three companies of the 23rd and proceed to cut off Jenkins' retreat. John Hall described what then transpired.

> About 2 o clock at night they reached Blacks or Howells cross Roads here the Ohio troops who had marched from Cole after 5 o'clock declared themselves Give out and beged for a couple of hours rest which was unwillingly given they started again at 4 oclock and reach the place whare they crossed the[re] about Two hours two late I am told the officers and men of the 13th was the worst disappointed set of men you ever heard of Jenkins had braged so much what he was going to do with them that they ware spilling for a fite and I have no doubt if they had been allowed to have went without the Ohio troops they would have wheped the Rebels but as I understand they are still picking up some stragglers if the Rebels accounts of the 13th be true our own farmer boys will fite The Rebels got here one box of amunition a goodeal of clothing and

they have stole about 200 horses from Union men our government want horses why not take all the Rebel horses in this country and pay for them at the end of the War. This would stop those raids otherwise if the union men are the only suffers they will certainly continue. They made the vote on the new state the test so let it be we ware threatened with this before the Election and I wrote to you[r] pl[ace] to that effect.

Remember me kindly to your family and to Governor Pierpoint. I am better in health since the operation on my eye.[397]

After the Point Pleasant fight Jenkins' band of men retreated east across the Kanawha River and moved into Cabell County to Howell's Mill on Mud River, where they encamped. The Wheeling *Intelligencer* reported that a gentleman, who had left Guyandotte on Wednesday, April 1, related that "Jenkins pickets were within three miles of that place" and that from there, he was supposed to be en route for "Dixie."[398] Another source, Captain J.V. Young wrote in connection to this phase of Jenkins' retreat that Jenkins "turned his course up the river, crossed at ten mile, and struck a bee line for Mud, but Captain [Isaac M.] Rucker, with his Home Guards, pitched into him and bushwhacked him to Mud River."[399] Young was with his company at Coals Mouth during Jenkins' raid and missed being "called on" by him.

The expedition in pursuit of Jenkins comprised of the detachments under Colonels Brown and Comly seems to have ended on the first or second of April. Comly returned to Charleston with Companies E and K.[400] Col. Brown returned to 13th headquarters by the 2nd. Brown, just in at regimental headquarters from the scout to Howell's Mills where Jenkins' had been reported, wrote a short note in response to Capt. Young's earlier correspondence describing the chase after Jenkins. Brown's note supports the gist of John Hall's letter and conveys how weary they all were. Brown and his detachment of 13th soldiers had come up with Jenkins at Howell's Mill and exchanged 'compliments' with Jenkins' pickets at Howell's during the early morning hours of April 2. Jenkins' men had been preparing to eat when firing began between Brown and the raiders.[401] The Confederates broke camp leaving their breakfast behind. Brown wrote to Young that Jenkins "had crossed and passed out about three oclock this morning [April 2], and my men were not in a condition to pursue him, he will probably pass over to the falls of Guyandotte and from thence to the Tug fork of Sandy where his wagons are."[402]

The Point Pleasant *Weekly Register* naturally had plenty to say about the "bootless attempt" to pursue the enemy. "[A]fter it was clear that Jenkins had given up all hope of taking the town of Pt. Pleasant there was a delay of 2 hrs. before the Victor No. 2 under command of Capt. F. Ford was permitted by Capt. E.P. Fitch, Chief Quartermaster at Gallipolis, to go up the Kanawha with the cannon and thus by shelling Jenkins army from the river, entirely prevent their recrossing to the west side of the Kanawha." Capt. Young, in command of his company (G) and Co. I of the 13ᵗʰ all posted at Coals Mouth, wrote to a friend in a disheartened way regarding the raid. He stated Col. W.R. Brown, cmdg. the 13ᵗʰ, had wished to pursue the raiders but unconcern for West Virginia among upper military had again foiled any possibility of making a decisive strike.

> The thief [Jenkins] came in with eight hundred but I don't think he got out with three hundred. He stole about one hundred head of horses and a great deal of other property and it all might have been saved. Jenkins might have been caught, but __ but __ but. Colonel Brown wanted to go but __. So, as it is, he is out and gone. The whole country full of squads of robbers. Three companies maintained their ground against six hundred. The firing commenced at 6 P.M. and lasted until half past eleven. (I mean at the Hurricane Bridge.) The whole country is in a state of excitement
>
> Lieutenant Brooks[403] has just returned from a two days scout in the Mud country, and says Jenkins is camped two miles above Barboursville on the Guyan river, and is reinforced by Clarkson.[404] They say that he has about eight hundred men, and when they rest he intends another attack on the 13ᵗʰ at this place and the Hurricane Bridge. I wish we were mounted and had a commander who cared for Virginia, or would let us go after them. But while we have commanders who won't let us drive the Rebels out of our country we might just as well quit, throw down our arms and make friends with the South. I have my company and Company I here, in all about one hundred men, but if old Jack does come I think he will have a good time of it.
>
> [...] Give my respects to Col. Oley[405] and tell him I want him in the valley, mounted. We want the 8ᵗʰ Va. here. Must have it. Captain, I am almost disheartened. But I know we are in a good cause.[406]

Kanawha Valley troops had not been deficient by any means, there just simply weren't enough of them to do the job that needed to be done. Between March 28 and April 15, "some five comp[anie]s" of the 23ʳᵈ Ohio Volunteer Infantry[407] "together with the 13ᵗʰ Virginia and the cavalry at Camp Piatt [detachments of the 2ⁿᵈ Loyal Virginia Regiment and other units] under Colonel [John] Paxton and and Gilmore"[408] "scoured the country"[409] in pursuit of Jenkins' men, killing and taking prisoner "about two hundred of Jenkins' marauders,"[410] stragglers mostly[411] "a[t] various points [...] who had been prowling about steeling evry thing they c[ould] carry off."[412] Colonel Paxton and Captain Gilmore went in pursuit of Jenkins "by different routes" ranging farther afield through Logan and Cabell counties "worrying him [Jenkins] badly and getting about forty prisoners,"[413] but the majority of the raiding party had made good their escape.

The "Union line" in the Kanawha Valley military district was extremely tenuous. So much so, that the defeat of Jenkins and his withdrawal from the area was widely considered a confluence of fortuitous events. Such luck could not be expected in case of future alarms. The line was too weak, and more of the same with more injurious results were expected at any time.

Suspicions were already afoot that Confederates were planning another advance into the Kanawha Valley. This sense of things kept Kanawha troops very much on the alert. Troops were shifted about and kept on rigorous schedules of picket duty. Given the tenor of the times, it is surprising that not more infantry were mounted. The floating question as to whether the 13th would indeed be mounted was finally resolved in the negative at the end of March. Brig. Gen. Scammon had wished to mount about 2,000 of his infantry—chief among these, the 13th Infantry Regiment. Maj. Gen. H.W. Halleck, however, did not approve "of that sort of mongrel force"[414] and so the 13th remained as they had been: a regiment of foot soldiers. Their rigorous duty in this rugged landscape of mountains and ravines, rivers and tributaries and bottomlands, in all kinds of weather, made of them something else ... what we might today call "special forces," especially tough and adept on the march and in a fight. Their stamina and military skill was honed and tempered at this time. Their time 'holding the thin line' would become indispensable to them as soldiers and to the larger armies they served with in the bitter campaigns of 1864, when they starved, marched in the hot Virginia sun and fought one battle after another in quick succession.

April 1863

Hard marches, scouting and skirmishing

Life in the Kanawha Valley military district continued much as it had previous to the engagements of March 28 and 30. April 1863, proved to be a time in which Federal and Confederate forces continued to shift about. Detachments moved along the fronts constantly looking for strategic advantage, here and there penetrating and gaining some temporary advantage. The men of the 13th West Virginia made hard marches, scouted and skirmished "frequently" with the enemy.[415] On April 6, 13th regimental headquarters was transferred from Point Pleasant to Hurricane Bridge, "a distance of 48 miles."[416] Headquarters remained at Hurricane for the month of April.[417] The regiment continued to be divided with posts at Point Pleasant, Mud Bridge, Coals Mouth and Hurricane Bridge. Field and staff officers and Cos. A, B, C, D and F were all stationed at Hurricane Bridge.[418] Co. E remained stationed at Point Pleasant.

At mid-month a shift of troops was ordered. On Friday, April 17, the 5th West Virginia Infantry, counterpart to the 13th in garrisoning this section of country, left Ceredo and took up post at Barboursville. Also this day (April 17), Col. Hayes wrote to Col. Brown to "direct Co. F of your Regt. to move to Pt Pleasant unless there be some special reason for sending another Co. instead."[419] Brown apparently had some reason for leaving Co. F in place at Hurricane Bridge. Instead, on April 19, Col. Brown ordered Lieutenant O.W. Griswold, cmdg. Co. H, to "report with your company for duty to Capt. J.D. Carter cmdg. Post at Point Pleasant, Va. You will be ready to march Tuesday morning the 21st inst."[420] Co. H arrived and remained with Co. E at Point Pleasant for the month of April. Young's Co. G continued to be stationed at Coals Mouth, probably still together with Co. I.[421] April 20, found the 13th divided in the following manner: Companies E and H were at "Camp Samuels," Point Pleasant; Companies B and G and "a detachment of recruits" (i.e., recruits belonging to Company I) were at Coalsmouth, Kanawha County; and Companies A, C, D and F were at Hurricane Bridge.[422]

On Saturday, April 25, 13th Companies at Point Pleasant and Coals Mouth were all ordered to report to Col. Brown at Hurricane Bridge. Upon their arrival, Brown directed "Lt. Col. Hall to take 3 companies of the 13th and locate at Mud Bridge"[423] (also called Mud River Bridge[424]) in Cabell County. Hayes had written that day to Lieut. Col. James R. Hall from 1st Brigade Head Quarters informing him that one company of the 5th Virginia Infantry had been ordered to relieve him at Coals Mouth and that when the 5th arrived at Coals Mouth, Hall was to report with his command (Cos. B and G, 13th Virginia) to Col. Brown at Hurricane Bridge. Brown would "designate three companies of the 13th Va. Regt" which Hall would then "take to Mud Bridge and establish a camp there." Two companies of the 5th Virginia would then report to Hall at Mud Bridge.[425] Only one company, that of Captain R.B. McCall, may have in fact reported to Hall at Mud Bridge. On April 28, McCall, commanding the detachment of 5th Virginia wrote to Hall: "In pursuance to orders from Hd. Qtrs. I am ordered to report to you with my Company for duty at Mud Bridge. I now await your orders at that place."[426] Thursday, April 30, Col. Brown issued Special Order No. 10 to Lt. Col. Hall:

> You will take three Cos of the Regt (Cos B G + E) and proceed to Mud Bridge on the Guyandotte Turnpike at which place you will find a detachment of the 5th Va Regt under the command of Capt R B McCall who will report to you for duty — You will select a suitable cite for a fortification — encamp there and fortify.[427]

Hall arrived at Mud Bridge with his command of three companies of 13th soldiers on April 30. The united command of 5th and 13th detachments "erected an earth fortification"[428] as ordered. R.B. Hayes had also written to Capt. J.D. Carter at Point Pleasant on the 25th regarding the move saying "[a] company of the 5th Va. Vols Inf has been ordered to relieve you at Point Pleasant. On their arrival there you will report with your command to Col. Brown at Hurricane Bridge."[429] The 13th Regiment drew clothing on April 30,[430] probably at Hurricane Bridge.

When the 13th moved they made "hard marches"[431] and when they stayed at post, they had long hours of picket duty. Such was the case with, 13th soldier, Mark E. Robison, of Co. E, who wrote a letter home on Monday, April 6, from Point Pleasant. At post there, he had been standing long hours on guard duty and going without sleep. His letter was written on patriotic Union paper with a biblical quote from Isaiah at the top. Robinson wrote:

April the 6 1863

Dear father and Mother I this After noon seet my Self to try to write you a few Lins to let you know that I am well at this present time in whitch I hope an trust you all ar in joying the same like Blessing I wood like to See you All but I cante tell when I shal Bee Bleest withe pleasure of seeing you all it may bee a long time and it mant Bee So long But i Shal trust an hope to God that I may have the pleasure of Seeing you All wonce more in Life I have a hard time her a standing Gard I have bin a an gard so mutch for the last weeke Or two an lost So Mutch Sleep that I feel striped at the present Company E is all that is her at this time and it looks like that wee ar going to stay her The Citsons [Citizens] has sign a Bill for to keepe us her Man to Man have Sign it So I cant tell whether wee will Stay or not I wish to know how you [f]elt after you got home a Bout hour Batle wee got 7 or of them it Done me good to Shoot At them I think I turned up Some of ther heels to Sun and Wee will try to do the same for them if tha Come back a gane I feel thankful to God that I got out as well as what I did I wood like to see pease so I could come home but I Doant see mutch Chance for it Soon I have nothing that is remaining to write to you At the present it Lookes Like ther is no Chance fore me to Come home ther fore I want you to write to me an Let me know how you ar all giting a Long I got a Letter from Moses the other Day and he Was well at the time he wrote the Leter he sed he did not know whether tha wood come Back or not But he thought that wood go to dixey Joel is Well at this time So I Shal Say no more at the present Ondly wishes you to remember me for ever as your treu son Affection son untill Death us Do part

Mark E Robison
Writing to his father
Mr John Robison

Please anser this soin Write to me as often as you can an let me know how you all ar giting a long Grandson B Moor is well at this time he is tending Van hauel[ ]ing At the hospital

So good by
For a while
Mark E Robison
To John Robison[432]

April 6, Hayes had again occasion to write to Colonel Brown regarding the tardiness of his reports:

> Your Tri monthly for the ten days ending March 26th has not yet been rec'd at these Hd. Qtrs. Their delay causes us much embarrassment. You will hereafter forward your Tri monthly reports promptly on the 6th, 16th, and 26th of each month and your Trimonthly report for March 26 and monthly report for March 1863 immediately.[433]

Major General A.E. Burnside, commanding the Department of the Ohio, wrote to General Halleck on April 17, proposing measures "to prevent further instances of what just occurred at Point Pleasant." Burnsides' perception was that the lack of mutual understanding between the military departments of Eastern Kentucky and West Virginia had made the region liable to raids like the one made on Point Pleasant. Burnside proposed that only that portion of West Virginia lying west of the ridge between the Little and Great Kanawha Rivers should be under Major General R.C. Schenk's jurisdiction, while that to the west should be consolidated with the forces of Eastern Kentucky. In this way, held Burnside, the present liability could be reduced by consolidating the forces of the two districts into one command. Had his suggestion in fact been heeded, it might well have alleviated some of the obvious difficulties in authority compromising the efficiency of the District of the Kanawha.

Burnside cited in support of his proposal that (1) the Confederate force in front of Newbern to the Tennessee line "acts as one body" and whether the enemy went on the offensive through Eastern Kentucky or West Virginia "they can only be properly met when there is an intimate connection between the troops on the Kanawha and on the Big Sandy." (2) The Kanawha Valley's connection with its headquarters in the east was "remote, roundabout, and precarious." Necessary stores were most conveniently drawn from the Department of Ohio (headquartered at Cincinnati) with its base of supplies located at Gallipolis, Ohio. (3) Thus, if the two districts were connected, establishment of a chain of outposts across the entire front would have been an easy matter, while supports at central points, like Charleston and Louisa, would be available from any point on the circumference.[434] Burnside would also have liked to have increased the efficiency of his department by having his troops "organized into a corps" with General George L. Hartsuff in command of it.[435] Unfortunately, nothing changed for the command in the Kanawha Valley. Things went on as they had been except perhaps just 'a little more so.'

Another Confederate incursion into West Virginia

Union troops in the Kanawha Valley did not have long to wait for the next Confederate raid. In addition to partisan attacks on scouting parties, the next major

Confederate incursion came with General John D. Imboden and General William. E. Jones. These forces raided northwards through Summerville with cavalry and infantry from April 20 through May 21. Their ultimate target was the Baltimore and Ohio Railroad in West Virginia.

The Jones-Imboden raid was one of the most comprehensive movements made against the railroad and its effects on the northwest were immediate. The goals of the raid were ambitious. They were to destroy all the bridges and trestling on the Baltimore and Ohio Railroad from Oakland, Maryland, to Grafton; "to defeat and capture Union forces at Beverly, Philippi, and Buckhannon;" to overthrow the government at Wheeling; to recruit men for the Confederate service; and to collect supplies, particularly horses, cattle and grain. Further, Confederate authorities were in hopes that the raid would enable their forces to (1) regain and hold Northwestern Virginia permanently, or at least for long enough to inflict great damage to the B&O railroad west of the Alleghanies and (2) with the shift of southern troops north, to draw the Federals out of Winchester and thus open the lower Shenandoah Valley to their own quartermasters and recruiting officers. The raiders fell far short of their goals but returned south by way of Summerville with a large train of supplies.[436]

Union forces all over the area held on tenuously. The raid caused considerable minor fighting. To counter this raid, "[r]egulars and home guards became more active, commands were shifted, and methods were altered."[437] At last, the need for highly mobile defense troops commanded by competent aggressive commanders became fully apparent. For the first time in West Virginia cavalry "became the chief arm in the defense structure. [...] Never again was the new State in danger of a general Confederate invasion."[438] One wonders if the question of whether or not to mount the 13th Regiment had been decided after the Jones-Imboden raid instead of one month before, if the 13th might not indeed have been mounted as intended by Governor Pierpont. Indications are that it would have been regardless of Halleck's condescending posture regarding putting such a "mongrel force" in the field. A reading of the Point Pleasant *Register* suggests that by end April the patience of private citizens had also been exhausted, and they were readily taking the law into their own hands to deal with guerrillas whose usual business was stealing, taking horses, bacon, meal, flour, salt, yarn, cloth etc.

At length, Jones and Imboden left as they had come, by way of Summerville laden with a large train of booty. Fearing, however, that the Federals would send a regiment from Fayetteville to intercept them, Colonel John McCausland (C.S.A.), formerly of Mason County, Virginia, was directed to make a demonstration with his command against Fayetteville to prevent Federal forces from detaching any of their force to go to Summerville.[439] On April 20, General Burnside would report that the disposition of the troops of his department were the following: his advance posts were "now on the Cumberland, from Celina to Somerset, with strong pickets as high up as Barboursville." There was now "no rebel force" in Kentucky north of the Cumberland, to speak of, except "some strolling bands and guerrillas."[440]

Promotions and orders

Beginning Wednesday, April 15, Col. Brown devoted considerable time to correspondence and issuance of orders. This continued for several days. He wrote recommending that John S. Cunningham of the 11th Virginia Regiment be appointed Adjutant General of the 13th to replace Emory J. Bridgeman, who was killed in action at Hurricane Bridge.[441] Lieutenant and Adjutant William I. Mathews had been filling in again since the death of Bridgeman but then on April 20 Brown issued Special Orders No. 6 and 7 providing that,

> 1st Lieut + Adjutant W.I. Mathews 13th Reg V.V.I. having rec'd from the Governor of Virginia a commission as Captain of Co. H. [13th] Regiment is hereby relieved from duty as Adjutant. He will take command of his Co. To supply his place 2nd Lieut J.S. Cunningham Co. G. is detailed and will report in person at this Head Quarters.[442]

And

> 1st Lieut and Adjutant William I Mathews 13th Regt Va. Vol. Infty have received a commission from the Gov of Va as Capt of Co H of that Regt is hereby ordered before the U. S. Mustering Officer at Wheeling Va for the purpose of being mustered out as 1st Lt and Adjutant and mustered into service as Capt of Co H 13th Regt Va Vol Inft.[443]

Cunningham was reported commissioned 1st Lieutenant and Adjutant of the 13th West Virginia as of April 2, 1863.[444]

April 15, Brown issued a series of special orders. Special Orders No. 1 provided that: "Commanding officer Co B 13th Regt Va Infty will detail a suitable Non Comd officer to take charge of and proceed to deliver to the Provost Marshal at Charleston Va Charles T. Carroll Prisoner captured by the forces of this command."[445] Special Orders No. 2 provided that, "Commanding officer Co. A will detail Milton Wilson (Private) for Nurse in Hospital."[446] Special Order No. 4 ordered "Lt Col. J.R. Hall 13th Regt Va Inft [...] to proceed to Charleston on business connected with the Regiment Transact it and return without unnecessary delay."[447] On April 18, Brown issued Special Order No. 5 from 13th Headquarters at Hurricane Bridge, directing that

The Non-Commissioned Officers and Privates belonging to the 13th Regt Va Vol Inft, who were taken prisoners by Genl Jenkins during his late Raid in the Kanawha Valley were paroled and released by him in Violation of the Cartel, and are therefore not bound by the Parole. They will therefore return to duty immediately as if no parole had been given.[448]

A question had arisen as to whether the parole given the 13th soldiers captured by Jenkins was legitimate and binding. Brown had written to Col. Hayes and Hayes in turn wrote to Captain James L. Botsford, Assistant Adjutant General, District of Kanawha, at Charleston, Virginia, on April 14, 1863, for confirmation. Hayes wrote:

I have received information that eight recruits for the 13th Va were captured by Jenkins during his late raid and immediately released on their parole. My understanding of it is they should be immediately released to their regiment for duty. Am I correct? Hayes[449]

Two days later Hayes wrote in reply to Brown that,

the men in question were paroled and released in violation of the Cartel. They are not therefore bound by the parole, and will be ordered by you to return to duty as if no parole had been given. This decision is according to General Order No 15 Current Series issued by Genl Schenck. I have not been furnished with the Orders or I would send it.[450]

On May 4, Captain Martin P. Avery, Assist. Adjutant General at 1st Brigade Head Quarters, Camp White, also wrote to Colonel Brown in response to a previous letter from him regarding the validity of the parole. Averell wrote and enclosed a copy of General Orders from corps headquarters which probably ended the matter:

Col. Brown

In reply to your letter of April 28 in regard to paroled prisoners of war. I call your attention to General Order H 16 from Hd Qtrs., 8th A.C. a copy of it enclosed. 'Men not paroled in accordance with the cartel are not considered as having been prisoners of war and if such men are about from their commands they will be reported as Deserters.' The Col. commanding this brigade directs that you arrest any paroled soldiers within your reach as Deserters who are absent without the proper authority and send them to the nearest Provost Marshal[451]

April 23, Col. Brown and regimental Quartermaster Stephen Comstock had occasion again to be censured for problems with their monthly reports. Brown was reprimanded by Colonel Hayes because irregularities in his tri-monthly reports consistently rendered them incomplete.[452] Hayes also admonished Brown for his failure to send in his lists of "Deserters or Absentees" as requested in a "Circular" from Brigade Headquarters (dated April 14) sent pursuant to General Order No. 16 received from Division Head Quarters.[453] Capt. J.D. Carter's letter to Adj. J.S. Cunningham two days later in which he acknowledged receipt of Cunningham's latest correspondence, "calling attention to certain orders + circulars in reply,"[454] probably referred at least in part to the irregularities in reports referred to by Hayes. Carter continued in his letter to Cunningham:

> I have never been furnished with said orders, and circulars, consequently cannot refer to them. Capt. Matthews has not said any thing to me in reference to the matter. I have no Deserters to report. I have never had a man to leave without my consent, we are all presented or accounted for.[455]

13th Quartermaster Stephen Comstock received similar promptings. April 23, Quartermaster William McKinley (in future, the 25th President of the United States) wrote to Comstock, reminding him:

> Lieut.
>
> You will forward to these Hd. Qtrs. on the 4th, 9th & 28th of each month your tri monthly report of Means of Transportation condition +c. You will also report on receipt of this what transportation you may need Quartermasters stores +c. I have some stationery send in your Requisitions, will ship to you.[456]

On the 25th, McKinley again wrote to Comstock instructing that: "In your tri-monthly reports to this office you will hereafter add the amount of forage also the amount and kind of tools."[457] After the shift of 13th detachments from Point Pleasant and Coals Mouth to Hurricane Bridge and Mud Bridge, Hayes again wrote to 13th Regiment headquarters regarding proper submission of reports to district headquarters and to Washington in the following memorandum (Order No. 19) to Col. W.R. Brown, Lieut. Col. James R. Hall and Col. A.A. Tomlinson, commanding 5th West Virginia Infantry Regiment. On Wednesday, April 29, Hayes wrote:

> You will make a consolidated report to those Hd. Qtrs. of the troops under your command on the 8th, 18th and 28th of each month. You will not wait for the reports of the detached companies

of your Regiment as they have been ordered to report direct. This will also apply to the monthly returns except the copy which must be sent to the Adjutant General at Washington.[458]

On April 23 and 24, Col. Brown issued orders regulating the order of the day in camp. Special Orders No. 8 directed that:

Until further orders Company drills will be from 9 to 10-1/2 O.C. A.M. and from 2 to 3-1/2 O.C. P.M. —

Dress Parade will be at 6 O Clock P.M.[459]

Special Order No. 9 issued April 24 directed that in regard to those in need of medical care:

There will be a sick call at 8-1/2 O clock every Morning at which time, the first Sergeant will make out a list of sick soldiers in their respective Companies and march them to the Hospital and report them, with a list of their names to the surgeon of the Regiment for examination. —[460]

On April 28, to curtail a problem with the indiscriminate discharge of firearms in camp, Col. Brown ordered that "any Officer or Soldiers who shall hereafter discharge Fire arms of any description within this command (unless at an enemy or in necessary self-defence) thereby creating unnecessary alarm within the same; shall be sent forthwith to Divis[ion] Head Q[ua]rt[e]rs to be tried and dealt with according to the 49th Artic[le] of War."[461]

Morning reports for the month of April 1863 indicate that Company A was stationed at Hurricane Bridge, Putnam County, for the entire month, and that the company had present and absent three commissioned officers and eighty-three enlisted men. On April 1, a new recruit joined the company. April 2, one enlisted man, George Walter Fitzwater, died in "Hospital Hurricane Bridge" of "disease &c." One drummer was reported present and on duty at camp for the month; one artificier was reported present from April 1 through 20; and a farrier or blacksmith was present from April 21 through to the end of the month. Two enlisted men were absent with leave April 1 and 2. No one was reported in arrest or confinement during the month and no one had special or extra duty. One to two privates had daily duty April 17 to 30.

The month of April saw a numerous details for detached service drawn upon Co. A. On April 3, C.C. Frame, Milton Wilson, Job Hall and William Compton were ordered to Charleston. On Monday, April 12, many details were made. One officer and seventeen enlisted men left camp. "Remarks," record another group on detached detail: Lieutenant Greenbury Slack with twelve men. These were ordered on a scout in the direction of Guyandotte. Other detachments included Sergeant George Danner and four [?] men, who were detailed to guard prisoners to Charleston. Groups of men were recorded returned from detached duty. These included William Compton, Job Hall and Milton Wilson returned, and Charles P. Quigley, Andrew Snodgrass, Daniel Snodgrass and Levi Hennings returned to camp from detached service. On April 13 and 14, one officer and thirty-six men were out on detachment. April 14, Christopher Frame returned from Charleston and Lieut. Slack came in with twelve men from scouting duty. On April 16, one Lemuel Anecomie (a member of the 13th W. Va. Regiment?) was ordered to Charleston with a prisoner. Sunday, April 18, 1st Sergeant George Danner returned from Charleston. Monday, April 21, Levi Hennings and Andrew Snodgrass reported at Hurricane Bridge. April 22, Charles P. Quigley returned to camp from his home at Charleston, and "Danil Snodgras" (Daniel Henry Snodgrass) returned from his home at Charleston. April 23, S.M. Comice (member of the 13th Regiment?) returned from Charleston, and April 24· Sergeant Amos Brown reported for duty.

Co. A had a minimum of six and a maximum of ten men sick during the month. Beginning April 25, there was a sharp increase in the number of sick men. Between April 25 and 29, fourteen to seventeen men were reported "present sick." Some of the men had of necessity (due severity of the illness and risk of contagion?) to be conveyed to other hospitals, leaving by April 30 just seven men sick in camp. The rest were reported as "absent sick."   April 1, Perry Gatewood was detailed for detached service at Regimental Hospital Point Pleasant and on Sunday, April 18, he was ordered from Point Pleasant to Hurricane Bridge Hospital. April 10, Ralph Pauley returned from home unfit for duty. April 16, Milton Wilson was detailed for hospital nurse. Monday, April 21, John Moor was absent with leave, sick at his home at Charleston. April 22, Elijah H. Newel was sent to Point Pleasant Post Hospital. April 25, George W. Ramey returned from home unfit for duty and Obediah Buckner was reported absent sick on the the 25th.[462]

Morning reports for Company B indicate that the company was stationed at Hurricane Bridge April 1 through 16. April 17 to 30, the company was reported at Coalsmouth. A small number of men were granted leave of absence during the month. One to three enlisted men had leave of absence April 1 and 15 and two to four enlisted men between April 25 to 30. No one was absent without leave during the month. One enlisted man was reported "missing in action." On April 18, Corporal Patrick H. Caldwell died in Hospital at Hurricane Bridge of "Typhoid Pneumonia"[463] brought on "by exposure."[464] Patrick had been present on duty from the date of muster-in (Oct. 8, 1862), until end of February 1863.

At some point thereafter, he had become sick. A special muster roll for April 10, 1863, indicates that he was at that time sick in hospital.[465] On April 20, Milton Stewart, Captain, Co. B, wrote with compassion and compliment to Robert Caldwell, father of Patrick:

Coals Mouth Va April 20th/ 63

Mr Caldwell

Dear Sir. The bearer of this will convey to you the sad news of the death of your son. He had been suffering for some time past with lung-fever — the skill of the surgeon was unavailing — and God took him away from us. We mourn the loss of a good soldier — a noble patriot, you the loss of a loving Son. Our heartfelt sympathies are with the bereaved,

Very Truly Yours
Milton Stewart
Capt. Co. B 13th Va. Vols.[466]

On April 25, 1863, Capt. Stewart again wrote to Robert Caldwell regarding the return of Patrick's possessions and how he might proceed to collect money yet due the soldier from the federal government. Stewart wrote:

Dear Sir:

Enclosed you will find $2.50 the amount of money Patrick had with him at the time of his death. I send you in a box to Point Pleasant all his clothing, consisting of 2 Blankets 3 prs Drawers, 1 pr Socks, 1 pr Shoes, 4 Shirts, 1 pair Trowsers, 1 Blouse, 1 Great Coat, 1 Comfort, Pair of Gloves, Hat and 2 Ambrotypes of himself. By sending to Eben Maloney Revenue Collector at Point Pleasant you will get the box. He has due him after deducting the balance, on his clothing account, — $30.50 as Corporals pay, and $75.00 due him on bounty, which will fall to you Any lawyer almost can collect this for you, R.L. Stewart of Gallipolis is engaged in that business I would respectfully ref[er] you to him. His papers will be made out once sent to Washington — you will have no difficulty

Respectfully
Milton Stewart
Capt Co B 13th Va Vols.[467]

Company B finished the month of April with three commissioned officers and eighty-nine enlisted men present and absent. None were designated for special or daily duty. Four privates were detailed for extra duty from April 1 through 16. One to six enlisted men were detailed for detached service each day of the month. No one was present or absent in arrest or confinement during the month. Co. B had from two to eight men present sick and from eleven to sixteen absent sick. The greatest number of men reported sick came at the end of the month.[468]

Morning reports for Company C show that the company was stationed for the duration of the month at Hurricane Bridge. No one had leave of absence. Private Abraham Asbery deserted the company at Point Pleasant on April 12. He remained in desertion through the end of April, "supposed to be in vicinity of the Ohio River 20 miles below Point Pleasant."[469] No one was present or absent in arrest or confinement. Co. C had eight to thirteen men present sick over the course of the month and none absent sick. At the end of the month, the company tallied present and absent three commissioned officers and seventy-seven enlisted men.

No one in Company C was detailed for special or daily duty. April 1 through 8, five privates had extra duty. One commissioned officer had detached duty for the entire month of April. Details of enlisted men on detached service were: one enlisted man April 1 through 8; four enlisted men April 9 through 25; and two enlisted men April 26 through 30. April 2, thirty-three men of Co. C "scout[ed] on Mud River, Virginia." This scout returned to camp at Hurricane Bridge on April 3. April 23, John P. Harmon and on April 25 James L. Woodyard reported to camp for duty.[470]

Company D reports for April indicate that the company was stationed at Hurricane Bridge for the month. Co. D had one drummer present for the entire month and one fifer present March 1 to 11 and March 14 to 22. One enlisted man was granted leave of absence April 6, April 14 and the 15th. One enlisted man was reported absent without leave April 8 through 10. Three enlisted men returned from desertion on April 1. The company finished the month with an aggregate of three commissioned officers and eighty-four enlisted men, present and absent.

2nd Lieutenant George Snowden was assigned to special duty on April 1 and 2. Each day of the month, one private had extra duty and two to four privates had daily duty. April 1 or 2, Lieutenant James Hanna and fifty-two enlisted men were among the "five companies" of the 13th W. Virginia ordered to go in pursuit of A.G. Jenkins after the battle of Point Pleasant. The detachment under 1st Lieut. Hanna reportedly "went in pursuit of General Jenkins forces towards Barboursville." April 20 to 22, one commissioned officer was detailed for detached service and April 22 to 30, one enlisted man was detached. The company had one to six men present sick during the month. Five to eleven men were absent sick, including Captain Simon Williams, who was sick for the entire month. Two privates were present in arrest or confinement April 1 through 19 or 20.[471]

While home on sick leave and visiting at the local store, Capt. Williams spoke about the 13th Regiment to those present. In particular, he speculated that the 13th "were to be moved soon to Tennessee if they were, he argued, they were not going to be home for a long time."[472] Lovina Burrows, wife of David Burrows, Company F, overheard Williams and wrote to her husband about such a possibility. David wrote in response:

> [...] About what Simon Williams up there is talking about the men, there is nothing to it. We don't know whether we will stay here very long or not and we don't know where we will go when we leave here.[473]

April 8, Captain J.D. Carter, Co. E, commanding the Point Pleasant military post, issued Special Order No. 39 ordering James H. Gunter, Co. E, to

> take charge of 3 stragglers from the 8th Va. Regt + 28th OVI and deliver them to the Provost Marshal at Parkersburg also 1 Rebel prisoner of w[ar] and 1 citizen and deliver them to Provost Marshal at Wheeling, Va. and return without unnecessary delay. [...] H.H. Bogess Captain + A.Q.M. will furnish transportation. Lt. Hill will furnish subsistence.[474]

April morning reports for Company F indicate that the company was stationed at Hurricane Bridge April 1 through 30. No one was absent with leave or without leave. One new recruit, Private Abner Johnson, enlisted in the company on April 2. Emory J. Bridgeman died on April 2 of wounds received in action on March 28 at the battle of Hurricane Bridge. At the end of the month the company numbered three commissioned officers and seventy-seven enlisted men.

No one in Co. F was detailed for special duty. Two privates had extra duty each day of the month and five privates had daily duty. One commissioned officer and fourteen enlisted men had detached service April 1 through 3 (detachment likely in the pursuit and mop-up of Confederate stragglers in the aftermath of the Jenkins' raid). Co. F was detailed for detached duty each day of April but the number of men on detachment stayed low with just one to two commissioned officers and three to five enlisted men on detached duty at any one time. Co. F reported no men present sick for the period of April 1 through 6, and just two to four present sick for the remainder of the month. Four enlisted men were reported absent sick for the period of April 1 through 27, and three were so reported April 28 through 30. No one was present or absent in arrest or confinement.[475]

On Friday, April 10, Brigadier General Scammon wrote to Col. Brown regarding trouble with Company G:

> The Co[mpany] at Coalsmouth under Capt. [John V.] Young is in a bad state of discipline. You must look to this and correct it. It would be well to send another Comp[any] there with a Superior officer or station one there in place of Capt. Young.[476]

April 14, Col. R.B. Hayes, cmdg. brigade, wrote to Capt. Young with orders to "direct a detail of two good men as teamsters for these Hd. Qtrs. to report to Lt. Wm. McKinley A.A.Q.M. immediately."[477] April 19, 1863, Capt. Young wrote to his wife, Paulina ("Marsh") from Coals Mouth. He had been in hopes of having opportunity to get leave to visit home

> but Col. Hall is going to the Point tomorrow and will be gone one week, and Gen. Scammon has turned over the command to me until he returns, which will keep me here one week longer.[478]

Young, who would likely have welcomed being superceded by another officer at this juncture as recommended by Scammon, was tired of the responsibility of the command of the Coals Mouth Post and its attendant problems and hardships as he explained in his letter to his wife:

> You don't know the trouble that it causes me. The she-devils are running to Headquarters reporting me every week — But thanks to General Scammon — he is not General Crook.
>
> Sam Wilson's negroes ran away last night — the last one, big and little, and two of Sam Rust's. Will let them go — it is their Southern rights. Mr. Lewis tells me this morning that the Secesh are cutting up on the lower side of Coal about their negroes, and say Captain Young won't give them any protection at all. Well, suppose I don't. I did not leave my happy home to protect rebels' property – won't do it.
>
> There are three companies here now and it will give me something to do this week. [...] There is no medicine here.[479]

The 13ᵗʰ West Virginia needs hospital staff and surgeons

Not only was there no medicine but 13ᵗʰ hospitals were understaffed and "outside hiring" had to be resorted to. To resolve this, John Hall, Esq., wrote to Governor F.H. Pierpont on Thursday, April 23, introducing and recommending Dr. A[bram] Williams, graduate of Jefferson Medical School, Philadelphia for service with the 13ᵗʰ

Regiment. Hall wrote that Williams had "considerable hospital experience" and was "a modest man of laborious habits [...] free from the Vises that distroy the usefullness of so many Physicians in the army." Hall continued writing that "Dr. [Samuel G.] Shaw as well as many of the officers of the 13 Va Volunteers have authorized me to say to you that the appointment of Dr. Williams as assistant surgeon would be acceptable. I am still deriving comforts from the last operation on my Eye." In post script he added that "[t]he Regiment is divided and it would be in saving to the Government to give it an assistant surgeon instead of out side hiring."[480]

April 28, George M. Kellog, Medical Director of the District of Kanawha, also wrote to Gov. Pierpont requesting an appointment for an assistant surgeon for the 13th. Kellog wrote:

> I have the honor to request of you the appointment of an assistant Surgeon for the 2nd Va Cavalry as also for the 13th Va Regt Vol Inf — the campaign being now about opening for the spring we need an immediate increase of our medical officers as the exigencies likely to arrive cannot be well met by the medical officers of this division.[481]

Abraham D. Williams, age twenty-three, was subsequently enrolled by Governor Pierpont and joined the 13th West Virginia for duty at Point Pleasant, on May 10, 1863.

Dr. Williams

Abram D. Williams was born in Williams, Indiana, a small town named for an early progenitor of the Williams family and situated on a branch of the White River in Spice Valley, Lawrence County, Indiana. Abram was born September 16, 1835, and was thus aged about 28 years when he joined the 13th Regiment as Assistant Surgeon.[482] He was born into a family of well-to-farmers and stock-breeders and was the nephew of Dr. Elkanah Williams, a pioneer physician of the specialty fields of ophthalmology and otology and occupant of the first chair of ophthalmology in a U.S. college. It comes as no real surprise that at the time of the taking of the 1860 Indiana Census (taken for Spice Valley Township, Lawrence County, on August 7, 1860), the 24-year-old A.D. Williams was listed as a student (probably a medical student embarked upon a 3 year course of study), still single with a surprising personal estate of $2,400.00 and rooming together dormitory-style with three other fellows.[483]

A.D. Williams began his medical studies at Miami College of Cincinnati, Ohio, where his uncle, Dr. Elkanah Williams, one of the early practitioners and teachers of ophthalmic and otologic medicine in this country (see *Appendix*, p. 390-92) also taught. Miami Medical College was absorbed into Ohio Medical College of Cincinnati in 1858. (It was re-established again as a separate college by 1865.) The Civil War interrupted Abram's studies and forced re-organization of many school curricula as in many colleges in both the North and the South, students and faculty joined the war effort. Professors, lecturers and medical "demonstrators" left to obtain commissions with army and navy hospitals. Having no teachers, medical schools reduced or suspended classes altogether. Abram was fortunate. He continued his studies—for a time at least—privately with his uncle.

The teacher and student relationship is obvious in a letter written by the elder man to his nephew. The letter was dated November 12, [1862], and is contained in a collection of Dr. E. Williams correspondence published by Ancestry.com. Dr. E. Williams and his wife, Sallie, had traveled to Seminary Hospital at Frederick, Maryland, to nurse and treat a young man named Eldridge. Dr. E. Williams wrote to Abram with vivid medical detail describing the condition of the dying man, together with a description of the effect of the "stimulate" administered by Dr. E. In passing, Dr. E. referenced the long interruption in Abram's lectures writing: "I am very sorry that you have been kept back from Lectures so long, but it has been unavoidable. We will try to make up for it in the future."[484]

Abram, however, determined to finish his course of study, transferred to the prestigious Jefferson Medical College, at Philadelphia.[485] Jefferson College neither suspended classes nor ceased graduating their students during the war or during any war for that matter, since its first class graduated in 1827. Williams graduated from Jefferson Medical College in 1863, and soon after enrolled in the 13th West Virginia Infantry to occupy the position of Assistant Surgeon.

January 31, 1865, Abram resigned from the army and went to Europe where he took post-graduate study in diseases of the ear and eye. The 1850s and 1860s brought exciting progress in modern otology and ophthalmology but one had to travel to Europe for study (the locus of the new sciences of opthalology and otology) not only with funds but prepared to take lectures given in the foreign languages of German and French. After completing his otological and ophthalmological studies, he returned to Cincinnati to practice and lecture in otology at Miami College of Medicine.

Sometime before 1870, Williams moved west. The 1870 U.S. Census for Kansas City, Missouri, Ward 4, records that A.D. was 33 years of age, a practicing physician and sharing living quarters with a grocer and a student dentist. Sometime between 1870 and 1873, Williams moved to St. Louis, Missouri, and set up a practice there. In 1872, he married Isabella M. Williams (born in Kentucky September 1836).[486] Isabella, called Belle, was a daughter of Senator John Williams of Kentucky. Abram and Belle had 2 daughters: Elizabeth ("Lizzie

Ann") born about 1872 and Annie P. born about 1877.[487] In 1873, A.D. Williams' authoritative and widely-read book, *Diseases of the Ear* was published.[488] St. Louis City Directories from 1863-1923 suggest that he resided and practiced in St. Louis from the 1870s to the early 1900s. For many years he conducted his practice in the diseases and surgery of the ears and eyes at Eighth and Chestnut Streets in St. Louis. As was typical for the period, he may have had his offices adjacent to his residence.[489]

Williams came to have a national reputation as an ophthalmologist and otologist. He practiced and taught "for many years" in St. Louis. In 1894, Williams was listed as one of the faculty at St. Louis Woman's Medical College and Hospital (in existence from 1883 to 1996). He taught ophthalmology and otology. In 1894, the college and curriculum was described in the following advertisement:

> Prominent among St. Louis' most important medical institutions is the Woman's Medical College and Hospital, Sixteenth and Pine streets. It was incorporated in 1891. The hospital is located in the college building, contains twenty rooms and wards neatly furnished and supplied with all the appliances necessary for the comfort of the sick. Besides having equal privileges at all the city hospitals, the students of this college have exclusive access to this. Faculty: G.F. Hulbert, M.D., president, gynecology; George Wiley Broome, M.D., secretary, practice of surgery and clinical surgery; M.L. Boas, M.D., treasurer, practice of medicine; C.F. Wilson, M.D., osteology and demonstration of anatomy; Given Campbell, M.D., nervous and mental diseases; Hugo Rothstein, M.D., diseases of women; C.O. Van Ness, M.D., chemistry; I.N. Love, M.D., pediatrics and clinical medicine; Paul Paquin, M.D., microscopic and analytical diagnosis; Hanau W. Loeb, M.D., diseases of nose and throat; P.F. Hellmut, M.D., D.D.S., dental surgery; Ellen M. Osborne, M.D., instructor in midwifery; A.D. Williams, M.D., ophthalmology and otology; W.C. Usser, M.D., physiology; F.P. Gillis, M.D., obstetrics; Henry Jacobson, M.D., genito-urinary and rectal surgery; John H. Duncan, M.D., diseases of the skin; Chas. W. Schleiffarth, M.D., pathology; Augustus C. Bernays, M.D., anatomy; Governor Chas. P. Johnson, legal medicine; Geo. W. Cale, M.D., principles of surgery and clinical surgery; Frank Parsons Norbury, M.D., cranio-cerebral surgery. The professors give lectures at the college every week on the special branches over which they are assigned. The course is three years, graded. The

instruction is didactic, clinical, demonstrations and recitations. The grades required from the students are higher than those usually demanded, and every effort is made to properly prepare and fit them for the responsibilities assumed after graduation. For further particulars as regards qualifications for matriculating, for graduating, methods of instruction, fees, text books, etc., our readers should write for the annual announcement of the Woman's Medical College and Hospital — an honor to St. Louis, the State and country.[490]

Dr. Abram D. Williams was also an honorary member of the St. Louis Medical Association. After the death of his wife Belle in 1910, he moved back to Williams, Indiana. The 1910 Indiana Census lists him as 74 years. Living with him are his two daughters: Elizabeth age 35 and Annie aged 32. He died two years later on August 6, 1912, at his home in Bedford, Indiana, at age 77. Cause of death was given variously as the result of "acute indigestion, following a long illness"[491] and due to complications of disease and old age. He was buried at Bedford, Indiana.[492]

Now that we have graduated, married and buried the good doctor, let us return to that point on May 10, 1863, when the young 28 year old Williams "joined for duty and enrolled" in the 13th West Virginia Infantry Regiment at Point Pleasant. He was appointed Assistant Surgeon and regimental returns indicate the following record of duty. He was present with the regiment May to September 1863. For October 1863, his presence or absence was not stated. November 1863 to April 1864, he was reported present with the regiment. Beginning May and June 1864, he floated between regiments and hospitals. During those two months, he was "absent detailed on special duty in the 15th (or 5th?) West Virginia Regiment. July and August of 1864, he absent assigned to duty in 5th West Virginia Regiment, and November and December 1864 he was "absent assigned to duty in the 1st Regiment Virginia Veteran Volunteers." January 1865, he resigned. His resignation was "accepted by Special Order No. 24 dated Jan. 31/65," issued by Major General George Crook.[493] During his military career with the 13th and other West Virginia regiments, he resigned twice and was twice granted leaves of absence.[494]

The trouble began end of September 1863, when Dr. Samuel G. Shaw resigned his position as Surgeon of the 13th West Virginia. Then, jockeying for position began. When Dr. Williams was passed over for promotion in favor of a "non-professional" physician (i.e., without medical diploma), he tendered his resignation (for the first time) in the following letter:

Hospital 13 Reg Va Vol Inft

Barbourville W. Va.
February 20, 1864

To
Thayer Melvin
Capt + AAG.

Sir:

I have the honor herewith to tender my resignation as Asst Surg 13 Reg Va Vols Inf. It is with reluctance that I state some of the many reasons I have for adopting the course which, I think, my self-respect and the regard which I have for the profession which I hold alike sacred demand.

First. On the 29th day of Sept 1863 the resignation of our late Surgeon was accepted by special order no 45 from Department Head Quarters.

Overlooking all the ordinary customs of promotion, the Colonel commanding and a majority of the officers of the regiment persisted in recommending Dr. Dalley 2nd Asst Surg 5 V.V.I. who has no medical Diploma and cannot therefore be a regularly authorized Physician and had him commissioned Surgeon of this regiment. All of which is alike disreputable to myself and the Profession which I represent and I cannot with honor to myself consent to serve in the connection which I am placed.

Second: The persistent efforts of some of the officers of the regiment to have a non-professional man commissioned Surgeon of the regiment over the claims not only of myself but also the heads of other ranking Asst. Surgeons in this Division of the army, have an injurious influence upon my further usefullness in the same and make it necessary for me to sever my relations therewith.

I regret that I deem it my duty to thus sever my connection with the regiment and the Service, but I consider that my honor, as a gentleman, and self-respect as a professional man, demand the course that I am, in justice to myself, compelled to pursue.

I am with much respect
Your obdgt Servant
Asst. Surg. 13 Reg V.V.I.[495]

Shortly after tendering this, his first letter of resignation, a leave of absence was granted Williams. The leave of absence was necessary for reasons of personal business—that he might settle a legal matter requiring his attendance at court. Then, his leave was extended upon request of Lieutenant Colonel James R. Hall, comdg. 13th W. Virginia, John Hall's son, that Williams might attend to the father, who was Williams' patient. John Hall, about this time (after 6 days of deliberations) had been found guilty of involuntary manslaughter by the Mason County Circuit Court, Point Pleasant. Hall the father had grit and stamina but it is no surprise that given his age and the rigors of jail and trial that he was at this time gravely ill and required Williams' immediate medical attention.

Extant documents tell the tale. A ten-day leave of absence was granted by Gen. Scammon (undated but certainly previous to February 3, 1864, when Scammon was captured by Confederate guerrillas). This leave of absence was granted that Williams might travel to Cincinnati, Ohio, to attend to "private business connected with the Civil Court."[496] Then, on February 22, 1864, Brig. Gen. George Crook now commanding the Kanawha Division (Scammon having been captured), granted Williams a seven day leave of absence to attend to "important private business."

> Head Quarters 3rd Division
> DEPARTMENT WEST VA.
> Charleston, West Va., Feby 22, 1864
>
> Special Orders,
> No. 37.
>
> Assistant Surgeon <u>Abram D. Williams</u> 13th Regt. Va. Vol. Infantry has permission to absent himself from the regiment for some (7) days from the 1st day of March 1864 on important private business.
>
> By command of Brigadier General Geo. Crook
> [by] E.W. Clark, Jr.
> Acting Assistant Adjutant General
> Comdg. O. 13th Va. Infy
> [illeg.] Hd. Quarters 1st Brigade[497]

A week later Lieut. Col. James R. Hall wrote to Col. Brown requesting that Williams' leave of absence be granted without delay and extended, it being of "utmost importance" to James' father (John Hall). Lieut. Col. Hall's correspondence complete with telegraph is as follows:

Point Pleasant West Va.
Feby 29th 1864
Col William R. Brown
Comdg 13 West V Infty

Dear Sir

Enclosed you will find a Telegraphic Dispatch I last night received from Brigd Genl George Crook comdg this Division

If Doctor Williams has not yet left Barbourville you will please hand him the Dispatch and start him off as quick as possible. Say to Doctor Williams that we will get the leave extended — Col you will please attend to this immediately as it is of the utmost importance to my Father —

Very Truly Your Friend
James R. Hall
Lieut Col 13 W.V.V.I.

Col I wish you would allow Sergt Love the Bearer of this to return as his Father is laying dangerously ill — He is a relative of mine and only comes because he knows how important that this reach you Jas R Hall Lt Col 13 W Va.[498]

U. S. Military Telegraph

By Telegraph from Hd Qrs 3d Div Dept W Va Charleston 186[4] to John Hall Asst Surg A D Williams has permission to be absent from the command for seven days from the first of March — Nearly the terms of the order

Signd By order
Gen Geo Crook
Gen M Kellogg
Surg + Med Dir[499]

As for Dr. Williams first resignation, it was not accepted. Instead, he became something of a roving doctor. While on the rolls of the 13th Regiment, he did duty away from the regiment at army hospitals and in the field with various other regiments such as the 5th and 1st Veteran Virginia Infantry Regiments. When Major General David Hunter's army returned to the Shenandoah Valley in July of 1864, after the disastrous raid on Lynchburg, Williams, who was on duty in the army hospital at Charleston, was ordered east with the Army of the Kanawha, as it was then designated, pursuant to the following order:

> Medical Directors Office
> Kanawha Valley Forces
> Charleston W. Va. July 17, 1864

Special Orders
No. 1

Asst. Surg. A. D. Williams 13th W. Va. I. now on duty at Post Hospital is herby relieved and will report to his regiment for duty

He will accompany the squad of men who leave hospital and convalescent camp tomorrow morning

> L.L. Comstock
> Surg. 7th W.Va. V. Cav.
> Act. Med Director
> Kan. Val. Forces[500]

September 18, a day before the battle of 3rd Winchester, Virginia (fought September 19, 1864), Williams was ordered from the 13th to the 5th West Virginia Infantry Regiment (both regiments in 1st Brigade, 2nd Division, Army of West Virginia) pursuant to the following order:

> Head Qrs. First Brigade 2nd Division A. W. Va.
> Summit Point W. Va
> Sept. 18th, 1864

Special Orders
No. 10

    II. First. Asst. Surgeon <u>Williamson</u> 13th Regt. West Va. V, I, is placed upon duty in the 5th Regt. W. Va. V. Inf. and will report to Lieut. Col. W.H. Enochs without delay.

> By Command of Col. R.B. Hayes
> R. Hastings
> Capt and A.A.A.G.
> Col. W.R. Brown
> 13th Va. Vol. I.[501]

With the new year and obvious end of fighting in the Shenandoah Valley, Dr. Willims again submitted his resignation. This time his reasons were more compelling and his resignation was accepted.

Hd. Qrs. 13 Va Vol Inft
Camp Hastings Md.
January 19, 1865

R.P. Kennedy, A.A.G.

Sir:

I have the honor herewith to tender my Resignation as Asst. Surg 13 V.V.I. for the following reasons.

1ˢᵗ It is very necessary for my professional interest as well as <u>public good</u> for me to return home to Cincinnati where I can do better for myself and <u>more</u> for the Service by way of <u>Special Hospital Duty</u> in a military <u>Ophthalmic</u> Hospital there. I have had several <u>urgent</u> Solicitations to come and assist in said Hospital by the Surgeon in charge.

2ⁿᵈ Very important private business, which has accumulated since I entered the Service, demands my <u>immediate</u> presence at home, in order to avoid pecuniary loss and to secure outstanding debts, which have been running for several years.

3ʳᵈ I have <u>very urgent</u> reasons, which it is not proper to state here, for tendering my resignation

For these several reasons I have the honor to forward herewith my resignation and <u>respectfully request</u> that it be <u>immediate</u> and <u>unconditional</u>.

There will be two medical officers[502] left with the regiment, which as an Organization, will be mustered out of the Service in six (6) to seven (7) months.

I am with much respect
Your obedient Servant
A.D. Williams
Asst. Surg. 13 V.V.I.

To R.P. Kennedy
Major + A.A.G.

I certify on honor that I am not indebted to the United States on any account whatever and that I have made my <u>last</u> and <u>final</u> Return of property to the proper Departments and have, in the proper way, accounted for all the property for which I have been responsible; and that I was last paid by Major Stafford — to include the 31ˢᵗ day of August 1864.

A.D. Williams, Asst. Surg 13 V.V.I.[503]

Dr. Williams was discharged January 31, 1865, pursuant to Special Order No. 24 West Virginia.[504]

Morning Reports for April 1863 Continued

Morning reports for Company H indicate that the company was stationed at Hurricane Bridge from April 1 to 19. On April 19, Col. Brown ordered Lieutenant O.W. Griswold commanding Co. H (Captain Taylor W. Hampton having resigned March 10, 1863), "to report with your company for duty to Capt. James [John] D. Carter comdg. Post at Point Pleasant, Va. You will be ready to march Tuesday morning the 21st inst."[505] April 20 and 21, the company was "marching." April 22 to 30, Co. H was stationed at Point Pleasant. One to four men were absent with leave April 1 through 25. Andrew J. Allen returned from desertion on April 1 after almost two months absence. Two enlisted men were reported absent without leave for the entire month of April. At the end of the month, Co. H had present and absent two commissioned officers and seventy-eight enlisted men. No one had special or extra duty. One private was detailed for daily duty for the month. No one had been ordered out on detached duty. The company had fourteen to nineteen men present sick over the course of the month, and four to eight men were absent sick at any one time during the month. On April 12, Jacob Shoemaker was wounded accidentally in the hand at Hurricane Bridge. No one was reported present or absent in confinement.[506]

Morning reports for Company I show that the company was stationed at Point Pleasant April 1 and 2. April 3 through 30, the company was stationed at Coalsmouth. 2nd Lieutenant W.E. Feazel, the only commissioned officer in the company, returned to the company on April 4 and was present for the duration of the month. Ten privates were present in camp April 1 through 4, and thirty-seven to forty-four privates were present April 5 through 30. On April 4, Sergeant Peter Darnel "returned from Elk River with 27 men a gain." No one was absent with leave during the month. Three enlisted men were reported absent without leave April 1 through 3. Samuel Hunter, Michael Bailey and Ezra Hansberry were dropped from the rolls for desertion. The company also gained recruits. Three joined the regiment on April 5; one joined April 23; and another joined April 30. At the end of the month, Co. I tallied present and absent one commissioned and forty-five enlisted men. No one had special, extra or daily duty. One commissioned officer and twenty-four enlisted men had detached service April 1 through 3, and twenty-four enlisted men were detailed for detached duty on April 4 and 5. Co. I had one to six privates present sick during the month. None were reported absent sick. No one was present or absent in arrest or confinement.[507]

May 1863

Still scattered by companies on the periphery of the military district

On the major military fronts, Union armies moved into new offensives. The vast Army of the Potomac, now under command of General Joseph Hooker, was positioned at Chancellorsville and at Fredericksburg. Friday, May 1, the battle of Chancellorsville began. It would continue until the 4th, when Hooker's forces fell back across the Rappahanock. Chancellorsville was a brilliant success for Confederate General Robert E. Lee, although it cost him the life of 'Stonewall' Jackson. The South grieved for the beloved West Virginia General but it was also much encouraged and emboldened by the victory won at Chancellorsville. Lee, full of confidence, determined to invade the North with his victorious Army of Northern Virginia. As he moved into Maryland and then deep into Pennsylvania, on all parts of the lines inside the Department of the Ohio the enemy became particularly active. U.S. Army General Ulysses Grant, still at Vicksburg, had moved closer in, below the city on Mississippi soil, and on Monday, May 18 the siege of Vicksburg began.

At the beginning of the month of May, the 13th continued to be divided at various posts. Co. A was at Coals Mouth with Co. I. May 1 through 9, Co. D was stationed at Hurricane Bridge. On May 10, Co. D was marching for Camp White. Beginning May 11, Co. D was stationed at Camp White, near Charleston, together with Co. F. Companies B, E and G were at Mud Bridge. Co. H was at Point Pleasant. Co. C was encamped at Hurricane Bridge until May 8, when they were ordered to proceed to Mud Bridge to dismantle Howell's Mill. This accomplished, they were on the march for Camp White. Having plenty to occupy them, the individual companies expected to stay where they were all summer and all winter.

Not only the 13th Regiment was scattered by companies in detachments at different outposts, all regiments of the Kanawha Division were scattered in this way at various outlying posts—all considerable distances removed from military headquarters at Charleston. As further raids, like the one just concluded by Jenkins, were anticipated, the troops of the department were stretched thin over the large distances encompassed by the department and called "Federal lines." This was a fiction of course, in reality this stringing out of troops with gaps between offered the Confederates opportunities to attack the Kanawha troops in detail.

The first attack came in the first week of May. On May 4, forces skirmished at Lewisburg. It appeared as if again this season, West Virginia would again be a battleground. Imboden, Jackson and Jenkins were massing their forces at Weston, and another attack on the railroad in the vicinity of Parkersburg and Clarksburg was anticipated. Regular forces which could be spared were sent to Clarksburg, and all Clarksburg militia were sent to protect the railroad.[508] On Tuesday, May 5, West Union, lying between Parkersburg and Clarksburg, was attacked by two Confederate cavalry regiments. Bridges were destroyed between West Union and Clarksburg and wires were cut. Colonel George Robert Latham commanding the defense of West Union asked for aid from Parkersburg. None, however, could be provided because Parkersburg had no reserve forces. West Virginians again looked to Ohio Governor David Tod for assistance. Gov. Pierpont at Wheeling was sent two thousand light rifles (short French muskets) and accouterments, five hundred Lindner carbines from Washington by quick conveyance and two hundred thousand cartridges from Pittsburg. Col. R.B. Hayes, at 1st Brigade Head Quarters Charleston, responding to this new alarm wrote to his commanding officers at Hurricane Bridge, Mud Bridge and Barboursville (respectively to Col. W.R. Brown; Lieut. Col. J.R. Hall; and Col. A.A. Tomlinson) as follows:

> Hd. Qtrs. 1st Brigade 3rd Div. 8th A.C.
> Camp White, Va. May 5, 1863
>
> Information deemed reliable by General Scammon has been received which makes it proper to direct that you place your post as soon as possible in condition to receive the enemy. Complete at once such defensive works as have been commenced or if not yet begun select the best point in your vicinity and make all needful preparations to hold it against any Rebel force that may visit you. Scout the country near you — guard against surprise and if the enemy approaches send immediate information here.
>
> By command of Col. R. B. Hayes
> M. P. Avery, Capt. and A A A G

To Col. W. R. Brown Comdg Co. 13th Va at Hurricane Bridge and Lt. Col. James Hall comdg. Detachment of 13th at Mud Bridge, and to Col. A. A. Tomlinson comdg. 5th Va. at Barboursville Va.[509]

J.V. Young, Captain commanding Co. G, referenced the heightened tenor of the times in a letter to his wife, Paulina, from Mud Bridge:

> We are fortifying here and that strongly and I think Jenkins or any other Rebel leader comes he will find something to do. I understand that there is a considerable Rebel force in North Western Virginia, and what has become of our men I can't tell. But I heard this morning by the doctor of the 4th Va. that the 8th is at Parkersburg but how true I can't tell. I shall move tomorrow to the fort, and fix myself to stay some time here, and when I get fixed and you get well I shall look for you down. [...] I am cooking for myself [...] Chickens and turkeys are plentiful and don't cost anything but an effort to get them. [...]
>
> Let me know how the Rebs are doing up there and what has become of old Bailey. Tell my friend Freeman to look after our pet Rebs; give them no peace. Keep me posted about your neighborhood.
>
> There is considerable excitement now at Headquarters but the cause is kept from us, for good reasons I suppose. But as I learn the facts I will let you know. I don't expect to get leave to come home until the excitement is over. [...] I look forward to when I can return home to the arms of my little family, an honor to them and my country. Surely I could not stay at home in peace and see my country struggling with traitors, and not raise my arm as well as my voice in the defense of the liberty of our depressed people; and our holy religion.[510]

Adjutant John S. Cunningham also wrote home regarding the anticipated Confederate incursion:

> Head Quarters 13th Va Regt
> Hurricane Bridge May 5th, 1863

> My Dear Little wife

> From information received it is probably that before many days we may be visited by a Rebel force how large I am not able to say therefore if you have preparations to make you will be forewarned —

I do not anticipate any large force so as to cause any of our forces to fall back from above but a raid similar to that of Jenkins about the 1st of April last. — The Rebels are trying to make a surprise party but preparations are being quietly made and we are confident of success —

If Father is at home you advise him to go to Wheeling as he has had it in contemplation for some time. Of course we do not know where the blow will fall but we are better prepared here than at Coalsmouth to fight the Rebels.

I do not wish you to give yourself any unnecessary alarm because I can take care of myself but I wish you to be relieved of the anxiety for your Father in case the rebels should make an attack upon Coal. The movements of the enemy are known to only a few and what I impart to you is to be kept secret for a while longer. [...] Please write to me all the news at Coalsmouth since the companies have left

Your Affectionate husband
J S Cunningham[511]

Orders were now issued at a rapid rate from division and regimental headquarters. On May 6, Col. Brown ordered Captain J.D. Carter to "proceed to Mud Bridge with your Co. and report for duty to Lt Col. James R. Hall commanding Post at that place."[512] Brown also ordered Capt. James W. Johnson, commanding Company A to "take a detachment (A + G) and proceed to Mud River as high up as Hamline and obtain all the information of the enemy you can, and on your return you will destroy all the Mill Burrs in the Grist Mills on Mud River between Hamline and Mud Bridge."[513]

On Thursday, May 7, Jenkins' men visited their old neighborhoods. Supplies were stolen. After nightfall, "a lot" of Jenkins' guerrillas, among them Henry and William Bryan and John Tafft of Mason County, who had been with Jenkins when he attacked Point Pleasant on March 30 went see what they could steal on the north side of the Ohio River in the neighborhood of Swan Creek, about seventeen miles below Gallipolis. They stole anything lying around loose: saddles, bridles, etc. The following morning, a party of eleven Union citizens started in pursuit of them and overtook them "within a circuit of about 6 miles." The Bryans and Tafft were caught with most of their stolen goods and jailed. "They were armed with breechloading carbines and revolvers."[514]

May 8, Colonel Brown issued Special Orders No. 13 ordering Capt. Van D. McDaniel, commanding Co. C, to take part of his

Company and proceed to Mud Bridge — thence down Mud River as far as Howells Mills and take the Mill Spindles (or such other machinery as can be easily placed back without injury, so as to disable the mills from ever grinding until ordered by Military authority) and bring the same to this Post.[515]

Col. Hayes had offered to send a man out to the 13th W. Virginia Regiment to drill artillery squads. May 8, Brown wrote in reply that he "would be glad to have a man for a week or two."[516] Adjutant J.S. Cunningham wrote to 1st Brigade Head Quarters on May 8, acknowledging that "Genl. Orders H 73, 85, 87, 89, 90, 91, 93 War Dept. Series of 1863" had been received at 13th Regiment Head Quarters at Hurricane Bridge.[517]

On the march to put out fires

The operations of Generals John D. Imboden and John Echols against the B&O Railroad in vicinity of Lewisburg, Virginia, brought a wave of new alarms. Military authorities at Charleston, anticipating trouble ordered a concentration of troops, which included the 13th Virginia. On the evening of Saturday, May 9, the 13th Regiment received orders from Brigade Head Quarters "to break up the encampment at Hurricane and Mud Bridges and to report at Camp White on the west side of the Kanawha River [near Charleston]."[518] Col. Hayes' orders to Col. Brown and Lieut. Col. James R. Hall were the following:

Hd. Qtrs 1st Brigade, 3rd Division, 8th A.C.
Camp White Va. May 9th, 1863

Colonel

You will break up your camp at Mud Bridge and move with the troops of your command to this camp without delay. Bring your Artillery and Camp Equipage with you and press into service any transportation you may require which is in your vicinity. In abandoning your Post you will not destroy any defensive works which you may have or any Government property which can be saved. Prompt compliance with this order is required.

By command of Col. R.B. Hayes
to Col. Brown commanding at Hurricane Bridge W. Va.
to Lt. Col. Hall commanding detachment 13th + 5th Va. V. I.
at Mud Bridge.[519]

On the morning of May 10, the 13[520] left camp. They marched from Hurricane Bridge to Camp White, a distance of twenty-four miles.[520] While on route for Charleston, Adj. Cunningham wrote to 1st Brigade Head Quarters from Coals Mouth informing that the 13th had arrived at Coals Mouth and that he (Cunningham) had "sent back terms for Lieut. Col. Hall." Cunningham also requested that a "boat may be sent down for Camp Equipage."[521] The regiment at length arrived at Charleston on the evening of May 10. They "went into camp two miles above the town on the west side of the river."[522] In *Annual Return of Casualties* is recorded that on May 11 the 13th moved "above Camp White 2 miles."[523] *Field History of the 13th Regiment* records that on May 11, "the 13th Regt. was encamped two miles above Camp White on the west bank of the Kanawha River."[524] The regiment was now stationed in "Camp near Charleston, West Virginia."

David Burrows, Private, Co. F, had became ill with fever on May 10 but had stood the march to Charleston nonetheless. "We left Hurricane Bridge on the 10th," Burrows wrote to his wife on the 15th, "and I was not very well then and marching made me worse. We are about one mile above Charleston on the lower side of the river. We expect to fight a fight here soon."[525]

J.V. Young also commented upon the concentration of troops at Camp White in his letter to his wife, Paulina, dated "Camp White Kanawha, May 1863." The letter seems to have been written sometime after their arrival at Camp White and before their departure for Fayetteville on May 18. Young wrote:

> I am here yet and well [...] Company G is all right, ready for a fight. [...] What we are going to do, or where we are going I can't tell. [...] The boys like this place very well and I believe would rather stay here than at Coal. [...]
>
> The drum is beating for guard mounting. The sound is heard far up and down the river. The 5th Va. is above us and the 23rd Ohio is below us. Witcher's Cavalry[526] is on our right and a heavy battery on the hill at the ferry branch, with one company, 1st Va. Cavalry, besides other troops in the neighborhood. Therefore you may see that we are not afraid of Jenkins or any other horse thief.
>
> The Rebels drove the 91 Ohio from Summerville night before last, and captured their trains, and it is rumored that they have with their army from 3,000 to 5,000 head of cattle, and 500 head of horses, all stolen from North West Virginia. What the country will come to I can't tell, while we have such commanders. [...]
>
> I wish you would come up and stay with me a few days for I don't know how soon we may be ordered away from here, and where the General only knows, for I don't. I have some things to

send home and can't send them well at present. You can ride up or come on a steamboat. They all know where the Thirteenth is camped. You can stay at Albert Tulley's who lives here. [...] My company is in better condition this morning than it has been for some time. We report 70 for duty this morning. We think Col. Brown and Col. Hall almost the best men in the world. They have the entire confidence of the officers and men of the Regiment, and I have no doubt but the officers and men would fight for them as long as one was left to raise an arm; or a voice to cheer them in their efforts to re-establish our Government. [...][527]

Correspondences continued to arrive at 13th Regimental Head Quarters from Col. Hayes at Brigade Head Quarters. These contained orders to the 13th that suggest that preparations were being made to meet the enemy. The tension occasioned by Hooker's defeat at Chancellorsville; the renewal of Confederate operations in Western Virginia; and the expectation that lines of communications between the Kanawha Valley and the outside world would soon be severed can be sensed in Hayes' orders. Underlying the language of the following correspondences is the sense of uncertainty and crisis which underlie these days.

Tuesday, May 12, was a rich day for orders and information. On the 12th, Hayes wrote as follows. To "Col. Brown at Camp White; [Lieutenant Colonel] Tomlinson of the 5th Va. Inf.; Capt. S.J. Simmones Battery Camp White; Capt. G.W. Gilmore 1st Va. Cav.; Capt. D. Delaney Co. A 1st Va. Cav. and Lt. John S. Witcher Co, 3rd Va. Cavalry," Hayes wrote:

> Hd. Qtrs. 1st Brig. 3rd Div.
> Camp White Va. May 12, 1863

> I am directed by Gen. Scammon to have all in readiness for an attack. I do not think it necessary to awaken the men at present but you will see that sentinels are on the alert and such preparations are made as will enable you to have your men ready for an attack on the shortest notice.

> Capt. Simmons and Lt. Austin in case of alarm will put their guns in position to bear on Charleston. By order of Col. R.B. Hayes.[528]

Hayes also ordered that soldiers held in arrest be employed on the breastworks being erected opposite the mouth of Elk River (Special Orders No. 27):

> Commandant of Regiments and Detached Companies of this Brigade stationed near Camp White will hereafter direct all men

held in confinement in their respective commands be sent with a sufficient guard to report to the officer in charge of the fortifications opposite the mouth of Elk river. The Guard and prisoners will bring with them their dinners.[529]

May 12, Hayes ordered Brown to "please direct a Detail from your command for Guard duty to-morrow May 13 of 2 Corp[ora]ls [and] 14 men. They will report to these Hd. Qtrs. [Hd.Qtrs. 1ˢᵗ Brig. 3ʳᵈ Div. 8ᵗʰ C.] at Guard Mounting 8 a.m."[530]

Wednesday May 13 was another day in which a flurry or orders were issued by Hayes. Hayes wrote to Brown this day with a number of directives. Brown was to

direct a Detail of 5 Privates to report to Capt. C. A. Sperry Prov[ost] Marshall at Charleston for duty as Provost Guard. The men should be selected from the best as this Guard patrols the town. They will take with them their accoutrements. They will not be relieved except they are found incompetent.[531]

Also on the 13ᵗʰ, Hayes ordered Brown to "direct a detail of 2 Blacksmiths from your command for duty at these Hd. Qtrs. [Hd.Qtrs. 1ˢᵗ Brig. 3d Div.] They will report immediately at the Blacksmith's shop at the Ferry."[532] Hayes also ordered Brown to "direct a Detail to report at these Hd. Qtrs. to-morrow at 9 o'clock A.M. for Brigade Guard of 1 Captain as Brigade officer of the day, 2 Corpls.; 13 men."[533] Thursday, May 14, Hayes wrote to Brown with the following: "You will direct a Consolidated Morning Report of the troops under your command be forwarded to these Hd. Qtrs. Immediately" and detail "Sergt, 2 Corpls., 13 Privates" for Brigade Guard to report to "these Hd.Qtrs." — 1st Brig., 3d Div. 8ᵗʰ A.C., Camp White on May 15 – "at 9 o'clock A.M."[534]

Sunday, May 17, Hayes had occasion to write letters home, seemingly pleased overall with his command and confident in their present situation. To his uncle he wrote:

We are in no danger here. We have built a tolerably good fort which we can hold against superior forces perhaps a week or two or more. We have a gunboat which will be useful as long as the river is navigable. My whole brigade has been here. The most of it is good and the rest is improving.[535]

To his wife, he wrote in greater detail, conveying his impressions of the 13ᵗʰ Virginia:

My whole brigade except two or three detached companies is now here. Delany, Simmonds, the Fifth and Thirteenth Virginia and a new cavalry company were sent for during the recent scare. We have nearly finished a tolerable fort, and have a gunboat. I have thirteen pieces of artillery.

I am most agreeably disappointed in my Virginia regiments. The Thirteenth is new and composed of West Virginians, but it has capital officers and they promise well in all respects. I reviewed them this morning. Their appearance would be creditable to an old regiment.[536]

Fighting erupts at Elizabeth Court House Ravenswood and Fayetteville

As fate would have it, military authorities at Charleston hadn't long to wait for the enemy to make his move. On Saturday, May 16, fighting erupted at Elizabeth Court House and at Ravenswood, Virginia, to the north of Charleston. May 17, Brig. Gen. John McCausland made a sudden attack on the Federal outpost at Fayetteville. McCausland, with "a few thousand Virginia troops" had come in on the Princeton road from the line of the Virginia and Tennessee Railroad and had "bombarded our earthworks until driven off by Col. [Carr B.] White,"[537] commanding in this engagement the 12th, 20th and 91st Ohio Volunteers and 2nd Virginia Cavalry.

Brig. Gen. E.P. Scammon anticipating the attack on Fayetteville wrote to Colonel [R.B. Hayes, presumably] regarding the "assault of Jones force." Scammon advised [Hayes] to

make all your arrangements with a view to defense first and rapid forward movement afterwards. If this movement be well met it will completely fatigue + demoralize the enemy — while we, not being worn down by long marches, shall be ready to go forward. While I am writing, I hear from Fayette that the enemy are in sight all along the front of our works + White says the ball is about to open. I learn also that the reinforcements sent him have arrived and I pray that all may go well. If as I suppose, they fancy they are to fight but a single regt. and are only prepared for that we shall be successful — otherwise Col. W. will have a hard time and we must be ready to succor him if possible. At all events, the programme previously laid down as to our movements in the Valley, must be observed to the letter. Come over to my old quarters this evg. At eight or between six and seven o'clock + we will talk over matters. There are some details which I would wish to talk rather than write. E.P. Scammon B.G.[538]

McCausland's advance was correctly thought to be a diversionary movement, a feint, ordered to occupy Kanawha Valley troops while Generals Imboden and Echols at Lewisburg, Virginia, operated against the B&O Railroad. It had nonetheless, to be countered. Gen. Scammon hurried reinforcements forward to Fayetteville and drove the attacking force beyond Princeton.[539] On May 19, Scammon "sent three regiments and some artillery, the rest of McMullin's Battery,"[540] to re-enforce Col. Carr B. White, commanding 2nd Brigade, Kanawha Division and the Federal military post at Fayetteville. J.M. Merrill (regiment not given) wrote the following in a letter to his wife, Lydia, dated "Camp Piatt, May 21st, 1863," regarding the fight at Fayetteville. "The 12th and 91st Ohio and 13th Virginia are at Fayetteville, with 13 days rations and about 400 of our regiment keeping open communication with them and Gauley Bridge."[541]

The 13th had been waiting in camp at Charleston, ready to move at a moments notice. The evening of Sunday, May 17, orders were received to report at Gauley Bridge, Fayette County.[542] Hayes wrote to Brown that day:

> You will immediately proceed with your regt. to Gauley Bridge A SteamBoat will be at your Landing for you at 1 P.M. — Take all the Guard with you that is on this side of the river.
>
> All extra baggage you may have had better be left here under a guard for the present.[543]

Andrew Stiarwalt, musician with the 23rd Ohio Volunteer Infantry, wrote in his diary regarding the transfer of the 13th upon the iron clad steamer, which embarked on May 17. He wrote:

> their was an Iron Cladd Steamer was put in motion on the Kanawha river it is maned by a Comp[any] of marine and mounted with 4.24 pound houtyers [howitzers] + 2-30 pound paritsguns [parrot guns] it is to operate up + down the river on patroll duty the 5th + 13th VA Regts passed Charleston they are to be stationed at different points along the Kanawha river to d[o] post or guard duty this 17th May/63.[544]

May 17 to 19, the 13th was on the march to Gauley Bridge and Fayetteville, a distance of forty-nine miles.[545] On May 18, Confederate forces under McCausland concentrated to attack Fayetteville and on May 20 a general attack was made.

The 13th reported at Gauley Bridge at 9 o'clock in the morning on Tuesday, May 19. They remained here until 12 o'clock noon. Col. John T. Toland, 34th Ohio Infantry Vols., commanding the military post at Gauley Bridge, ordered Brown to send a detachment of his men to the rifle pits for fatigue duty. Toland wrote:

You will please direct a detail from your Regt. for fatigue duty as follows:

1 Lt., 2 Non Comm. Officers, 60 Privates.

They will be provided with tools and will report to Maj. J.W. Shaw, 34th Regt. at Gauley Bridge who is charged with superintendence of work on Rifle Pits — at 11 o/c. A. M. this day.[546]

Adj. John S. Cunningham found a moment to write a brief note to his wife:

Head Quarters 13th Regt Va Vol Infty
Gauley Bridge May 19, 1863 12 oclock M

My Dear little wife:

We have just arrived at this place and the men have rested about two hours We have just received marching orders for Fayettville — We are well and in good spirits and will proceed to Fayetteville in about an hour as soon as we arrive there I will write to you again[547]

At twelve midday, Colonel Brown received orders to proceed to Fayetteville without delay. Col. Toland's orders to Brown were as follows: "You will march with your command to Fayetteville, immediately. You will proceed with caution, so as not to fall into an Ambuscade of the enemy."[548] At 2 o'clock the 13th "was again in motion"[549] and reached Fayetteville the same evening.[550] 13th Regiment *Field History* records that it arrived at Fayetteville on the 19th, at "9 o'clock P.M." and was "assigned to its position in line of battle for the next day."[551] The regiment bivouacked for the night at 9 o'clock "expecting an attack every minute from the enemy whose campfires could easily be seen from our position."[552]

Early in the morning, at day break, the enemy opened their batteries upon the 13th (and other troops drawn from the Kanawha Division for this expedition) with shot and shell.[553] The Confederates had a total of four pieces of artillery trained upon Scammon's troops. "Their fire was reported as being accurate, but doing little damage."[554] The 13th West Virginia's part in countering the attack on Fayetteville and maneuvering involved in pursuing the retreating enemy were described in several sources. The author of 13th *Regiment Field History* records in his typically spare style (a style lean on the kind of detail and that might brighten a historian's heart) that

we lay under their fire until 1:30 P.M., when the enemy retreated. The Commander of the Regt. received orders from the Brigade Commander, Col. White, to be, ready at 10 o'clock P.M., with three (3) days' rations, to march in pursuit of the retreating enemy.

The rebels were pursued as far as Raleigh C.H., being fired upon by them once on the way, doing but little injury. We encamped in sight of Raleigh C.H. for the night, after having fired a few shots at the retreating enemy.[555] At 12 o'clock M. the following day orders were received to march back to Fayetteville, at which place the Regt. arrived at noon, May 23d, and remained two days to rest.[556]

Regimental Return 13th Va. Vol. Infantry for May 1863 has the following more detailed record:

We lay under their fire until about 2 o'clock when the enemy retreated. Col. Brown then received orders from Brig. Commander Col. White to be ready to move in one hour in pursuit of the retreating enemy immediate preparations were made the troops being provided with 3 days rations.[557] We followed the rebels as far as Raleigh Court House being fired upon once by them on the way doing but little injury. We encamped in sight of Raleigh for the night after having fired a few shots into the rear guard of the foe. The next morning for Fayetteville. Again bivouacked on McCoys farm first night reached Fayetteville next morning at 9 o'clock remained here to rest for 2 days then rec'd orders to return to Camp White near Charleston, Va. marched thence on the evening of the 25th reaching Kanawha Ferry same evening distance 12 miles. From thence marched to Camp Piat, reaching there on the afternoon of the 27th. Next morning left on steamer for Coals Mouth arriving there about 10 o'clock where we remained until the end of the month, whole distance marched 190 miles.[558]

While resting in camp at Fayetteville, John S. Cunningham wrote the following home on May 23 describing the formidable force assembled to advance into the Kanawha Valley; their march to the threatened point; and the severity of the forced march after the enemy.

In Camp Near Fayetteville Va
May 23rd 1863

Mr Dear Little wife

I wrote you two little notes in pencil one from camp near Charleston and the other from camp Gauley informing you of our departure from the latter place. At that time there was a considerable force of Rebels investing Fayetteville and the forces had been fighting two days previous to our being ordered up to relieve them—

We made a forced march[559] to this place expecting to fight our way through to the Garrison but was agreeably disappointed. We left the Ferry at the Falls of Kanawha at 4 P M Tuesday May 19th and arrived at the fortifications at 10 P.M. and our positions were assigned for battle immediately on our arrival — at early dawn the morning following the rebels opened their batteries upon us and the fight continued until 1-1/2 [1:30] P.M when the rebel force was compelled to retreat towards Raleigh C.H. with a loss of some men (artillery men) —. Immediately after the retreat a Brigade composed of the 12th + 91st Ohio and our own Regt (13th Va) with a battery of 6 field pieces commenced the pursuit all night and the next morning at 8 oclock we brought them temporarily at bay and exchanged a few shots and it is thought the rebels sustained a loss of some wounded men — the 91st only had one wounded slightly we continued the pursuit closely to the vicinity of Raleigh C H where we arrived at dark on the evening of the 22nd. The Rebels had planted their batteries ready to receive us — Our battery immediately opened upon them and we formed in line of battle — but the enemy immediately fled without returning a shot it was now very late at night and we rested until morning — at 6 o clock we were in the town — the rebels had been retreating the balance of the night very much exhausted the men scattering to the woods — at 12 M we commenced our return march to Fayetteville and arrived at this place at 10 A M after pursuing the enemy 30 miles — It was a very forced march and our regiment is very much fatigued and foot sore having marched since May 10th 133 miles — we tried very hard to capture the field pieces from the enemy but did not succeed — At all events there will not be an expedition sent against Kanawha valley again very soon in any large force as it is a complete failure — The 13th Va behaved nobly they were under fire of the rebel batteries about 9 hours — and the rebel force was quite formidable under the command of John McCausland — It was intended by the rebels to concentrate their force under McCausland Jenkins and Imboden and make a raid into the Kanawha this time but there was a failure as you now know Jenkins + Imboden are called off into eastern Virginia

You can inform all concerned there were no casualties in the 13th Va and all the men from Coalsmouth are well and hearty and are now sitting around the camp fires pounding coffee for their supper. I am now sitting under a tree writing this note on a board in my lap.

[...] I enclose to you a rebel newspaper found at Raleigh CH. some of the boys got papers as late as the 18th from Richmond. Your Affectionate husband

J S Cunningham

It is probable we may be moved away from this place soon but write to me direct the letter to follow the Regiment — we may return to Gauley. J.S.C.[560]

May 24, Col. Carr B. White (commanding 2nd Brigade and the recent defense of Fayetteville) requested Brown to "please forward to these Head Quarters immediately a detailed report of the action taken by your command in the events of the past week."[561] Col. Brown responded with the following report.

Head Quarters 13th Regt. Va. Vol. Inf'try
Camp near Fayetteville, May 24, 1863

Colonel:

I have the honor to make the following report of the part taken by this Regiment in the events of the last week connected with the attack of the enemy upon this post [Fayetteville], and his retreat and our pursuit of him. —

On Tuesday May 19th at 12-1/2 o clock I received marching orders from Col. Toland Commanding at Gauley Bridge to report without delay to Col White commanding at Fayetteville, I was also cautioned to guard against surprise as the enemy had made an attack on Fayetteville and the place was very probably invested by them, and that I would most likely have to cut my way through in order to relieve the Garrison

This Regiment crossed the river at the ferry below the falls of Kanawha at 4 o clock P.M., leaving our train on the north side of the river, and succeeded in entering Fayetteville at 9 o clock P.M. without having met or seen any signs of the enemy.

> Upon our arrival I was immediately assigned a position in line of battle, and we lay on our arms during the night.
>
> At daylight May 20ᵗʰ the enemy opened his battery and kept up a vigorous fire during the forenoon and closely shelled the battalion of this Regiment under Lt Col Hall which was stationed on the slope of this hill south of the town. —
>
> When the enemy retreated and it was decided to pursue, this Regiment was assigned its position in the Brigade and at 10 o clock P.M., we started in pursuit.

Indeed, Hall's battalion stood up to a rapid artillery fire early in the day. Confederate artillery officer, Milton W. Humphreys, who was on duty with his gun 'Maggie,' recorded that on the morning of May 20 he fired "a total of 65 rounds" at the Federals. Confederate troops then retired 12 miles and bivouacked for the night.[562]

Col. White's orders in regard to the May 20ᵗʰ pursuit embodied in General Order No. 3, indicate that the 13ᵗʰ was marching in rear of McMullen's Battery, which was behind the 91ˢᵗ. This arrangement seems clearly to suggest that the 13ᵗʰ was so placed as to be within supporting distance of the 91ˢᵗ Ohio, which had "the advance of the Infantry"[563] with McMullen's Battery following in rear of the 91ˢᵗ. Gen. Order No. 3 outlining the order of march on the pursuit was as follows:

> This command will leave camp at 10 O.C. this P.M. in the following order Major McMann with Detachment of cavalry will constitute the advance guard.
>
> Col. Turley with 91ˢᵗ Regt O.V.I. the advance of Infantry
>
> Capt McMullen with Battery in rear of the 91ˢᵗ O.V.I.
>
> Col Brown with 13ᵗʰ Regt Va Vol Inft V. V. I. in rear of Battery
>
> Lt Col J D Hines with 12ᵗʰ Regt V V I rear guard
>
> The commanders will see that there is no straggling in their commands.

Brown continued his Report:

> — At 8 o clock A.M. on the 21ˢᵗ the enemy was brought to bay for a short time and some shots were exchanged between the advance of this Brigade and the Rebel artillery and rifle men. This Regiment was ordered to deploy on the hill to the right of the road and to be

held in readiness to support the 91ˢᵗ Ohio in case the rebels should advance in force, this did not however become necessary as the rebels were soon discovered to be again in full retreat — We again pursued, occasionally skirmishing with the rear of the enemy for 10 miles to within one mile of Raleigh C.H.[564] when we were halted at sunset, and two pieces of cannon were brought to the front and shelled the town, the enemy immediately left without returning a shot. —

When the artillery was brought to the front, our positione was assigned on the left of the Brigade, here we rested on our arms during the night. Early on the morning of the 22ⁿᵈ Companies B. F. + G of this Regiment under command of Lt Col James R Hall were deployed as skirmishers and entered the Town of Raleigh which they found evacuated by enemy, the rebel force having left immediately after the first shell was thrown into the town the night previous. —

At 12 o clock, M[idday] I received orders to return to Fayetteville and arrived at this place [Fayetteville] with the Brigade at 10 o clock[565] A.M. May 23ʳᵈ —

I have no casualties to report in my Regiment.

I am Colonel very Respectfully
William R. Brown
Colonel Comdg 13ᵗʰ Regt Va Infty[566]

Milton W. Humphreys, artillery officer with McCausland's command (C.S.A.), provided insight on events at this juncture, which as is often the case, fills in gaps in the reports of Union sources. Early in the morning of May 21, wrote Humphreys, just as the Confederates were resuming the march, Scammon's troops appeared in their rear. A few shells were fired upon the Federal advance "from the bronze rifle," which caused the Federal van guard to fall back. The Confederates marched on experiencing no further molestation until late that afternoon. They had halted at Raleigh Court House (now Beckley) when their scouts came in and reported that Federal troops were approaching. McCausland turned around taking three companies of infantry; one cavalry company; and one howitzer (Humphrey's own gun, Maggie) back to meet the Federal advance. McCausland positioned his troops at Raleigh Court House, a small hamlet at that time, posting Maggie "in front of a hotel in the south-west angle between the main and the Logan roads." The Federals planted a battery of their own "on a mountaintop mile away, where they already had a fortification, and opened on us, their first projectile (seemingly a percussion shell) [...] exploding within ten feet of [Maggie.]" McCausland galloped up to Humphreys and asked if Maggie could reach the Federal battery. Upon being told by Humphreys that he had not the proper fuses to effect this, McCausland gave orders to "limber up and gallop off, which we did ignominiously, while the enemy threw shells after us with remarkable precision [...] [W]e marched a few miles and found the army at what we called 'Camp Piney.' "567

Losses sustained on the expedition to Fayetteville

There had been no infantry fighting per se in front of Fayetteville and hence few casualties. Col. C.B. White reported overall for the brigade just two killed, seven wounded, and nine missing.568 Col. Brown reported that the 13th sustained no casualties. W.W. Harper related in his letter to the county newspaper that the men of the 13th returned from Fayetteville "well but considerably worn down" as they had "traveled nearly 100 miles in five days." Harper noted that one man was "slightly wounded."569 Captain A.F. McCown, commanding Co. F, reported that one of his men, William Daniel, had been captured by the enemy "having straggled from his company, while near Raleigh on the 22nd."570 There was other trouble as well. 2nd Sergeant Thomas M. Hackett of Co. H, "having on one occasion absented himself without leave from his company and on the recent march from Gauley to Fayetteville, Va. [having] acted in a manner unbecoming a soldier" was ordered "reduced to the ranks" (subject to the approval of Col. Wm. R. Brown), pursuant to Company Order No. 2 issued by Capt. W.I. Mathews, cmdg. Co. H, on June 1.571 Hackett's reduction was approved by Brown. He was "reduced to the ranks for negligence of duty" on June 3, 1863.572

Soldiers' letters describing the recent operation at Fayetteville

During the layover at Fayetteville, the men took the opportunity to write home. On May 24, W.W. Harper , Sergeant Major, 13th West Virginia, wrote to the editor of the Point Pleasant *Register* describing the expedition to relieve Fayetteville.

> *Mr. Editor:* — This Regiment received orders at Hurricane Bridge in Putnam county, on the evening of the 12th to strike tents and report without delay at Charleston, Kanawha county, and in obedience to orders we were under way by six o'clock, a.m. We reached the mouth of Coal River by 12 o'clock — left this point at 2 o'clock and reached Charleston at 10 p. m., bivouac[k]ed for the night. Next morning we went into camp about two miles above the town pitched our tents and remained there until 12 o'clock Sunday the 17th when we received orders to report immediately to Gauley Bridge. The excitement then

commenced and our cotton houses were soon demolished rolled up and packed into wagons. The train started up the river on its west side, the troops embarked upon the steamer Ingomar with the expectation of being landed at Loup Creek, but the boat could not stem Calvin Creek Shoals and consequently we were landed at this place about 11 o'clock at night, here we bivouac[k]ed for the night, next morning started for Gauley, got within 4 miles of the place the same day, bivouac[k]ed for the night in an open field on the bare ground, our covering the blue Heavens. In the morning we marched to Gauley reaching there at ten o'clock the 19[th]. Went to work immediately pitching our tents, but, alas! how uncertain and impermanent are the things of this world, at 2 p.m. orders were received to report immediately to Fayetteville, distance 12 miles we reached there about 9 o'clock. Our forces were then put in position for battle with the expectation of being attacked at any time, for the fires of the rebel camp could be distinctly seen from our position. The enemy made no demonstrations during the night, but the first intelligence we got from them in the morning was the bursting of a shell over one of our fortifications, thrown from a rebel gun. This aroused our forces and it was quickly answered from two of our guns, the firing thus commenced continued until 3 o'clock, when the rebels retreated, when this fact became known preparations were immediately made to follow. A detachment of the 2d Virginia Cavalry under Major McMahan was to take the advance, the 91st Ohio, Col. Turley, next; then a section of McMullins Battery; the 12th Ohio and 13th Virginia brought up the rear.[573] We traveled until 2 o'clock a.m. and halted for two hours to rest, at 4 o'clock made a forward movement, at 8 o'clock we came so near the rear of the fleeing rebels that they planted their cannon in or close by the road and with their infantry deployed on the right side of the road they opened fire upon us which checked our advance for a short time, but before we got everything ready for the attack upon them they fled and left the way open to us. Again we moved forward as fast as the men could safely travel, at 6 o'clock we came up with them again at Raleigh, we immediately got two of our guns in position and opened upon them when they fled in great confusion and judging from the suddenness of their flight they were very much alarmed; we remained here all night, next morning some skirmishing parties were sent forward to search the town to learn whether there were any of the enemy there; your correspondent was one of the party and was the second man to enter the town, but no rebels were to be found. — The pursuit here ended and we left again for Fayetteville which we reached at 10 o'clock Saturday the 23d. All well but considerably worn down, having traveled nearly 100 miles in five days. We had one man slightly wounded. Two of their men were killed on Wednesday from our cannon. The object of the raid is not known, but one thing is certain it cost them more than it come to. W.W.H.[574]

Captain John V. Young, commanding Co. G 13th West Virginia, wrote home to Paulina, from camp at Fayetteville on May 24.

> I am well at present and Co G is all right. We left Charleston last Sunday and have been marching ever since, night and day. We ran the rebels to Raleigh Court House, shelled them out, and returned here last night tired and sore feet, without the loss of a man, but one got his finger shot off.
>
> I had the biggest company in the Regiment and was sent out to take the town, which we did with Company A. But when we get there the rebels had left and we were all broke down and could not follow them any further. I tell you Colonel McCauslen did good running this time. On Wednesday we lay under his cannons all day, the shells bursting over our heads, but nobody hurt. Co. G. made a great deal of sport about their bad shooting.
>
> I can't write much this time. I have no place to write but on my knee. We are camped out in the field without tents and knapsacks at Gauley. Where we will go next I can't tell [...] I must stop. The mail goes out at twelve and I have no time. I am officer of the day and will be busy, so goodbye.[575]

While at Faytteville ...

While the Regiment as a whole rested at Fayetteville, Col. C.B. White, commanding the brigade of troops to which the 13th momentarily was attached (i.e., 2nd Brigade, Kanawha Division), designated the 13th for certain duties contained in the following orders. On May 24, Col. White wrote to Brown directing him to "please direct a detail from your command for fatigue duty 1 Sergt; 10 Privates. They will report at these Hd. Qtrs. immediately."[576] The same day, Brown was ordered to detail from his command "for Guard duty One (1) Captain Three (3) Sergeants Three (3) Corporals Twenty seven (27) Privates" to "report at the usual place of Brig. Guard Mount at 8 oc AM tomorrow May 25, 1863." Brown was also to "direct a detail for fatigue the same as to day the Sergeant in command will receive his instructions from the Sergt now on duty."[577]

On Monday, May 25, the brigade was inspected by the departmental Inspector General. A circular had been issued to that effect from 1ˢᵗ Brigade Head Quarters at Camp White. Addressed to Col. Wm. R. Brown et al., the Circular provided as follows:

> Hd. Qtrs. 1ˢᵗ Brig.3rd Div.
> Camp White May 25, 1863
>
> Circular:
>
> The troops of this command will hold themselves in readiness for inspection by the Inspector General of the Middle Department at 1 o'clock p.m.[578]

Also this day, Brig. Gen. E.P. Scammon issued Special Order No. 34 directing that the 5ᵗʰ and 13ᵗʰ West Virginia Regiments return to their former posts:

> 1. The 5ᵗʰ and 13ᵗʰ Regiments Va. Infantry will return immediately to the position formerly occupied by them at Hurricane Bridge Mud Bridge Barboursville Point Pleasant and Mouth of Coal River — each post and detachment to be proportioned as to amount of forces same as before.
>
> The Chief Quartermaster of the 3ʳᵈ Division will furnish necessary river transportation upon the requisition of Regimental commanders.
>
> 2. Witcher Cavalry will report to Lieut Col Hall at Mud Bridge[579]

In obedience to Scammon's orders, Col. Hayes wrote to Col. Brown directing that "[i]n pursuance of orders from Brig. Genl Scammon you will return with your command to these Hd. Qtrs. by easy marches."[580] Brown received these orders to return to Camp White on May 25. He had orders from Col. White this day as well. White, still in command of the brigade, wrote to Brown: "You will please direct a detail from your command for fatigue duty one Serg[ean]t 10 Privates. They will report immediately to Lt. Engle A.C.S. at the Commissary B[ui]ld[in]g."[581] White also ordered Brown: "You will please order 1 Company from your command to proceed immediately to outside cavalry post near Sargers — the 'officer of the day' will direct their movements."[582]

Thus ended the latest Confederate attempt to re-enter the Kanawha Valley. The incursion had been for naught. As Hayes wrote in a letter home, the enemy had "bang[ed] away" at Fayetteville for "three or four days doing nothing."[583] That the enemy might not repeat such an assault in this part of his district, General Scammon established a camp at Flat Top Mountain—"a very steep and rugged elevation of some mile."[584] From Flat Top, he sent out various scouting expeditions.[585]

May 26, 27 and 28, the 13th Regiment returned to Coalmouth on the Kanawha River. They marched on the return a distance of fifty miles.[586] The regiment was back in its old bailiwick, where it's services were badly needed. The "entire region" between Eastern Kentucky and the Kanawha River was "filled with secesh," "numbering 3,000 to 4,000." "[T]he Guyandotte and Sandy Rivers [we]re fordable at many places,"[587] permitting the secesh easy travel across country.

The 13th arrived at Camp White on the evening of May 27. Upon their arrival there, they were ordered to Coalsmouth. Andrew Stiarwalt, musician, 23rd Ohio, observed on the 28th, that "the 5th + 13th VA Volunteers went down the Kanawha river."[588] The 13th arrived at Coalsmouth on the evening of May 28. The regiment remained at this post until June 1, when orders were received for Companies A, C, D, F and H to march to Hurricane Bridge. These companies returned to Hurricane Bridge on June 2, marching a distance of twelve miles.[589] Companies B, E and G were at the same time (presumably June 1) ordered to Mud Bridge in Cabell County, to their old encampment. R.B. Hayes wrote to his uncle, Sardis Birchard, from Camp White on June 2: "We are stronger here than we were. I have now a full brigade, four regiments infantry, a battery, and three companies cavalry. We fortify all points deemed important."[590]

The 13th Regiment companies remained at their Hurricane Bridge and Mud Bridge camps until Monday, June 29.[591] On June 29, the 13th marched twelve miles marched to Coalsmouth.[592] June 30, the 13th "marched to Camp White 15 miles."[593]

In camp at Coals Mouth, Col. Brown issued new orders. On May 30, General Order No. 9 authorized the formation of a "council of Administration" for the regiment. Lieut. Col. James R. Hall, Capt. J.W. Johnson and Capt. Van D. McDaniel were appointed council members. They were to "proceed to organize themselves immediately for proper action—appointing Capt. A.F. McCown for Secretary."[594] May 31, Brown had General Order No. 10 issued providing that: "The Commanders of Companies will have their tents struck at daylight tomorrow morning (June 1) preparatory to an early march."[595]

Private David Bailey, Company B, is killed by a citizen on Election Day

Forty-year-old David Bailey of Leon, Mason County, enlisted in the 13th Regiment at Point Pleasant, on August 5, 1862. Prior to enlistment, he had worked as a farmer. He had a wife and children. His eldest son, John W. Bailey, aged twenty-one, joined the regiment with his father the same day, eventually becoming bugler for Co. B. After muster-in, David was detailed for detached service with the regimental hospital. He had duty as an ambulance driver.[596] Most recently, from January through April (1863), he had been detailed for extra duty as cook for the hospital.

May 28 had been designated as election day by the most recent Constitutional Convention which had convened at Wheeling on February 19, 1863. On this day, the first election for State and municipal officers was to be held in all counties of the newly formed State. May 28, Private David Bailey was in Leon, his home town. He was killed. A contemporary news report of the incident recounted that:

> during the day [he] became engaged in a discussion in regard to the Government policy with a number of Copperheads from the vicinity of Ten Mile, which resulted in an altercation, the soldier proving to be the best man. Later in the day, William Shields, Louis Shields and Columbus Greenlee, waylaid Bailey on his road home. William Shields struck the soldier on the left temple with a rock, or some other hard substance, and fractured his skull. The murderer escaped during the confusion, and is now in all probability within rebel lines. His accomplices were arrested and are now held for trial.[597]

The events leading up to his death as preserved in family lore are presented quite differently. Bailey was indeed in Leon on election day and after voting, he went to the store there hoping to hear the news. Leon, located at the mouth of Thirteen Mile Creek (founded between 1835 and 1840), had once been a friendly town. Unknown to Bailey, the storekeeper had senselessly vowed to kill the next Union soldier that he saw. With no inkling that he might be attacked, Bailey approached the store (later known as Burdette Hardware). Some men had congregated there and as he entered, he was struck down with a scale weight. He died a short time later. His body was taken and buried but the location of the grave was forgotten and lost.[598]

Miscellaneous regimental records May 1863

Extant regimental records indicate that Company A was stationed at Hurricane Bridge, until May 10. May 11, Co. A was transferred to "Camp near Charleston, V[irgini]a." May 29 through 31, the company was at Coals Mouth. On June 1, the company returned to Hurricane Bridge. Co. A reported that it had marched over "two hundred miles" during the month.[599]

Morning reports indicate that at the beginning of May the company reported present and absent three commissioned officers and eighty-three enlisted men. There were no changes to the company (such as enlistments or discharges) reported for the month. Detached service was reported only for the period of May 4 to May 13 (i.e., the period prior to the expedition to Fayetteville). A minimum of two to a maximum of ten men were absent on detached service during this period. May 4, 1st Sergeant George Danner and three enlisted men took ten prisoners to Charleston and Danner returned from Charleston on May 9. May 6, Captain Johnson "went with five men on a scout up Mud." For May 10 and 11 is the following entry: "This morning we rec'd an order to report at Charleston without delay. We left Hurricane Bridge 8 o'clock p.m. and arrived at Charleston 2 o'clock A.M." Remarks entered in the pages following morning reports describe particulars of their expedition to Fayetteville:

May 17: We left camp near Charleston at 2 o'clock and Steam boat to Pain[t] Creek.

May 18: We march from Pain[t] creek to camp Hodelsen.

May 19: We arrived Gauley Bridge at 2 o'clock p.m. Left Gauley Bridge at 4 o'clock.

May 20: March to Fayetteville. W[oodson] B. Hall wounded accidentally shot his thumb off.

May 21: Arrived Fayetteville 9:00 p.m. Left Fayetteville at 10 o'clock p.m.

May 22: March towards Gauley to Plak [Peak?] Farm.

May 23: arrived Fayetteville at 10 o'clock.

May 24: at 9 o'clock a.m. left Fayetteville, at 4 o'clock march to the Falls of Kanawha.

May 26: Left Falls of Kanawha and march to Clifton.

May 27: Left Clifton 5 o'clock a.m. March 15 mile to Ba _ _ s Town.[Brownstown?]

May 28: We arrived Coals Mouth at 10 o'clock a.m.[600]

Morning reports record that during the month Co. A had a minimum of three and a maximum of eleven men present sick and seven to twelve men absent sick. May 1, Corporal A.J. Davis and Michael Baxter were "absent with leave from the Surgeon at Regimental Hospital, Hurricane Bridge." On May 2, F.M. Cobb was reported absent at home, "with leave from the Surgeon at Hurricane Bridge Regimental Hospital." May 9, William Withers returned to the company from home and reported for duty. May 11, Capt. James W. Johnson was reported "[a]bsent sick at Mouth of Coal," and "Samuel Snodgrass report[ed] to his company." May 13, William P. Compton was reported "absent sick at Post Hospital," and Sergeant M. Grinstead and V.B. Edens both "report[ed] for duty." May 14, John Moore returned from home and reported for duty. Samuel Teel was noted "absent at home." May 16, Corporal Andrew J. Davis returned from home and reported for duty. One enlisted man had leave of absence on May 29. On the 30th, John Tully and Philip Wintz were absent with leave. May 31, Philip Wintz was reported absent "at home, at Charleston." On May 29, the day after their return to camp from the expedition to Fayetteville, seven enlisted men were absent without leave. Some must have returned as just two enlisted men were absent without leave on May 30. No one was reported in arrest or confinement.[601]

Morning reports for Company B show that the company was stationed at Mud Bridge, May 1 through 9; May 10 and 11, the company was on the march; and May 12 through 17, the company was "in Camp near Charleston." May 18 through the 23, the company was again marching; May 24, it was at Fayette Court House; May 25 through 28, again on the march and May 29 through 31, it was at Coals Mouth. At the end of the month, Co. B numbered present and absent three commissioned officers and ninety enlisted men. One man enlisted in the regiment on May 1. One enlisted man returned from being missing in action on May 5. May 1, "John Waugh returned to his company at day." May 28, enlisted man David Bailey was killed by a citizen in Mason County. The company had from one to three men absent with leave May 1 through 17. No one was reported absent without leave. No one had special or daily duty. One to five privates had extra duty May 6 through 31. Two to ten enlisted men had detached service for all but two days of the month (May 12 to 13). Co. B reported one to four men present sick (including Capt. Stewart sick May 7 and 8) and approximately thirteen to fifteen men absent sick (original manuscript is not entirely legible). No one was present or absent in arrest or confinement.[602]

Morning reports for Co. C indicate that the men were stationed at Hurricane Bridge until ordered to dismantle Howell's Mills on Mud River. On May 8, Col. Brown issued Special Order No. 13 to Capt. Van D. McDaniel, cmdg. Co. C, sending him and the company on this errand. Brown directed McDaniel as follows:

> You will take a part of your Co. and proceed to Mud Bridge —
> thence down Mud river, as far as Howells Mills, and take the Mill
> Spindles or such other machine as can easy be placed back without
> injury — so as to disable the mill from ever grinding, until ordered
> by Military Authority, and bring the same to this post.[603]

Co. C was reported "on the march" May 10 and 11. May 12 through 16, the company was recorded as being "in Camp near Charleston." May 17 through 28, Co. C was again marching. May 29 through 31, the company was reported at Coals Mouth. Company musicians were present to signal the order of the day, their tunes and rhythms imparting energy and lightening the burden of the company's exhausting schedule of movements. No one was absent with leave. Thirty-year-old Private Bazalleel Meek deserted on May 4 from Hurricane Bridge.[604] At the end of the month, Co. C reported present and absent, three commissioned officers and seventy-seven enlisted men. The company reported no details for special, extra or daily duty, perhaps due to being so frequently on the march. One commissioned officer and two enlisted men were on detached service for the entire month. The record suggests that Co. C enjoyed good health in May. There was a total of only one to five men "present sick." The record of these five soldiers is following:

> May 3rd, James Cox was sick in the Regimental Hospital at Hurricane Bridge. He returned to duty May 10th, 1863. May 4th, Thomas E. Randle (Randall) was sick in Post Hospital at Point Pleasant and returned to duty May 20th, 1863. May 5th, Ira Hilton was admitted to Regimental Hospital at Hurricane Bridge. He returned to duty May 8th.

Records at the Adjutant General's Office indicate that on May 16 or 17 George Burris had a wound dressed at "13th Virginia Regimental Hospital."[605] Absent sick included David Eads, who was admitted to "U.S.A. General Hospital Point Pleasant," on May 19 for "Sequ of pneum" (pneumania and repercussions of pneumonia?) He returned to duty on June 9.[606] May 29 through at least May 31, Francis A. Windon was sick in hospital at Coals Mouth. No one of Co. C was present or absent in arrest or confinement during the month. Three to five enlisted men were reported "Retained by Civil Authority."[607]

The following series of entries (contained in "Remarks" following McDaniel's morning reports) so remarkably rich in detail in comparison to the usual bill of fare for this regiment, chronicle the 13th Regiment's part in the expedition to Fayetteville:

> May 9: The regiment received orders at Hurricane Bridge, Va. on the evening of May 9, 1863 from Brigade Head Qtrs. to break up camp and report at Charleston and on the morning of the 10th, we left Hurricane Bridge and arrived at Charleston. On the evening of the same day, went into camp 2 miles above the town, on the west side of the river. The next day remained here until Sunday, the 17th, when orders were received to report at Gauley Bridge. We reached there on the 19th at 9 o'clock A.M. Remained there until 12 o'clock, when we were ordered by Col. Toland to proceed to Fayetteville without delay. At 2 o'clock P.M., the regiment was again in motion and reached Fayetteville same evening at 9 o'clock P.M. Bivouacked for the night, expecting an attack every minute from the enemy whose camp fires could easily be seen from our

position. Early in the morning, the enemy opened upon us with shot and shell. We lay under their fire until about 2 o'clock, when the enemy retreated. Col. Brown then received orders from Brigade commander, Colonel White, to be ready to march in one hour in pursuit of the retreating enemy. Immediate preparations were made, the troops being provided with 3 days rations. We followed the Rebels as far as Raleigh Court House, being fired upon once by them on the way, but doing little injury. We encamped in sight of Raleigh for the night after having fired a few shots into the rear guard of the foe. The next morning started for Fayetteville, bivouacked on the McCoys Farm first night. Reached Fayetteville next morning at 9 o'clock A.M.; remaining here to rest for two days, then received orders to return to Camp White near Charleston. Marched thence to Camp Piat, reaching there on the evening of the 27th. Next morning left on steamer Victress for Coals Mouth and there we remained until the last of the month.[608]

Morning reports submitted for Company D also record that a strenuous program of marching was ordered for the regiment. Co. D was in encamped at Hurricane Bridge May 1 through 9. "Remarks" indicate that the company had been ordered to report to Camp White on May 10. The company commenced marching on this day as ordered. They arrived there on May 11 and "halted for the night." May 12, Capt. Simon Williams, James Robinson, Jacob Tucker, Marshall Smith and Henry Hoffman were all sent to "Post Hospital at Point Pleasant Virginia." No one was reported present sick after May 21. Apparently all sick men were sent away in anticipation of the counter-movement to Fayetteville. Six to at least thirteen men were reported absent sick for the month.[609] No one was present or absent in arrest or confinement. On May 17,

> Co. D with the Regiment was ordered to report to Gauley Bridge.
>
> May 18: Reached Gauley Bridge on the 19th ordered from there the same day to report at Fayetteville, arrived at Fayetteville same day.

Ordered in to reinforce Fayetteville, Co. D was again marching on May 19. May 20 through 21, the company was at Fayetteville. May 22 through 28, they were again marching. May 29 through 31, the company was at Coals Mouth. One drummer was reported present with the company each day of the month. A fifer was reported present March 1 through 11; and March 14 through 22. Nineteen-year-old Samuel D. Hannah filled the position of drum major at this time. He would soon to be promoted to the position of principle musician for Co. D and would so remain, until the regiment was mustered out.

Highly mobile as the company was at this time, no one was granted leave of absence. Two enlisted men were reported absent without leave on May 4, and one was so reported on May 5. The company had no loss or gain during the month and finished the month with present and absent three commissioned and eighty-four enlisted men. As to duty, no one was reported detailed for special duty. One private had extra duty on May 1, and two privates had daily duty all month. May 1 through 12, one enlisted man was detailed for detached service. May 13 through 31, two enlisted men had detached duty. Co. D reported one to three men present sick May 1 through 21. 2nd Lieutenant Charles T. Latham was reported sick May 4 and 5. Captain Simon Williams, who apparently suffered from bouts of ill-health, was again sick from May 6 through 10; and 13 through 17.[610]

Morning reports submitted for Company F show that the company was stationed at Hurricane Bridge May 1 through 9. Co. F like Co. D probably received orders to proceed to Camp White on May 10, and they "[m]arched pursuant to orders. Crossed Coal at 12 o'clock m., arrived one mile below Charleston at 10 o'clock P.M. and bivouacked. May 11: 12 o'clock M. [Midday] ordered up the river marched 2 miles and went into Camp [nea]r Charleston." May 12 through 17, Co. F was in "Camp near Charleston, Va." May 15, David Stevenson deserted from camp at Charleston. May 18 through 28, the company was again marching. May 29 to 31, Co. F was reported "at Coals Mouth." The company had one drummer present each day of the month and one fifer present May 10 through 18 and May 31. No one was absent with leave. At the end of the month, Co. F numbered three commissioned and seventy-seven enlisted men. No one was reported detailed for special duty. One to three privates had extra duty each day of the month and four to five privates had daily duty. Three to five men (one commissioned officer and enlisted men) had detached assignment each day of the month.

According to reports, Company F enjoyed fairly good health. No one was reported present sick May 18 through 21; and May 26 through 28. Just one to four men were reported present sick all other days. Two to five men enlisted men were reported absent sick each day. Among the sick was David Burrows, who became ill with fever on May 10. No one was reported present or absent in arrest or confinement. Remarks accompanying McCown's reports provide interesting details about the regiment's part in the expedition to Fayetteville:

May 17: Ordered to Gauley at 3 o'clock P.M. Marched on board
steamer Ingomar. She took us to Cabin Creek, bivouac[k]ed.

May 18: Marched at 8 o'clock, arrived in 4 miles of Gauley Bridge at 6 o'clock P.M. and bivouac[k]ed.

May 19: Arrived at Gauley Bridge at 9 o'clock A.M. Put up tents and at 3 o'clock P.M. was ordered to Fayetteville, where we arrived at 9 o'clock P.M. and found the place invested by the rebels, we lay on our arms all night.

May 20: The enemy opened their artillery on us at daylight and kept up a brisk fire until 2 o'clock P.M., when they withdrew their forces in the direction of Raleigh, at 10 o'clock P.M. our regiment, the 91st and 12th OVI under Col. White started in pursuit.

May 21: All day on the road, skirmishing occasionally with the enemy, at dark, we shelled the town of Raleigh.

May 22: 22 men of the company under Lieut. [Timothy] Russell were sent on a reconnaissance about Raleigh, returned to this Co. at 12 o'clock and were then ordered to return to Fayetteville it being certain that the enemy had effected a safe retreat.

May 23: Arrived at Fayetteville at 10 o'clock A.M., lay in the shade all day waiting for orders, bivouacked that night.

May 24: Remained at Fayetteville all day resting, bivouaced again at night.

May 25: Left Fayetteville at 3 o'clock P.M. marched to the Falls of Kanawha, where we camped for the night.

May 26: Marched to Clifton where we again bivouac[k]ed.

May 27: Marched to Brownstown and stayed all night waiting for steamers to transport us to Coalsmouth.

May 28: Marched on board steamer Victress, arrived at Coalsmouth and once more put up our tents after an absence of 3 weeks. We lost one man, William Daniel, having straggled from his company, while near Raleigh on the 22nd inst. was captured by the enemy.[611]

Morning reports for Co. H record that the company was stationed at Point Pleasant May 1 through 6. On May 7, the company was ordered to Hurricane Bridge. The company straightway started marching and arrived at Hurricane Bridge that day.

They remained at Hurricane Bridge May 7 through 10. On May 10, the "Regiment received marching orders for Charleston. Left Hurricane 11th May reaching destination same day, distance marched 24 miles." They remained in camp with the rest of the regiment at Camp White May 11 through 17. May 11, Captain W.I. Mathews issued Company Order No. 1 giving notice that "Pursuant to Special Order No. [left blank] issued from Regimental Head Quarters, the undersigned [Mathews presumably] assumes command of Co. H, 13th Regt."[612] "Remarks" indicate that on May 17, the regiment received marching orders to proceed to Gauley. The men "left by boat at 2 P.M." May 18 and 19, the company was en route, marching. May 19, the 13th Regiment arrived at Gauley at 9 A.M. and went into camp. At 12 midday, they received orders to march for Fayetteville. May 20, they were at Fayetteville. They had marched a distance of fifteen miles from Gauley to Fayetteville. May 20 and 21, the 13th marched a total of twenty-nine miles. The " had a fight on [May] 20th to Raleigh CH 29 miles."[613] On the 20th, "the enemy shelled us without effect." May 21 through 23, they were on the march.

> May 21: At 3 P.M. of 20th rebels retreated. Pursuit commenced at 9 P.M. Marched til 3 A.M. then halted til daylight.
>
> May 22: Marched again and overtook rebels at 9 A.M. a few shots [fired] rebs retreating.
>
> May 23: Pursued til evening. Arrived at Raleigh. Discontinued pursuit. 23 May arrived at Fayetteville.

Making the return trip to Fayetteville on May 22 and 23, the company marched twenty-nine miles.[614] May 24, Co. H left Fayetteville. May 25, the company with the regiment returned to Gauley "marching twelve miles." The 26th through 28th, the command continued marching and arrived at Coalsmouth on the evening of May 28. May 29 through 31, they remained at Coalsmouth.

Co. H reported no buglers, farriers, blacksmiths nor artificiers present. They had one drummer to sound out the daily rhythm, he was present May 1 through 20 and May 24 through 31. One fifer was present May 27 through 31. Just who these musicians were is open to conjecture. Albert C. Jameson and William T. Shaver, who had been serving as company musicians respectively, since September and December 1962, were no longer with the company. Fifteen- year-old A.C. Jameson had been in poor health and unable to stand the rigors of army life and had been absent sick for some months. He would soon be discharged from the service upon a writ of *habeus corpus* brought by his father. William T. Shaver had deserted March 6, 1863, at Hurricane Bridge.

One enlisted man was reported absent with leave on May 10. Two to four enlisted men were absent without leave May 1 through 31. Capt. Taylor W. Hampton, who had resigned his commission was replaced by William I. Mathews, superceding Lieutenant O.W. Griswold. Two enlisted men were reported deserted on May 10. At the end of the month, Co. H had present and absent a total of three commissioned officers and seventy-six enlisted men. One private was delegated for extra duty May 1 through 23. Two privates had extra duty May 27 through 31. Over the course of the month, two to sixteen men were reported present sick with a greater number of men sick during the first week of May. Numbers of present sick fell as the month progressed. The sick men may have been sent away to hospital or recovered as the month wore on. The company had between at least four and upwards of seventeen men absent sick each day. No one was present or absent in confinement.[615]

Morning reports for Company I are at best skeletal. 2nd Lieutenant William Feazel had difficulties keeping abreast of his reports. They do, however, provide some basic information. For the first time, non-commissioned officers were reported, in the number of three sergeants and four corporals all present May 1 through 31. Twenty-nine (or twenty-eight) privates were in camp for the month. No one was reported absent with or without leave. There was a loss to the company of one enlisted man, a recruit, John W. Howard, rejected for disability and dropped from the rolls on May 7 of 8. At the end of the month, Co. I reported present and absent one commissioned officer and a total of forty-four enlisted men. The company had four to five men present sick including "Lieutenant Feazel sent to Hospital Charleston, Virginia," on May 12. Feazel was reported sick May 13 through 18. He returned to the company on May 18, was sent to Gallipolis, Ohio, on May 21 and remained absent through the 31st for some unspecified reason. Five enlisted men were absent sick. No one was present or absent in confinement.[616]

A humorous note was sent by Brig. Gen. E.P. Scammon to his subordinate officer, Col. Hayes on May 29, which leaves much to the imagination of the reader. Scammon wrote:

> At Mr. Jeffries May 29.
>
> Col. R.B. Hayes Comdg. Brig.
>
> Dr. Colonel:
>
> Will you do me the favor of <u>intimating</u> to Col. Brown that he is under your Command and that when he wishes to have muskets discharged he must get your permission? It would not surprise but w[ould] greatly <u>annoy</u> me to hear from Cincinnati that we had been attacked by the enemy. The sooner you let Col. Brown <u>at</u> [illeg.underlined words] understand that you are their actual <u>commander</u> the better for all concerned. Very truly E.P.S.[617]

June 1863

"Squaley Times"

During the first half of June, the continued dry weather had begun to take a fearful toll upon agricultural products all over the countryside. "Wheat had been seriously injured, oats about all done in, corn and grass [were] very short."[618] Rebels in the Kanawha Valley had continued to be troublesome since McCausland had been repulsed at Fayetteville.[619] On June 15, U.S. President Lincoln issued a proclamation announcing that "the armed insurrectionary combinations" in several of the States were threatening to make inroads into Maryland, West Virginia, Pennsylvania and Ohio, which required additional military force for the U.S. service. Ten thousand more volunteers were needed from West Virginia to serve for 6 months.[620]

Also on this day (June 15), Gov. F.H. Pierpoint issued an ominous proclamation directing "Commandants of Regiments and Companies of Virginia Militia" to immediately call together their companies and regiments to hold themselves in readiness to go to the field "at an hour's warning." Arms and equipments would be furnished at the several places of rendezvous. Pierpont continued.

> The enemies of our liberty and prosperity are again threatening our peaceful homes. Citizen soldiers stand by your firesides and defend them against the common foes of a free government. Make every available spot a rifle pit from which to slay the enemy. You know the roads and the passes.[621]

The amended West Virginia State Constitution having been adopted by the citizens of the new State at Spring elections this year, the government of West Virginia was officially organized on June 17. On June 19, the U.S. Congress voted that the forty-six counties of Western Virginia be admitted as a new State. Saturday, June 20, Lincoln issued a proclamation under which West Virginia was formally admitted to the Union as the 35th State, and Arthur I. Boreman, West Virginia's first elected governor, was sworn into office on this day. Former Gov. Francis H. Pierpont delivered a speech in the streets of Wheeling upon the occasion of Boreman's inauguration. He exhorted the people to acknowledge that in light of recent events the fight was not over: "lose not your sacred liberties. Fight as long as a mountain presents a battery — or a grotto remains to serve as a rifle pit. [...] Never yield the right of a freeman."[622] On Monday, June 29, pursuant to Lincoln's June 15th call for more volunteers, Gov. Boreman issued a proclamation of his own regarding the escalation of Confederate operations into Maryland, West Virginia, Pennsylvania and Ohio that necessitated additional military force be raised. Boreman urged persons liable for military duty to come forward and volunteer. "The State is not considered safe from the invasion of the enemy," explained Boreman, "and it is known that some sections are infested with bands of guerillas, murdering and robbing the people. This call is deemed necessary to repel invasion and for the protection of the people." The militia troops recruited under this call were to rendezvous at Wheeling, Clarksburg, Parkersburg, Point Pleasant, Charleston or Ceredo and would receive subsistence from the nearest U.S. Quartermaster.[623]

On Wednesday, June 24, the Department of West Virginia was organized pursuant to General Order No. 186, issued by the War Department at Washington.[624] The troops within the limits of the department consisted of E.P. Scammon's division, formerly 3rd Division, 8th Army Corps, Middle Department, now known as Third Division, Deparent of West Virginia. Brigades comprising the division were commanded by William W. Averell, Nathan Wilkinson, James A. Mulligan and Jacob Campbell. Brigadier General Benjamin F. Kelley had command of the Middle Department. Scammon's division consisted of two brigades: First Brigade, to which the 13th Virginia belonged, commanded by Rutherford B. Hayes, had its headquarters at Camp White, Charleston. Second Brigade, commanded by Carr B. White, was posted at Fayetteville.

The newly organized department labored to prepare itself and fortify for the next Confederate incursion. Towards the end of June, it was reported that soldiers stationed along the B&O Railroad were busily building heavy hewn timber block houses for the defense of bridges and trestlework.[625] The troops of Scammon's division continued to be deployed along a wide front (from the Ohio River to Fayetteville, West Virginia), and they set to work to fortify their positions and to scout from posts along the rivers and tributaries towards the interior.

Regimental reports and correspondence dated June 1863 indicate that after the regiment returned to Coals Mouth at the end of May, it was again dispersed to various points. On June 1, Companies A, C, D, F and H were ordered to Hurricane Bridge and Companies B, E and Young's Company G were ordered to Mud Bridge.

These companies remained stationed at their respective posts until Monday, June 29, when the regiment was ordered to report to Camp White, Charleston. On the 29th, the 13th encampments at Hurricane Bridge and Mud Bridge were broken up and the regiment proceeded by way of Coals Mouth to Camp White. Companies A and H and a squad of recruits remained at Coals Mouth.

On June 1, Col. Brown issued Special Order No. 14 providing that: "2nd Lt John S. Cunningham of this Regt is hereby ordered to proceed to Wheeling Va to be mustered into service as Adjutant of this Regt."[626] On Thursday, June 4, Col. Hayes again had occasion to reprimand Brown on the "irregularity" of his reports:

> Sir: Your attention is again called to the irregularity in your tri monthly Reports. The last regular report received from you was for May 10. Our report of your Regiment for the 20th and 30th of May had to be made up from the irregular report required of you at this camp on the 15th ultimo.
>
> You will please have the reports for May 20th and 30th made out and forwarded to these Hd. Qtrs. without delay. Hereafter your reports must reach here by the 10th, 20th, and 30th of each month without fail. By command of R. B. Hayes.[627]

June 5, Hayes sent the same message to Brown, a duplicate of the order of June 4.[628] On the 5th, additional orders were received at 13th regimental headquarters from Colonel Hayes at Camp White, directing that scouting parties be sent out and in what direction that communication might remain open with Union troops on Big Sandy River. Hayes ordered Brown to "direct your scouting expedition towards the right so that this Brigade can keep the communication open between it and the forces of Brig. Genl. White at Louisa, Ky."[629] Also on the 5th, Hayes wrote to Lieutenant Colonel Hall, cmdg. at Mud Bridge, with orders tending to the same end:

> You will direct your scouting expedition towards Barboursville hereafter and inform Lt. Col. Tomlinson com[man]d[in]g at Barboursville of any movement of the enemy in that direction as well as reporting it to these Hd. Qtrs. forthwith. The design is to keep up communication with Brig. Genl White on Big Sandy by having the forces of this Brigade scout towards their right.[630]

Saturday, June 6, Col. Brown wrote to Charleston from 13th Headquarters at Hurricane Bridge requesting that he be furnished with "Genl Orders No. 58 and 113 War Dept 1863."[631] June 7, 1863, a letter was received at 1st Brigade Head Quarters from Gen. Scammon, which conveyed information obtained from a deserter from Wyoming County, Virginia, that "the enemy in that direction and most probably their intention being to reach the Ohio and capture the 5th and 13th Va. V. I. to be notified to be on the 'que vie.' "[632] On Thursday, June 11, Brown sent to 1st Brigade Head Quarters at Charleston his Tri-monthly Reports for May 20 and 30 with a word of explanation as to why they had been delayed.[633]

On June 11, John V. Young, cmdg. Co. G (13th detachment at Mud Bridge), wrote home from camp:

> [...] Colonel Hall is fortifying here just as though he was going to remain here during the war. But I think it very uncertain how long we will stay here. However, I shall keep myself in readiness to gather up when I am ordered.

> [...] All the troops are out on inspection and general review, but I have not gotten able to attend. [Young was in bed suffering from rheumatism in his back.] Lieutenant Reyburn takes charge of my company. There are not more than a dozen sick men left in camp, and Colonel Hall left me in charge of the Fort, and gave me orders that if I was attacked to fight until the last man was dead; while he would not be out of sight.

> [...] My books, orders, men and everything else had been inspected this morning by a general inspecting officer of General Schencks staff, Colonel Jones, but they are all right.[634]

That another attack was anticipated at this point in time is clearly evident. Captain Young was sent out from Mud Bridge with his men on June 15 "on a scout to Mud and Guyan River."[635] Two days later, Col. Hayes wrote the following to Col. Brown:

> Hd. Qtrs. 1st Brig. 3rd Div. 8th AC
> Camp White Va. June 17, 1863

> Col. Brown comdg. 13th

> Have your command in readiness to march at a moments notice with 3 days rations. Scout the country in your front thoroughly and report any indications of the enemy to these Hd. Qtrs. by the shortest possible means. Carry out these instructions without exciting suspicion of intended movement.

> By command of R. B. Hayes[636]

The scout embarked as ordered on Thursday, June 18, when Lieut. James R. Hall, commanding a detachment of 13th and 5th Virginia Volunteer Infantry at Mud

Bridge, sent "Capt[ain J.V.] Young with a squad of men on a Scout toward the right of Mud River."[637]

Young wrote a letter to his friend, Captain John T. Bowyer,[638] from Mud Bridge on June 26 thanking Bowyer for his commission and relating particulars of the scout and his perceptions regarding the imperative need for mounted troops to scout the area; the local population's resistance to the new State government; how these close ranks to protect "the Rebs;" the prevalence of horse thievery; that Union folk were losing confidence in the power of their new government to protect them and their property; and that they were leaving.

> [... A]s you up there appear to be much excited and have not time to write and let a body know what you are scared about, I will try and give you a hint of our troubles down here.
>
> Well in the first place, the Rebs are stealing all our horses, and if the Governor of Va. don't do something for us on this side of the Kanawha River we might just as well hang up the fiddle, and I assure you that nothing but mounted men will do any good here. Now, Sir, I have been scouting this country hard for two years, and you must know that I understand the country and the people by this time. If you think the people are willing to submit to the new Government of Va. over here you are mistaken. Although they claim protection from us.
>
> Last week it was reported here that there were Rebs upon Mud, and Col. Hall ordered me to take thirty men and look after Jeff's horse thieves. Well, I proceeded up Mud to the mouth of Middle Fork, then up the Middle Fork one mile to Mr. Eagle-sons. Here I captured two rebel soldiers, Joseph Tinsley[639] who commanded the squad of horse thieves about a mile above this place. An attempt was made to rescue the prisoners. A squad as large as our own fired on us. Three of the party shot at me, which came very near hitting me, the rest fired on the company, but as Providence would have it, none were hurt.
>
> On last Monday I was sent out again after other horse thieves that had been stealing the neighbors horses out of the plow. This time I found their Camp where they had kept their stolen horses for more than 12 months. Not one man or woman would tell me one word about those thieves, but denied that they had ever been there, when the fact is that they fed and harbored them all the time, and those men and women say they claim protection from the northern government. I struck the trail of the stolen

horses, 8 in number, which had started for Dixie the day before I got there, but I back trailed them to their camp near one Peter Burns, whose wife stated that they had been there and boarded at her house and cut grass in their meadow for their horses. (This I forced out of her by telling her I would burn — —) I saw where they cut the grass, where they had carried it to the horses, but for all this Peter Burns denies that he knows anything about them or who they were. I have two Burns in the Guard House here but I think Col. Hall will release them. I have taken several horses from these kind of men, but they are given back again. But when the rebels get a Union man's horses he is gone forever from him — Therefore the loyal citizens here are losing confidence in our Government. They say, and truly too, that the Secesh is protected while they are unprotected just because they did not rebel against their Government and for this they are leaving this part of Virginia — and they think it is very hard, indeed that they have to leave their homes and the graves of their fathers and everything that is near and dear to them just because they are loyal and the Secessionists are permitted to remain at home and enjoy their property and be protected too by their sons who flew to their country's call to put down this infernal rebellion.[640]

All along the front from Fayetteville to the Ohio River, Union army detachments stayed at the ready, fortifying, scouting and taking prisoners. A detachment including the 2nd Virginia Cavalry (U.S.A.) scouted from Fayetteville to the Big and Little Coal Rivers June 18-19 and to Loup Creek on June 26. A detachment of the 5th Virginia Infantry holding the country to the west of the 13th at Barbourville, on the Ohio River left camp to scout to Logan County on June 20. One soldier correspondent with the 5th calling himself "Grotius" wrote to the Gallipolis *Journal* describing the scout, "one of a number of scouts sent out" by the 5th but included here as representative of the kind of duty the 13th was doing in the same kind of interior country against an enemy operating in the same way with varying amounts of success.

Camp at Barboursville, Va.,
June 26th, 1863.

Dear Friend Harper: Sergeant Fuller, of Company H, with a detachment of thirty men has just returned from a very successful scout into Logan county, a district of country where the rebels have been in the habit of conducting affairs pretty much in their own way.

The Seargent left camp on Saturday morning last [June 20th]: his instructions were to scout the country as far as he went. Capt. Walker, of the Confederate States army, was also on the scout, and had stolen several horses and saddles, and had taken some prisoners, belonging to the home guards of Logan county. The rebels had started to return, when the Sergeant was informed that they were in the neighborhood, and immediately rallied his men and started in pursuit. Marching all night, they overtook the rebels snugly ensconced at the house of a 'warm friend,' quietly sleeping. Their picket, at the advance of our boys, threw down his gun and started on the double quick to Camp but there were some among our boys who could run too, and arrived at the quarters as soon as the reb did; the boys knowing that their enemies were greatly superior in numbers, would not risk a 'general engagement,' but rushed upon them with a yell which aroused them from their peaceful slumbers, and put them all to fight. The Sergeant demanded them to surrender, upon which, Capt. Walker called out: 'I surrender, by G—d.' The boys marched off with their spoils, consisting of the rebel Captain, his Lieutenant, and a small number of privates, fifteen stand of small arms, five horses and saddles, and had the satisfaction of releasing the home guards, besides capturing a deserter who left our Regiment last winter. Nobody was hurt on our side, and but two rebels, both killed. Besides this, the Sergeant sent into Camp at different times, eighteen prisoners. This is but one of a number of scouts sent out since our return to this place. Not a day passes but more or less prisoners are sent into Camp, by our scouts.—

'Success to the sang trade.' Respectfully, yours, &c. GROTIUS.[641]

On the 18th, Colonel R.B. Hayes ordered that "Commandants of Regts and Detached Companies of this Brig. will make immediate application to Lieut. Col. Chesebrough A.A.G. 8th A.C. to have all 2nd Lieutenants of their commands promoted from Enlisted men discharged in order to be mustered by the Apt Commissary of Musters."[642] On the 18th, Quartermaster Stephen Comstock had his May returns corrected by 1st Lieutenant and Quartermaster William McKinley and he returned for resubmission. McKinley returned Comstock's returns with the following note appended: "Your return[s] for month of May 1863 are respectfully returned for correction. See Forms 23 and Abstract M Camp and Garrison Equipage transferred. Errors are marked in pencil."[643]

Friday, June 19, Col. Brown wrote from Hurricane Bridge requesting 1st Brigade Headquarters to appoint a "Board of Survey" "to examine into the Loss of Clothing issued to 13th Regt. Va.V.I."[644] Also this day, Brown issued General Order No. 13, approving certain reductions and promotions in rank:

> II Order No [left blank] issued from Head Quarters Co A reducing Sergeant George W. King to the ranks for disability and promoting Corporal R.H. Davis to sergeant, vice George W. King reduced, — and promoting private James H. Tully to Corporal, vice R.H. Davis promoted, — is hereby approved. They will be obeyed and respected accordingly
>
> II II Order No [left blank] issued from Head Quarters Co D reducing Sergeant Julius Hatcher to the ranks for disability, and promoting corporal A.W. Darnel to sergeant, vice Julius Hatcher reduced, — and promoting private John C. Kimes to Corporal, vice A.W. Darnel promoted, — is hereby approved. They will be obeyed and respected accordingly.[645]

June 20, Col. Brown devoted at least a portion of his day to his administrative duties. He wrote to Charleston acknowledging the "receipt of General Orders No. 106, 127, 125, 132, 135, 139" and reporting that he had "not received Nos. 52, 54, 55, 73. 84, 111, 126, 141."[646] Also on the 20th, Brown wrote to Brigade Head Quarters Charleston forwarding a "requisition for arms with instructions from A. Buffington relating to the manner of making out the returns." Brown added "[i]f an inspector is needed to Condemn the arms no one [is] on hand [and that] an inspector be appointed."[647] New orders for the day (June 20) included, Special Order No. 15 authorizing a detachment from Co. A to escort of prisoners to Charleston:

> Corporal [Cornelius] Page + 4 men of Co. A is hereby detailed to take charge of and will proceed to deliver to the Provost Marshal at Charleston West Va Nathaniel Birchfield Co D and Sampson Smallridge a Rebel Soldier. — He will take Receipt for the same and return to these Head Quarters without unnecessary delay.[648]

The 13th Regiment West Virginia scouts along the Kanawha River in search of rebel mail stations

On June 19, another scout was ordered. Col. Hayes wrote to Brown ordering him to search and shut down certain suspected Confederate mail stations:

Hd. Qtrs. 1ˢᵗ Brig. 3ʳᵈ Div. 8ᵗʰ AC
Camp White June 19, 1863

Col. Brown commandg
at Hurricane Bridge

Information which is deemed reliable has been received here that there are 'rebel' mail stations at the following named houses John Blackwell 3 or 4 miles from James Lenhams opposite Buffalo. Wm. Frazier 4 or 5 miles further on the River, A.W. Handley[649] in Teayses Valley.

The Col. comdg. directs that you have these houses searched on Monday night June 22 at 12 o clock — And if anything is found which is suspicious arrest the parties and send them here under Guard with the evidence against them. By order of Col. R.B. Hayes[650]

This scout and search took place as ordered on June 22. June 24, a report was submitted. The document reporting the results of the search is only a fragment and damaged at that, the upper and lower portion bearing the signature being torn away. Subsequent correspondence from Charleston, however, supports the inference that the report was written by Col. Brown. The body of the report is nonetheless complete and is reprinted as follows:

Sirs

I have the honor to forward herewith to you a report of the search for rebel Mail made June 22ⁿᵈ inst., —

On the 20ᵗʰ inst I received orders from Brigade Head Quarters to search the following houses which from information deemed reliable were stations for rebel mail. — John Blackwell 3 or 4 miles from Jones Lenhaws opposite Buffalo, Wm Frasier 4 or 5 miles further on the river, A W Hanly in Teays Valley, —

Parties were detailed at the Head Quarters of this Regiment in time to search the houses at the time specified viz 12 o.c. on Monday might June 22ⁿᵈ inst — At the houses of John Blackwell + William Frasier there was no evidence found of any mail or communication — at the house of AW Hanley there was found two Rebel papers. The Lynchburg Daily Republican dated June 1ˢᵗ — the Lynchburg Daily Virginian dated June 5ᵗʰ inst — a proclamation and circular letter issued by John H Reagan P[ost]

M[aster] General of the C.S.A. addressed to the Post Master in the southern confederacy You will find the papers accompanying this

No other evidence was found at the house of direct communication with the rebels.[651]

Hayes responded to Brown the same day (June 24) writing: "Send A.W. Hanley the man at whose house the Lynchburg papers were found to these Head Qtrs under guard."[652]

The regimental paper trail indicates that towards the end of June, Col. Brown turned his attention to optimizing and tightening the efficiency of his medical system. June 21, Brown issued General Order No. 14 from headquarters at Hurricane Bridge regarding the misuse of ambulances as general conveyances. This was not a unique problem. The perversion of ambulances from their proper use was army wide. Army ambulances were made as two-wheeled and four-wheeled carriages. Two-wheeled carriages were the more numerous but four-wheeled carriages were actually best for transporting wounded or sick men. The 13th probably had two but no more than three ambulances. In all likelihood, they were carriages of the two-wheeled variety. Brown continued:

> It is hereby ordered that hereafter no Officer Private Soldier nor Citizen will be permitted to ride in the Ambulances belonging to this Regiment unless ordered so to do by the Surgeon of the Regiment certifying to the fact that the man is sick or unable to travel in any other way
>
> Ambulances are intended only for the conveyance of the sick and wounded and we cannot expect them to be properly respected by the Enemy if they are constantly used as ours have been heretofore for other purposes than those intended And any Driver having charge of an Ambulance who will suffer this Order violated by any of the class of persons above referred to will be severely punished[653]

On June 22, Brown administered another downstroke writing to Hayes from Hurricane Bridge reporting that it was "desirable that the Hospital be removed from Coals Mouth to Mud Bridge as there are good buildings there." Further, he asked for Hayes' advice regarding the removal of the hospital to that point.[654] Hayes responded the next day writing that "[i]t is desirable that your Regimental Hospital should be as near to your command as possible. After consulting your Surgeon you will locate it where it will best accommodate the sick."[655]

Morning reports for the month of June indicate that on June 1 Company A was ordered away from Coals Mouth to Hurricane Bridge. The company remained at Hurricane Bridge until at minimum June 19 (or more likely, through to the end of the month). By order of Gen. Scammon, Co. A "left Coals Mouth June 1ˢᵗ a.m. and arrived at Hurricane Bridge 2 o'clock p.m." The company had present and absent, three commissioned officers and eighty-three enlisted men. June 1, Corporal Robert H. Davis was promoted to 3ʳᵈ Sergeant, and James H. Tully (one of the company's musicians) was promoted to corporal. One drummer (probably fifty-year-old, tall, grey-haired, Samuel Snodgrass) was present with the company for the month. June 1 and possibly June 2, one fifer and one farrier or blacksmith were also reported present. June 1 to June 15, one artificier (the term for a skilled laborer; a military mechanic) was reported present, and two artificiers were reported present from June 16 to 30. On June 1, Irdell Harel (or Hurell) was detailed for daily duty at Post Hospital, Coals Mouth. Detached duty was heavy. The following is the record of men detailed for detached duty from Co. A:

June 2 to 8, two enlisted men

June 9 Lieutenant Greenbury Slack and eight men, detailed from Co. D, were ordered to scout on Mud River. These returned from scouting on June 11.

June 11 to June 20, two enlisted men

June 21, Corporal Cornelius Page, James Light, James Spradling, William Gray, George W. King were sent as guards to escort three prisoners to Charleston. This party did not return as a group to camp. On June 25, James R. Spradling was reported to have returned from Charleston. Then June 26, Corporal Page, George W. King, James Liyth (Light) returned from Charleston. William Gray, who had failed to return, was reported absent without leave.

June 21, Lieut. Slack was ordered on a scout with twenty-eight men. These left camp at 4 o'clock p.m. on June 21. Slack returned to camp with his men at 5 p.m., on June 23.

June 23, seven enlisted men and one commissioned officer had detached service.

June 24 through 30, two enlisted men

June 26, Lieutenant Samuel S. Mathers and five men were ordered on a scout.

June 14 through 19, Capt. James W. Johnson was absent with leave. Johnson returned to the company in the "P.M.," on the 19th. June 1, Philip Wintz, who had been absent at home at Charleston, returned to Co. A. The following is the record of those absent without leave: June 5 through 10, Private Andrew Snodgrass. Snodgrass had been absent at his home in Charleston and returned to camp on June 11. June 23, Andrew J. Cobb was reported absent without leave. He had returned to his home on June 14 and not returned. He may have been absent until June 26. June 26, two enlisted men were reported absent without leave, and June 27 through 30 one enlisted man was reported absent without authority.

June 5, William P. Compton "repor[ted] for duty" and June 24 Sergeant Grinstead reported for duty, both presumably returning from sick leave. Co. A had one to five enlisted men present sick for the month of June with a decrease in the number of present sick at the close of the month. One commissioned officer was present sick June 4 through 7 and June 14 and 23. Many were absent sick during the month. On June 1, seven enlisted men were reported absent sick. June 2, five more men were reported absent sick (total of twelve). Elevated numbers of sick persisted for the rest of the month. On June 3, William Riley, George W. Ramey, Ref (Ralph) Pauly, "B. Hall" (Woodson B. Hall) and Francis M. Cobb were all reported "sick in Hospital at Coals Mouth." June 8, "H. George" (William H. George) and "W. Ramey" (George W. Ramey) returned from hospital to report for duty. Beginning June 20, Joseph Scott was absent, sick at home at Charleston. No one was reported in arrest or confinement.[656]

Morning reports for Company B show that the company was marching for Mud Bridge on June 1. They arrived there on the 2nd and remained stationed at Mud Bridge until June 30, when they were again marching. Just one to two men had leave of absence. These were absent June 1 through 3 and 8 through 21. No one was absent without leave. One man enlisted in the regiment on June 22. The company finished the month with a total of three commissioned officers and ninety-one enlisted men present and absent. Five to nineteen men were delegated for detached service each day of the month. A large group, comprised of two commissioned officers and seventeen enlisted men, was detached June 9 and 10. Co. B had two to eight men present sick for the month and six to thirteen absent sick. Higher numbers of men were reported sick at the beginning of June. No one was present or absent in arrest or confinement.[657]

Christopher Columbus Barnett, Private in Co. B (about seventeen to eighteen years of age), wrote the following letters home to his parents, James and Rebecca A. Barnett, during the month of June:

Mud Bridge
June the 8, 1863

Dear father and mother

it is witch plesur that I take mi pen in hand to write yo a few lines
to let you know that I am well at presant am hope when those few
lines comes to hand that thay may find you all well and injoying
god helth We were paid off the other day and thay [subtracted the]
clothing Bill out [so] i have none to send home tis time but i have
15 Doless i have not moe to rite this time i wont you to rite as
soo[n] ase you ge[t] this[658]

Mud Bridge Cabell Co.
June the 23 1863

Dear father

it is witch the grates of plesur that i take mi pen in hand to rite you
a few lines to let you know that i am well at present ant i hope that
when those few lines comes to hand thay may find you all well and
in joying goo helth as for gamling i never dun much but i have quit
playing atol i don't play atoll for nothing i think that [illeg.] pay
day that i can send forty dolers home next time i recived yoles leter
and i was glad to [illegible].[659]

Morning reports and "Remarks" submitted for Company C show that the regiment left Coals Mouth at 7 a.m. on June 1. They marched a distance of twelve miles
and reached Hurricane Bridge at 12 noon. Co. C was stationed at Hurricane Bridge
until June 29, when they received marching orders and were in motion at "2 P.M."
They arrived at Coals Mouth at "8 P.M." having again marched the distance of twelve
miles. In the morning, on June 30, Co. C started for Charleston at 7 a.m. They
marched a distance of twenty-five miles and arrived at Charleston at 6 p.m.

Musicians Marcus L. Jones and George E. Warner were present each day of the
month. No one absented themselves from the company with or without leave. The
company sustained no loss nor gain. At the end of the month, Co. C had present
and absent, three commissioned officers and seventy-seven enlisted men. Two to five
men were detailed for daily duty and one commissioned officer and two enlisted men
were assigned detached duty each day of the month. No one was reported present
sick and just three to five enlisted men were reported absent sick during the month.
Among the sick were Richard T. Ellis and James L. Woodyard, sick in "Regimental
Hospital at Coals Mouth" beginning June 3. Francis A. Windon, who had been sick
in hospital at Coals Mouth beginning May 29 returned for duty on June 7. June 8,

William McDaniel reported sick at Point Pleasant and continued sick through the end of the month. On June 11, Richard T. Ellis reported for duty "from Hospital." John Gormon entered 13th Regimental Hospital on June 19 and remained there at least through the end of the month. June 24, Private George W. Barnet was reported sick in Post Hospital at Point Pleasant. On June 26, Barnett was admitted to "U.S.A. General Hospital Gallipolis" with the complaint: "old age." His age was given as forty years. He was assigned to the Invalid Corps and returned to duty on October 12, 1863.[660] No one was present or absent in arrest or confinement.[661]

Company D morning reports indicate that on June 1, the company moved with the regiment from Coalsmouth to Hurricane Bridge. The company then remained at Hurricane Bridge until June 29, when the 13th Regiment was ordered to report at Charleston. Co. D arrived at Charleston and pitched tents next morning on June 30. Drum major Samuel D. Hanna was reported present with the company June 1 through 24 and June 28 through 30. One fifer was present each day of June. No one was absent on leave. One enlisted man (probably Private Nathaniel Burchfield) was reported absent without leave for the period of June 9 through 12. Burchfield was reported "deserted" on June 13. By June 17, he appears to have been apprehended and by June 22 was reported in confinement at Charleston, where he remained June 23 through 25. Sheldon Gibbs and John Carr were reported absent without leave, and June 26 through 28 the two enlisted men were reported "lost deserted." June 20, Sergeant Julius H. Hatcher was reduced to the ranks for disability. Corporal Arthur W. Darnel was promoted to sergeant in his place and John C. Kimes was promoted to corporal. The company finished the month with an aggregate present and absent of three commissioned and eighty-two enlisted men.

Two privates were detailed for daily duty and two enlisted men had detached service every day of June. One of those on detached duty was Henry C. Williamson, who on July 2 was reported "left on detached duty" as "Ordnance Sergeant." Included in "Remarks" is reference to the scout of June 9 led by Lieutenant Greenbury Slack (Co. B) on Mud River for which eight men of Co. D were detailed.

One to five men of Co. D were reported present sick June 7 through 30. This number included 2nd Lieutenant George Snowden, sick June 7 through 11 and 13 through 15. On June 15, Lieut. Snowden was sent to "Post Hospital" at Point Pleasant. Captain Simon Williams was also again reported sick on June 30 along with one to three non-commissioned officers also reported present sick. Co. D had nine to thirteen men absent sick during the month. One enlisted man was absent in confinement June 22 through 30.[662]

On June 20, Mark E. Robison (Co. E) wrote home about the "squaley times" in their neck of the woods:

Camp Mud bridge
June:20; 1863

Dear father and mother

i seat my self to let you no that i ame well at this time and hope
when thes few lines gets to hand they may find you ingoying the
Same blessing i wold like to cume home but cant at present times
her is so Squauley that we cant leave Camp the repoart is 2000
Coming doon geyand [Guyan] but i be leave that hit tis turnd out
to bad 3 breach whackers [bush whackers] and they is few upe on
mud so i don't thik they is eny danger of tham we ar going to take
a scout upe one mud in few days and over one to Coal so i think i
will be ta home by the last of next weake if we don't get the scout i
cant tell when i will be at home but i think we will get hit lutenent
has got the promes of hit as qick as this litle exsite ment [excite-
ment] gets over times her is prety good her corn her lokks verry
well and wheat a hit is a nise plase her so no moar at present your
true and a fectnent Son un tell death

Mark E Robison to Mr John Robison

i have got some money to bring home if i dont get a chance to
bring hit i will rite to you in the next leter[663]

June morning reports for Company F noted the change of station from Coals
Mouth to Hurricane Bridge. June 1, the company was on the march for Hurricane
Bridge. They marched a distance of twelve miles to their new camp. They remained at
this new station from June 2 through 29. On June 5—a momentous occasion—the
company was paid by U.S. Paymaster Major Cowan for the period of December 31,
1862, to April 30, 1863.[664] June 29, Co. F "marched from Hurricane Bridge, crossed
Coal river near the mouth, [and] bivouaced at Coalsmouth." On the 30th, the company was again marching. They commenced the march at 5 a.m. for Charleston and
"arrived there at 5 P.M. and pitched [their] tents on the south side of the river."

One drummer (James King?) was present on duty with the company for the
entire month. June 1 through 8 and 21 through 30. Fifer James Edwards was present
on duty. One commissioned officer was absent with leave June 27 through 30. No
one was absent without leave. There was loss to the company of two enlisted men.
One was reported "deserted" on June 6, and the other "captured by the enemy" on
June 10. At the close of the month, Co. F numbered three commissioned officers
and seventy-five enlisted men. One to three privates had extra duty each day of the
month, and four privates had daily duty. Two to three enlisted men were detailed

for detached service each day of the month. On June 11, Private John M. Young was "ordered to his company from recruiting and reported for duty." Three to five men were reported present sick June 23 through 30. A low of four to a high of seven enlisted men were reported absent sick each day. These numbers improved towards the end of the month. During the last nine days, just four were absent sick. No one was present or absent in arrest or confinement.[665]

Morning reports for Company H record the transfer of the company from Coals Mouth to Hurricane Bridge, where the company remained stationed June 2 through the 30. The company had a full complement of commissioned and non-commissioned officers. Sergeant Thomas M. Hackett was reduced to the ranks on June 1 for an unspecified cause, and Private William Shannon was promoted Sergeant in Hackett's place on June 2. One commissioned officer was absent with leave June 14 through 19. Two enlisted men were absent without leave June 1 through 8, and one enlisted man was or continued absent without leave June 9 through 20. On June 21, John E. Paul and Harrison Thacker returned from absence without leave. There was a loss to the company of one enlisted man, fifteen-year-old company musician, Albert C. Jameson (Jamison). He was discharged "by order of Civil Authority" on June 20. On June 3, Col. Brown wrote to 1st Brigade Head Quarters requesting that the discharge papers of Lieutenant William I. Mathews and William Perdue "be forwarded that they may be mustered into service in their new promotion."[666]

At the end of the month, Co. H had present and absent three commissioned officers and seventy-five enlisted men. Two privates had extra duty each day of June, and four enlisted men were detailed for detached duty on June 23. Over the course of the month, the company had five to ten men present sick each day and ten to eighteen men absent sick. "John Snyder and James M. Drake returned from Hospital for duty June 23, 1863." No member of Co. H was present or absent in confinement[667]

The Case of Albert C. Jamison

There were of course minimum standard requirements for the enlistment of volunteer soldiers. These had been set by the War Department to guide recruiting officers and if wantonly disregarded, carried some severe penalties for recruiters. These standards included the requirement that at the time of enlistment recruits must have reached a minimum age of eighteen years. Males younger than eighteen could only volunteer with the consent of a parent or guardian. There was also a height requirement, which set a minimum standard of five feet. There were also physical and health requirements although it is anyone's guess what these actually were though it seems likely that they had chiefly to do with having a complete set of arms and legs and the ability to see.

That these rules were flagrantly broken with an easy grace on repeated occasions was an all too frequent occurrence in the recruitment of West Virginia regiments. This generalization cannot be overemphasized most particularly in connection with regiments recruited later into the war (such as the 13th), when most able-bodied men of requisite age had already joined the armies of the North and South. There are several instances of censure from civilians leveled at officers responsible for recruitment of the 13th companies but there is not a scrap of evidence that anyone was ever called to account for ignoring the minimum age standard or any other enlistment standard for that matter.

The recruitment of the later companies of the 13th had much in common with the practice prevalent in the Confederate States: any one who wished to shoulder a musket and take a shot at the enemy was signed up and sworn in. A frightful number of minors enlisted in the 13th served until the arduous campaigns of 1864, when they broke down entirely but there is again, not a shred of evidence among military records to indicate that this was considered anything out of the ordinary and very few parents, to my knowledge, or guardian of any of these minors sought recourse through the legal system to reclaim their sons. Moses Jamison turned to the courts to reclaim Albert, who had enlisted against his father's wishes.

Albert C. Jamison, aged fifteen or sixteen years, had joined for duty and been enrolled in the 13th Regiment on September 20, 1862, at Kygerville, Gallia County, Ohio, by Taylor W. Hampton. Hampton had recruited Co. H and been commissioned Captain but had by the time of Jamison's discharge, resigned his commissioned over some dissatisfaction related to back-pay and was completely out of the picture at the time of the following events. Jamison was duly mustered into service by 2nd Lieutenant Richard R. Crawford, Mustering Officer, on December 11, 1862. Jamison was appointed musician for Co. H.[668]

Notations entered into 13th Regimental Order Books provide some context for the case of Albert C. Jamison. Shortly after the soldier was mustered in, his father, Moses P. Jamison, appealed to the Kanawha County Circuit Court seeking to have his son discharged by civil authority. In this he was successful. On December 31, 1862, Judge James H. Brown presiding at Circuit Court, then in session at Charleston, directed Col. William R. Brown

> that the body of Albert C. Jamison detained in your custody as it is said, together with the day and cause of his being taken and detained by whatsoever name, he may be called, in the same — you have under safe and secure conduct before the Judge of our Circuit Court of Kanawha County immediately after the receipt of the "writ" or before any other Judge of a Circuit Court in the 9th Judicial district of Virginia to do submit to and receive, all and

Singular, those things which shall then and there be considered of him in this behalf —[669]

Lieut. Col. James R. Hall granted permission for Albert to be taken from his company and in obedience to the above writ, Moses P. Jamison, Albert's father,

> appeared before the Judge of the Kanawha Circuit Court [on June 6th] bringing with him the body of the said Albert C Jamison and no return being endorsed on said writ — the Court proceeded to hear the evidence in the case — and it being proved to the satisfaction of the Court that said writ was served upon Col Wm. R. Brown to whom it was directed, who had the said Albert C. Jamison in his custody — and that he detained him as a volunteer Soldier in the said Regt and that he Commit[t]ed him to the custody of the said Moses P. Jamison, Father of the Prisoner to be produced before the Court in obedience to said writ and that the said Brown detained him for no other cause and it was further proved that the said Albert C. Jamison is a minor under the [law] for absence without leave; is hereby approved. By order of James R. Hall, Lt Col Commanding, by John S. Cunningham, Adjutant.[670]

At Circuit Court held at the Kanawha Court House on Saturday, June 6, John Slack, Clerk of the Court, made a transcript from court records of the hearing of evidence in the case of Jamison. Albert C. Jamison, wrote Slack, was aged

> fifteen years and was induced to volunteer in said Regiment without the Knowledge or consent of his said father Moses P. Jamison and against his will in the year 1862 — that the said Albert C is a boy of feeble constitution and delicate health, and unfit to endure the life of a soldier and has been until recently unable to perform duty as a Soldier for a period of six months and desires to be discharged also his Father desires his discharge.
>
> And thereupon it is considered by the Court that there is no lawful cause for the taking, and detention of the said Albert C Jamison as afore[said]
>
> It is therefore ordered and adjudged that he be discharged and Liberated from the Custody aforesaid ——
>
> Virginia Kanawha Circuit Court Clerks Office

I John Slack clerk of said Circuit Court do hereby certify that the foregoin[g] is a true transcript from the records of said Court. In witness whereof I hereunto set my hand and offer the Seal of said Court this 10th day of June 1863. Teste Jno Slack Cl[er]k.[671]

On June 7, A.C. Jemison or Jamison was delivered over to civil authorities on a writ of Habeas Corpus pursuant to General Order No. 12, issued by Col. Brown on June 6, 1863, from 13th Regiment Headquarters at Hurricane Bridge. The order read as follows: "Capt William I Mathews Comp (H) will deliver into the hands of civil authorities Albert Jemison (Musician) in your Comp. A Rit of Habeas Corpus having been served on me for that purpose."[672] In obedience to the writ, Jamison was detained and committed to the custody of his father, Moses P. Jamison, "to be produced before the Court in obedience to said writ and that the said Brown [Colonel Brown] detained him for no other cause and it was further proved that the said Albert C. Jamison is a minor under the [law] for absence without leave; is hereby approved."[673] Albert was retained by civil authority from June 7 until June 20, 1863, when he was discharged by civil authority effectively ending his military career.

Levett Perdew, another young Private of Co. H, wrote to his parents from Hurricane Bridge, suggesting on the other hand that army life was not half bad.

> Hurricane Bridge
> Putnam County
> June 10th 1863

> Dear Father and Mother I take my pen in hand to inform you that I am well at this time and truly I hope thes few lines will com to hand and find you enjoying the same Blessing

> Father I have nothing of interest to writ you at this time only we are a laing here a doing nothing only ateing Hard Bread and sault meat and agiting as fatt as Hogs of a most year and the lasist fellows you evy seed and I beleave some of the Boys wold fight at the drop of a hat we have jest got in from off a scout and as luck wold have it we did not see eny rebs and we was glad of it for the weather is to worm to kill meat as fatt as we are at this time a part of Company B of the 5 Va was with us but they don't look so neat and sleck as we do I will say no more abot this at the presant I sent you 20 dollars by Gaurdon Bi[illeg. letters] which I guess you got it I have nothing more to write at this time only I want you to writ to me oftoner then you do I have wrot several letter to you and

recives no ancer from them so I will close by saying I remain youre affectionat son untel Death

Levett Perdew[674]

Morning reports submitted for Company I record that the company was stationed at Coals Mouth for the month of June. Co. I had present and absent 1st Lieutenant William E. Feazel, cmdg. company; four sergeants; four corporals and forty-two enlisted men. On June 1, Elijah E. Riley was promoted to "3rd Duty Sergeant." One enlisted man was absent with leave, June 1 through 10. One enlisted man, Alexander Craig, was reported absent without leave June 22 through 29, having deserted from camp at Coalsmouth. He was dropped from the rolls for desertion on June 30. Co. I gained recruits during the course of the month. One man enlisted in the regiment on June 8; two more joined on June 11; two were added June 12; two more joined June 22; and another man joined on June 25. At the close of the month, Co. I had present and absent one commissioned officer and fifty-two men. A letter of authority, given by J.T.B. (J.T. Bowyer?) on June 27 notified of or granted permission to Henry Stump, a Private in the 9th Va. Infantry, who desired permission to recruit for Company I, which was not yet full.[675] The company reported none absent sick but one to seven men were reported present sick during the course of the month. This number included two non-commissioned officers reported sick June 12 through 21. No one was present or absent in confinement.[676] Though under-sized, Company I was mustered into the United States service on June 30 (or July 26, 1863, at Charleston, West Virginia), by Mustering Officer for 3rd Division 8th Army Corps, Lieutenant Cyrus S. Roberts.[677]

Probably some time during summer but no later than September 1863 recruits for the 13th (probably belonging to Co. I) were lodged and boarded by one Henry Snyder. It was a common practice for troops to be quartered in private dwellings before leaving for camp. William E. Feazel, who made the contract and certified the account for Snyder, forwarded the account to Captain W.C. Thorpe Disbursing Officer, at Wheeling. For "a little more than one year," however, Thorp had failed to pay the account. On September 9, 1864, Feazel, who felt honor bound to see Snyder paid, wrote to F.H. Pierpoint to enquire as to what had become of the account and whether Captain Thorpe had yet drawn the money or not.[678]

Correspondent for the 13th begins his series of letters to the county newspaper

On Tuesday, June 23, William H. Harper, Sergeant Major, Company F, wrote to the editor of the Point Pleasant *Weekly Register* from Hurricane Bridge. Although he already had been sending intermittent letters to the *Register* giving news of the regiment, with this letter he officially commenced a series of letters on behalf of the men to friends and families at home. He would continue to serve as regimental war correspondent until his resignation from the service March 2, 1865. Though often given to sermonizing on the evils of secession, his descriptions of life on and off duty, in camp and on the march, are colorful and often amusing. Harper commenced with the following missive:

> *Mr. Editor:*— It may afford the readers of the *Register* who may have friends and acquaintances in this Regiment some satisfaction to hear from us once in a while, it is, however, a very difficult matter to keep up a very interesting correspondence for any length of time when the Regiment is not moving about; you can find but little from which to write that is of an incidental character, and especially is this true of us in our present locality. Here we are as the boys sometimes say, outside of God's world and on the very borders of wide spread ruin and desolation; it is true we are encamped on the south-eastern extremities of a most beautiful Valley, drained by the tributaries of Hurricane Creek, which carries its water Northward into the Kanawha and those of Mud river running in a more Westerly direction, and, yet, wherever you turn your eyes, you behold starring you in the face the inevitable results of the rebellion. Large meadows and grain fields are thrown out and wiped to the earth by roving herds of half starved cattle and sheep, and as though heaven itself designed to inflict a still more fearful and complete destruction upon the people, has withheld and continues to withhold its genial showers in consequence of which vegetation of all kinds is perishing and the ground is beginning to present the appearance of a barren waste. There has been but very little rain here since the 1st of May, and we would not be astonished beyond measure were the fruit itself turned into the apples of Sodom.
>
> Is all this because of the crime of secession? Let some wiseacre answer the question if he can.
>
> The Regiment is divided at present, into two detachments, one is at Mud Bridge, in Cabell county, commanded by Lieut. Col. Hall, the other at this place, commanded by Col. Brown. We have less sickness in the Regiment at present, than ever before, there being but few of our men unable for duty. We are doing a considerable amount of scouting at present, and are snaking into camp and forwarding to headquarters a goodly number of the Butternut fraternity. A scout went out on Mud River on the 21st inst., in pur-

suit of some horse thieves that were said to be prowling about in different sections of the country gathering up horses. Some of this same gang no doubt passed within a short distance of our camp on the night of the 16th inst., and stole three horses from a Union man, who lives within some six or eight miles of this place, and went with them unmolested. It was this gang and others like them that our scout was sent out after. They succeeded in finding their camp, but the rebs had made good their escape. Another party was sent into the neighborhood bordering on the river to search for rebel mails, it having been reported that there were several prominent secessionists engaged in that work. Their houses were searched but nothing could be found.

The boys have a great many different kinds of amusements,[679] some of them very funny; among them, we might name the following: They catch some poor forlorn dog, and then tie a coffee pot or tin pan to the posterior elongation of his spinal column, and then the poor irrational animal is turned loose amid the roars and shouts and yells of the gathering crowds of mischievous Yankees. It becomes at once hideous and one would suppose from the pitiful shrieks of the poor dog and the velocity with which he moves over the ground, that he thought Jeff Davis and all the Southern Confederacy were just at his heels. Here we end for the present. We hope to do better next time.[680]

On June 23, Col. Brown wrote to 1st Brigade Headquarters inquiring "whether a Mustering officer has been appointed for the Division."[681] Col. Hayes wrote to Brown in response, on Thursday, June 25, that "[t]here is a mustering officer appointed and permanently attached to this Division. He is present at Fayetteville and will on his return most probably visit you."[682] Correspondence from Gen. E.P. Scammon to Hayes suggests that in a matter of days, Scammon hoped to launch an offensive. On June 23, Scammon wrote to Hayes the following:

Confidential
Head Quarters 3rd Division 8th Army Corps
Charleston June 23, 1863.
Colonel R.B. Hayes
Comg. Brigade.

I hope to receive such information as will justify my moving to the front within a week at farthest. Be ready to move at the end of three days from this date — And, for once I hope to be able to move Secretly.

You will see that all preparations are made without imparting this information or any hint thereof to any person whomsoever. Exercise your own ingenuity as to the mode — but be sure that all is ready and that no rumor or <u>Surmise</u> is caused by your preparations. If necessary, forbid all visits to town and keep citizens out of Camp. Let purchases for these few days be made by trusty officers O U[or N].C.S. — otherwise the rumor is here to get wires. Very truly E.P. Scammon B.G.[683]

Orders given by Quartermaster William McKinley at Brigade Headquarters to regimental quartermaster Stephen Comstock suggest that at this point in time, the 13th West Virginia had to supply itself from the country. On June 23, McKinley wrote to Comstock: "You will retain all the teams now in your possession until otherwise especially ordered. The fact of having to supply yourselves warrants you in keeping all your transportation."[684] June 25, Col. Brown forwarded "a box of books and blanks" to 1st Brigade Headquarters.[685]

Another nervous shifting of forces took place at the end of June and beginning of July. Sunday, June 28, Col. Hayes wrote to Col. Brown, at Hurricane Bridge and to "the Commanding Officer at Mud Bridge" (i.e., Captain Milton Stewart) with new orders. To Capt. Stewart Hayes wrote: "Direct the 3 companies of 13th Va at your post [Companies B, E, and G] to report without delay to Col. Brown at Hurricane Bridge, Va."[686] Brown was informed that:

The three (3) companys of your regiment at Mud Bridge have been ordered to report to you at Hurricane. As soon as they arrive you will march with your regiment for this place — At the Mouth of Coal River leave two companys with a Good Officer to command.

Send your Artillery to Mud Bridge.[687]

Hayes also wrote to Lieut. Col. Tomlinson, cmdg. 5th West Virginia Vol. Inf., on the 28th ordering him to "send one company of your regiment to enforce the two companies at Mud Bridge, Va."[688] June 29, the 13th camps at Hurricane Bridge and Mud Bridges were broken up and the regiment was ordered from these places back to Camp White. Companies A and H and a squad of recruits remained at Coalsmouth.[689] June 30, the 13th marched from Hurricane Bridge to the mouth of Coal River. From Coal they embarked on transports and went up the Kanawha River to Charleston. They remained stationed near Charleston until July 9.[690] The regiment drew clothing at

Charleston on June 30.[691] Hazard Farley, principal musician and drummer for the 13th deserted on the 30th. He remained absent until August 31, 1863.[692]

That forces were being concentrated in preparation for a forward movement was not lost on Andrew Stiarwalt, musician for the 23rd Ohio Volunteer Infantry, who remarked that

> on July 1st the 13th VA Regt arive + pass up the River and in camped about two miles above Charleston on the 5th July the 9th VAI moved up the River on the 8 July the 13th VAI moved further up the river it is rumered that there is to be forwadment on too Newburn Station Nine Comps of the 23d OVI Regt shiped on Steem Boats on the 9th July and moored up the Kanawha River on the 10th July they arrived at the mouth of Loop Creek Seven Comps started up in Rout for Fayetteville and Raleigh and incamped about 4 miles from the other two comps followed[693]

July 1863

Grim realities

July 1863 was marked by a series of Federal victories. July 1 through 3, the great battle at Gettysburg was fought and won. July 3, surrender of Vicksburg was being negotiated at Vicksburg, and the campaign in Tennessee was being successfully concluded by Gen. W.S. Rosecrans, who had forced Gen. Braxton Bragg out of most of the State. The news from Gettysburg, Vicksburg and Tennessee would certainly have heartened Union citizens and soldiers in the Kanawha Valley and been celebrated on the Fourth, the national holiday, but the news of these victories had not yet reached the Kanawha Valley. West Virginia's status as buffer between the Northern and Southern States in rebellion was taking its toll on civilians and soldiers in garrison in the Kanawha. There was little open celebration of the national holiday. Remarks made by newly elected West Virginia Governor Arthur I. Boreman in his inaugural speech shed light on the state of things generally at this point in the summer:

> It seems to me that the position of our people in the beginning of the troubles, and their condition since, have not been understood by our friends around us. In the commencement of these difficulties we were part of a Southern State, whose convention passed an ordinance of secession and this fact caused many to sympathize with the South without reflecting whether it was right or wrong. We were situated between the South and the North, and in case of a collision it must necessarily result that ours would be contested territory; that if we adhered to the Union the South would deal with us much more severely than if we were a part of a Northern State, or of one that had [made] no attempt to secede; and that we would be, what we have since been so truthfully called by many, the great 'breakwater' between the North and those in rebellion in the

186

South. All these matters were weighed and considered by us, but we determined, with a full belief of what would occur, and what has since occurred, that the Government was too good to be lost, and the rights and immunities which we knew we were enjoying were too precious to be surrendered on the uncertainty of the results of experiments in the future. We thus took our position with our eyes open; knowing what civil war had been, and what it could only be again if once commenced: and we have not been deceived. Our State has been invaded by traitors [...] they have applied the torch to public and private property; they have murdered our friends, they have robbed and plundered our people; our country is laid waste, and to day gaunt hunger stares many families of helpless women and children in the face. This picture is not overdrawn. It is a simple statement of the facts. Yet notwithstanding all this, the Union men of West Virginia have not looked to the right or the left, but through all these difficulties and dangers they have stood by the Government.[694]

Despite the grim realities alluded to by Gov. Boreman some positive results had been achieved by the government. This is apparent in a letter written by Judge H.J. Samuels to General George Crook in which he stated that on his arrival at Guyandotte, Cabell County, "in July, 1863, I found society so far protected and the people feeling safe that I held courts in this county and Wayne during that year without interruption."[695]

It will be remembered that on June 29 (with the exception of Cos. A, H and some recruits left behind at Coals Mouth) the 13th West Virginia had broken up their camps at Hurricane Bridge and Mud Bridge and marched via Coals Mouth to Camp White. Immediately upon the transfer of the regiment to Charleston, a problem surfaced at 1st Brigade Headquarters at Camp White in connection with Dr. Samuel Shaw, the regimental surgeon of the 13th, Gen. Scammon inquiring of "by what authority surgeon Shaw 13th Va V.I. grants furloughs to the men."[696] The same day (Wednesday, July 1), Col. Hayes wrote to Col. Brown saying: "The General Commanding directs you to make inquiry by what authority Surgeon Shaw of your command grants furloughs."[697] It is within the realm of possibility that Dr. Shaw, a doctor of many years experience, sent sick soldiers home, now that the regiment was back behind Union lines, where they might recover under better conditions.

On July 2, Company B was ordered to Point Pleasant. They would remain stationed there until the whole regiment was united to go in pursuit of John Hunt Morgan and his cavalry division. July 3, Hayes ordered a "Detail from 13th W. Va. for fatigue duty in Charleston to report to Lieut. Maurice Watkins Ordnance Off.

at 8 o'clock A.M. July 4, 1864."[698] July 6, John S. Cunningham declared that he had been performing the duties of 1ˢᵗ Lieutenant and Adjutant since April 20, 1863. He was mustered into the service at this grade.[699] On the 6ᵗʰ, Co. F was ordered to Loup Creek.

Scammon shifts his troops to Cotton Mountain, Fayette County

On July 7, Rebel cavalry ("companies or bands of horse thieves and bushwhack-ers," according to J.V. Young in his letter to daughter Emma, written on July 10, from his station at Cotton Hill) attacked Federal trains "along the Cotton Hill road" on Cotton Mountain. The raiders, continued Young, "took all the teamsters prisoners, and the best of the horses, and made their escape. [...] [T]hey are very troublesome to our transportation."[700] In consequence of this nuisance, on July 8 General Scammon sent orders to 1ˢᵗ Brigade Head Quarters indicating that the "13ᵗʰ Va.V.I. be placed in readiness to move to the South side of Cot[t]on mountain. A boat will carry them to South Creek."[701] July 9, the "remaining part of the [13ᵗʰ] regiment" received orders to march to the south side of Cotton Mountain in Fayette County. The "remaining part" consisted of Cos. C, D, E and G. These companies arrived at Cotton Mountain on July 10 and remained there until the 12ᵗʰ.

Transfer of the 13ᵗʰ West Virginia to Cotton Mountain was part of a larger shift of troops. Col. R.B. Hayes' journal entry for July and a history of the 23ʳᵈ Ohio Regiment in James Comly Papers indicate that the Kanawha Division left Charleston by steamboat and moved up the Kanawha River in the latter part of the day on July 9. The Kanawha troops spent all day of the 10ᵗʰ at Loup Creek, and on July 11 moved to the foot of Cotton Mountain on the Fayetteville side.[702] On July 10, Company F (13ᵗʰ Regiment), still at Loup Creek, received orders from Col. R.B. Hayes to cross the river and join the rest of the regiment on the south side of Cotton Mountain, Fay-ette County. Co. F arrived there at 6 o'clock p.m. on the 10ᵗʰ and joined that part of the regiment already at Cotton Mountain camping at the old camp of the 89ᵗʰ Ohio.

Before departing Camp White for Cotton Mountain, Sergeant Major William Harper found time to write to the home front to convey news of the 13ᵗʰ Regiment. Since June 23, the date of his previous correspondence the 13ᵗʰ had been further divided, related Harper. From two detachments, the regiment was now divided in four detachments. One detachment was at Coals Mouth, one at Point Pleasant, one at Loop (Loup) Creek and another at Charleston. These detachments were proba-bly serviced by three sutlers — Will Sherwood, Columbus Shrewsbury and perhaps also one "Wilber." The camps at Coals Mouth, Point Pleasant, Charleston and Loup Creek were all located on the highly navigable Kanawha River, tributary of that supe-rior conduit, the Ohio River. Sutlers servicing these camps would likely have been able to receive their goods regularly from their wholesale suppliers. This stood in contradistinction to camps whose sutlers had to transport goods overland. Overland conveyance was more problematical because of the poor state of transportation, lack of wagons, horses, mules and the likelihood that wagons conveying sutlers' goods could be set upon by guerrillas.

Some information is known about the sutlers servicing the 13th Regiment in 1863. Columbus Shrewsbury, for one, had formerly served as Commissary Sergeant and then 2nd Lieutenant of Company A 4th West Virginia Infantry. He resigned his lieutenant's commission May 26, 1863,[703] to serve as sutler to the 13th together with Will W. Sherwood. Sherwood had, at some point in time, obtained official authority to serve as sutler to the 13th. See illustrations L5Ba and L5Bb, both 5 cent brass tokens and L25B, a 25 cent brass token (in David E. Schenkman; *Civil War Sutler Tokens and Cardboard Scrip,* Post Office Box 155 Bryans Road, Maryland 20616: Jade House Publications, p. 74). The brass tokens were cut and stamped by John Stanton (Cincinnati, Ohio) with Sherwood's name and the name of the unit (the 13th Regiment), all script for use in purchasing from Sherwood's tent of supplies.

Columbus Shrewsbury was an avid correspondent and his letters (preserved in the Roy Bird Cook Civil War Collection at West Virginia University, Morgantown) provide information about the regiment and the business of sutlering. Columbus was born in Kanawha County, West Virginia, on June 5, 1832. He was the son of John C. Shrewsbury and Elizabeth (Farley) Shrewsbury. In 1856, he became a resident of Mason County, West Virginia. He was married to Cynthia A. Jarrett (born in Kanawha County, September 19, 1838) in Meigs County, Ohio, on December 28, 1859. Fannie E., the first of the couple's children was born in Mason City, Mason County, on October 13, 1860. It is to Fannie that Columbus' more interesting war time letters were addressed. Columbus enlisted in the Union army in 1861 as a private and served with the 4th Virginia Volunteer Infantry. In 1863, he was promoted to second lieutenant and discharged for disability in the same year.[704]

Shrewsbury wrote to his wife and little daughter Fanny (letter dated "Coals Mouth Sunday July 5th 1863") as follows. Shrewsbury penned his letter on a humorous piece of printed patriotic stationary. His last line, "Eyes to the front!" refers to a line drawing—a cartoon—printed on the stationary, depicting a group of soldiers supposedly standing at attention but in reality gawking at a buxom and comely lass walking in front of their line. Shrewsbury wrote:

> Dear Wife
>
>   My health is good the hot weather goes very hard with me. I have a good house to Board at plenty of Onions and Lettuce and other vegetables; The fleas are plenty heare they anoy me so that I can scarcely sleep I have sold a good many goods since the first of

the month. Will Sherwood is up at Charleston on the opposite side of the River with the bala[n]ce of the Regiment Wilber has gone home to [s]pend the 4[th] of July I have nothing new to write all is quiet so far as I know If you will write me what kind of Bark you want I will try and get them if that Medicine that you got of the Doctor does you any good get some more and try and get well. Fany you must go to Sunday Scool and see the little Baby.

Eyes to the front

C Shrewsbury[705]

W.W. Harper also wrote this month to the county paper describing in considerable detail recent events in the life of the regiment. These included the accidental wounding of Corporal Logg (James Legg, also called Junius Legg or Logg) of Co. E; the Fourth of July; and a "sumptuous dinner" given by Lieutenant Lemuel Harpold of Co. C. Harper wrote as follows:

> *Mr. Editor:*— Since my last, this Regiment has moved from Hurricane Bridge to this place [Charleston]. We stated then that the Regiment was divided into two detachments, it is now divided into four detachments one at Coalsmouth under command of Capt. Johnson, one at Point Pleasant, under command of Capt. M. Stewart, Co. F under command of Capt. A.F. McCown, left this afternoon for Loop Creek, where it will be stationed for sometime perhaps, to guard the Government stores; the balance of the regiment is still at this place, but how long it will remain here is not known. We are ready, however, for any movement that may result in good to our glorious cause, and that will prove disastrous to the enemy.
>
> The 9th Virginia left here on Sabbath last for Fayetteville, where it is probable it will remain for some time, that depends, however, upon the motions of the rebels.
>
> Junius Logg, a Corporal in Company E, met with a very serious accident on the 4th. He had taken his gun and went a short distance into the country to hunt some game, and he was passing through the woods with his gun slung over his shoulder, the butt behind and his left hand a hold of and over the muzzle, the cock caught the brush and caused the discharge of the piece, the contents entering close to the heel of his hand and passed out at the knuckle joint of the middle finger, making it necessary to amputate the finger. This is another warning to those men who are so careless

in handling their firearms. We have had more men injured in this Regiment by the careless use of firearms, than by the enemy.

The Fo[u]rth of July was a rather dull day with us in these parts. Captain Simmon's Battery fired some 70 shots, but as for anything else all was quiet and seemed to wear the insignia of mourning rather than rejoicing and perhaps it were all the more proper, for it can trully be said that this nation is today the house of mourning and the judgements of the Almighty are abroad in the land, and on that day subsequent events show that two armies were engaged in bloody conflicts, and upon that day many a noble spirit took its flight from that bloody field of death to that better and more peaceful region we trust, where the clatter and din of war is forever hushed.

The body of a soldier belonging to the 91st Ohio, was found floating down the river on Monday. We made him a box, rough it is true, put him into it and buried him on the bank of the beautiful Kanawha. We felt sorrowful when we reflected that another patriotic soldier is gone. His name is James Baice. May he rest in peace.

A very sumptuous dinner was given to the officers of the Regiment today at Company C's headquarters, superintended by Lieut. L. Harpold. I could not help reflecting upon some of the good dinners eaten upon other occasions and under quite different circumstances.

Lieut. Harpold is a free hearted, kind officer, and I presume this act of kindness on his part, will not soon be forgotten by those who participated in the dinner. W.H.H.[706]

A few blithe spirits celebrate the Fourth in Charleston

Yes, this first celebration of July 4th, as West Virginia became the 35th State of the Union, was a surprisingly quiet one in the Kanawha Valley. Harper and other writers had similar observations on the day. Diarist Henrietta Fitzhugh Barr, a southern sympathizer living in Ravenswood, remarked in her diary: "[p]eople had too much on their minds to celebrate" and the 4th "passed off quietly."[707] Army correspondent in camp at Charleston ("D.S.,") observed that the heat had also been well nigh intolerable exhausting all and that the city was overall lacking in "patriotic demonstration" although there were "many Union people in Charleston."[708]

Charleston is a "haughty town," wrote D.S., over which, as he listened to the cannons boom a salute he saw just one American flag floating and that was "on a Govt. boat lying at the wharf. Even headquarters can produce only <u>two stars</u> on such a day as this — and a staff (can't say its <u>all standing</u> though.)"[709] The monotony of camp life at Charleston, was felt by all on the Fourth. This sameness was relieved during the night when Federal pickets were fired into by bushwhackers and on the 5th, when "some movements" were afoot, "which whether they result in anything or not are at least salutatory in the way of breaking the monotony of camp life."[710] The rebels around Charleston were in the main trying to keep U.S. troops just sufficiently occupied that no troops would be withdrawn to be sent elsewhere.[711]

Despite the general somber feeling, some of the more blithe spirits in the Kanawha Division celebrated the "glorious Fourth." These hoped that a favorable outcome to the east and west would soon relieve them from the tedium of garrison duty, and they would soon move to push against the enemy in his own territory. The 9th Virginia had been transferred from Charleston to Fayette and the hope prevailed that more troops would be forwarded for this purpose. R.B. Hayes wrote home in high spirits: "We had a good Fourth [...] A good deal of drinking but no harm. We let all out of the guard house."[712] Some 13th soldiers, among them Nathaniel Burchfield and Sheldon Gibbs, were among those released from the guard house in the spirit of the holiday.

To the southside of Cotton Mountain; to Fayetteville and Raleigh Court House to fight McCausland; and back up the Ohio in pursuit of Morgan's Raiders

July 8-9, the 13th marched a distance of forty-five miles to Cotton Mountain in Fayette County.[713] Upon their arrival here, they camped at the old camp of the 89th Ohio. On July 10, Co. F rejoined the regiment from Loup Creek and the regiment, such as it was—Cos. C, D, E, F and G—was ordered to the south side of the mountain. They arrived there on July 10 and remained there until July 12.

On Saturday, July 11, orders were received at 1st Brigade Headquarters that Scammon desired the troops of the 1st Brigade to move to Fayetteville on July 12.[714] Col. Hayes, commanding 1st Brigade, did not "fully approve" of the expedition and had a bad feeling about it. He wrote in his diary for July 12:

> We are starting on an expedition to Raleigh County and perhaps further. I do not fully approve of the enterprise. We are too weak to accomplish much; run some risks; and I see no sufficient object to be accomplished. [...] Dear boys, darling Lucy, and all, good-bye! We are all in the hands of Providence and need only be solicitous to do our duty here and leave the future to the Great Disposer.[715]

In compliance with Scammon's orders, the 13th West Virginia Regiment was ordered from the south side of Cotton Mountain near Fayetteville on July 12. The same day, they made the march to Fayetteville, a distance of six miles.[716] Then, Monday, July 13, the regiment was ordered to march for Raleigh Court House.[717] The small army comprised of three brigades of Scammon's Kanawha Division, cavalry and artillery marched for Raleigh this day. The countryside through which the troops passed was "frightfully wild and rugged country."[718]

The 13th, marching with Hayes' 1st Brigade, reported in its regimental returns that on the march to Raleigh, they "[s]kirmished some with the enemy on the way."[719] James Comly noted in his journal that 2nd Brigade was in advance on the march and had some "little skirmishing" and "1st Brigade had no fighting at all. Nobody hurt here." Scammon's cavalry (3rd Brigade) "was sent to press the enemy's rear."[720] The 13th arrived at Raleigh Court House about 12 noon on the July 14. Co. F was selected among others for the probe forward into the town, and they "drove in the enemy's pickets" through the town and camped in Raleigh C.H. at 12 o'clock m[idday]."[721]

The Confederates, for their part, had been fortifying beyond Raleigh "about 5 miles distant [from the town], on the south side of Piney creek."[722] They had, as R.B. Hayes remarked in a letter home, been "annoying" Federal military "a good deal from there."[723] Indeed, they were found to be waiting for the Kanawha troops behind strong fortifications there. Scammon, with the concurrence of his officers, deemed the works too costly to storm by frontal assault in terms of loss of life. A plan to turn their position was proposed by Lieut. Col. James Comly, cmdg. 23rd Ohio V.I. The Kanawha troops would "draw in after" the enemy and by so doing compel Confederate troops at Lewisburg to make an advance on Gauley and Charleston in rear of the Federal troops at Raleigh. 1st Brigade would "be immediately sent back to meet such an advance."

Comly's plan being approved, 1st Brigade was ordered to move accordingly at 2 p.m. The 2nd Brigade was in the meantime charged with the task of destroying the Confederate works at Raleigh. First Brigade marched to the camp occupied by them "on Monday night."[724] It became apparent that as a practical matter attempting to turn the position would require too much time and be altogether too costly an undertaking when weighed against the advantage gained and Scammon resolved to retire without attempting to take the works. He decided to withdraw his troops the next morning and then, encamped for the night in position at Raleigh. It was the cavalry feint pressing on the Confederate rear which ultimately and unexpectedly had the desired effect: the Confederates pulled themselves back from their position during the night (night of July 14). At about 2 a.m.,[725] they vacated their fortifications at Piney leaving some stores behind. Scammon had a portion of his troops destroy the vacated rebel works on Piney River, ammunition and etc., left behind by them there. The rebels had left before the Kanawha troops could themselves fall back. The enemy having retreated, the 13th remained in Raleigh until the 15th, and then return to Fayetteville. They arrived there at "12 o'clock July 16."[726] From Fayetteville to Raleigh, the 13th Regiment had marched a distance of twenty-nine miles.[727]

A soldier correspondent calling himself "Twenty-third" (from the 23rd O.V.I.?) wrote in humorous vein regarding the expedition to Raleigh. "The Rebels," wrote Twenty-Third,

> had a force strongly fortified, near Raleigh Court-house, which was superior in numbers to all the forces which General Scammon could 'rake and scrape' for the expedition. In this country, the roads are not like the Miami Valley, [Ohio] and an animal which cannot climb a tree, is confined in his movements to roads which go through the mountain passes. Unfortunately one of these passes, from which roads radiate in every direction, was occupied by the force at Raleigh, and it was necessary to pass through, or very near there, no matter what route might be afterward. So an infantry force moved up so as to make a feint against the enemy's position, and hold him to his works, while the cavalry passed to the rear. The cavalry expedition was to make a feint of going in the direction of Princeton, and afterward debouch in the direction of Wyoming C.H., and strike rapidly for the railroad. This plan succeeded admirably in all except one thing: the force at Raleigh took our feint for a real attack, and ran away in night. They were 'felt of' as tenderly as possible by Colonel White, with a small portion of his brigade— the 1st Brigade being kept at least three miles off—but they left. So we came back and went after Morgan, and the cavalry proceeded on their way [to fight at Wytheville].[728]

Adj. John S. Cunningham wrote to his wife relating details of the expedition of July 8-14:

> Head Quarters 13th Va Regt Vol Infty
> Camp Laurel Creek South side Cotton Mountain July 14 [1863]
>
> My Dear little wife
>
> Last Wednesday (June 8th) I bid you good bye at Mr Rands in the evenings in one hour afterwards we were on our way to Gauley River or rather to the Falls of Kanawha. At Camp Piatt some part of the machinery broke + we had to lay by until morning — Early the following morning the transport started for Loup Creek Shoals at which place we arrived at 9 A. M. — The Regiment debarked on the west side of the Kanawha River and immediately took up the line of march to the Falls — at which place we arrived at noon — Dined on a cracker + marched to this camp by sunset — There-

fore you will see by this date this Regt has been in quarters one day — There is no doubt these bushwhackers have been plentiful as 7 waggons were captured within 400 yds of the camp the day before this Regt arrived at this post —

All the movements indicate an onward move on the part of the troops in this Department therefore you may be prepared to hear from me from some other point — I do not know this to be so but believe it to be so.

When we left Charleston We had only 4 cos with us since then one company has joined us one co is at Pt Pleasant, 2 cos + Detach of Recruits are at Coalsmouth, thereby reducing our effective strength very much — Whilst I am writing this the 1st Va Cavalry, Simmonds Battery + the 23rd Ohio are passing us — the 23rd will encamp near us for the present Genl Scammon passed us within the past half hour — There is a constant stream of waggons passing us all the while [...].[729]

Inasmuch as the Federal expedition to Raleigh had been concluded, Gen. Scammon sent back his 1st and 2nd Brigades to Fayetteville and sent cavalry forward to the railroad at Wytheville.[730] Scammon returned with the "Twelfth and Twenty-third Ohio Volunteer Infantry, Thirteenth and Ninth Virginia Volunteer Infantry, and McMullins battery of 6-pounders"[731] to Fayetteville. The 13th West Virginia Regiment remained at Raleigh on July 15. At one o'clock p.m., that day, they received orders to march back to Fayetteville. The troops began the march to Fayetteville on the morning of the 16th. The 13th arrived at Fayetteville "at 12 o'clock [noon] July 16"[732] ate their the midday meal and then continued the march in the heat of a hot summer's day until at about 6 o'clock p.m. Thereafter, the 13th with the rest of the troops "halted for the night, after a march of 17 miles, at a point about 8 miles from Fayetteville."[733] The 13th, 23rd Ohio (and presumably other Kanawha Division troops on the retrograde) "bivouac[k]ed at Cotton Hill."[734] At 6 o'clock p.m. during the evening meal at Fayetteville, the entire 13th "regiment was ordered to Gallipolis, Ohio, to go in pursuit of the Rebel General John Morgan."[735]

Brigadier General John H. Morgan
is reported dangerously close to Gallipolis

Col. Hayes and his command of "about 1200 infantry"[736] with the remainder of Scammon's troops sent on the expedition to Raleigh Court House had been completely cut off from the outside world for over a week. In that time, no news had been obtained or received regarding Vicksburg and Gettysburg. To hear the latest dispatches, Col. Hayes obtained leave from Gen. Scammon to ride to Fayetteville, while the men encamped to learn what news there was at the nearest telegraph station. At the station, Hayes learned from the telegraph operator about the Union victories at Vicksburg and Gettysburg and that Confederate cavalryman General John Hunt Morgan had been raiding on grand scale virtually unopposed through southern Indiana and Ohio.[737] He was at this time reported at Hillsboro, Ohio,[738] below Chillicothe, traveling in the direction of Gallipolis and getting dangerously close to it. Morgan was expected to arrive at the latter place on July 18.

Hayes was startled by this news. Not only was the infamous raider entering country for which the Kanawha troops (chiefly Hayes' command, 1st Brigade, Kanawha Division) were responsible but it appeared as if Morgan would escape through Hayes' district with ample opportunity to ransack Gallispolis, a prosperous town and military hub of supplies and men for the tri-State area of Ohio, Kentucky and West Virginia. Immense quantities of government stores estimated in the neighborhood of one million dollars in value[739] were kept at Gallipolis. These, Morgan could destroy or plunder before making good his escape across the Ohio River into Kentucky or West Virginia. Hayes also learned from the telegraph operator that Morgan was "hard beset by Union cavalry." General Edward H. Hobson was a day's march in the rear. It was obvious that Morgan was almost certainly seeking to escape from the North by crossing the Ohio River at Gallipolis, "where there was no adequate force to dispute his passage or to protect large quantities of supplies which had been collected there."[740]

Ever quick on the uptake, Hayes apprehended that if Morgan was indeed making for Gallipolis, he was 'within range.' Hayes formed a plan. Morgan was close enough to be struck by Scammon's troops but only if the men were started immediately and by forced march were pushed forward to Loup Creek, there to meet steamboats if any could be procured. If boats could be sent immediately, the men could be embarked by early the next morning and reach Gallipolis in time to head off Morgan.

Hayes quickly telegraphed to his adjutant at Charleston, asking if there were any steamboats there. The answer was yes. Hayes immediately ordered them to be sent to Loup Creek, "the highest navigable point on the Kanawha."[741] The steamboats at Charleston were immediately started up the river. They were due to arrive at Loup Creek by early next morning. Having secured these preparations, Hayes galloped the fifteen miles back to camp.

Hayes arrived back at camp by nightfall. He laid his plan before General Scammon. Scammon, however, was reluctant and there was "sharp controversy."[742] Scammon did not believe that what Hayes proposed could be done given the condition of the troops and the time constraints. The troops were already tired out from a seventeen mile march in the heat of a hot July day. Col. Hayes (a practiced attorney before the war) brought "all his powers of persuasion to bear"[743] and succeeded in obtaining Scammon's permission to push forward with two regiments from his command, 1ˢᵗ Brigade. He took with him the 23ʳᵈ Ohio and 13ᵗʰ Virginia "and a section of artillery"[744] (McMullin's Battery) to go in pursuit of Morgan. Scammon left the 9ᵗʰ Virginia Infantry at Fayetteville and the 5ᵗʰ Virginia remained where it was at Charleston. Not to be outdone by his subordinate officer, Scammon himself with members of his staff accompanied the troops on their expedition after Morgan and directed their movements. Scammon took with him to Gallispolis in addition to the 23ʳᵈ Ohio and 13ᵗʰ Virginia also the 12ᵗʰ and 91ˢᵗ Ohio Infantry Regiments.[745]

Hayes' men were cooking their evening meal when he gave them their orders. He lay the situation before them and explained the need for an immediate start. Though worn out by the day's march, they received the orders with "wild hurrahs"[746] and soon commenced their night march to Loup Creek. The men of the 13ᵗʰ Virginia coming from an area where folk were "not unused to raids,"[747] they likely took a keen interest in Morgan's approach and in what part they might play in the break up of this notorious command. J.Q. Howard, Hayes' autobiographer in post-war years, wrote of this juncture.

> In 30 mins after the orders were read to the soldiers, the column was on its march. The road was mountainous, the darkness dense, the route almost impassable, but the Kanawha river was reached at the break of day. The steamers were both in sight, and on these the eager men and the artillery were embarked. By daylight the next morning this timely succor was at Gallipolis.[748]

An anonymous correspondent from Columbus, Ohio, to the *New York Times* wrote

> In half an hour [Hayes'] little column was in motion, groping its way along the rough mountain road. The night was moonless, and the darkness sometimes so intense that the regiments were compelled to halt until the clouds cleared before they could go forward. All night the weary march was continued, and just as dawn began to streak the summits of the mountains, the column, reaching a high point over looking the Kanawha Valley, near [Gauley Bridge] saw the two steamboats rounding a bend and coming up the river. The troops and the boats reached the wharf almost simultaneously, and within an hour the whole command had embarked, and the steamers were under full headway down the Kanawha, their decks strewn with tired and sleeping soldiers.[749]

Another account, extracted from a history of the 23rd Ohio, relates that upon hearing of the proposition to go after Morgan at Gallipolis, the men responded "with cheers" and

> cries that they were off for God's country; and soon after the two regiments started. They marched until near midnight, and rested for a couple of hours, and a little after daylight, as they approached Loup Creek, they saw the two steamers just rounding the bend. The boats and the troops reached the landing almost simultaneously. Without losing any time the regiments embarked, and were soon on their way down the river. The tired soldiers stretched out on the deck and slept nearly all day.[750]

Thirteenth regimental records reveal that on July 17 the 13th marched to Loup Creek. At 12 o'clock noon, they embarked on the transports, *Victress, Victor No. 2* and *General Meigs* "to go in pursuit of Morgan in Ohio, arrived at Point Pleasant at 8 o'clock p.m."[751] "on the evening of July 17th."[752] Lieutenant Colonel James Comly, commanding the 23rd Ohio Infantry, noted in his diary that when the soldiers arrived at Loup Creek, they boarded the *Victor No. 2*, then when they met the *B.C. Levi*, "the two boats were lashed together" and in this way the boats made their way to Camp Piatt, their next destination. Charleston, wrote Comly, was reached at four o'clock in the morning, rations were obtained and they started immediately for Gallipolis.[753]

The 13th traveled from Fayetteville to Loup Creek, a distance of forty-five miles.[754] 13th companies which had not been ordered forward to Fayetteville the previous week were picked up by the steamers as these made their way down the Kanawha from Loup Creek to Point Pleasant. Cos. A and H and probably also the "squad of recruits" (Co. I), all stationed at Coals Mouth (St. Albans today), under command of Capt. James Johnson and Co. B (stationed at Point Pleasant, Capt. Milton Stewart, cmdg.) joined the 13th companies which had been sent to Fayetteville. As the transports traveled down the Kanawha and approached Coals Mouth and Point Pleasant, they turned in to shore to pick up the detached companies. At last, all arrived at Gallipolis at 10 o'clock p.m. the evening of the 17th.[755] The distance traveled from Loup Creek to Gallipolis was ninety-two miles.[756]

Militia had been assembling at Gallipolis for several days amounting to a considerable force and some heavy guns were in position.[757] A part of Morgan's forces had reportedly camped the night of July 17 "fifteen miles" from Gallipolis.[758] By daylight on July 18, Hayes'

> troops disembarked and took positions to defend the town [of Gallipolis], but Morgan had been advised by spies of their approach when six miles away, and turned his column northward toward

Pomeroy [...]. Colonel Hayes instantly re-embarked [his men on the boats] and steamed up the river to overtake him.[759]

Scammon received word that the raiders had veered off away from Gallipolis and were instead pushing by to the east "up the Ohio as if to cross at Pomeroy,"[760] Ohio, another fording place on the Ohio River. Scammon left behind a small detachment of the 23rd Ohio Infantry with 2,500 militia to secure Gallipolis[761] on the 18th, under Captain A.A. Hunter, commanding Post at Gallipolis. Scammon then re-embarked his troops and continued up the Ohio River towards Pomeroy, where it was anticipated that Morgan would attempt to cross at a ford into West Virginia.

A post war correspondent to the Cincinnati *Times* noted that the 23rd Ohio and 13th Virginia Volunteers had arrived just in time to succor the town of Gallispolis. On the morning of the 18th, began correspondent, there had been no other trained troops at Gallipolis to stop Morgan from entering the town. All day on the 17th and during the night, the steamers transporting the 23rd Ohio and 13th Virginia Volunteers were forced forward. At daylight, on July 18, they reached Gallipolis

and they were not a moment too soon, for John Morgan was within six miles of the town, and there were no troops there to dispute his entrance. Hayes moved forward at once with his forces, and he had not proceeded a mile when he sighted the skirmishers of Morgan. Morgan's men halted, and the officers could be seen holding a council. They were evidently taken aback at the sight of an army, and Hayes had taken care to display his troops in the largest proportion possible. Morgan evidently thought Scammon's whole force was there to capture him. If he had known there were but twelve hundred men it is entirely probable he would have consolidated his four or five thousand men, and tried to cut his way through. And Hayes felt his disparity of numbers, but hoped with rested troops, who were wrought up to the highest purpose, to defeat the great raider. He could at least, hold the town until succor arrived. But Morgan did not pause to offer battle Hobson was in his rear, and a fleet was on its way from Cincinnati to prevent his crossing, so he turned quickly to the eastward, and made haste for Pomeroy. Hayes, with equal celerity boarded his boats again, and arrived in time to head Morgan off again.[762]

An unpublished history of the 23rd Ohio contains the following corroborating account of the arrival at Gallipolis and departure for Pomeroy of Scammon's men. When the boats transporting the 23rd Ohio and 13th West Virginia Regiments reached Gallipolis, the troops immediately disembarked and formed in line of battle to receive Morgan, who was reported five miles from town. Scouts were sent out, and it was soon reported that Morgan had veered off, and was making for Pomeroy. Col. Hayes ordered the troops on board the steamers and started for Pomeroy.[763]

Regimental reports indicate that upon their arrival at Gallipolis in the morning of the 18th, once it was learned that Morgan had skirted Gallipolis, forces were disposed from Hayes' command to secure the safety of town and the balance of Scammon's troops were ordered to board the steamboats again.[764] These proceeded hurrying to push up the Ohio River in hopes of overtaking Morgan and his raiders at Pomeroy. 13th reports indicate that by transport the regiment was conveyed from Gallipolis to Pomeroy a distance of twenty miles. Pomeroy was reached "at 12 o'clock M."[765] At Pomeroy, the 13th "had a skirmish."[766]

While waiting at Gallipolis for their next orders, the men of the 23rd Ohio and 13th Virginia may have had occasion to witness the following martial display, which, for all it being peripheral to the story of the 13th Virginia, is too delightful to omit or consign to a footnote. Harrison G. Otis, then serving in the rank of Captain in the 23rd Ohio Volunteer Infantry, published the following anecdote of the Morgan raid, which was related to him by an officer, whose regiment passed with Hayes up the Ohio on their way to intercept Morgan.

The officer had strolled toward Gallipolis town center while the transport he was on was "coaling." There, on the public square (still today where it has always been) was the drill ground where the militia, "consisting of old and middle-aged men, and of boys," gathered from "the back counties," had been mustered for organization and drill in anticipation of hostilities with Morgan's raiders. The officer related that

> Disorder was the order of the day. An officer of volunteers was assigned the exasperating duty of forming them into companies and regiments. Company and regimental officers were elected by the rank and file. The exigencies of the occasion which had called them into the field would not allow of time for instruction, yet in a fit of desperation copies of tactics were issued to the field offi-cers. Morgan was reported to be rapidly approaching. Drilling was vigorously practiced from day's dawn until dark, the drill ground being the public square. [...] All was activity of the most intense military disorder. Soon, however, several persons on horseback, with swords, any one of which, in the matter of rustiness and avoirdupois, might have been the sword of Bunker Hill, or of ye bold buccaneer, began to ride frantically up and down, shouting

themselves hoarse and very red; while others on foot made equally as much noise. As a consequence of this, one long irregular line of men, and several short ones were formed. The long one the writer ultimately discovered was a regiment, while the shorter ones were companies—which astonished him very much. They had not yet received their arms, the maneuvers which the writer witnessed being the 'setting-up' process. The party who shouted the loudest, (a fat steamboat captain of vast dignity,) and who, consequently, succeeded in getting the greatest number of people together, it came to be known was the Colonel. Riding to the front, accompanied by the Lieutenant Colonel and Major, he faced about, and in an awful voice commanded, *'Aatention, battalion ! Right face! Forward by file right, double quick, march!'* Not a man stirred. The only effect this order seemed to produce upon the men to whom it was given was to cause them to expectorate a little more copiously than usual. The Colonel with increased dignity repeated the command, with the same result. Directly the Lieutenant Colonel was observed to approach him, when a brief council of war was held, which, ended, the former officer—the most Virginianist West Virginian it has been my fortune to behold—rode a few paces nearer to the line, and prefacing his observations with a volume of tobacco juice which made a great blotch upon the plain, exclaimed, *'Look wild, thar, youns! Prepar to thicken 'n march endways—thicken! Right smart, git!'* Whereupon the whole line formed double files, and marched off by the flank at double-quick time in good order. The Colonels order, as he had learned and delivered it, was Greek to them; but when rendered comprehensive by a translation into the native English, it was very promptly and intelligently executed.[767]

Pursuit of John Hunt Morgan: Engagement at Pomeroy

"July 18ᵗʰ, the Regiment debarked at Pomeroy and had a skirmish with General Morgan's force, about two miles in the rear of Pomeroy, when Morgan's men retreated into the country, in the direction of Buffington Island."

— *Field History of the 13ᵗʰ Regiment West Virginia Infantry*, W. Virginia State Archives.

To bring the reader up to date.
From the Confederate side and other fine points

On July 2, Brigadier General Morgan had crossed the Cumberland Mountains and launched a raid with his division, known as the "Ohio raid" or "the great raid." The object of the expedition according to then Colonel, Basil W. Duke, Morgan's second in command, was

> to save Gen. Bragg's army in its retreat from Tullahoma from annoyance by the Federal cavalry and draw as large a force of the enemy as possible after himself, so that it might not be able to take part in the battle which Gen. Bragg intended to deliver after he had crossed the Tennessee River. [Morgan] started with 2,460 men, rank and file.[768]

The "great raid," coincided with the Vicksburg and Gettysburg campaigns and ended for all intents and purposes at Buffington Island on July 19 although Morgan and the last remnant of his command did not surrender until July 26. Morgan cut through central Kentucky, southern Indiana, the suburbs of Cincinnati and southern Ohio. As he drew closer to West Virginia, he entered country where civilians and troops were not unfamiliar to raids. They did not panic and were in fact accustomed to employing what in the modern era we might call "counter-insurgency" tactics. No surprise that it was here in southernmost Ohio and West Virginia country that Morgan's command was at last broken up and largely apprehended.

Pomeroy was in its appearance a typical small Ohio river town. Much more of a contender in economic terms in those days, than now, it supported four churches, one newspaper office, a machine shop and about a dozen stores. It had grown quickly due in large part to the abundance of coal in the region. The first coal mine had opened in Pomeroy in 1819. By 1846, the town had a population of about 1,600 people. Col. W.R. Brown, commanding 13th Virginia Volunteers, had been a resident of Pomeroy (at least by 1860) and was a machinist[769] by profession. Before the war, he worked as a formeman in a machine shop in Pomeroy.[770] Pomeroy's buildings were closely built huddled together one on the other and edging the several streets that had been fitted within the narrow interstice between the Ohio River bank and the hills. The hills were a prominent feature of the town scape. They rose behind the town almost like a canopy in some places, and they defined its northern perimeter.

Early Saturday morning (according to various reports to the Cincinnati *Commercial*), Morgan moved his forces, numbering in all about "4,000 men, well-horsed and equipped with revolvers, saber and carbines" toward Pomeroy, "the ford below Buffington Island, on the Ohio River."[771] Morgan approached the Ohio River but "found all the roads guarded."[772] Morgan halted his men "some 3 miles distant," from Pomeroy, "near R[o]cky Spring."[773] Here "the militia, well secured in position, tried at long range the effect of squirrel rifles and other arms at hand, and seven mounds attest the accuracy of their fire."[774] Likely too, was that here, Morgan also learned that Fitch's gunboats were waiting for him. Thus, it was that the militia held Morgan's men "in check until the 23d Ohio and 13th Virginia landed at Pomeroy and marched to the scene of action."[775] Morgan turned his course "east up the road back of Pomeroy."[776] At some distance beyond Carrs Run, Morgan's men divided. One group proceeded in the direction of the Ohio River to strike the ford at Buffington's Island, and the other group took up the line of march towards Chester, Ohio.[777]

The push to arrive at Pomeroy ahead of Morgan paid off. When the 13th Virginia and the 23rd Ohio reached Pomeroy aboard the steamers at twelve noon, they found "the militia waiting in position for Morgan." The 23rd and 13th were straightway disembarked and deployed to intercept the raiders. Two howitzers were taken from the steamboats and loaded with grape and cannister. "About noon" Morgan came. The 23rd Ohio went out to meet him and found Morgan present "in force." The 13th was sent for and lines of battle were formed and skirmishers thrown out. "Morgan's men dismounted and did the same."[778] "Morgan's officers were not long in discovering that something tougher was in front of them than militia regiments, they suddenly drew off remounted, and made still further up the river."[779]

In his official report to Adjutant and Captain James L. Bottsford at 3rd Division Headquarters, Hayes reported that at Pomeroy

> By direction of Genl Scammon the 23rd Ohio under Lt. Col Comly and the 13th Va. under command of Col. Jones of Genl Scammon's staff,[780] were marched to the roads on which Morgan was attempting to reach the river. On the approach of the Rebels these Regiments formed in line of battle and with four companies thrown forward as skirmishers, advanced to meet the enemy who were dismounted and also formed in line of battle. A short skirmish ensued, when the rebels retreated, hastily mounted their horses and pushed on up the Ohio. Our loss was one wounded Pvt. Corporal Clemens 23rd Regiment O.V.I. The enemy lost 5 killed and 16 wounded upon the field.
>
> The troops were immediately embarked on transports and steamed up the Ohio to Buffington Island [...]
>
> My command it is believed prevented the Rebels from crossing the Ohio at Pomeroy and gave important aid to those engaged in intercepting them at points higher up on the River. It is proper to add that General Scammon and staff were with the advance during

the whole of the expedition and that all the movements referred to in this report were directed by the General.[781]

13th regimental records, (Record of Events Field and Staff) confirm the skirmish back of town: after disembarking at Pomeroy, the 13th skirmished with the Rebels "back of the town" or in rear of the town, "about 2 miles."[782] One source characterized the skirmish engaged in by the 13th as "severe."[783] Morning reports for Co. A carry: at Pomeroy, "We march[ed] of[f] the Boat in the Hills 4 miles from Thoun [Town] and pursued the Rebels. After 3 [hours], we march back on the boat again."[784] "Remarks" in Co. F reports note that on July 18 Co. F with the regiment "[s]teamed up to Pomeroy and disembarked went out on the hill attacked Morgan driving him off."[785] Capt. J.V. Young, Co. G, wrote in a letter to his wife:

> The 13th Regiment did not lose a man, notwithstanding we made a charge on the gray backs back of Pomeroy with a yell that made them fly in every direction. We charged down a steep hill right on them. If you ever saw devils doublequick, you may imagine their speed.[786]

Without more information it is impossible to state which 13th companies were engaged at Pomeroy and which held in reserve. It is distinctly possible that all companies were engaged in back of Pomeroy, except Co. D, which got onto shore at Pomeroy, "[b]ut did not have to fight."[787]

Skirmishing between the 23rd Ohio, 13th Virginia and Morgan's advance[788] subsided when the "Confederates retreated and commenced to move up the country."[789] The raiders moved rapidly in the direction of Buffington Island.[790] The 13th Regiment meanwhile, with the 23rd Ohio, returned to Pomeroy and the transports. Aboard the transports the 13th and 23rd followed the long bend in the river and proceeded up as far as Buffington's Island. Gen. Scammon, who with his staff had general command of the expedition, stated in his report that after "a slight skirmish," his "troops were immediately re-embarked, and steamed up the river to Buffington Island."[791] They arrived at Buffington at daylight "on the morning of the 19th."[792]

Remarks made by Morgan's men regarding the resistance encountered as they passed Pomeroy are all in agreement. Major James McCreary, 11th Kentucky Cavalry, C.S.A., wrote in his diary for July 17 that they halted for the night at Pomeroy. "The enemy are in considerable force in front," continued McCreary on July 18. "We attacked them[793] and drove them from our front and then moved rapidly in the direction of Buffington, where we intend to cross."[794] Basil W. Duke, Morgan's second in command, stated that

In passing on the 18ᵗʰ near Pomeroy, there was one continual fight, but, now, not with militia only, for some regular troops made their appearance and took part in the programme. The road we were traveling runs for several miles at no great distance from the town of Pomeroy, which is situated on the Ohio river. Many bye-roads run from the main one into the town, and at the mouths of these roads we always found the enemy. The road runs, also, for nearly five miles through a ravine, and steep hills upon each side of it. These hills were occupied, at various points, by the enemy, and we had to run the gauntlet.⁷⁹⁵

Colonel [Warren] Grigsby took the lead with the Sixth Kentucky, and dashed through at a gallop, halting when fired on, dismounting his men and dislodging the enemy, and again resuming his rapid march. Major [Thomas] Webber brought up the rear of the division and held back the enemy, who closed eagerly upon our track.⁷⁹⁶

About 1 o'clock of that day we reached Chester and halted, for an hour and a half, to enable the column to close up, to breathe the horses, and also to obtain a guide, if possible (General Morgan declaring that he would no longer march without one). That halt proved disastrous—it brought us to Buffington ford after night had fallen, and delayed our attempt at crossing until next morning.⁷⁹⁷

Allan Keller, a contemporary historian, described the gauntlet referred to by Duke as a five mile long narrow road, which "debouched into an area of open land near the Rock Springs fairground, where a side road followed Kerr's Run to the river above Pomeroy." According to Keller, here, Union troops (regulars and home guards) concentrated their firepower.⁷⁹⁸

Anna Starr, resident of the Mason City area (West Virginia), wrote to her husband, William C. Starr, 9ᵗʰ West Virginia Infantry, from Pomeroy on July 27. Making reference to remarks made by Basil Duke, who had been held prisoner at Pomeroy, she wrote that,

[...] Basil Duke told several of our citizens here, that they had met with more obstructions, bushwhacking and serious opposition in Meigs county than they had, all put together before the Pomeroy Militia captured 79 of them and kept them in the courthouse here five days they all told the same tale that they knew as soon as they came into Meigs county and were saluted the first thing with a bridge destroyed in front of them they were sure their doom was sealed; these prisoners said they fully intended to burn Pomeroy and destroy all the saltworks but the fear of our gunboats and the spirited opposition of the Pomeroy Militia who fought them several hours out back of Mr. Heckards saved the town.[799]

The foregoing opposition and the encounter with Hayes' forces seems to have persuaded Morgan of the wisdom of crossing the river into West Virginia farther upstream. The successful blocking of Morgan at Pomeroy may appear as just the most recent in a series of measures and strikes to "bag" the Confederates but the interception at Pomeroy may well have been the critical stroke in bringing down Morgan and his elusive rough riders. Morgan was being pursued by Brigadier General Henry M. Judah with his force of eleven hundred men—just a portion of Judah's command, the 3rd Division, 23rd Army Corps. These eleven hundred consisted of the "5th Indiana, 11th Kentucky, and 14th Illinois Cavalry, and two sections of artillery and four Mountain Howitzers, under Captain Henshaw." Also in pursuit of the raiders was Brig. Gen. Edward H. Hobson and his command of three brigades from his division of Kentucky troops and on Morgan's right was Lieutenant Leroy Fitch, commanding tinclads from the 6th Mississippi Squadron of Federal gunboats.

Up to the time Morgan drew near to Pomeroy, he had, particularly in Ohio, been doing "all the damage he chose to."[800] So rapidly had he been moving since commencing this lightning raid that no one had been able to catch up to him. He had been well nigh unstoppable. After doing considerable damage to the Ohio railroads at Jackson and Berlin, he then passed through Vinton, located 5 miles north of Centreville and kept "on east thence to a point two miles back of Pomeroy," where he was at last intercepted, by Scammon's troops.[801] After this encounter Morgan's men concentrated their efforts on effecting an escape. The destruction fell off.

Hayes' regiments were not first upon the scene of action at Pomeroy but Morgan's encounter with them seems to have been a determining factor. First of the major Federal players to arrive at Pomeroy was Lieut. Leroy Fitch's gun boat (probably his flagship the *Moose*), which was at Pomeroy by Friday, July 17. Fitch, with the gunboats under his command,[802] had followed Morgan constantly up the Ohio River for about 450 miles since he had crossed over into Indiana. In addition, about 200 "poorly armed" Home Guards had also assembled at Pomeroy, and roads leading to Pomeroy "for 15 miles out" had been blockaded with felled trees.[803] Anna Starr wrote in another letter to her husband (dated "Mason City July 20· 1863"), that gunboats were at Pomeroy waiting for Morgan and fearing these, he "slipped by Pomeroy on some back road, and stopped at Rock Springs to feed his horses."[804] Mrs. Starr's letter is of interest as it also makes mention of "three steam-boat loads of soldiers," Scammon's forces. Her letter is the following:

I <u>hear</u> that Gen. Scammon with a part of the forces under his command is marching across to Parkersburg to head off Morgan and then <u>again</u> I hear that Gen. S. was with the 23rd Ohio and 13th Va who came down the Kanawha and passed up the Ohio, yesterday on steamboats in pursuit of the famous John. [...] We fully expected a big battle in Pomeroy yesterday so the boys and I crossed over to Mason City [...] intending to take to the hills when the cannonading commenced by Morgan fearing the gunboats we had here slipped by Pomeroy on some back road, and stopped at Rock Springs to feed his horses he is skedaddling up the river some where and our gunboats and three steam-boat loads of soldiers are pushing up alongside to prevent his crossing.[805]

These last mentioned would have been Hayes' regiments. When Hayes' troops engaged Morgan in rear of Pomeroy, he gave up on the idea of accomplishing anything at Pomeroy. Morgan was forced farther up river with his exhausted men thereby giving Union forces converging upon him time to catch up and close in.

In a letter home to his wife, Lucy, R.B. Hayes emphasized the importance of the interception of Morgan's men at Pomeroy, writing: "<u>We</u> can truly claim that Morgan would have crossed and escaped with his men at Pomeroy if we had not headed him there and defeated his attempt." Had Morgan been able to cross the Ohio and escape into West Virginia at Pomeroy, it would have been almost impossible to apprehend him. Not only because of the difficult terrain through which the Federal troops would have to scour for Morgan's men but because Morgan would have found many sympathizers willing to offer aid and refuge.

After the engagement at Pomeroy, as indicated above, Hayes' troops were immediately put aboard the boats and for the duration of the pursuit the men of the 13th and 23rd "were quartered on steamboats." Hayes regarded the chase as a happy adventure. He stated in a letter home to his wife that on board the boats

> men were singing, bands playing. <u>Our</u> band [the 23rd Ohio regimental band] was back and with us, and such lively times as one rarely sees. Almost everybody got quantities of trophies. [...] Morgan's raid will always be remembered by our men as one of the happiest events of their lives.

As Hayes observed, Morgan's men seemed "only anxious to get away." "We were at all the skirmishes and fighting after he [Morgan] reached Pomeroy," wrote Hayes but "[i]t was nothing but fun — no serious fighting at all. I think not over ten killed and forty wounded on our side in all of it."[806] Morgan's men were so jaded and worn-down from hard-riding that there was no fight in them, no grit in their attacks.

As for Federal losses at Pomeroy, the 23rd had one man wounded in the hand. Lost horses stolen by the cavalrymen along the way were re-captured. The 13th reported no casualties at all. Gen. Scammon reported that the raiders lost five killed and sixteen wounded left on the field at Pomeroy.[807] Citizens reported that the Confederates had three killed and sixteen "wounded and left at roadside houses."[808] Co. F 13th W. Virginia Regiment reported that it had captured "9 horses."[809]

Up the Ohio to Portland, Buffington Island and to Big Hocking River

From Pomeroy the conveyance of so large a body of Federal troops as Scammon's command was did not go unnoticed. Henrietta Barr at Ravenswood, a devoted secessionist, noted in her diary, for Saturday, July 18, that after 9 p.m. Scammon's force, "several thousand men" loaded aboard "six gunboats" went up the river, past Ravenswood.[810] On July 19, the 13th "[r]eached Buffington [Island where they] found Morgan endeavoring to cross."[811]

> "Old Morgan is done ...
> a long and tiresome march both by land and water"

—Captain J.V. Young to Paulina, his wife, dated "Point Pleasant, July 24, 1863, 3 o'clock 20 min. P.M.," Roy Bird Cook Coll. WVU, Morgantown, trans. p. 77.

"... Skirmishing with the rebels back of the town [Pomeroy] about 2 miles. They retreated and commenced to move up the country. Regiment returned to Pomeroy and from this place went up the Ohio River as far as Buffington's Island. Arrived there early on the morning of the 19th, where an engagement took place in the village of Portland between our forces and Morgan's resulting in the defeat of Morgan and the capture of some 800 of his men. This Regiment then moved up the river in pursuit of the flying rebels as far as Big Hocking River. The next day [July 21] returned to Point Pleasant. Left P[oin]t Pleasant on Aug. 2nd for Charleston, W.Va. and remained until [Aug.] 5, 1863, thence ordered to Coals Mouth, remained until Aug. 5, 1863, ordered to Barboursville, arrived Aug. 7th. Distance marched 200 miles. By transport 288 miles. Total 488 miles."

— *Field History of the 13th Regiment West Virginia Infantry*, West Virginia State Archives at Charleston and Regimental Return 13th Va. Vol. Infantry for August 1863, National Archives, Wash. D.C.

"The most successful and jolly little campaign we ever had."

— Col. R.B. Hayes, diary entry dated "July 22, 1863."

"The loss of Morgan's command ... was a heavy blow to the rebellion."

— Report of Maj. Gen. Ambrose E. Burnside, U.S. Army, commanding Department of the Ohio, of operations March 25-August 10, 1863, p. 14.

After Pomeroy, Hayes' command kept pace with Morgan's progress through Ohio, and arrived in time to be in at the surrender of a large part of Morgan's force at Buffington Island. Then moving up and down the Ohio river, Hayes' men, with other regiments of the Kanawha Division, prevented Morgan from crossing at several points.

— Paraphrase of R.B. Wilson, "Kanawha Division: Its Campaigns. [Part] I—continued," *The National Tribune*, February 18, 1897, p.2.

While Hobson's and Judah's cavalry pursued overland, gunboats under command of Lieutenant Leroy Fitch, patrolled the Ohio River to prevent Morgan from escaping over the Ohio River into Kentucky or West Virginia. As Morgan raided through Indiana and Ohio, he had succeeded in escaping at every juncture before pursuing forces could be gotten into position. He had encountered opposition at Pomeroy that had slowed his pace. As he passed through Gallia and Meigs County, local militia were able to retard his progress; Scammon's troops and militiamen had been in position to make it hot for him; and other land troops and gunboats on the river were at last able to close in and come within striking range of him. At Buffington Island, where he intended to attempt another crossing into West Virginia, Morgan's luck ran out. His men were exhausted from their many days in the saddle and the Ohio River was high and swift-running for that season of the year, making the river crossing a dangerous undertaking. It was no longer a matter of "fording" but a swimming of men and horses across fast moving water.

The burden of the main fighting was born on the Ohio side by Hobson's and Judah's cavalry troops supported by Fitch bombarding with his gunboats (warships from the 6th Mississippi Squadron) from the Ohio River. The cavalry pressed the raiders on flank and rear and the tin-clads fired from the river until the raiders were forced into a bottleneck at Buffington. Hayes' 23rd Ohio and 13th Virginia came into action at what can only be described as the very exciting termination of the raid, when Morgan's division dissolved and scattered to escape from the bottleneck.

The 23rd Ohio and 13th Virginia disembarked and embarked again in the Hocking River area going where the heavy tinclads could not go. Their job was multi-fold: to scout and guard fords and penetrate back into the rough countryside to capture prisoners. They established and manned prisoner depots; escorted prisoners down river; and they likely performed anything considered necessary at the time in 'their' bend of the river between Belleville, West Virginia, and Hocking River, Ohio, in the way of mopping up the 'great raid.'

Infantry regiments were critical components of the 'mopping up' as the Confederates had become much scattered over as much as fifty or sixty miles of country. It was only the infantry which could get back into and scour the difficult country back from the Ohio. Ohio regiments seem to have scoured the Ohio side and West Virginia regiments the West Virginia side.

There are, as always with the Loyal 13th, difficulties in gaining a clear image of the sequence and locus of events as references to geographic locations in contemporary accounts were fairly general. Buffington, Hocking and Belleville were the only notable place names in the vicinity, and they seem to have been used interchangeably to refer to the same twenty or so miles of land on either side of the Ohio River in the two different States (Ohio and West Virginia). In addition, Col. W.R. Brown, commanding the 13th, did not it seems, submit a report to brigade headquarters summarizing what part his command played in the chase and mop up. Neither is there specific reference to the regiment in Col. Hayes' report, which in any event creates some confusion inasmuch as Hayes obviously had his dates wrong by a day.[812] With this apology for introduction, the following is my reconstruction of the part played by the Kanawha Regiments in the termination of Morgan's raid.

The land battle in the valley adjacent to Buffington Island

From Pomeroy, Morgan moved still farther east, towards Buffington Island, where he planned to re-cross the Ohio into Virginia (West) and to safety in the Southland. The race was on for the well-known but little guarded crossing below the village of Portland, Ohio, at Buffington Island. Morgan's command, numbered now something under "nineteen hundred strong."[813] In hot pursuit, were Brigadier General Edward H. Hobson, in overall command of operations with three brigades from his division of Kentucky troops and Brigadier General Henry M. Judah with 3rd Division, 23rd Army Corps. These forces were upon Morgan's left flank and rear. On Morgan's right was the fleet of gunboats under Lieut. LeRoy Fitch.[814] Gen. Scammon's troops (among these Hayes' regiments, the 23rd Ohio and 13th Virginia) were also making their way up the river on steamboats with as much speed as the tricky navigation of the shoals and deep fast water above Pomeroy would permit.

Up to now it had been one great horse-race. Horse stealing extraordinaire had kept the raiders ever freshly mounted and able to gallop for hundreds of miles through the country and ever just out of reach of their pursuers.[815] The stress of the raid was taking its toll. Although still outdistancing his pursuers, Morgan had met greater resistance in Ohio, his men were spent, the result of many days in the saddle and the command was weighed down with a heavy train of loot. If he could maintain his lead to cross the river into West Virginia all might still be well. His concern was for the river crossing and that his command arrive at the ford before the gunboats could arrive. The river crossing would be dangerous enough an undertaking without having to make it under fire. It had not been a normal year, however. Usually with a late spring one could expect a drop in water level making the Ohio River in eastern Ohio fordable on foot or on horseback but this was not so in 1863. Due to a late melt of snow and ice in the mountains, the water level in the river was unseasonably high, at flood stage and fast moving. This rise in the river cut against the raiders in two ways. It was the reason that the tin-clad gunboats of Fitch's 6th Mississippi Squadron and steamer transports, bearing the 23rd Ohio and 13th West Virginia, were able to range farther upriver than usual and the reason that it was no easy matter for the raiders to cross the river at Buffington and escape into West Virginia.

The raiders passed through Chester and reached Portland, Ohio, at "about 8 p.m."[816] on the night of the 18th. Portland was a small village located on the Ohio River bank "a short distance above Buffington Island"[817] and opposite the head of the island. "[T]he night was one of solid darkness"[818] and the river so high that Morgan and his officers were faced with a considerable dilemma. Several obstacles presented themselves. Morgan and those with whom he advised fully appreciated that if the command was not crossed this night, "there was every chance" that on the next day (July 19), they would be attacked "by heavy odds."[819] Col. Basil W. Duke wrote of this juncture:

> The troops we had seen at Pomeroy were, we at once and correctly conjectured, a portion of the infantry which had been sent after us from Kentucky, and they had been brought by the river, which had risen several feet in the previous week, to intercept us.[820] If transports could pass Pomeroy, the General knew that they could also run up to the bar at Buffington Island. The transports would certainly be accompanied by gun-boats, and our crossing could have been prevented by the latter alone, because our artillery ammunition was nearly exhausted—there was not more that three cartridges to the piece, and we could not have driven off gun-boats

with small arms. Moreover, if it was necessary, the troops could march from Pomeroy to Buffington by an excellent road, and reach the latter place in the morning. This they did.[821]

To cross at the ford—the only feasible fording place in this part of the river—, Morgan had first to take an earthwork thrown up to guard the ford. The work was defended by "about three hundred infantry—regular troops—and two heavy guns [...] mounted in it."[822] To cross at this point, the earthwork which commanded the ford had to be taken. The high water level, even if the taking of the ford were attempted was a huge concern. Wrote Basil Duke, "the river usually fordable there at this season of the year was high. The June rise produced by the melting of the snow in the mountains, had come this year a month later than usual, and the waters were up."[823] The surge in the river was estimated at five feet, more or less. To complicate matters the night was "thoroughly dark."[824] Morgan had been unable to procure a guide and knew nothing of the ground and in the solid darkness the officers and men could only guess from which direction to move on the attack. It was feared that they would surely run afoul of one another in the darkness. With rested troops this movement would have been difficult enough but with more or less exhausted men, Morgan judged it to be beyond the pale. Ultimately, Morgan determined not to hazard a night attack on the fieldwork but "to take the work at early dawn and instantly commence the crossing, trusting that it would be effected rapidly and before the enemy arrived."[825]

In preparation, Morgan ordered Duke to place two regiments of his brigade near the fortification and to attack it at daybreak. Accordingly, Duke placed the Fifth and Sixth Kentucky (each about 250 men strong) about four hundred yards from where, in the darkness, he supposed the work to be located. Two parrot guns were also placed in supporting distance of the two regiments on a tongue of land projecting northward from a range of hills, which ran parallel to the river. Morgan's division "encamped in the cornfields, at the end of a private lane running parallel to the [river] road."[826] Morgan's decision to delay at this point would permit Union forces converging upon the cavalrymen time to catch up. By morning, the Confederates would find themselves completely surrounded.

Setting the scene of action

A short description of the lay of the land at the Buffington battlefield, description of the adjacent river channel and disposition of forces on the morning of July 19 would be helpful to the reader in visualizing the next sequence of events. The orientation of the Ohio River channel at Buffington Island was then, as it is today, roughly on a north-south axis. A slim chute of river flows between the island and the Ohio State shore. The field of action at Buffington was a valley of river bottom land in the shape of an elongated triangle situated between the Ohio River and a range of hills. This range of hills formed the valley's western boundary. The hills ran parallel to the river at its southern boundary but at the valley's northern terminus, both hills and the river inclined toward each other, creating a bottleneck where the hills then ran to the water's edge. Adjacent to where the hills reached the river was a precipice. There, the land fell abruptly away into a steep ravine.

The valley was roughly a mile long and its southern extremity from the hilly ridge to the river was "about 800 yards wide."[827] In the immediate vicinity of the fording place itself, "[t]he rugged precipitous hills which follow the river for hundreds of miles, often at a distance of only a few hundred yards, fall back considerably."[828] The river, flowing through this rugged terrain had created high river banks obstructing the view for soldiers aboard the warships and transports and made for uncertain aim when it came to concentrating firepower towards the land from the river. The open and fertile river front property, however, was prime for cultivation and for crossing a small army such as Morgan's.

On the morning of July 19, Morgan was attacked by three thousand of Hobson's cavalry[829] and, wrote Basil Duke, "by several thousand infantry which had been brought up the river on transports, while three river gunboats steamed up and took part in the fight."[830] Judah, Hobson, Fitch and Scammon's troops all converged at the fording place below Portland, where they supposed Morgan was. Neither Judah, Hobson, nor Fitch, however, knew of the proximity of the other and it was with uncanny timing that their attacks on the Confederates began almost simultaneously.

Judah's forces commanded the southern extremity of the valley (the base of the triangle) and he attacked from there. Approximately equidistant from either end of the valley lay the Chester road down which Hobson was spurring his cavalry toward the sound of Judah's guns. Fitch was in position at the foot of Buffington Island with his three gunboats making his way upriver and the transports containing Scammon's men (among these the 13th Virginia) were down river from Fitch but in proximity to him, following him. Indeed, these transports may very well have followed the *Moose*, Fitch's flagship up from Pomeroy.

At first both brigades of Morgan's Division—2nd Brigade commanded by Basil W. Duke and 1st Brigade commanded by Colonel Adam R. Johnson—confronted Judah but then under pressure from Hobson's assault on his rear, Johnson's command had to re-position themselves to form at right angles to Duke's line to face into Hobson's advance. Judah blocked the river road at the southern extremity. Hobson blocked the Chester road leading back to the Ohio interior and Fitch blocked at the river. The only other way out of the valley was the road at the northern apex of the wedge, an access that was hardly more than a precipitous mountain path from which the land fell away sharply toward the river forming a steep ravine. This exit, although raked by fire from the gunboats, was the Confederates' best hope for escape. As the raiders began crossing the river they were caught at last and forced into a fight. The valley would prove an unfortunate place for the outnumbered rough riders, essentially a trap. They were caught in the cross-fire of Judah, Hobson and Fitch.

The chase on the Ohio River

Despite the difficulties the *Moose* had in pushing eastward from Cincinnati into more mountainous country (i.e., swimming up river against the river's flow and navigating shoals with the necessity of periodically having to warp the boat over shallow areas), the *Moose* came up with Morgan at Buffington in good time and without being sighted. "A few hours before the Confederates reached the crest of the nearby shore the *Moose* was anchored off Sand Creek Bar, below Buffington Island."[831] The other gunboats of the 6th Mississippi Squadron were down river from the *Moose* in blockade.

Earlier that week, as Morgan was observed to be moving still eastwards from Cincinnati, Lieut. Fitch determined to establish a blockade of some forty miles in length around Pomeroy. The "numerous shoals and shape of the river"[832] required this measure. With the exception of the *Moose*, Fitch's flagship and the *Imperial*, his dispatch steamer, the remaining tinclad warships of the 6th Mississippi Squadron were distributed in this stretch of river where the water level "was so low and the fords so numerous"[833] as to offer many an opportunity for escape to the cavalrymen. "Four major fords and a number of minor fords" were patrolled.[834] The patrol boats remained in that part of thie river. The sole representatives of the U.S. Navy squadron at Buffington were the *Moose*, the *Imperial*, and a makeshift gunboat, the *Allegheny Belle*, which was piloted by John Sebastian. The *Belle* had been outfitted several days before with cotton bales and several field guns pursuant to orders from Maj. Gen. Burnside, cmdg. the military department. These boats also carried Scammon's infantry regiments.

At about two o'clock, Sunday morning the 19th, Lieut. Fitch, anchored with the *Moose* and *Imperial* off Sand Creek Bar below Buffington Island, decided to see if he could maneuver himself closer to the foot of the island. Fitch had the "*Imperial* tow him up slowly. Unfortunately a dense fog set in and the two steamers were forced to reanchor." At about seven o'clock in the morning, the sound of musketry firing "a little ahead off the port bow"[835] reached the ears of those aboard the *Moose*.

Meanwhile, the raiders broke camp. After a rest of some hours, Morgan's men were again on the move in the early hours of Sunday, the 19th. They reached the ford at Portland at three a.m.[836] "All was quiet, a dense fog wrapped this woodless scene."[837] Not long afterwards, Morgan started his troops: Col. Duke was ordered to attack the works commanding the ford at Buffington and Col. Abner W. Johnson ("Stovepipe

Johnson") was ordered to attack any forces which came up the river to dispute the crossing over the Ohio. Johnson's men positioned themselves a short distance from Buffington Island and entrenched. The artillery positioned on the "first rising ground to the left of the river road" supported the troops at the ford and those who would first attempt the river-crossing.[838] Morgan took position on a hill (perhaps with the artillery pieces) to have perspective on proceedings. With Morgan were the reserves.

At dawn, the Fifth and Sixth Kentucky moved to attack the earthwork but they found that it had been evacuated during the night. Duke immediately sent Morgan word of this revelation and that it was safe to cross the troops over into West Virginia.[839] Supposing that the Federal garrison had retreated down the Pomeroy road, Duke also directed Col. Smith to take the 5th and 6th Kentucky Regiments and move some four of five hundred yards further down the Pomeroy road to follow them up.[840] Another courier reached Morgan about the same time as he received Duke's message. This courier informed Morgan that a Federal gunboat had approached to within close proximity of the Confederate battery but being fired upon the boat "had retired precipitately."

After verifying these two reports, Morgan, decided to commence crossing his command over into West Virginia. At 4 a.m., two companies were "thrown" across the river and were "instantly opened upon" by Federal forces.[841] In minutes, Duke heard the "rattle of musketry" from the direction that the Fifth and Sixth Kentucky Regiments had moved. Smith and the Fifth and Sixth Kentucky had unexpectedly come upon General H.M. Judah accompanied by his staff and escort conducting an advance reconnaissance.

Judah had moved up the river road from Pomeroy and arrived at Buffington with his troops at about 5 a.m.[842] A "very dense fog" blanketed everything. Judah halted his force and proceeded in person with a small advance guard, consisting of his staff and a seventy-five-man cavalry escort, to reconnoiter the river bottom at the bar. It was now about 5:30 a.m.[843] and still not yet dawn. (Modern day calculations show that sunrise occurred on this date at Buffington Island at 5:53 a.m.) Visibility was so limited by the fog[844] that after proceeding "down a road surrounded by inclosed fields"[845] for about a quarter of a mile they inadvertently stumbled into Col. Duke's men dismounted and in ambush.

Judah's party reached the point where the stage road emerged from a woods and where the Ohio River and Buffington Island were within view with Portland nearly a mile to their left.

Judah's men were subjected here to a severe fire on front and flanks and nearly surrounded. "The storm of bullets [...] produced a confusion in the staff and advance, resulting in a stampede."[846] Finding it impossible to resist the Confederates and taking casualties in killed and wounded; losing forty to fifty in prisoners; and losing one piece of artillery, Judah fell back upon the main body of his troops. "Order was soon restored," wrote E.H., a correspondent to the Cincinnati *Commercial* (reporting from aboard the *Imperial*, Fitch's dispatch boat) and Judah had his artillery "brought to the front."[847]

Having successfully intercepted and dispersed the Federal advance, Duke instructed Col. Smith to take his men and fall back to the ground where they had originally formed to attack the earthwork. Duke then rode to Morgan to received instructions. It was now about six o'clock in the morning and starting to get light. The battle of Buffington Island was poised to open.

Smith in the meantime dismounted and posted his men at the southern extremity of the triangular valley to hold that point against Judah's forces when they advanced to attack. Judah had reformed his troops since falling back upon the main body and now brought his batteries forward. He commenced firing upon Morgan's guns and then advanced to attack. Smith's undersized regiments were not up to the force of Judah's assault. Judah was superior in numbers to Smith. When Duke returned from his consultation with Morgan, he found Smith's regiments in full retreat. Someone had ordered Smith's men to "rally to horses" and in the process of doing this they were charged by three regiments of Judah's cavalry. The 5th and 6th Kentucky received orders to face about and drive back the enemy cavalry, which was somehow accomplished despite their being outnumbered and very low on ammunition if not out altogether. In less than half an hour, these movements had all been made.

Smith's desperate counter-charge availed little, however, as Judah had in the initial assault driven Smith back upon Gen. Hobson, coming in upon the Chester-Pomeroy road. Hobson had recaptured his own as well as two Parrot guns positioned by the enemy to command Buffington ford. Further, in the course of the Confederate counterattack, a portion of the 5th Kentucky, was cut off from the rest of the regiment and captured. After the counter-charge, what remained of the 5th and 6th regiments were again dismounted and formed in line across the valley. Duke sent couriers to Morgan requesting that the 2nd Kentucky be sent to him to strengthen the thin line his two regiments presented to the enemy and to post men along the ridge on their right flank. Col. Johnson's rear vedettes posted on the Chester road had also by this time been driven in by Hobson, and Johnson formed his brigade to resist the enemy's assault from that direction. Johnson's line was at right angles to Duke's. Almost "at the same time" that Hobson and Judah attacked, "the gunboats steamed up and commenced shelling us without fear or favor,"[848] wrote Duke. With the rising of the sun, the fog lifted, showing the gunboats at the river, and for Morgan, all hope of fording the entire command at Buffington Bar dissipated.

The battle at the river

Let us take a step back, to pick up the thread of events at the Ohio River, as they occurred from the onset of hostilities that morning and were experienced by soldiers and sailors on the boats. As the sound of musketry fire became more intense a short distance up river to his landward side, Lieut. Fitch sprang to action. Cincinnati correspondent E.H. described the action:

> In the meantime, the volleys of musketry had reached the ears of [Lieutenant Fitch commanding the Federal fleet on] the gunboat Moose or '34,' [...] was lying, wrapped in a dense fog, a mile below the foot of the Island. Three other boats of the fleet, the Reindeer, Victory and Springfield, were lying further down, being unable to go higher on account of the shallowness of the river above. As soon as the musketry was heard, the Moose started up, moving very slowly necessarily, and being worked with difficulty through the shoals. When she had a gained a position between the Island and the Ohio shore, she opened with shell from her port and how[itzer] guns upon the cavalry and artillery of the enemy.[849]

Taking soundings as they probed forward, the navy-men aboard *Moose* and *Imperial* carefully navigated their boats over the bar and up the channel between the island and the State of Ohio. Myron J. Smith, in his excellent reconstruction of the naval battle at Buffington wrote that while the steamers progressed 'fairly into' the chute, they were hailed from shore.

> Easing in, the *Moose* picked up Captain John J. Grafton, a member of Judah's staff, who had lost his way. The soldier was able to give Fitch his first indication of the impending fight as well as the relative positions of the opposing forces as the infantryman last knew them. While the two men spoke, the fog suddenly started to dissipate. Moving ahead once more, the *Moose* yawed just enough to open fire over the high river bank with her three port broadside guns, which had been elevated to their maximum angle.[850]

Fitch was ignorant of the army's maneuvers on shore at this, his first position, and his artillerymen—never before engaged in battle—could only estimate where to lob their twenty-four pound shells. The roar of these shots overhead created confusion in the ranks of Federal and Confederate troops alike.

The Confederates, meanwhile, thronged up river at the water's edge attempting to cross. "The river was very full in consequence of a heavy rain away up the river,"[851] wrote Major James McCreary, 11th Kentucky Cavalry, and while Morgan's men were "trying the river" to determine if it was indeed fordable, "the gunboats steamed up."[852] As they tested the waters, "the discharge of artillery from down the river"[853] could be heard. During the next moments, the Federal transports "landed their infantry,"[854] and "a heavy drumming sound of small arms in the rear and right" could also be heard from the banks of the river as "three black columns of infantry" fired upon the Confederates, who were then "in close column, preparing to cross."[855] Then, wrote McCreary,

> thousands of cavalry moved down upon us, and the artillery commenced its deadly work. We formed and fought here to no purpose. [...] Shells and minie balls were ricocheting and exploding in every direction, cavalry were charging, and infantry with its slow, measured tread moved upon us, while broadside after broadside was poured upon our doomed command from the gunboats. It seemed as if our comparatively small command would be swallowed up by the innumerable horde.[856]

In the valley, Morgan's entire division found themselves exposed to the crossfire coming now from three directions. The valley was level and completely open and fragments of shell filled the air. Even the 2nd Kentucky and 9th Tennessee (of the 1st Brigade) and the 8th and 11th Kentucky (of the 2nd Brigade), which had not thus far been engaged, were completely under fire and began to suffer demoralizing effect.

Meanwhile, in rear of the lines of his fighting men and amid escalating confusion, Gen. Morgan was withdrawing regiments (belonging to the 2nd Brigade) through the bottleneck at the northern perimeter of the valley despite the many stragglers, who were galloping about wildly away from every shell. Suddenly, the long train of Confederate wagons and ambulances dashed to the north end of the valley the only obvious avenue of escape. They became entangled with each other. Horses broke loose and plunged wildly. Some were caught up in the battery of howitzers attached to the 2nd Brigade in position at the northern end of the valley and guns and wagons upset and fell into the steep ravine there.

In the intervening time, the left flank of Duke's line, held by the 6th Kentucky in position roughly three hundred yards or more from the river, was bearing the brunt of the heavy attack on that point, when it was completely turned. The 6th was almost surrounded. Although the battle was going badly for the raiders, Colonels Duke and Johnson continued fighting for more than half an hour to give Morgan and the rest of the division time to exit the valley through the bottleneck at its northern perimeter. Observing that Morgan had in that time succeeded in extracting his men from the scene of action, Johnson and Duke "determined to withdraw simultaneously."[857]

Their men were remounted and with the 6ᵗʰ Kentucky at the rear fending off Union cavalry, Johnson's and Duke's men fell back up the valley for a mile or so "in column of fours, from right of companies"[858] "in fairly good order," "keeping up at the same time a sort of resistance for some distance."[859] Then, however, panic overtook the men and as Johnie Chandler observed from his vantage point with the 7ᵗʰ Ohio Cavalry (Hobson's command), "the end" came about in "a matter of minutes."[860] Pressure from Hobson attacking from over the hill on the Chester-Pomeroy road and the gunboats raking the northern terminus of the valley with grape-shot at the moment the 7ᵗʰ Michigan Cavalry (Hobson's command) charged into the crowd of fugitives turned the Confederate withdrawal into a route.

Dispatches to the Cincinnati *Daily Commercial* relayed that by the time Confederates had fallen back toward the island in response to pressure from Judah and Hobson, Federal forces from the rear had also

> moved up, and they, in combination with gunboats, again began the battle. Morgan replied very feebly. It quickly became evident that his ammunition was nearly exhausted, and all were confident of his immediate capture. After killing about forty of his men, the balance grew desperate and made a grand rush, breaking through our lines and scattering in all directions, seeming as if each man was taking care of his individual self.[861]

The raiders, perceiving that there were just two roads of escape[862] from the valley, broke ranks and rushed for these exits. Almost immediately they thronged and blocked the exits. It was over. This charge forced Duke and "some fifty other officers and men [...] into a ravine on the left of the road," where soon afterward they were captured.[863] Myron J. Smith Jr. wrote in his analysis of the battle at the river's edge[864] that a large group of Confederates, finding themselves encircled, overwhelmed, bombarded and stalled at the bottleneck at the end of the valley, made a break for it by descending a steep ravine to the river in an attempt to cross here. Screened by the Confederate rear-guard still holding Federal cavalry at bay and covered by the two guns on the north bank, the men quickly descended the ravine. Coming under fire from the warships, the raiders scattered in all directions. They found no escape from the ravine but by the way they had come and those who were able to, scrabbled out and scattered to the woods. So quickly did the end come that Morgan's men were able to fire off just a couple of rounds when their lines were broken.[865] So quickly had the battle proceeded and then concluded that Hayes' regiments arrived at the scene of action in time only to see the end. By 10 a.m., Morgan was defeated at Buffington, and the raid was over.

Morgan's men scatter

Caught in the crossfire of artillery and musketry coming from three directions and being forced back to the river's edge by Federal cavalry charges, the Confederates made desperate last attempts to scatter and escape. In one instance, they detached two of their twelve pound guns and drew them to Swan Bar, a mile or so above the head of Buffington Island. From there they intended to bring them to bear upon the *Moose* that some relief might be had on that flank. Lieut. Fitch countered these efforts, however, by quickly moving the *Moose* and the *Allegheny Belle*, carrying Hayes' regiments, up to Swan Bar and throwing shells in amongst the raiders manning the guns. After a few shots from the bow pieces, the two guns were abandoned.

Another segment of the Confederate command, a portion of the column had managed to make it's way still farther up the river bank and was in the process of crossing the river. Fitch next turned his attention to this column, which had reached about a third of the way across the river. He shifted his fire to the men in the water dispersed and drove the men back to the adjacent "ravines and hills."[866] Well knowing that "General Judah would look out for those left in the rear,"[867] Fitch continued upriver traveling "[a]bout 15 miles"[868] (roughly, by my calculations, to the area of Sugarcamp Run, Ohio, between Belleville, West Virginia, and Hockingport, Ohio).

The sharp engagement at Buffington Island had lasted less than half an hour.[869] A number of eyewitness sources published at the time stated that about half of Morgan's command was captured or killed at Buffington. Gen. Burnside wrote in his official report to the War Department that an estimated "two-thirds"[870] of Morgan's force and all his artillery and supplies had been captured. Duke claimed that Morgan reached Buffington with 1,800 to 1,900 men,[871] and that Morgan retreated from Buffington with "[b]etween eleven and twelve hundred men [...] closely pursued by Hobson's cavalry."[872] The actul number of losses will never be known, certain only is that after the fight at Buffington Island, Morgan's Division no longer existsed.

Morgan's men scattered on both sides of the river. It was supposed, perhaps it was wishful thinking, that few raiders had made it to the Virginia side from Buffington. It would have been harder to apprehend them in that countryside due to the rough nature of the ground and because their many friends across the river would have facilitated their escape. Hopes ran high in the Union army and civilian population that in the twenty-four hours after the battle the whole force would be captured.

After Buffington, the battle disintegrated into a kind of ragged running fight up river and into and over the adjacent ravines, hills, woods and underbrush. Morgan and his men, now broken into larger and smaller groups, unencumbered by wagon train, wounded, artillery and plunder, proceeded as each thought best. Some headed for the deeper fords up river to arrive, they hoped, before Federal forces came near. The main group with Morgan turned back from the river, left the main road and headed for the interior, "keeping at a respectful distance from the river."[873] They moved in the direction of Marietta, Ohio. Judah's cavalry went immediately after Morgan with a brigade of General Hobson's cavalry not far behind. Lieutenant Fitch in the *Moose* and John Sebastian piloting the *Allegheny Belle*, supported by General Scammon's troops (in transport steamers? Aboard the *Allegheny Belle*?), continued on, traveling up the Ohio to frustrate any further attempts of the cavalrymen to cross the river.

Weaving together framentary 13ᵗʰ West Virginia sources

The part played by Kanawha regiments is recorded in *Field History of the 13ᵗʰ Regiment*, 13ᵗʰ regimental returns and company morning reports. These records provide the following fragmentary information. We know that the 13ᵗʰ arrived aboard transport, at the head of Buffington Island early "on the morning of the 19ᵗʰ" as it was becoming daylight.[874] They had traveled a distance of thirty-six miles from Pomeroy to Buffington Island.[875] They arrived in the midst of the engagement taking place "in the village of Portland" between Morgan's cavalrymen and Federal troops. This engagement, resulted in the defeat of the "rebels under Genl. Morgan and the capture of some 800 of his men."[876] Hayes' troops arrived in time to witness the defeat at the river and to skirmish with the enemy and prevent his reaching the Virginia side.[877] From Captain Albert F. McCown's Company F records, we learn that when the 13ᵗʰ reached Buffington, they "found Morgan endeavoring to cross," and that Co. F "disembarked" probably to the Virginia side of the river and "skirmished with him driving him back into Ohio."[878] Capt. Young with Co. G wrote to his wife that

> After a long and wearisome march on land and water, we have all returned to this place safe and sound. We were in some skirmishing with Morgan and were in sight of the battle fought at Buffington's Island but were not engaged in it. We were drawn up in line on the Virginia side to prevent his crossing, which we did, all but Morgan and about eighty of his men made their mistake in Va. We captured nearly one hundred prisoners, one inspecting General captains and several Lieutenants. About three thousand and [illegible word or words] of his men and horses are now in our hands with all his artillery, small arms, wagons, buggies etc. Old Morgan is gone up sure.
>
> [...]
>
> You must do the best you can and watch your horses. Morgan's men are scattered all over the country and must have horses.[879]

Co. B muster roll records that the company was present at Buffington Island when Morgan attempted to cross the Ohio but that they "were not called into action."[880]

Ohio sources provide additional context

Ohio sources provide additional perspective. Andrew Stiarwalt, musician 23rd Ohio, recorded that the ball got rolling at Buffington at

> 7 o clock it was so vary foggy that our progress was vary slow but we arived in time to assist Gen Hobson in capturing about 1700 men and all of Morgans artilery and in preventing the Rebs from crossing the Rive[r] we ware under charge of Gen. E.P. Scammon [...] the attack took place abou 7 A M a perfect Rout took place among Morgan's men So much so as that they scattered all over the country being perfectly discourged and fagged out with fatigue of night and day marching to keep out of the way of our persuing Troup who determined to Capture them if possible[881]

Colonel R.B. Hayes, who traveled aboard the transport with his troops, wrote in his private journal that Morgan had been found at Buffington Island at daylight. "He was here attacked by General Judah's cavalry and the gunboats. Not much fighting by Rebels, but great confusion, loss of artillery, etc., etc. On to Hockingport."[882]

Hayes seems to refer to that point in the battle when having fallen back up the valley for about two miles the Confederates in their panic made a break for it descending the ravine to the water's edge. One third of Morgan's command tried to cross here. Lieut. Col. James Comly, cmdg. the 23rd Ohio, noted in his journal that the Kanawha troops had landed at Buffington but the enemy escaped and made their way upriver before Scammon's forces "could be brought within reach" and "into action."[883] E.H., writing for the Cincinnati *Daily Commercial*, indicated that the 23rd and 13th arrived in the midst of the battle and landed at Buffington Island but not soon enough to be in the fight proper.[884] What Scammon's men witnessed at that juncture, wholly or in part, was described by Lieut. Fitch in his official report. Morgan's men were attempting to ford the river at Buffington

> just above this island, making for the river and attempting to ford. I at once engaged him, drove him from the banks, and captured two pieces of his artillery, a portion of his baggage train, horses, small arms, etc. During this time General Judah was pressing his rear.
>
> He did not engage us over an hour, when his forces broke in the utmost confusion, throwing away their arms and clothing and taking to the hills. A portion, however, moved up along the bank in

hasty retreat, but I followed them so closely that they soon broke
and disappeared up the ravines and over the hills. In this column
moving up along the bank were several buggies and carriages,
which were abandoned to us. [...] The road along the bank was
literally strewn with his plunder, such as cloth, boots, shoes, small
arms, and the like, but I had not time to land and take possession
of these things, as I wished to keep up the river with the remnant
of his scattered band, knowing that General Judah would look out
for those left in the rear.[885]

The sight which opened into view at the river's edge was doubtless exciting and
satisfying but also strange and wonderful. "A Cincinnati gentleman" wrote in 1865
that Hayes and his men

reached the bend just in time to see the head of Morgan's column
attempting to cross, attacked it, and captured a large part of the
force with their enormous pillage; and such a spectacle was never
exhibited before. The stores had been gutted for three hundred
miles, and vehicles on the road captured to hold and convey the
plunder. Besides, the horses were loaded with calico, blankets, mus-
lims, shoes, ribbons, glassware, crockery, hats, caps, stockings and
every article that could be attached to the saddle, bridle and other
harness; and this for an army of four thousand horse in superb sav-
age raggedness, yet radiant with their bountiful spoils.[886]

The sheer amount of plunder taken in Ohio and Indiana exceeded anything
that even Col. Basil Duke, Morgan's second in command confessed to having seen
before, "so great was the men's disposition to punish the North."[887] They had stolen
on an unprecedented scale and it must have been a sight for the Kanawha troops to
see. Correspondent E.H. elaborated with detailed description of the panorama to the
Cincinnati *Commercial*:

The immense amount of plunder which had been gathered in Ohio
and Indiana, was also retaken in the wagons which had been stolen
to transport it. There were thousands of dollars worth of groceries,
provisions, and dry goods of every variety. There were silks, and
satins, cotton and linens, vestings, broadcloth—in short, an abundance of everything to eat, drink or wear [...]. The most valuable of
the recaptured plunder, however, was the horses, over a thousand
of which were taken, together with a train of buggies, carts, market
wagons and vehicles of every description. [...]

A great portion of the plunder of the battlefield was picked up by the people around, and many on the boats have taken away swords, carbines, etc., as trophies and momentoes. As soon as the battle ceased, hundreds of claimants of horses made their appearance. No doubt some were entitled to them—others were not. However, the great abundance of these animals caused their disappearance to be a subject of but little remark. So with other articles of plunder dropped by the guerrillas in their headlong flight.[888]

Fouteen miles Hocking River to Hocking Port, Lee's Creek, Belleville and Hocking

The Kanawha troops had not arrived in time "to render important assistance"[889] at the battle but there was important work still to be done to intercept that part of Morgan's troops which had escaped and were presumed to be headed farther up river to attempt crossings at fords higher up. These fording places were deeper and within relatively short distances of one another (between fifteen to twenty miles apart). Regimental returns indicate that immediately after the battle at Buffington, the 13th "pursued Rebels" and was ordered and "then moved up the river in pursuit of the flying rebels as far as Big Hocking River"[890] or Hocking River as it is known today.[891] Pursuant to these instructions, the 13th was conveyed by transport up to Hocking River fourteen miles.[892] Hayes' regiments seem then to have proceeded to Hocking Port, either by transport and/or on foot to guard "the fords over the Ohio at Lee's Creek, Belleville, and Hocking."[893]

After the battle had ended and Morgan's men turned back from the river, Fitch immediately continued upriver without securing his captures of guns, men and materials, trusting to those in the rear to attend to captures. At Gen. Judah's suggestion, Scammon and his men were sent up river "under convoy of a gunboat, to Blennerhassett's Island."[894] Kanawha troops would scout for prisoners and guard prisoners taken in the countryside. There was also the problem of communications to attend to. All had gone down so quickly. Lines of communication had been interrupted by the raiders so that communication was difficult. A communications station was set up at the foot of Blennerhassett's Island.

There were also problem with the boats themselves. The *Moose*, heavier than the transports and *Allegheny Belle* due to its armature, could only with difficulty negotiate the river through the mountainous country all the way up to Blennerhassett's Island. Scammon's men, however, aboard the lighter *Allegheny Belle* were able to sail all the way to Blennerhassett's. A note from Col. William Wallace commanding the 15th Ohio Volunteer Infantry, reporting to Gen. Burnside from the ford at the "Foot of Blennerhassett's Island," verifies that Scammon and troops stopped first at Blennerhassett before returning to Hocking River pursuant to their original orders. Wallace wrote on the 19th, that he had his "troops and artillery on Virginia side in good condition." He continued:

> Was detained on the bar [sandbar presumably] several hours. General Scammon came up from below with transports and troops. Ordered me to remain here, as his force was sufficient for the enemy, below, and departed down again with his troops for Hocking.[895]

This stop at Blennerhasset's Island is not mentioned in 23rd Ohio or 13th Virginia records. It seems the men were not permitted to disembark but simply were turned around and conveyed back downstream to Hockingport. There, they were set ashore, where they guarded ferries and probably prisoners. Col. Hayes wrote in his report to 3rd Division headquarters that from Buffington his command went up the Ohio to Hockingport,[896] and in his private journal, Hayes wrote that he and his men left Buffington and moved "[o]n to Hockingport; [where they] guarded the ferries over the Ohio at Lee's Creek, Belleville, [both in West Virginia] and Hocking [in the State of Ohio]."[897] Comly noted that after embarking on their boats again at Buffington Island, they "moved up to Hockingport, 15 m[iles] from Parkersburgh, where we again landed." Co. A (13th W. Va.) morning reports record that at Buffington, they "pursued the Rebels," and then "left Buffington south on up to Hawkings point [Hockingsport]."[898]

Newpaper accounts in the form of dispatches, letters, regimental histories and diaries also provide particulars and mention Hayes' men and the scenes at the river's edge as Morgan's men attempted to escape capture. To enjoy the history preserved in these bits however, one must tolerate some vagary. A linear construction of time and specific place are beyond my abilities. Confusing is that it is not altogether clear if chroniclers actually refer to Buffington Island or Craig's Bar opposite Belleville Island farther up river or some other fording place. Both Buffington and Belleville Island were locations where the fleeing raiders made significant attempts to cross the river into Virginia and where Union forces engaged and effectively countered. Ohio and West Virginia shorelines on this part of the river had a sameness to them and by this time and at this juncture all involved were weary and it is difficult to know what spot is specifically being referenced. From above Portland below to Blennerhassett's Island above, it was a landscape comprised of small towns and villages, small islands, shoals, tributaries and fording places in close proximity. This stretch lies on either side of Hocking River, a major tributary of the Ohio and consequently geographic features may have been simply lumped together under the town name 'Hockingport' or 'Hocking River' for convenience sake to indicate the larger area of operations in which men engaged in a kind of running fight to capture and deliver up prisoners. Bearing such qualifications in mind, consider the following.

One of the things Hayes and his command accomplished was putting a stop to the ferrying over of the raiders to the Virginia side of the river. This was an important stroke. It has been often said that about half of West Virginia was pro-South. The defeated raiders reaching the Virginia side could easily elude capture in the rough terrain of that countryside and move south to safety aided and secreted by sympathic pro-South private citizens. A correspondent to the Cincinnati *Commercial* signing himself simply "H.," reported that during the fight at Buffington a scow passed repeatedly "to and fro,"

> carrying horses and men for Morgan safely to the Va shore. The gunboats were there, as was apparent from their smoke-stacks just visible below the island, but they seemed to ignore the existence of the scow. [...] The scow continued its operations until it was captured by some forces on the Va shore under Gen. Scammon.[899]

Capt. Young, Co. G 13th W.Va., writing home from Point Pleasant on July 24 noted that he had seen a group of raiders "on the va. side but whether Gen. Morgan was with them or not I can't tell but one thing I do know they are bringing his [Morgan's] men in here every day. Two have just come."[900] "History of the 23rd Ohio Regiment" preserved in James Comly's papers at the Ohio Historical Center, Columbus, contains the statement that Hayes' troops reached Buffington Island "at daylight" on July 19. There they "found that Morgan with about 300 of his men had succeeded in crossing the river, when he was attacked by Gen. Judah's cavalry, which at that point had caught up with him, and also by the gunboats. Gen. Morgan returned to his command on the Ohio side of the river, and they then scattered through the country."[901] An Ohio account (orig. 1876) referenced by Harrison G. Otis in his newspaper, the *Santa Barbara Daily Press* and corrected by him, also seems to reference this incident. Otis indicated that Morgan was perceived ferrying men over. He had "seized a steamboat and,"

> had ferried over about three hundred of his men, when Colonel Hayes arrived, seized the boats, and put a stop to any further progress in that line. Morgan himself had crossed the river, but seeing that his main body was about to be cut off, he recrossed, and remained with his soldiers to share their fortunes. After, some fighting he drew off again, and made for other points up the river. But the last opportunity for escape had passed, and the Confederate raiders, hardly beset by Gens. Hobson and Shackleford, were speedily driven to the wall and forced to surrender.

The surrender of Morgan and his main body took place many miles farther up the river, in Columbiana County, after some hundreds of his men had successfully crossed into Virginia and made good their escape.[902]

E.H., our Cincinnati correspondent, also served up valuable details about Belleville although he seems to misidentify Hayes' regiments. The 12th Ohio and 17th Virginia, wrote E.H., with Scammon followed behind the *Moose* up and back as far as Hocking Port on the Ohio side fourteen miles above the Island, where they were disembarked to co-operate in the pursuit. E.H. reported that from Buffington, the raiders

fled in every direction, our cavalry pursuing them along the river, east and west and over the hills north, while the gunboat [the *Moose*] continued [to fire on] those who attempted to swim the river. Working her way up, fast as the shoal water would permit, accompanied by our pursuing cavalry, along the bank, [the *Moose* fired upon about 200] rebels attempt[ing] to swim their horses across at Craig's bar opposite Bel[le]ville, 10 miles above Buffington.[903] About 30 were killed or drowned here, [...] the majority were driven back into the hands of our cavalry.[904] The Moose then continued up the river, 6 miles higher, when, on account of shoals, she was obliged to turn. She had been followed thus far by the transport, bearing the 12th Ohio and 17th Virginia Regiments of Infantry, [the 23rd Ohio and 13th West Virginia?] which, under General Scammon, had come down the Kanawha, from Virginia, but had not been landed in time for the fight. These followed the Moose up, and back as far as Hocking Port, which is on the Ohio side, fourteen miles above the Island, where they were disembarked, to co-operate in the pursuit. As the Moose was coming down, her officers were sitting on the hurricane deck, watching the Ohio side, unsuspicious of danger, when, at the mouth of Lee's Creek,[905] a sudden volley of musketry was poured into the boat by about 50 rebels,[906] who had crossed below, hastened up, and ambuscaded themselves in the bushes at the mouth of the creek. In answer to this, a few rounds of shrapnel were thrown at the bushes, by which eight were killed and several wounded. The boat then returned to the battle-field, where squads of prisoners were being brought in.[907]

A reporter to the *Cincinnati Daily Commercial* writing from Columbus, Ohio, traced out additional details as to what transpired after Buffington Island. Morgan left the main road and continued roughly in a parallel direction with the river, staying behind the hills adjacent to the channel and away from the gunboats, began correspondent. He only hazarded to approach the river when he had reached Belleville. There, opposite Belleville, at Craig's Bar, distant about "ten miles above Buffington," was another ford. Supposing he had outdistanced the gunboats, Morgan approached the citizens in the vicinity and demanded their help in crossing the river. In response to the raiders' threats, the citizens "began to bring forth old flat boats and scows, seeming to be in tumultuous hurry, but in reality causing all the delay they possibly could."[908] At length, Morgan's advance, about three hundred men at Craig's Bar, filled the boats and another two hundred were near the riverbank, attempting to swim their horses over from this point. The boats carrying Morgan's advance were

> just passing out from shore, when a gunboat appeared coming round the bend a short distance below. It came up in few moments but not in time to prevent the boats from reaching the opposite shore. The gunboat destroyed the flats, which the escaped rebels had set adrift, and then ran across to the Ohio side, keeping up a brisk cannonade, shelling the woods on the Virginia side. Another gunboat arrived [the *Allegheny Belle*?] just then and opened on the rebels remaining on the Ohio side. Morgan fell back out of reach of the cannon, and drew up his forces, seeming to wait for a land attack from the crew of the gunboats. Hobson and Judah were a mile or two in the rear moving up swiftly.[909]

Shells from the gunboats had effectively terminated the crossing at Belleville although Col. Adam R. Johnson succeeded in reaching the Virginia side with over three hundred of his command (2nd Brigade) by swimming the channel before the boats had maneuvered within range. Those yet midstream, who had not been killed or wounded when the boats had opened fire, were forced to return to the Ohio side. Among those midstream was General Morgan himself, who was "standing with his staff about one hundred yards out in the Ohio River, when a part of his command swam that river at Belleville "just twelve miles above where [they] had a fight at Buffington Bar, Ohio."[910] When fire from the gunboats grew too hot to withstand, Morgan returned to shore with those men stopped in the escape midstream. This group fell back to join those on the Ohio side.

Skirmishing with Federal forces subsided and the Confederates again took flight. The cavalrymen fell back setting upon a zigzag course northward into the Ohio interior. Some of these seemed to be trying to escape over eastern Ohio into West Virginia or Pennsylvania and others doubled back continuing west. From this point on, "portions of the command were captured at different places."[911] The main body of men with Morgan were followed up by General James M. Shackelford commanding about "500 men of Hobson's command" and other Federal cavalry—"some freshly remounted in Cincinnati [who] were placed on cars and unloaded, ultimately check[ed] Morgan at Salineville."[912]

Captain Sidney P. Cunningham, 2nd Brigade Morgan's Division, stated in his report that it was one "ironclad and two transports" (likely Scammon's command is referred to in the transports following in rear of the *Moose*) that had moved down upon them as they attempted to swim the river at the Belleville ford. Cunningham wrote that

> [a]fter leaving the road at Portland, the command was marched to Belleville, some fourteen miles, and commenced fording, or rather swimming, at that point. Three hundred and thirty men had effected a crossing, when again the enemy's gunboats were upon us one ironclad and two transports. Again we moved up the river.[913]

As daylight faded, those who had effected the crossing with Cunningham left the river. They impressed guides and collected together the other "360 men who had made the crossing" and marched out towards Claysville, West Virginia.[914]

Allan Keller also referenced the landing of troops at Belleville in his book published in 1961. He wrote (without citing sources) that

> a small party of Rebels rode hurriedly upriver to the vicinity of Hockingport. The residents had listened to the sound of cannonading during the battle at Buffington Island, and then noted, with growing fear, that the firing was getting closer as the *Moose* went into action at Belleville Ford. [Infantry, Scammon's men presumably were put] ashore at Indian Run,[915] and they headed into the woods. Then came the rattle of musketry [...].[916]

Correspondence to the Cincinnati *Gazette*, reprinted in the Baltimore *American,* reported that from Buffington Island Hayes' troops were sent in pursuit of a remnant of Morgan's command, which had escaped to the Big Hocking River. The transports containing Hayes and the 23rd Ohio and 13th West Virginia "were then run up to a point between H[oc]kingsport [and illegible place name] and landed on the Ohio shore, to intercept the Rebel[s]."[917] The evening of July 20, it was reported at Cincinnati that Scammon's men "had captured the force [yet at large] or compelled it to surrender."[918]

A later Jackson County history, without any reference to sources, suggests that the 13th Virginia was in no position to have been involved in blocking Morgan's attempt at crossing the Ohio at Belleville and that in fact, it was a mismanagement that they failed to be in position to prevent the crossing before the enemy could be driven back to Ohio by the gunboats. The writer explained adding that Gen. Scammon's troops, "which had been guarding the West Virginia shore" were "brought across to help guard the prisoners on the Ohio side, leaving the way clear for Morgan [...] Scammon's men , who might have barred the crossing [at Belleville], were on the wrong side of the Ohio, and more than a dozen miles downstream."[919]

Lieutenant Fitch's own report relates that he traveled upriver about 14 or 15 miles when he "again fell in with another portion of [Morgan's] forces fording." "The current," continued Fitch, "was so very swift and the channel so narrow that it was sometime before [he] could get within range of them." Gradually, he maneuvered closer to the cavalrymen. When he could get within striking distance, he opened fire on them throwing shells in their midst. Some of the fugitives had already made it across. "[M]ost" were driven back and again fled upriver. Several of those midstream were killed, "25 or 30" men were wounded and drowned and "15 or 20 more horses" standing on both banks were captured although Fitch himself did not have time to stop for them.[920] After Belleville, stated Fitch, he continued to push upriver, where he sighted another portion (25 or 30 men) of Morgan's command crossing opposite Lee's Creek. Due to very shallow and swift running water, however, he was unable to maneuver in close enough to get them within range and to drive them back and the Confederates were entirely successful in crossing. Having reached as high as it was safe for him to travel "at this stage of water, and the river still falling," he dropped down below Buffington Island.[921] By ten o'clock a.m., the Moose had "returned to the battle-field, where squads of prisoners were being brought in."[922]

Hocking River

Scammon's Kanawha troops were put ashore to patrol between Hocking River and Belleville, where due to the mountainous nature of the countryside, the *Moose* could not operate effectively. The terrain was rugged, the riverbed was rough, the water level shallow and falling in consequence of the subsidence of the recent surge. Hayes' men were disembarked at Hockingport with orders to guard fords over the Ohio at Hocking River, Ohio, Lee's Creek and Belleville, West Virginia.[923] Between Hockingport and Belleville was just under five miles of country. Col. Carr B. White's command (2nd Brigade, Kanawha Division) was concentrated in vicinity of Belleville. 1st Brigade, Kanawha Division (Hayes) seems to have ranged on the Virginia shore within a five mile stretch nearby, preventing crossings from that shore.[924] The presence of the Kanawha Division in proximity to the ford crossings at Belleville and that 5 mile stretch of river was of critical importance. This stretch of river was beyond the reach of the tin-clads, which required deeper water to maneuver, and Hobson's and

Judah's cavalry were occupied elsewhere. In this five mile or so, the Kanawha troops were all that stood between the Confederates and escape into West Virginia.

One story involving the 13ᵗʰ Regiment "at Belleville" survives into the present by virtue of Herbert Roush's book, *If Thou Wilt Remember* (based on the letters of 13ᵗʰ West Virginia soldier David Burrows and Roush family lore). Roush related that after being taken by transport to Belleville, the 13ᵗʰ was fired upon by the *Allegheny Belle*. The *Allegheny Belle* was patrolling "up and down the river shooting everything that moved," when she "fired on the 13ᵗʰ, thinking they were Confederates. After this incident, the 13ᵗʰ established a prisoner compound under the river bank."[925]

Wherever prisoners were detained, some delay and hardship was inevitable. There were insufficient boats to transport the prisoners and there were no rations for men or prisoners at Buffington. Boats and rations had to be requested of Gallipolis and Portsmouth and then forwarded to Buffington. Likewise, there was probably no medical care but what a regimental medical officer might provide from what he had carried with him.

Both Scammon and Hayes reported that their commands performed effectively in preventing escapes into the West Virginia interior. 13ᵗʰ regimental records, however, reveal little of the mop up. They are abbreviated to the point of being telegraphic. Co. A muster rolls for July and August indicate that after Pomeroy, Co. A boarded a transport traveling "to Hockingport, Ohio." There, we are told, they "engaged in the capture of Morgan's forces and returned to Coalsmouth Aug. 8, 1863. Travelled over 200 miles."[926] "Remarks" entered with Co. H morning reports recorded that after leaving Pomeroy, the company "moved up and down the [river] until July 20ᵗʰ, when [they] moved down the river."[927] "Remarks" with July morning reports for Co. D provide a bit more information. These indicate that the 13ᵗʰ Virginia was also detailed in detachments by company. Co. D arrived at Buffington Island on July 19 "at which place Morgan made a stand and got defeated" and then, wrote the reporting officer:

> We went on to Hockingport and then this Co. changed boats and went to Parkersburg and from there we came back and started for Cincinnati with a boatload of rebel prisoners.[928] We arrived at Cincinnati July 23ʳᵈ. Delivered our freight and started [for] Pt. Pleasant, W.Va. Arrived at Point Pleasant and awaited orders July 26, 1863.[929]

Orders were issued in the field and save one, no written orders for detachments survive for other companies One order only, issued by Scammon to Hayes from Hockingport on the 19ᵗʰ, survives among a file called Miscellaneous West Virginia Papers. Scammon ordered Hayes to "guard the fords between this place and Belleville — If Steamer Baltimore and Victor No '2' arrive during the night you will detain them with your command unless otherwise ordered."[930]

One fares little better with contemporary unofficial sources.[931] *13ᵗʰ West Virginia Field History* ("13ᵗʰ Regiment Memoranda") which was never intended to be sent to the War Department at Washington but was compiled and sent to the State government of West Virginia also offers little but the bare facts of the regiment's involvement in ending Morgan's raid. We read here, that the 13ᵗʰ pursued Morgan and the remnant of his command to the Big Hocking River and then returned to Point Pleasant on July 21.[932] Another unofficial source states that at Hockingport, the 13ᵗʰ "skirmished."[933] Cincinnati newspapers also reported that on Monday, July 20, a portion of Morgan's command skirmished near Hockingport and at Coal Hill near Cheshire, Ohio, before turning northward away from the river. Twentieth century West Virginia historian, Boyd B. Stutler, describing the crossing at Belleville (without referencing his sources for so believing), wrote that so relentless was the Federal pursuit prosecuted that by the time "the 13ᵗʰ West Virginia Infantry landing from the steamboats and other troops moving down from Parkersburg, the raiders had no time to rest and refit themselves."[934] Captain J.V. Young, Co. G 13ᵗʰ West Virginia, wrote from Point Pleasant to his wife that he was waiting at Point Pleasant with all but two companies of the 13ᵗʰ West Virginia—the "two companies" having been "left up the river to guard the prisoners." He speculated that once the companies were collected and together again "I think the Regiment will be ordered to Coal or Winfield."[935]

Hayes wrote in his official report that "[d]uring the night of the 19ᵗʰ the two Regiments [the 23ʳᵈ Ohio and 13ᵗʰ Virginia] were deployed on the Virginia shore for a distance of 5 miles to prevent the enemy from crossing. It is believed that no Rebels succeeded in crossing during the night although attempts were made to do so."[936] Once prisoners had been captured, a secured holding area had to be established with an appropriate number of guards. Guards were also necessary to secure transport steamers while loading and when prisoners had to be moved they required an escort. That the 23ʳᵈ and 13ᵗʰ made a successful job of guarding this stretch of river day and night is attested to by the fact that they collected over 200 prisoners (see list of prisoners captured *infra*). Col. Hayes, in his modest way, assessed the contribution of his regiments in his official report to division headquarters as follows:

> My command it is believed prevented the Rebels from crossing
> the Ohio at Pomeroy and gave important aid to those engaged in
> intercepting them at points higher up on the River.[937]

Back Down to the Kanawha

By Monday, July 20, the crisis was winding down. The day, however, was otherwise an eventful one of departures and captures. Gen. Judah left Buffington the morning of the 20th. General Hobson remained behind to command all forces at Buffington. With the exception of a small force employed at Buffington to clear out captured property, Hobson had his troops out scouting in every direction. Five hundred more Confederates were reportedly captured on the 20th.[938] Also on the 20th, "seven hundred of the prisoners captured on Sunday were started down the river" for Cincinnati, including Colonels Richard C. Morgan (John's brother), Smith, Duke, Ward and Captain Thomas H. Hines.[939] Colonel Duke recalled in his post-war account that the prisoners "were marched down the river bank some ten miles to the transport" and crowded into the boat. The steamer left for Cincinnati as soon as they were all aboard.[940] They were "some three days in making the trip." They traveled down river "as rapidly as the low stage of water, and the speed of the little boat, [...] would permit."[941]

Inasmuch as Co. D 13th West Virginia, was among those escorting prisoners to Cincinnati, it is not amiss to include Colonel Duke's complementary words regarding the kind treatment accorded him and the rest of the men by their guards during the trip. The guards were inclined "to ameliorate our condition as much as possible," wrote Duke. They were extended "that courtesy, which characterized the men of both armies, who had served at the front, during the first two years of the war."

> The soldiers who guarded us from Buffington to Cincinnati were characterized by this spirit in an unusual degree, and carried out this practice, [...] I must say, more than any troops I had even seen. We met with treatment so different, afterward, that we had occasion to remember and compare.[942]

Tuesday morning, July 21, the *Moose* and *Imperial* left for Cincinnati. The blockade established by Lieut. Fitch around Pomeroy, employing tinclads of the 6th Mississippi Squadron, was maintained until news was received that Morgan and the last of his men (ca. 300 in all) had surrendered near Salineville, Ohio, on July 26. On July 21, it was reported that the remainder of Morgan's forces, now only in small detachments, were "scattered considerably over the country for 15 miles or more back of Buffington and Pomeroy"[943] and that they were in no condition to inflict further damage to the national interest or her people. Stragglers were being picked up "hourly"[944] by Federal regulars and militia. The Kanawha troops returned by steamboat "to their old camps on that river."[945]

July 20, the 13th W. Va. Regiment received orders to return to Point Pleasant.[946] The regiment was set in motion that day for the return trip. From their location on the West Virginia side (probably about half way between Lee's Creek and Hocking River), they had to march to Belleville; a distance of "three miles."[947] At Belleville they boarded transports and were then conveyed to Hockingport, where "after laying at

Hockingport all day the 20th"[948] they moved down river and arrived at Point Pleasant on the 21st.

Bits of information in 13th West Virginia regimental records give some indication of the return trip begun on the 20th. Co. A reports preserve that the company was "ordered back to Bellvill, Va. Sleep 1 night and 1 day. From Bellvill were ordered back to the U.S. Steamer. William Gray left at Bellvil."[949] Co. F reports record that on the 20th, the company "[s]tarted down the river on our return, on board steamer Emma Graham."[950]

From Belleville the regiment traveled to Buffington a distance of "ten miles."[951] It was probably at Buffington that the Kanawha soldiers picked up the "160 prisoners out of a total of 2,373 captured."[952] These 160 they escorted down the Ohio to military authorities at Gallipolis. From Buffington, the 13th "[r]eturned by transport to Gallipolis + Point Pleasant 60 miles."[953] The boats patrolled as they proceeded back down the Ohio to their debarkation points. Scammon's troops kept a watchful eye on the adjacent countryside for signs of Morgan's men. Andrew Stiarwalt wrote that "on 20th to the 22nd we patrolled the Ohio River to the mouth of Kanawha River on the 22d July/63 we arrived at Camp White," at Charleston, West Virginia.[954]

Having delivered the prisoners in their custody to military authorities at Gallipolis for safekeeping until they could be transferred north,[955] Scammon's men headed up the Kanawha. From Gallipolis, the 13th West Virginia was ordered to Point Pleasant.[956] The 13th was "left at Point Pleasant;" the 23rd Ohio was left at Camp White, Charleston,[957] arriving there "at 10 A.M.;"[958] and the Fifth Virginia Infantry was "sent to Gauley Bridge."[959]

One story told about 13th soldiers guarding Confederate prisoners down river on July 20 was related by George Wilding, staff member of the Mason City (or New Haven) Home Guard Company. Wilding's story was quoted by Midred Gibbs in her book about the town of Mason City. Wilding stated that some prisoners conveyed down the Ohio River on the 20th were guarded by a section of the 13th. The boat containing prisoners and 13th soldiers halted at the town of Hartford but didn't land. The boat stopped long enough for it to become known that "[m]en and prisoners were both half starved. Word was sent around by 'grapevine telegraph' and soon after, 'all the cooked and baked food in town was hurried to the boat.' "[960] Herbert Roush refers to this incident in his book as well. Roush stated that while the *Moose* picked up a load of prisoners and took them to Hockingport, other prisoners were loaded on a boat heading down-river. These latter prisoners were guarded by a part of the 13th Virginia.

The boat bearing prisoners halted at Hartford City. The guards were

> confined to the boat, for fear some would try to get off for a visit.
> The word swept through town that soldiers and prisoners were half

starved. Hartford City responded by hurrying all the cooked and baked food in town to the boat.[961]

With the exception of Co. D (still making for Point Pleasant from Cincinnati) all other 13th companies had been disembarked at Point Pleasant by the evening of the 21st. They remained at this post until August 1 when the regiment was ordered to Charleston.[962] Co. F records indicate that on July 21 the company "[a]rrived at Point Pleasant, W.Va. at 12 o'clock m. and went into camp;"[963] that on July 22nd "1st Sergeant A[llen] C. Mason captured two of Morgan's men at Ten Mile[964] (Mason County); and that on the 22nd Mason "captured 2 more together with their horses and arms."[965] Co. D, which had been detached at Hockingport on the 19th to proceed to Parkersburg to escort a boatload of "rebel prisoners" from there to Cincinnati,"[966] returned to Point Pleasant and arrived there on July 26th. July 26 through 31, the 13th Regiment rested at Point Pleasant and awaited orders.[967] At Point Pleasant the 13th received orders to return to their former station at Coals Mouth.

Prisoners Taken

The question as to what captures were made specifically by the 13th during the mop up of the Morgan raid will probably never be answered.[968] It was a running fight and inconsistencies in accounting for prisoners and booty cover the field of extant materials. Correspondent E.H. reported that once the battle at Buffington was over the national forces "were immediately busied in securing and bringing into camp the many prisoners who had been taken," while "[a] couple of the best regiments of cavalry, with 2 or 3 pieces of artillery, were dispatched north after Morgan, who, at the head of 1,500 of his men, whom he had rallied together, had succeeded in making a disorderly retreat." A total of 1,700 prisoners had been mustered into camp up to 6 o'clock, on the evening of the 19th. The night of July 19, the prisoners that were collected at Buffington encamped on the field with a portion of the Federal forces charged with the task of guarding them.[969]

Co. F (13th Regiment) reported that "on the 19th we skirmished with Morgan's forces again, at Buffington, Ohio, where we took several prisoners." Regimental Returns for the 13th (submitted by the ever breviloquent Col. Brown) noted that the engagement at Buffington ford resulted "in the capture of some 800 of his [Morgan's] men" but Brown claimed no credit for any part of that number.[970] Col. Hayes, commanding 1st Brigade, wrote in his official report that "the total number of prisoners reported to these Hd. Quarters captured by the troops under my command was 208. [...] No full report of captured horses has yet been made — the number will exceed 50. A quantity of arms and cavalry equipments were also captured." Scammon, who wrote up his report with full figures of the enemy's losses, stated in his report to departmental headquarters that "First Brigade, under Colonel Hayes, reported 208 prisoners, including 6 commissioned officers, and upward of 50 horses captured; also a quantity of arms and cavalry equipments."[971] Capt. Young, whose company was still a part of the 13th West Virginia Regiment, wrote home that '[w]e captured nearly one hundred prisoners, one inspecting General captains and several Lieutenants. About three thousand and [illegible word or words] of his men and horses are now in our hands with all his artillery, small arms, wagons, buggies etc."[972]

A list of captures made by Hayes' regiments on July 19 at Buffington Island and up river "near the mouth of Big Hocking" was written in what appears to be John S. Cunningham's (Lieutenant and Adjutant of the 13th West Virginia Regiment) own hand. It is reproduced here. It is a list of captures made by the 23rd and 13th on the 19th which amounted in all to six officers and fifty-seven men. "Numerous" additional captures were made by Hayes' regiments as Comly tells us in his official report. These captures were made "up to the time when we entered the Kanawha, on our return." These were "properly accounted for elsewhere."[973]

[List compiled by John S. Cunningham, Lieutenant and Adjutant, 13th West Virginia Infantry], List of Prisoners of War Captured by the 23rd Ohio + 13th Va at Buffington + near the mouth of Big Hocking on Sunday July 19th 1863

F.M. Hare Capt + A.C.S 5th Arkansas Infty

D.K. Morton Brvt 2nd Lt Co E 8th Ky Cavalry [from] Missouri

E.W. McClean Capt Co B 2nd Ky Cavalry [from] Logan Co Ky

Wm. Hayes 1st Lt Co B 2nd Ky Cavalry [from] Covington Ky

John B. Cole, 2nd Lt. + Inspector Genl 1st Brigade Genl M[o]rgans Division [from] Lexington Ky

Wm. W. Page 1st Lt Co. C. Rich[ar]d Morgans Regt (Ky Cavalry) [from] Allen Co Ky

[Enlisted men:]

1   Wm. W. Thomas Co D 9th Ky Cav

2   Frank H. Grant Co D Wards Regt Ten Cav

3   Hugh Leggett Bymes Battery

4   Jos. R. Wheatly [Whealty or Wheatty]

5   Willis Colyer Bymes Battery

6   L. F. Mulligan Bymes Battery

7 John R. Walker Bymes Battery

8 William Warner Bymes Battery

9 Frank Stone Bymes Battery

10 Wm. Williams Bymes Battery

11 Dan Mathews Bymes Battery

12 Iolin [? John] Myone [?] Bymes Battery 3 miltrs. [?]

13 Wm. Pennington Co. A 5th Ky Cavalry

14 A. C. Smith Co. A 5th Ky Cavalry

15 Thos. Blincoe Co D 2 Ky Cavalry

16 Wm. P Still Co I Wards Regt Tenn C[av]

17 Johnson West orderly sergt Co B 2nd Ky Cav

18 Otway B. Norvell Co B 2nd Ky Cav[974]

19 A. Jordan Co I Wards Regt Ten. C[av.]

20 D. C. King Co A Wards Regt Ten. C[av.]
    Robert Andrews Q M Sergt Wards Regt

21 Jacob Swango Co D 5th Ky Cav

22 Wm. McQuny Co I 3rd Ky Cav

23 Elkanah Quesembury Co E 8th Ky Cav

24 Sidney Maupin Co E 5th Ky Cav

25 John Minny [Muny?] Co K 2nd Ky Cav

26 Rufus Tully 1st corporal Co B 2nd Ky Cav

27 R. J. Jeffries Co C 2nd Ky Cav

28 O. B. McCulley Co I 10th Ky Cav

29 G. Janbert Co B 2nd Ky Cav

30 Robert H. Chism Co C 5 Ky Cav

31 A. C. Piersall Co C 5 Ky Cav Chenaults[975]

32 J. W. James Co D 5 Ky Cav Smiths

33 T. C. Master Co D 4 [?] Ky Cav Clukes [?]

34 Benjamin Marshal Co A 8th Ky Cav

35 R. E. Crocker Co B, 2 Ky Cav

36 J. M. Gibbs Co B 2 Ky Cav

37 W. M. Bagwell Co B 2 Ky Cav

38 B. W. Porter Co B 2 Ky Cav

39 Jas. H. Baxter Co I 3 Ky Cav

40 J. W. Hite Co B 2 Ky Cav

41 Henry Traughber Co B 2 Ky Cav

42 S. W. Burnett Co B 2 Ky Cav

43 Ch[arle]s W. Haddox 3rd Sergt Co B 2 Ky Cav[976]

44 Wm. Sugg Co B 2 Ky Cav

45 A. J. Therman corporal Co. I 3rd Ky Cav

46 George Adams Co B 2nd Ky Cav

47 Frank Adams Co B 2nd Ky Cav

48 Alonzo Camp Grigsbys Regt 14 Ky

49 Wm. Hant [Hunt?] Co G 6th Ky Cav

50 Milton Parrish Co H 5th Ky Cav

51 T. J. Black Co F 3rd Ky Regt

52 J.M. Eddings Co B 2nd Ky Regt

53 J. K Watson 2nd Corporal Co B 2nd Ky Regt

54 Wm. D. Rose Co B 8 Ky Cav

55 William A. Rose Co B 8 Ky Cav

A. J. C. Robbins Orderly Sergt Co A 6th Ky C[av]

[Co ?] A [13th?] 1 Sergt 3 men

[Co ?] B [13th?] 4 men

[Co ?] C [13th?] 6 men

[Co ?] E [13th?] 6 men

[Co ?] F [13th?] 1 Sergt 5 men

[Co ?] G [13th?] 6 men

[Total] 2 [officers] 30 [men][977]

The cavalrymen listed above were almost all from Col. Basil W. Duke's brigade (1st Brigade, Morgan's Division) which was at the time comprised of the 2nd, 5th, 6th and 9th Kentucky Cavalry and Captain Edward P. Byrne's Battery of horse artillery.[978] So many men from Duke's brigade were lost or captured on this raid that the regiments to which they belonged were never again reorganized. It will be noted that not all prisoners were listed above with their military affiliation. Of those identified by organization, we find the following tally. At least one prisoner, Capt. F.M. Hare, belonged to the 5th Arkansas Infantry ("the Fighting Fifth"); at least twenty-two were from the 2nd Kentucky Cavalry (mainly from Co. B); four prisoners belonged to the 3rd Kentucky; one was from the 4th Kentucky; seven belonged to the 5th Kentucky; two belonged to the 6th Kentucky; five were from the 8th Kentucky; six to the 9th Kentucky; one to the 10th Kentucky (formerly Col. Adam Johnson's command—at Buffington Island, he commanded the 2nd Brigade, Morgan's Division); one to Col. J. Warren Grigsby's regiment; and nine were captured from Capt. Edward P. Byrne's[979] Battery, a four piece horse battery which included two parrot guns.

Miscellaneous information from regimental reports and official and unofficial correspondence of July 1863

Company A morning reports for the month of July indicate that the company tallied three commissioned officers and eighty-three enlisted men present and absent. The company gained one new recruit reported enlisted at Coal's Mouth station. He was present with the company from July 1 to 5, but thereafter he disappears from the record. On July 1, Samuel Midkiff returned to the company. He had been taken prisoner at Charleston in September 1862, "by the Rebels." Also on the 1st, Andrew J. Davis "being unfit for Corporal" was returned to the ranks. July 2, James A. Means was promoted to Corporal in Davis' place. The following men returned to the company from sick leave and reported for duty: July 2, Woodson B. Hall and Obadiah Bukner (Buckner); July 4, Robert H. Snodgrass; July 8, Andrew J. Cobb; July 12, Joseph Scott and John Teel and July 22, John Moore reported for duty. July 1, Perry Gatewood and Irdell Huerl (Hurel) were detailed for daily duty at "Regimental Hospital." July 3, Ralph Pauly was detailed for extra duty at "Regimental Hospital at Coals Mouth."

Detached service was detailed as follows: July 1 through 8, two enlisted men and July 10, three enlisted men, among these was William M. Glover detached as waggonier on July 10 to July 24, when he returned to the company. July 11, ten enlisted men were detached, among these were 1st Lieutenant Greenbury Slack and seven men who were ordered to scout on Mud River. Between July 12 and July 30, two to five enlisted men under command of a commissioned officer were detached, among these were Sergeant Miletus Grinstead and Thomas Pruitt detached on July 13 to guard ten prisoners to Charleston. Sergt. Grinstead returned to the company from Charleston on July 14. July 30, Sergeant Robert H. Davis, Corporal Andrew J. Davis, George M. Davis and John Hunter were detailed to "guard prisoner up to Charleston, Va." July 31, eight enlisted men and one commissioned officer were detailed for detached service.

Co. A had from one to five enlisted men present sick for the month and one non-commissioned officer was reported present sick from July 13 to 31. There were from seven to twelve enlisted men absent sick from July 1 to 25. The last week of the month just six enlisted men were absent sick. Elijah F. Newell was transferred to the Invalid Corps on July 11.[980] On July 15, John G. Moore and John F. Teel were reported sick at "Post Hospital, at Point Pleasant." On July 20, while the 13th was wrapping up their pursuit of Morgan, William Gray was left at Belleville, presumably due to sickness. He returned to the company from Belleville on July 27. Captain James W. Johnson had leave of absence July 8 to 31.[981] Beginning July 10, 2nd Lieutenant Samuel Mathers was absent with leave. One enlisted man was absent without leave for the first three weeks of the month, then two men were reported absent without leave and by July 28 just one was absent without leave. Among those reported absent without leave was Robert Gray on July 16, and Andrew Snodgrass reported absent on July 28. Snodgrass returned to the company on July 31. None in Co. A were reported in arrest or confinement.[982]

1863 was not only a pivotal year in the war but for Black Americans as well it was a watershed year. On January 1, 1863, President Lincoln's Emacipation Proclamation went into effect. In May 1863, with establishment of the Bureau of Colored Troops by the War Department, Black Americans were officially allowed to enlist in large numbers and U.S. Colored Regiments were organized. For the first time since the American Revolution, Black Americans were used as combat soldiers.

Federal authorities had until now been extremely reluctant to employ "negro troops" and put them into battle side by side with whites because of the widely held belief that the black man was an inferior being. By 1863, however, it was obvious that more manpower was needed to prosecute the war and from the War Department down to the privates in the field all realized that a black man could stop a bullet just as easily as a white man. As a result, recruitment for Black Americans was systematized, training camps were established and procurement of officers—all White men—was regularized. Captain James W. Johnson of Co. A was one appointed to command one of the new Black regiments. Johnson resigned his commission in the 13th West Virginia and was promoted and transferred to command the 3rd Regiment U. S. Colored Troops by order No. 335, issued by the War Department, on July 20, 1863.[983]

Morning reports for Co. B are incomplete but provide the following information. The company was stationed in camp near Charleston on July 1. By July 4, the company had transferred to Point Pleasant. July 19 through 21, they were on the march after John Hunt Morgan. By July 22, they were back at Point Pleasant.

Drummer James McDermitt was present every day except July 25. At the beginning of July, the company reported three commissioned officers and seventy-five enlisted men present for duty. At the end of July, there were reported three commissioned officers and sixty-seven enlisted men. No one had leave of absence. One enlisted man was reported absent without leave on July 25. No one was detailed for extra duty. July 27, Lieut. Col. Hall issued Special Order No. 18 from regimental headquarters at Point Pleasant, directing that "James McDermott of Co B is hereby ordered on Special duty for Ten days from this date. — He will report immediately to Capt Wolfe 3rd Va Cavalry."[984] Two privates were detailed for daily duty July 1 through 3, and five enlisted men had detached service each day of the month. Co. B had two to three men present sick during the month. No absent sick were noted.[985]

One of the men sick at this time was Private Christopher Columbus Barnett. Father James Barnett had gone to the military hospital where his son was a patient to care for him. Lieutenant Colonel J.R. Hall wrote the following kind letter to Columbus' mother Rebecca A. Barnett.

Barboursville West Virginia
July 20th 1863

Mrs Barnett

At the request of your Husband I write in regard to your Son Columbus

He is still very sick — but I hope the issue may be as you as a Mother may desire — You may rest assured that whatever can will be done for him and we will keep you informed of his condition — Mr. Barnett has written to you twice since he came down but sent his letters by mail this will probably reach you before his does

Very Respectfully
James R. Hall[986]

Thursday, July 2, 1863, Col. Hayes at Hdqts. 1st Brig. received orders from Brig. General Scammon, commanding 3rd Division at Charleston, conveying that "Capt. Compstons [Alfred F. Cumpston] Co 5th Va.V.I. to be relieved from the Post at Pt.

Pleasant Va Capt Stewart 13th Regt. Va. V. I. recommend as fit for the position."[987] The same day, Hayes wrote Capt. Stewart directing that "[i]n pursuance of orders from Division Head Quarters you are to proceed without delay with your company to Point Pleasant, Va. and relieve Captain Cumpstin [Cumpston] 5th Reg. Va. V.I. commanding Post."[988]

On July 23, Sergeant Samuel C. Love of Co. B was married to Miss Olivia D. Boggess. The couple were married at the bride's home by the Reverend S.K. Dix. The groom had just been promoted to the rank of Sergeant on June 12, 1863.[989]

Morning reports for Company C indicate that the company was stationed near Charleston July 1 through 8. July 9, Co. C was marching; July 10, they were at "Laurel Crick;" and July 23 through 31, the company was at Point Pleasant. At the end of July, the company had present and absent three commissioned officers and seventy-seven enlisted men. George E. Warner was one of two musicians present with the company. July 28, Sergeant William B. McCauley and John B. Patton were reduced to the ranks. July 28, Private John Plants was promoted to Sergeant in McCauley's place by order of Colonel Brown.

The company reported no one present sick. Eight enlisted men were reported absent sick July 1 through 8, and ten enlisted men were absent sick July 9 through 31. James S. Wood, David Eads and Markis D. Noble were reported "sick in Post Hospital, at Point Pleasant" on July 1. Eads was admitted to "General Hospital Point Pleasant" on July 1 for "General Debility." He returned to duty on August 9, 1863.[990] July 2, Asa S. Johnson was reported sick in "Regimental Hospital," and he was removed on this day to "Post Hospital" at Point Pleasant. July 6, George Mourning was sick in "Regimental Hospital." Mourning returned for duty on July 9. July 9, Francis A. Windon was reported sick in "Post Hospital, Point Pleasant."[991]

Morning reports for Company D record that the company was stationed in "Camp near Charleston, West Virginia," July 1 through 8. On the 8th, the company with the regiment was ordered to report to the south side of Cotton Mountain. July 9, the company was marching. They marched a distance of forty-two miles and arrived that day at Cotton Mountain, "a long and steep mountain running down to the Kanawha River"[992] July 10 through 12, they were at Laurel Creek. July 13, the company with the regiment was ordered to report at Fayetteville. On the 14th, the regiment was "ordered to report with Division at Rolley, [Raleigh] Va. in Pursuit of Mcoslin's forces." July 16 Co. D was

> ordered with the regiment to go in pursuit of Gen. John Morgan, who was at that time making a raid through the State of Ohio, arrived at Gallipolis and from there to Pomeroy, where we go off on the shore But did not have to fight. 19th we arrived at Buffington Island, at which place Morgan made a stand and got defeated. We

went on to Hockingport and then this Co. changed boats and went to Parkersburg and from there we came back and started for Cincinnati with a boatload of rebel prisoners. We arrived at Cincinnati July 23ʳᵈ. Delivered our freight and started [for] Pt. Pleasant, W. Va. Arrived at Point Pleasant and awaited orders July 26, 1863.[993]

July 26 through 31, Company D remained at Point Pleasant awaiting orders. The company reported a hefty record of enlisted men absent without leave for the month of July. Nine were reported absent without leave on July 26; 12 were absent July 27; 11 were absent July 28; 7 on July 29; 4 on July 30; and 3 on July 31. John Carr, who had absented himself from the company beginning June 23 was reported "lost deserted" on July 6. The company concluded the month with three commissioned officers and eighty-three enlisted men present and absent.

Co. D reported no details for special or extra duty. Two privates had daily duty July 1 through 4, and five privates had daily duty July 5 through 14 and 26 through 31. No one had any of these duties on the march. Two enlisted men had detached duty July 1 through 3 and one enlisted man was on detached service on July 4. One private was reported present sick on July 12. July 26 through 31, the number of present sick jumped to between eleven and thirteen. July 1 through 14, the company had from fourteen to seventeen men absent sick. July 26 through 31, just three men were reported absent sick. On July 10, Privates John Killingsworth and Henry Hoffman were sent to "Post Hospital" at Point Pleasant, presumably from Cotton Mountain. Private Henry Hoffman had received a gun shot wound to the head. He was admitted to "General Hospital Point Pleasant" on July 11 and returned to duty July 17, 1863.[994] July 1 through 4, two enlisted men—Nathaniel Burchfield and Sheldon Gibbs—were absent in confinement. Gibbs was reported in confinement at Charleston, Kanawha County, July 1. July 4, Nathaniel Burchfield and Sheldon Gibbs were reported "released from the Guard House at Charleston and brought to their company and was restored to duty by order of Col. Wm. R. Brown."[995]

James Legg, Corporal, Company E, was admitted to "General Hospital at Point Pleasant" on July 5, 1863, with "[g]un shot wound hand." He was transferred to "General Hospital Gallipolis" on March 3, 1864.[996]

July reports for Company F show that it was stationed in "camp near Charleston" on July 1. On July 7, Brigadier General Scammon sent orders to 1ˢᵗ Brigade Head Quarters directing that "One Company 13ᵗʰ Va. V. I. proceed to Loup Creek to guard stores."[997] July 7 or 8, pursuant to Scammon's order, Col. Brown issued Special Order No. 17 to Capt. Albert F. McCown, cmdg. Co. F, directing him to furnish his company "with three days rations and [be] ready to proceed by transport to Loup Creek to Guard Government stores, where you will remain until further orders."[998] At 5 o'clock p.m. on July 8, Co. F marched from Charleston to board the transport steamer that would take them to Loup Creek. The company boarded the *General Meigs*.⁹⁹⁹ At 4 o'clock p.m. on July 9, Co. F arrived at Loup Creek. They "threw out pickets" and guarded government stores as ordered. They encamped at Camp Laurel Creek on July 10. On the 10th, David Stevenson, who it was presumed had deserted, reported and rejoined the company. He had been sick and not been able to report in sooner. July 10, Co. F received orders from Col. R.B. Hayes to cross the river and join the rest of the regiment on the south side of Cotton Mountain. Co. F arrived at Cotton Mountain at 6 o'clock p.m. on the 10th. July 12, they marched for Fayetteville, where they arrived at 6 o'clock p.m. and bivouacked. July 13, the company was marching. The 13th Virginia started for Raleigh Court House together with the 12th and 91st Ohio Volunteer Infantry Regiments and 9th Virginia Volunteer Infantry. They were to "to drive away the rebels" at Piney Creek. July 14, Co. F arrived at Raleigh Court House and "drove in the enemy's pickets and camped in Raleigh C.H. at 12 o'clock m[idday]." "[T]he enemy retreated." On the 15th, the company (with the regiment) "started on its return to Fayetteville at 1 o'clock P.M., the rebels having evacuated Piney and part of our forces destroyed their works."

July 16, Co. F and the regiment arrived at Fayetteville and "at 12 o'clock m[idday] got dinner and at 6 o'clock p.m. started for Gauley, bivouac[k]ed at Cotton Hill." July 17, Co. F "[m]arched to Loup Creek and embarked on boats to go in pursuit of Morgan in Ohio, [they] arrived at Pt. Pleasant at 8 o'clock p.m." July 18, the company with the regiment "[s]teamed up to Pomeroy and disembarked went out on the hill attacked Morgan driving him off."¹⁰⁰⁰ Company F had been "ordered forward and skirmished with Morgans forces near Pomeroy, Ohio July 18, where we [they] captured 9 horses." July 19, they "[r]eached Buffington [where they] found Morgan endeavoring to cross, [Company F] disembarked, [and] skirmished with him driving him back into Ohio," "several prisoners" were captured here by Company F.

July 20, the company "[s]tarted down the river on our return, [from pursuit of Morgan] on board steamer Emma Graham." July 21, they "[a]rrived at Point Pleasant, [...] at 12 o'clock m. and went into camp." July 21, orderly sergeant, "1st Sergt. A.C. Mason captured two of Morgan's men at Ten Mile and on the 22nd he captured 2 more together with their horses and arms."¹⁰⁰¹ Co. F was ordered back to Loup Creek to guard stores again, as embodied in Special Order No. 18, issued by Lieut. Col. Hall from Point Pleasant on July 27. It would seem that this order was not carried out as the company remained at Point Pleasant. July 31 was a day of note: Co. F (and with it probably the entire regiment) was paid for the period of April 30, 1863 to June 30, 1863, by Paymaster Major Cowan. On August 3, Co. F was transferred to Coalsmouth.

July reports show that Co. F had a drummer present July 1 through 31. July 1 through 7 and July 27 through 31, one fifer present.¹⁰⁰² No one was absent with leave. Three enlisted men were absent without leave July 27 and 28; two again were absent without leave on July 29; and one man was so reported July 30 to 31. July 4, William Daniel returned to the company "from capture." He had 'straggled' from his company while near Raleigh on June 22 and was captured by the enemy. One man returned from desertion July 10. The company gained in new recruits. Three new recruits were enlisted by Captain McCown at Point Pleasant. Oliver P. Roberts joined on July 15; Archibald Woodrum on July 27; and Frank E. Barrett enlisted in the company on the 31st.

There were losses as well. Private Franklin Hayse died "of disease &c." in hospital at Point Pleasant on July 13. Another enlisted man was reported deserted July 14. At the end of the month, Co. F reported a total of three commissioned officers and seventy-five enlisted men. No one was assigned to special duty. One non-commissioned officer had extra duty July 1 through 9. Private David Forbes, a blacksmith by trade, was ordered on extra duty with "A.Q.M." (Assistant Quarter Master) on July 7.[1003] Forbes continued to perform this duty for the rest of the month. Four privates had daily duty each day and two enlisted men had detached duty each day of the month. July 1 through 9, one non-commissioned officer was reported present sick. July 10 through 12, five to six privates were reported present sick. No fewer than seven enlisted men were reported absent sick during the month with a marked rise (from seven to fourteen) in that number during the second half of the month (i.e., July 13 through 26). No one was reported present or absent in arrest or confinement.[1004]

Captain J.V. Young's Company G 13th West Virginia, was stationed at Camp White "two miles above Charleston on the old Four Farm," where there were "tolerable good pastures." On Monday, July 6, he wrote to his wife, Paulina, with a richness of detail and emotion characteristic of the man.

> [...] I am well. But there is some sickness in my company at this place. There is no excitement up here we are all watching with breathless emotion the struggle of the Army in the East and West [...]
>
> I hear from Colonel Brown that there are some Rebs down on Mud, and in the Valley. I expect Company G will have to be sent back again, and the devil help them if I am sent back. You may say to the Rebs in your neighborhood that Colonel Brown says if the Rebels take one dollar's worth from you, he will send my Company G down there and turn them loose in the valley on the Secesh, and make them tired of bringing horse thieves into the neighborhood. Tell them this, and tell them that they shall make all the Union losses good. Their protection wont do them any good if they don't stop their friends from stealing. It is believed here that they are the cause of it. Mr. Handley[1005] need not boast of his protection for all the written that he could carry could not save him if my family is

again interrupted; and you may send him word of the fact, for all the officers up here believe that he and Jim Gray and J. Seasholes are the head and front of all the mischief done in the valley. There is strong talk up here of sending many of the Putnam Secesh to Dix, both male and female. I think it would be a good thing. You had better take care of your horses, and if you think proper, you may send me one and I think you can take care of the other.

We are two miles above Charleston on the old Four Farm. There is tolerable good pastures here. The 9th Va. has gone to Fayetteville. We expect to stay here some time.[1006]

Having some leisure time on his hands, on July 10 Young again wrote home. This time, with the 13th Regiment in position at the foot Cotton Hill "half way between Gauley and Fayetteville, where we expect to stay for some time to guard the wagon trains."[1007] To his wife, he wrote, "we are here at the foot of this miserable hill, six miles from the river and 6 miles from Fayetteville, for the purpose of guarding the transportation to Fayetteville."[1008] To his daughter, Emma, he wrote a much more detailed letter saying:

[...] This is a lonesome place, nothing to eat except what we brought out here with us. My Lieutenant and I have taken board with a farmer but he has not meat but plenty of milk and butter. They are good cooks, just from Buckingham County, old Virginia.

On the 7th of this month [July 7] the Rebs attack our trains on Cotton Hill and took all the teamsters prisoners, and the best of the horses, and made their escape. There are several companies, or bands, of horse thieves and bushwhackers in this county, and they are very troublesome to our transportation. I am willing to deed all this country to Jeff Davis if he will stop his 'Gray Backs' from stealing, and if he doesn't stop them I reckon we will have to do it, or stop them from breathing, but we will have to catch them first.

I feel very uneasy about home. [...] I am afraid the Rebs will get your horses and if they do I don't know how you will make out to live. But we must be willing to sacrifice everything for our Government and our holy religion. I mean the religion of our fathers — in other words, old Methodism. [...]

I don't think it will be long before the rebellion is put down. Lee's army is whipped to death; Rosecrans has driven Bragg out of Tennessee; Grant has taken Vicksburg, and you can see that rebellion is much crippled. [...]

Colonel McCausland (Rebel), I understand, is at Raleigh with considerable force, but we don't know how many. But if he wants to know how many there are of us, all he has to do is to come and see. We don't want Raleigh; and if he wants the Kanawha Valley he will have to come and get it. But I assure you, Emma, he will have a sweet time of it when he does come. There are three regiments at Fayetteville — the 12th, 11th and 9th Va. The 13th is at the foot of Cotton Hill, the 34th at Camp Piatt, the 23rd at Charleston; one battery of 12 guns at Fayetteville; another of as many more at Charleston. You can see by this that there will be some fighting done before they get the valley. I forgot the 2nd Va. Cavalry and part of the 1st, and also the 5th Va. at Mud Bridge, and Barboursville, which would round up some eight of ten thousand troops in the valley, enough to hold it from the Rebels. [...]

Direct your letters to Cotton Hill, in care of Colonel Hall, 13th Regt. Va. Vol. Inftry.[1009]

July 17, Burton Hamrick, Private, Co. G was admitted to "General Hospital Point Pleasant for an accidental[?] gun shot wound to the foot." He was transferred to "General Hospital Gallipolis" for the same complaint on March 3, 1864, and returned to duty June 28, 1864, not to the 13th Regiment but to the 11th Virginia Regiment as Young's company was not permitted to remained attached to the 13th by this date in 1864.[1010]

Back again at Point Pleasant after the "great raid," Capt. Young wrote home on July 24 from Point Pleasant. He expressed his hope that after the prisoners captured from Morgan were delivered at Cincinnati the 13th West Virginia Regiment would be gotten together again and stationed at one place "at Coal or Winfield."[1011] Happy that that Morgan and most of his men had been "gathered up," he wrote on an optimistic note to his wife: "I don't think there is any force of rebels in W.Va. None but a few scattering horse thieves, and they finally will be gobbled up."[1012]

Morning reports for Co. H indicate that the company was stationed at Coals Mouth July 1 through 16. July 17, it left Coals Mouth and arrived at Gallipolis remaining there until morning. July 18 through 20, Co. H was on the march. The morning of the 18th, the company left Gallipolis and arrived at Pomeroy, Ohio. They "then moved up and down the [river] until the July 20, when they moved down the river and arrived at Pt. Pleasant on the morning of the 21st after laying at Hockingport all day the 20th." They were reported at Point Pleasant July 22 through 31.

Co. H reported present and absent a full staff of commissioned and non-commissioned officers. A drummer was present July 1 through 18 and July 29 through 31. 1st Sergeant Harry Dunkel was absent on furlough for ten days (July 10 through

20). Sergeant William Shannon commanded in Dunkel's place until his return. 2nd Lieutenant William Perdue was absent without leave from July 21 to 25, when he returned to the company. Enlisted man Thomas M. Shiverent (Cheuvront or Cheveront) was reported absent without leave July 26. Cheveront returned to duty the next day (July 27). July 29, Francis M. Crofoot was reported absent without leave. Crofoot returned to the company on July 31. At the end of July, Co.H tallied present and absent three commissioned officers and seventy-four enlisted men.

Co. H was assigned detached duty each day of the month. The following details for detached service were ordered: two enlisted men July 1 through 17; eighteen enlisted men on July 18; sixteen men July 19 and 20; seven men detailed July 21 through 25; and two enlisted men detached July 26 through 31. Co. H had five to nine men present sick during the month including Captain W.I. Mathews, who was taken sick upon the return of the regiment to the Kanawha Valley on July 21. He remained sick through the 31st. The company had nine to twelve men absent sick each day of July. July 5, Sergeant Stephen Hooper, Private James Abbot and Private John E. Paul were sent to "Post Hospital" at Point Pleasant. Oliver Taylor returned to his company for duty from "Hospital, on July 12." July 25, John H. Duke was sent to "Post Hospital at Point Pleasant." Francis M. Barbour "returned from Point Pleasant Hospital" for duty on July 23. On July 24, George W. Green returned to the company for duty from "Hospital at Gallipolis, Ohio." Michael Stump was sent to "Hospital at Point Pleasant" on July 27. Private Jarrot F. Rigg was absent in arrest at Charleston July 19 through 31 for "sleeping on post."[1013]

Morning reports Company I record that the company was stationed at Coals Mouth July 1 through 11. July 12, Lieut. William E. Feazel, commanding, marched the company to Charleston and reported to Col. R.B. Hayes for duty. July 12 to 28, the company station was given as Charleston. July 26, Co. I was mustered into the United States service at Charleston by Lieutenant C.S. Roberts "to serve for three years."[1014] "Co. was organized by Lieut. W.E. Feazel at Charleston, W. Va. in July 1863 and as detachment was marched to Camp. Organization [of the Regiment] completed on Feb. 28, 1864."[1015] Another source records that instead of Charleston Co. I was mustered in at Guyandotte, on the 26th.

Colonel commanding brigade, R.B. Hayes, sent a despatch with orders on July 26 to the "Commanding Officer Co. I 13th Regiment V. I." from 1st Brigade Head Qtrs. Camp White, ordering him to move his "command up Elk river on the Charleston side to the Bluffs and picket the road leading up the river. This evening if possible."[1016] July 28, Sergeant Peter Darnel, commanding Co. I (Lieutenant Feazel was sick), left Charleston for the mouth of Big Sandy on Elk River. Darnel and company arrived at Big Sandy on July 29 and encamped. They remained there at least through July 31. The company tallied present and absent: one 1st lieutenant present July 1 to 20; four sergeants and four corporals present each day of July; and thirty-seven to forty-one privates. One new recruit was reported gained on July 6. No one was reported absent with or without leave. At the end of July, Co. I had a total present and absent of one commissioned officer and fifty-two enlisted men. It had three to seven men present sick during the month. The number of sick increased as the month progressed. 1st Lieut. Feazel was reported absent sick for the period of July 20 through 31. July 21, Lieut. Feazel was sent home to Gallipolis due to illness. No one was reported present or absent in arrest or confinement.[1017]

On July 16, Colonel A.A. Tomlinson (5th Virginia Infantry Volunteers) in command of the Post at Charleston, sent a telegram with the following order to Lieutenant Feazel stationed with Co. I at Charleston. The transport of troops indicated by Tomlinson's telegram was of course made necessary by the proximity of General John H. Morgan's raiders. According to Tomlinson's telegram, a similar order was sent to all commands stationed at the Charleston military post. It read:

> I have received a telegram from Col Hays ordering me to have all the officers and men on duty at this Post and be ready to march at noon on 17 inst. You will therefor have Your company at the wharf ready to be placed aboard of boats by the Specified time[1018]

Back in the Kanawha Valley

After the Morgan raid, the troops of the Kanawha Division resumed their former positions. Second Brigade, commanded by Col. C.B. White, took position at Fayette Court-House, and First Brigade headquarters were located at Charleston on the Kanawha River, while her regiments operated dispersed as before. The 13th West Virginia Infantry Regiment would remain in the Kanawha Valley until May 22, 1864.[1019]

For the moment, the 13th West Virginia was gathered at Point Pleasant. Here, they waited for their next orders and the return of Co. D from Cincinnati (and the return perhaps of another company from guarding prisoners). A certain amount of agitation related to reports of Morgan's men looking to steal horses prevailed at this time although for the most part, it was a period in which the command rested and refit. Capt. Young wrote to his wife from Point Pleasant, on the 24th:

> [...] would just say to you, take care of your horses for the woods are full of Morgan's men. They have all left their horses and all trying to make their way to Tennessee or Ky., and if they can steal horses they will do it. [...] There is quite an excitement now in this place [Point Pleasant]. They have discovered a squad of rebs eight miles above here. Company F and Company C are just now starting after them.[1020]

On July 26, Colonel Brown was absent from the regiment "after arms" to "bring them with him."[1021] July 27, Lieut. Col. James R. Hall issued General Order No. 15, which provided that

> II No [Order No. Left blank] issued from Head Quarters Co B reducing Corporals Wm Eckhard + William Ball to the ranks for disability; is hereby approved. —
>
> II II Order No [left blank] issued from Head Quarters Co C reducing Sergeant Wm B McCauley + Corporal John B Patton to the ranks for absenting themselves from the Company without leave; is hereby approved. —
>
> II II II Order No [left blank] issued from Head Quarters Company D reducing Corporals Richard W Meeks + George Miller to the ranks for absence without leave; is hereby approved.[1022]

July 28 Lt. Col. Hall ordered that a "Sergeant" take 3 men and take charge of and deliver to the Provost Marshal at Charleston the following deserters

1   Solon Cook Co D 12th Regt O. V. I.

2   Tilman Nichols 23rd Regt O. V. I.

3   Bernard Branor Co C 34th Regt O. V. I.

4   David Heltin Co H 34 Regt O. V. I.

5   John McNerland Co D 91st Regt O. V. I.

6   William Aleshire Co F 2nd Va Cav

7   Frances Kiser Co A 5th Va Infty

You will return without unnecessary delay[1023]

The troops of the Kanawha Division spent the remainder of campaigning season in the mountainous country adjacent to the rivers of the Kanawha Valley — in the vicinity of Coal, Elk, Gauley, New, and Greenbrier Rivers. "[T]he force of Echols, McCausland, Jenkins, and other rebel commanders"[1024] were in their front though the active operations had been reduced to small scale guerrilla operations.[1025] The troops of the Kanawha Division were actively employed in various ways. They spent their time in strengthening their positions, and they were detached on numerous expeditions of hard marching into the mountains beyond the rivers after guerrilla bands. Confederate partisan operations in the Kanawha Valley could not be entirely halted.

The extreme difficulty for infantry to cope with such enemy operations in a landscape which leant itself to irregular warfare prompted Col. Carr B. White (cmdg. 2nd Brigade, Third Division, 8th Army Corps) to conceive the idea of organizing a special force of hand-picked men from every regiment in his brigade "a corps of *elite* troops for scouting service."

> [S]uch a corps was accordingly created under special order issued about September 13. It was composed of the best shots, the best woodsmen, the best marchers, and the most dashing fellows that could be found in the three old regiments. The corps was organized under the designation of Independent Scouts, and ordered to 'make headquarters in the brush,' and thence sally out and 'engage at every opportunity' the bushwhacking companies that had long been annoying our front, attacking our picket lines, and interrupting communications with the rear.[1026]

The move to organize what we today would call a 'special forces' unit to cope with West Virginia partisans came about as a result of this unrelieved and exasperating problem together with the emboldening of Federal commanders in the West Virginia theater in the wake of recent Federal victories. With the opening of the spring campaign of 1864, scouts were taken (upon recommendation of the commanding officer) from every regiment in the division. This of course included the 13th W. Virginia. The requirements were rigorous and not a few of the men recommended by Col. Brown were returned to the ranks of their companies. So frequently did this occur in fact, that Gen. Crook reprimanded Brown with a warning that only the best woodsmen and soldiers be sent. The story of the Independent Scouts recruited from the 13th West Virginia in Spring of 1864 is told in *Beyond God's Country Vol. III; Tried in No Common Crucible Vol. IV; Halltown Races to Fisher's Hill Vol. V;* and *Cedar Creek to Muster Out Vols. VI,* of this series.

August arrived. The people supporting the Union cause were optimistic and confident. Morale had received a terrific boost during the month of July. Lee had been defeated at Gettyburg and had fallen back to deep within Virginia. Vicksburg, Mississippi and Port Hudson, Louisiana, had fallen. General Joseph E. Johnston, outnumbered and outmaneuvered, abandoned Jackson, Mississippi, to Gen. W.T. Sherman on July 16. General Rosecrans had driven General Bragg's army. Morgan's raiders, who had attempted a diversion in Bragg's favor by invading Indiana and Ohio, had been broken up and were now largely in prison. Federal forces had advanced across the Rappahanock. The all-important Mississippi River had been cleared of any significant Confederate presence. Emboldened by recent successes, Federal military authorities in West Virginia went on the offensive. A full-on purge of the land of its partisans and destruction of the Virginia & Tennessee Railroad became major objectives. Beginning August 5, U.S. Brigadier General William W. Averell led a cavalry raid through West Virginia, which would continue to the end of the month. Thursday, August 6, following Lincoln's proclamation, the North observed a day of thanksgiving for the recent victories. Business was suspended and church services were held.

On the other hand, supporters of the Confederacy were down-hearted and discouraged and their leaders called for renewed commitment and effort. Troops in the Kanawha Valley kept wary eyes on their district anticipating trouble at any moment. What cavalry the Kanawha District had was kept on the go. Col. R.B. Hayes wrote in a letter to his uncle on August 25: "I keep my cavalry moving as much as possible. The infantry has little to do."[1027] Towards the end of the month on August 21, Brig. Gen. Scammon sent a cavalry force — the 34th Ohio Mounted Infantry (Col. John T. Toland, cmdg.) and the 2nd Virginia Cavalry (Col. William H. Powell, cmdg.) to cut the Virginia & Tennessee Railroad at Wytheville. They "captured 120 prisoners, 2 pieces of artillery and 700 stand of arms, and as the citizens fired on the soldiers, destroyed the place."[1028] Guerrillas were still making scattered raids in the Kanawha Valley although not with the same spirit as before. August 27, the "notorious Bill Smith, who [wa]s slippery as an eel," made a raid on the village of the Guyandotte, where he robbed a store and captured a number of horses in the surrounding area. He was driven from that locale toward Trout Hill by one Major Thomas commanding two companies.[1029]

A trail of archival materials reveal the part played by the 13th Regiment within the larger context. On August 1, orders arrived directing the 13th to proceed Charleston. On the next day (August 2), they set out for Charleston by transport, distance traveled from Point Pleasant to Charleston sixty miles.[1030] They arrived at Charleston and encamped on the east bank of the Elk River in back of the town. August 4, Brig. Gen. Scammon wrote from 3rd Division Head Quarters, Charleston to 1st Brigade Head Quarters informing that a report had been received there that "half a regiment of Cavalry had been observed moving in the direction of Coals Mouth and he directed Col. Hayes, cmdg. 1st Brigade, that "[f]our companies of the 13th Va.V.I. be ordered to Coals Mouth to encamp one mile below in a good defensive position."[1031] That night, pursuant to Scammon's orders, Hayes sent per despatch the following to Col. Brown:

> Hd. Qtrs. 1st Brig. 3rd Div. 8th AC
> Camp White Va. Aug. 4, 1863
>
> Col. Brown
>
> You will direct four companies of your command with Lt. Col. Hall to move to Coalsmouth tomorrow morning. The steamer Victor No. 2 will transport them. Have them ready to load on at 5 o clock. They will camp one mile below the mouth of the river on the farm of Mr. Thompkins, selecting a good position that would

be easily defended in case of an attack by a superior force. Direct the officer in command to report promptly any information he may learn in regard to the movements of the enemy. By command of R.B. Hayes[1032]

Extant reports from this period indicate that after some reflection, it was determined that the entire 13th regiment be sent to Coals Mouth and not just the four companies requested by Scammon. Consequently, all 13th companies (i.e., the four companies with Lieut. Col Hall and the four with Col. Brown) received orders to proceed to Coalsmouth.[1033] Hall and his companies may have been first to set out for Coals Mouth (at five o'clock a.m. on August 5). Brown set out later that day. On Wednesday, August 5, Col. Hayes wrote to Colonels Brown and Comly, commanding the 23rd Ohio Volunteers in place of Hayes, commanding 1st Brigade, ordering each to "direct a detail of one Corpl and 6 privates for Provost Guard Duty to report to Capt. Williams Provost Marshal immediately."[1034] Again on the 5th, Hayes reiterated his orders to Brown regarding the move to Coals Mouth and detail of men for the Provost Marshall under the auspices of "Special Orders No. 56" directing that

1   Col. William R. Brown, Comdg 13th Regt. V.V.I. will move with the four (4) Companies of his command to Coals Mouth, and join the forces under Lt. Col. Hall.

2   Col. Wm R Brown, Comd'g 13th Regt Va. V. I, will direct a detail of two (2) Corporals and twelve (12) Privates for Provost Guard to report to Capt Williams Provost Marshall.[1035]

On July 29, Lieut. Col. Hall issued General Order No. 16 from regimental headquarters at Point Pleasant providing that:

So much of General Order No 15 as relates to the case of Corporal George Miller of Company D reducing him to the ranks is hereby re[s]cinded. — He is restored to his rank as of his original date — because detached by special authority and out of the limits of this Department — He will be obeyed and Respected accordingly.[1036]

On July 30, Col. Brown issued Special Order No. 20 directing that

Corporal Jacob C May Co G 13th Regt V.V.I. will report to Capt Warren Judge Advocate at Charleston West Va without delay. Private William Paul will accompany you — and after the transaction of all business required by the Judge Advocate yourself and William Paul will immediately return to your Company[1037]

At the end of July, the troops of Scammon's Division were stationed at Camp Piatt, Camp White, Coals Mouth, Fayetteville and Gauley Bridge. Division headquarters continued to be located at Charleston during the months of July and August. Headquarters for 13th West Virginia was at "Camp near Coalsmouth, West Virginia."[1038] Extant Company Muster Rolls for July and August indicate that Companies A, B, C, F and H were in "Camp near Coals Mouth." Co. F had transferred to Coalsmouth on August 3. Companies D, E and I were reported as stationed at "Camp Defiance, W. Va. near Coals Mouth." Company A reported that they returned from Hockingport to their old station at Coals Mouth on "Aug. 8, 1863. [...] Travelled over 200 miles."[1039]

On August 5, the entire 13th Regiment, pursuant to orders, left Charleston by transport for Coals Mouth. They arrived there that day after traveling west on the Kanawha a distance of twelve miles.[1040] They did not remain at Coals Mouth but immediately moved to Barboursville pursuant to new orders issued on the 5th. They arrived at Barboursville on August 7.[1041]

In the evening of the 5th, the 13th was in camp at Coals Mouth at an encampment they called "Camp Defiance." This Camp Defiance should not to be confused with the Confederate camp by the same name, which had been established on the eastern summit of Sewell Mountain in the fall of 1861 by Brigadier General Henry A. Wise (Governor of Virginia 1856 to 1859) although, as will be observed in W.W. Harper's letter below, even for him there seems to have been some confusion. On Saturday, August 8, 13th correspondent William W. Harper of Co. K, put pen to paper to tell people back home in "old Mason" about what their regiment had been up to. He wrote regarding their transfer on the *B.C. Levi* from camp on Elk River near Charleston to Tompkin's Farm, a mile from the Kanawha River near Coals Mouth.

> *Mr. Editor:*— This Regiment reached Charleston, W.V., on the 3d of the month, and went into camp the same day one mile up Elk river; some with tents and some without them — for some of our tents were left on the south side of Cotton mountain on our last trip to Raleigh. Here we remained part of the time without any protection from the burning rays of the sun, and part of the time wading through mud and water and sleeping upon brush to keep our weary selves from being swamped beneath the accumulated slush that had floated into many of our sleeping apartments. On the night of the 4th we received an order from headquarters to have four companies ready to move down the Kanawha under command of Lieut. Col. Hall, with instructions to encamp on the Tompkin's farm one mile from the river. At 12 o'clock a second order was received for the balance of our regiment to follow. Our

tents (what we had) were soon struck, and we again on our way down the beautiful Kanawha on the B. C. Levi and arrived at camp that night. I think I can safely say that this location exceeds by far in beauty and comfort any place we have ever been at, Pt. Pleasant not excepted, and that is a very comfortable place to be in.

Here we are encamped around what was once a very fine brick building, but is now in a state of dilapidation and ruin, but under the shade of a most beautiful grove of locust trees. This property belonged to rebels the greater proportion of whom are now away in Dixie in search of their "rights;" we hope they will not be disappointed, for if they find them we need never fear any more trouble from that source.

It was at this place where the rebels hords under the eccentric H.A. Wise, were encamped in July, 1861, and the headquarters of the hated Yankees are now in the very room then occupied by this same hair brained, bragadocia, and from the eastern steps of this building, he made his last traitorous speech in the Kanawha Valley, in which he uttered the following sentences: "Come with me to the border and we will make their towns and villages suffer. [Cheers] But first of all, let us attend to the traitors at home. Those who will not fight, we will take their arms from them and give them to those who will, and we will seize all their horses, their asses and their cattle, and drive them into camp, and we will feast upon the enemy. [Loud cheers]." This is the language of a man against those people who were trying to defend that Government under which he had always lived and which had secured to him all the blessings and privileges that mortal man could desire, but which he was then stretching every faculty of his being to destroy. Poor Henry, if we are not very much mistaken, his ears will never again be greeted with the melancholy notes of the whipporwill, or, the more discordant gibberage of the Katydid, in this beautiful Valley.

But I must close. We are all as well as could be expected, and are as ever ready to fight or die "for the land we love the best." More anon. W.W.H.[1042]

Captain J.V. Young, Co. G and Private David Burrows, Co. F, wrote home to their wives, like Harper, commenting on the pleasant aspect of their campsite at Camp Defiance. Young wrote on August 6:

You spoke of wanting bacon and flour. I would just say that in a few days we will have everything you want here, and you can send a wagon here and get it. [...]

The whole Regiment is here, and encamped where the Rebels camped 2 years ago. It is such a nice place, cool and shady. I hope we will remain here some time. [...] Wm. Racer has just reported himself to camp. The whole Regiment is in fine health and spirits.[1043]

Burrows wrote to his wife Lovina on the 17th, saying that after leaving Point Pleasant the regiment went to Charleston, and after staying one night there they returned to Coalmouth and camped "below Coal. We are camped in the prettiest place we was ever camped in. I would write for you to come up but there is no place for a woman to stay close to camp." He added that he had no chance to obtain a furlough but that he was having "pretty good times. We have as pretty a place to camp as ever anybody camped in. [...] There is nothing going on up here."[1044]

The Confederate States military authority had not transfered any of their forces away from West Virginia despite recent successes to the Federal arms on the major fronts elsewhere. Recent Federal movements against the Tennessee Railroad had caused the Confederates in West Virginia to strengthen their posts. Hayes wrote to his uncle, Sardis: "we have twice our numbers watching us. To keep <u>them</u> out of mischief, it is more likely that our force will be increased rather than diminished."[1045]

A report that the enemy was "in force at [the] marshes of Coal" reached 3rd Division headquarters—probably by August 6. In response to this news, Brig. Gen. Scammon wrote to Hayes at 1st Brigade Head Quarters giving orders "to have everything in readiness also to send word to Col. Brown Comdg. 13th Va. to be on the 'que vie.' "[1046] Hayes in turn, sent word to Lieutenant John S. Witcher, commanding Co. G 3rd Virginia Cavalry, to "[s]end scouts in the morning to the forks of Coal River with orders to report here [at Camp White, Charleston] in the evening and to [m]ake a detail of four men to report to Col. Wm. Brown comdg. 13th at the mouth of Coal river for couriers. By command of R.B. Hayes"[1047] Hayes then wrote to Brown:

Hd. Qtrs. 1st Brig. 3rd Div. 8th AC
Camp White Va. Aug. 6, 1863

Col. Brown

There is a report of the enemy being in considerable Cavalry force at the Marshes of Coal. We send you herewith four Cavalry men for Couriers — you will report by them any important information you may learn in regard to their movements. By command of R.B. Hayes[1048]

On August 9, Brown reported to Hayes regarding rebel activity in the area around Coals Mouth. Brown informed that he had received "information that a Heavly (rebel) 'citizen' was carrying salt from Ventricians stone [store?] below — and arrested them and brought them into camp." August 10, Brown reported that there was "no enemy within 18 miles around the country." August 17, Brown, reported to Hayes, that there "was no enemy near Coals Mouth."[1049]

Beginning about the 9th of August, a concerted effort to recruit more men for the 13th was made. On the 9th, Col. Brown issued a number of special orders to his companies detailing men for recruiting service. Special Order No. 23 detailed "Sergeant Grinstead and four men of Co A [...] to proceed to Charleston W Va for the purpose of making an effort to raise some recruits for this Regt." They were to report back to 13th Head Quarters at Camp Defiance "on Wednesday evening next." Special Order No. 25 authorized "Sergeant J R Walkup Co F [together with] private Leoni Clagg to go to the lower part of Mason County for the purpose of raising Recruits for this Regiment — he has permission to be absent six days at which time he will report at these Head Quarters." Special Order No. 26 detailed "Lt George Snowdon Co D and — Darnold Kimes and H F Sherman" for recruiting service to "be absent six (6) days at which time they will return and report at these Head Quarters." Special Order No. 27 authorized "Lt Bromley Co F" to be detailed "on Recruiting service for six (6) days" and to "report at these Head Quarters [13th Head Quarters at Camp Defiance] on the 15th inst." Special Order No 28 also issued on August 9 ordered "Lt Harpold, Sergt Wm McDaniel, John Claunch + Andrew J Long" to be "detailed on recruiting Service for this Regt for six days. They will report at these Hd Quarters on the 15th inst." Special Order No. 29 authorized "Sergt [Philip W.] Nicholson" of Company F to be "detailed on recruiting Service for ten days." He was thereafter to "report at these Hd Quarters on the 19th inst."[1050] On August 10, Brown issued Special Order No. 32 detailing "Sergt Van B Morris, Corp Lewis N [or V] Carter and three privates of Co G of this Regt," (Captain John V. Young's Company) "on recruiting service for ten days — they will report to these Hd Qrts on the 20th inst."[1051]

On Monday, August 10, Gen. Scammon ordered Charles Dally of the medical staff (13th Regiment) to "proceed to Point Pleasant to relieve acting Assistant Surgeon Dr. Strickland now in charge of Hospital at that place. Strickland to turn over all public property to Dally and to remain on duty with Dally."[1052] Also on the 10th, Captain Isaac M. Rucker (cmdg. Co. A 181st Regiment West Virginia Volunteer Militia) wrote to Col. Brown asking for his help in flanking a detachment of "horse thieves" operating near the Lincoln and Putnam County border. Rucker wrote the following from Winfield, Putnam County:

Dear Sir—

I am at this place with thirty eight mounted men — <u>without uniform</u> — And will proceed to-night to Col Hills on the Guyandotte Turnpike — and will start from Hills at daylight for Hamlin — with the intention of flanking a detachment <u>of Rebel Horse Thieves</u> —- stationed in that vicinity — and would be glad if you could <u>cooperate with me</u> — from any point you may think best —

I thought it best to inform you of my intentions lest you might have forces in that section who might mistake us for rebels — and to get your assistance if possible. I succeeded on yesterday in <u>Killing</u> the Notorious John Chapman[1053] on the Waters of Mud River — Please inform me if you can aid me — by the bearer of this —

> Very Respectfully Your obedient servt
> Isaac M Rucker
> Capt. Cmdg Co A. 181ˢᵗ Regt
> W. Va. Volunteer Militia[1054]

The fragmentary nature of regimental records do not permit one to determine whether Brown was able to send aid to Rucker although perhaps one of the several recruiting companies sent out by Brown at this time could have supported Rucker in his efforts to capture the partisans at Hamlin.[1055] As might be expected, requests for aid in apprehending other secessionists were received by Brown at this time from other sources as well. August 23, John Bowyer, U.S. Commissioner for Putnam County, wrote to Brown requesting that he arrest Jacob Grass and Thomas Paul and that he send John Young's Company to do it. Bowyer wrote:

> Sometime last Spring Jacob Grass + Thomas Paul who were Prisoners at Camp Chase were Brought to Wheeling and there released upon condition that they would on their arrival here that they would Give Bond + Security to the U. S. for their Good behavior and were allowed 20 days after their release to Execute the Bonds +c they were also to pay their fare home neither of which conditions have they performed. I am the U.S. Commission here for Putnam county Western district of Virginia Maj. Darr Pr[o]v[ost] Marshal General at Wheeling wrote to me that I should have them in case of failure as aforesaid and rearrested and sent to Head Quarters at Pt. Pleasant from which Place they would be sent on to Wheeling + Kept untill the conditions aforesaid were performed

Would it be Proper for you to cause their arrest + to send them to Wheeling or Lodge them in our Jail + then I could attend to them as Directed

Capt. Young + his company know them + they are not far off + could soon capture them I have no better means to arrest them + am disposed to do my full duty. Through Mr. Shaw or otherwise Please Inform me. [...] The above men are bad Rebels I know them well.[1056]

Note appended on reverse dated "Hd. Qtrs. 13th Regt. V.V.I., Camp Defiance Aug. 24, 1863," says: "The Above named men were notified and they went with own accord to the commission and gave satisfactory explanation to the commission and bonds."[1057]

August 13, Hayes wrote Col. Brown regarding his paperwork saying "Your attention is again called to the Circular [illeg.] from this Hd Qtrs July 29, requesting the Monthly Report for July of O[rdance] & O[rdnance] stores of your command."[1058] On August 16, Hayes left Camp White "to inspect the Thirteenth at Coal's Mouth." He planned to "take the band along for the fun of it."[1059] August 18, Brown wrote to Hayes "transmit[ting] blank receipts for ordinance and ordnances stores at Charleston W. Va. to be syned by Capt. [M.P.] Avery or some responsible person."[1060] August 20, Col. Brown issued Special Order 33 authorizing the formation of a court martial:

A Regimental Court Martial is Hereby Appointed to meet at these Hd Quarters on the 21st day of August 1863 at 10. O.C. A.M. or as soon as practicable for the trial of such prisoners as may be brought before it

Detail for the Court
Lt Col James R Hall
Capt Milton Stewart
Capt Albert F McCown
Capt John D Carter[1061]

The infantry had "little to do" wrote Hayes to his uncle. Indeed, soldiers of the 13th West Virginia had so little to to that they were turning to vice. On August 25, Col. Brown issued General Order No. 22 prohibiting gambling and recommending instead that the men study war "Tactics."

I Owing to so much discord derived from the indulged habit of card playing, the Col. Commanding deems it necessary to issue the following order that no gambling will be allowed within the limits of this camp by Officers or Non-commissioned officers, also that more attention be paid to the study of Tactics.

II II It is further ordered that gambling for Checks or money must cease through this command and only be allowed for pastime.[1062]

In connection with Special Order No. 33 regarding the convening of a regimental courts martial, Brown issued the following on August 27:

II The findings and sentences of the Court Martial are approved, and the Company Commanders are hereby directed to see that the sentences are duly executed. —

III The Regimental Court Martial of which Lt Col James R Hall of the Regt is President is hereby dis[s]olved[1063]

On August 27, Brown issued Special Order No. 36 and No. 37 (duplicate orders) appointing Captains Milton Stewart, John D. Carter and Albert F. McCown as "a board of survey to examine into and report on the condition of certain Commissary Stores now in the hands of S. Comstock Lt. + R. Q. M. 13th Regt Va. V.V.I."[1064]

The 13th Regiment (including Captain J.V. Young's Company G, which it seems was in the process of being reclaimed by the 11th Virginia Infantry) remained at Camp Defiance until September 3 when orders were received to proceed to Barboursville without delay and camp at that place.[1065] On Monday, August 31, the 13th marched for Barboursville and drew clothing.[1066] Co. D was issued clothing for the first time.

August 1863

Morning reports, orders
and other correspondence

Company A was at Point Pleasant on August 1. It left Point Pleasant on August 2 at 5 o'clock p.m. and arrived at Charleston early August 3 at 4 [or 1] o'clock a.m. August 3, Sergeant Robert H. Davis, [William] M. Davis, Andrew J. Davis, John Hunter and Jerry Jobling reported to the company. August 4, Sergeant Thomas Moore was left at Point Pleasant for duty. Moore returned to the company on August 8. August 5, Co. A left Charleston in the morning and proceeded to Coals Mouth where it remained until the end of the month. August 5, Robert Gray returned to the company. John Baxter was reported returned from Charleston and reported for duty August 8.

Co. A had two recruits. Thomas Witherew enlisted on August 5 at the Coals Mouth camp. Charles Eastwood enlisted on the 15th but failed to be mustered. Eastwood was discharged for disability by the regimental doctor on August 21. Capt. J.W. Johnson resigned July 14 or 20, 1863, and was transferred to the command of the United States "3rd Regiment Colored troops."[1067] The following promotions ensued in consequence of Johnson's resignation. Lieutenant Greenbury Slack was promoted to Captain, 1st Sergeant George Danner was promoted to the rank of 1st Lieutenant and Sergeant Miletus Grinstead was reported promoted to 1st Sergeant on August 24 (as embodied in the following Orders) to date it seems from July 20.

General Order No. 21:

I From special trust and confidence reposed by our Governor of West Va. in the good conduct of 1st Lieut. Greenbury Slack and Orderly Sergt. George Danner, he doth commission the 1st as Captain and the 2nd as 1st Lieut. of Co. A 13th Va. Vol. Inftry. They will be obeyed and respected accordingly.

II II Sergt. Miletus Grinstead is hereby appointed orderly Sergeant of Co. "A" vice Danner promoted and will be obeyed and respected accordingly.[1068]

One enlisted man of Co. A was discharged by promotion on August 22. One enlisted man was transferred out of the company on August 9, and one enlisted man was reported deserted on August 17. August 12, Michael Baxter died of disease at his home at Charleston. The number of present and absent for the company at the end of the month was reported as three commissioned officers and eighty-three enlisted men. One drummer was reported present for the month. One artificier was reported present from August 3 to 31. 1st Lieutenant Greenbury Slack was reported absent on leave from August 1 to 24. Two commissioned officers were reported absent with leave for four days between August 19 and 23. Enlisted men were absent without leave everyday of the month. Three, four, five and six at a time were absent from August 6 through 15. August 6, Andrew Snodgrass, William P. Compton, Philip Wintz and John Tully were first reported absent without leave. August 8, Sergeant Thomas Moore, John Moor, and David Smith were reported absent without leave. Richard George left the company without leave on August 12. Two men were absent on their own recognizance August 23 to 31. For the remaining days of the month, one enlisted man was reported absent without leave.

One private had daily duty each day. August 10, Sergeant Miletus Grinstead, Corporal James A. Means, Calvin Vance, Job Hall, John Hall and John M. Naylor had extra duty recruiting in Kanawha County.[1069] Abraham Chandler enlisted in the company on August 10. John M. Naylor was to have been absent from the company for just four days but he remained absent without permission until at least August 23. One commissioned officer had detached service everyday and anywhere from one to nine enlisted men had detached service during the month. August 9, Corporal James H. Tully, John Green, Daniel Snodgrass and William George were sent up the Kanawha River after a deserter. August 11, Corpl. Tully, William George and John Green returned with John Tully, William Gray, Philip Wintz and William Compton from Charleston. August 12, Sergt. Grinstead and Corporal James A. Means returned from Charleston. Alfred Jones and Thomas Hughes "returned from recruiting up Elk Aug. 25, 1863."[1070] William Hurel and William A. Means returned from Charleston, on August 31.[1071]

At most, ten men from Company A were present sick during the month. Between August 3 through 6, no one was present sick. There were up to thirteen men absent sick with improvement in that number at the end of the month. At that time, the number of absent sick dropped to eight. Francis M. Cobb returned to his company from "Post Hospital Point Pleasant" on August 10. August 12, a considerable list of sick soldiers was reported. Samuel Snodgrass and Nathan Snodgrass were reported sick in "Post Hospital Gallipolis, Ohio." Samuel Snodgrass would report back for duty on August 26. William Riley, John Newell and John Teel were reported sick in "Post Hospital Point Pleasant," and Ashar Ramey, George W. Ramey, Michael Baxter, Samuel Fluck and John Thomas were reported absent sick at home in Kanawha County. August 17, Robert H. Snodgrass was admitted to "General Hospital Point Pleasant" with enteric fever (typhoid fever). He died just two weeks later on September 2.[1072] August 20, John T. Newell was sufficiently recovered to return from "Post Hospital" and report for duty. Nathan Snodgrass reported for duty from "Post Hospital Gallipolis" on August 23. George W. Ramsy returned from Charleston unfit for duty on August 26. August 27, he left camp "for the Post Hospital." August 26, James Toothman, on duty as wagoner, reported sick. Jab [Job] Hall was detailed for wagoner in his place.

No one was reported in arrest or confinement. August 1, George W. King returned from Charleston. August 14, John Thomas and Michael George returned from home for duty. David Smith reported to his company on August 15, and on August 16, Andrew Snodgrass returned to the company. He had been absent since August 5. Job Hall returned from Charleston on August 16. Calvin Vance returned from Charleston on August 18. August 24 or 25, Milton Wilson returned from Charleston.[1073]

Morning reports for Company B note that the company was marching August 3 to 5. The command left Point Pleasant on August 3, destined for Camp White, Charleston, and arrived at Charleston at 5 a.m. on August 4. Next day (August 5), Co. B departed Charleston at 6 a.m., for Tompkins Farm with the note that a Union army camp had been established at Tompkins Farm on Cotton Hill, Fayette County. Captain Milton Stewart with six men was detailed for special duty in Mason County. August 7, Co. B was at Camp Defiance. They continued encamped there until at least the end of the month. One drummer was present every day. One man enlisted on August 6; two more enlisted August 11; and one man enlisted on August 21.

Between August 22 through 31, one, two and four enlisted men were absent with leave; one to three enlisted men were absent without leave August 7 through 12; and one enlisted man was reported absent without leave August 27. Total present and absent at the end of the month were three commissioned officers and ninety-seven enlisted men. One to four privates were reported on daily duty August 1 and 2 and August 8 through 31. Over the course of the month, two to thirteen men were detailed for detached service. Co. B reported three to five men present sick and six to nine men absent sick for the month. Private Charles Carroll, who was reported "absent in confinement at Charleston," was released on August 29.[1074]

Company C recorded that the company were stationed in Camp Barboursville (also named "Camp Defiance" in "Remarks" following reports) from August 1 through 31. August 1, Colonel W.R. Brown issued Special Order No. 17 from regimental headquarters at Point Pleasant regarding recent reductions and promotions in Co. C. John Plants was "promoted to the Rank of Sergeant in place of Wm B McCauley reduced and Andrew B Shobe Private Co C is also hereby promoted to the Rank of Corporal in place of John B Patton reduced they will be respected and obeyed accordingly."[1075] Co. C reported their two musicians (Marcus L. Jones and George E. Warner) present. No one had leave of absence. Two men were reported absent without leave August 9 through 12; one man was reported absent without leave on August 14; and one man absented himself August 26 through 29. The company gained seven enlisted men. John T. Burris enlisted on August 3 at Point Pleasant; two enlisted men had joined on August 18; one enlisted August 20; two enlisted August 22; and one enlisted August 26. At the end of the month, Co. C reported present and absent two commissioned officers and eighty-three enlisted men.[1076]

One commissioned officer had detached duty August 1 through 8, and from two to six enlisted men were detailed for detached service each day of the month with the greatest number of men detached at the end of August. Men were authorized to go on a recruiting tour pursuant to Special Order No. 28. This order had been issued by Col. Brown on August 9. It directed that "Lt Harpold, Sergt Wm McDaniel, John Claunch + Andrew J Long are hereby detailed on recruiting Service for this Regt for six (6) days. They will report at these Hd Quarters on the 15th inst."[1077]

Company C had five men present sick August 1 through 6, and two privates present sick August 18 through 31. There were ten to thirteen absent sick reported for the month. On August 8, David Eads returned for duty from "U.S.A. General Hospital Point Pleasant." He had been admitted to the hospital for "General Debility" on July 1.[1078] August 9, George F. Rollins was sick in "Post Hospital" at Point Pleasant, and August 17, Captain Van D. McDaniel was sick at "Point Pleasant Post Hospital." He remained there through August 31. One private was reported present in arrest or confinement for the period of August 22 through 31.[1079]

Company D reported that it was stationed at Point Pleasant on August 1. August 2, the company with the regiment was ordered to report at Charleston. They arrived at Charleston on the 2nd and remained there for two days (August 2 through 4). Then, they proceeded on to Coals Mouth, where they arrived "and pitched tents on Thompkins farm one mile below Coals Mouth." From August 5 through 31, the company was reported stationed at "Camp Defiance, W. Va." A drummer was present August 1 through 23 and August 25 through the 31, and a fifer was present August

1 through 14. No one was absent on leave. Two to nine enlisted men were reported absent without leave from August 1 through 8, and one enlisted man was reported absent without leave August 9 through the 18.

Numbers received a boost as four new recruits joined. These were James Williamson, who enlisted in the company August 12, and Nelson Flinn, who enlisted on August 15. John Piatt and Thomas W. Harshey enlisted on August 20. One enlisted man returned from desertion on August 28. Private James Burnett was reported to have drowned at Gallipolis sometime around August 17 or 19. Subsequent regimental records suggest that he was, instead, probably absent without leave. Sergeant John M. Graham was reduced to the ranks for disability on August 10. On August 11, Henry C. Williamson was promoted to fill the vacancy left by Graham.[1080] The company finished the month with present and absent three commissioned and eighty-six enlisted men. On daily duty were: five privates August 1 through 17; one non-commissioned officer August 17 through 31; and four privates August 18 through 31. Four to six enlisted men were detailed for detached service on August 8 and 9; and 13 through 31.[1081]

August 7, Col. Brown issued Special Order No. 21, authorizing Private Henry Williamson of Co. D "to take charge of two men" and "proceed immediately to Charleston W Va" where he was to "open and clean up Three boxes of guns belonging to this Regt and transact whatever other regimental business may be committed to his care — and then he will return without unnecessary delay and report at these Hd Qtrs."[1082] One commissioned officer and seven to nine enlisted men of Co. D had detached duty August 10 through 18. Among these a group detailed for recruiting service pursuant to Special Order No. 26. This order directed that "Lt George Snowdon Co D and — Darnold Kimes[1083] and H F Sherman are hereby detailed on recruiting service — they will be absent six (6) days at which time they will return and report at these Head Quarters."[1084] On August 13, Lieutenant Col. James R. Hall issued Special Order No. 33 detailing "Sergt Michael Roseberry of Co D [...] to proceed to Point Pleasant and transact such Regimental business as he is charged with and return by the next Government transport and rejoin his Company."[1085]

Twelve men of Co. D were reported present sick at the first of the month but these must have been conveyed to hospital elsewhere on August 2 or 3 as thereafter no more than seven men were sick in camp at a time. Three to as many as seventeen men were reported absent sick for the month. These number peaked during the period of August 3 through 13. August 18, Isaac Dingey was reported to have returned to the company from general army hospital. One musician and two privates were reported present in arrest or confinement on August 8. There must have been a spree as seven privates were present in arrest or confinement August 28 through 31.[1086]

August 7, Col. Brown issued Special Order No. 22 directing "Second Lt John H. Rosler Co E 13th Regt V. V. I. [to] take charge of 6 men and proceed to Charleston West Va on Special business connected with the Regt and for the purpose of recruiting. — He will report to these Hd Quarters August 13/63."[1087] August 21, Brown issued General Order No. 19 authorizing the following reductions and promotions in rank in Co. E:

> General Order No 1 issued from Head Quarters Co. E. reducing Corporal James Legg and Joel P. Thomas both being incapacitated from further active duty, the first from a wound — the second by being attached to the Invalid Corps is hereby approved. Marshal T Munday is hereby promoted to Corporal vice James Legg reduced and George W Smith is promoted to Corporal vice Joel P. Thomas Reduced.—

They will be obeyed and respected accordingly—[1088]

Company F reported that it started from Point Pleasant on August 2 and arrived at Charleston on the morning of the 3rd. The company remained at Charleston until the early morning of August 4 when it left Charleston and arrived at Camp Defiance near Coals Mouth that morning. Co. F had one drummer present each day of the month and one fifer present August 1 through 9. No one was absent with leave. One enlisted man was reported absent without leave August 14 and 15. Five new recruits were reported for the company. Among these were Thomas Roush, who enlisted August 1; Alexander Boles and Smith Tillis, who enlisted August 20; and Henry C. Casdorph, enlisted August 21. One unidentified man was reported as having enlisted on August 5. Corporal Russell B. Shrewsberry died in the Hospital at Point Pleasant on August 3 of typhoid fever. Private John M. Young was promoted to corporal on August 23.[1089] Frank E. Barrett deserted the company from Camp Defiance on August 16. At the end of the month, Co. F numbered three commissioned officers and eighty enlisted men. Two to three privates had extra duty each day of the month and three privates had daily duty each day. Two to seven men, enlisted and commissioned officers, had detached duty each day of the month.

On August 9, a number of orders were issued. Col. Brown issued Special Order No. 24 detailing "Sergeant Darnold Co F" [Peter Darnell] to "take charge of one man private James H Goal [Goad] Co I and proceed to Letart Falls Mason Co and arrest Irwin Lowe a deserter from Said Co [Company I] and will report to these Hd Quarters without unnecessary delay." Also on the 9th, Brown issued Special Orders No. 25, 27 and 29 authorizing groups of men from the company for recruiting service. Special Orders No. 25 detailed "Sergeant J R Walkup [and] Private Leoni Clagg to go to the lower part of Mason County for the purpose of raising Recruits for this

Regiment — he has permission to be absent six days at which time he will report at these Head Quarters." Special Order No. 27 detailed Lieutenant Joseph Brumley for "Recruiting service for six (6) days." Brumley was to "report at these Head Quarters on the 15th inst." Special Order No. 29 detailed Sergeant Phillip W. Nicholson for "recruiting Service for ten days. He will report at these Hd Quarters on the 19th inst."[1090] Special Order No. 34 authorized "Sergeant Darnold" (Darnell) of Company F "to take charge of Twelve (12) men from Co (I) and [...] proceed up Elk River Kanawha Co. West Va on a recruiting expedition Sergeant Darnold will see that his men are kept together as fast as is possible and will report weekly either in person or by letter to these [13th] Head Quarters"[1091]

A party from Co. F consisting of one commissioned officer and from four to six enlisted men were detached for the period of August 10 to 18. Co. F had very few sick in camp, just one to two at most, for twenty-one days out of the month. There were, however, between ten to sixteen enlisted men reported absent sick each day. Captain McCown seems to have seen to it that sick men were removed from camp to the overall benefit of his company and the regiment. No one was reported present or absent in arrest or confinement although authority and orders to arrest one and possibly two enlisted men of the company were given to Col. Brown in the following correspondence. Thursday, August 27, Brigadier General E.P. Scammon caused a letter to be written to Brown informing him that:

> Complaints have been entered in these Hd Qtrs against Wm Daniels of 13th Va Infy. who in company with Jno Daniels stole two horses from Elias and Rufus Lively of Fayette Co. One was a large gray horse with a lump on his back and another lumps on his hind foot just above hoof, the other was a sorrel horse, small. [illeg. word] men passed through Fayetteville with the horses
>
> You will arrest the said Daniels, and see if either horse can be found in Your command. If so send it to this post. Send Daniels under Guard.[1092]

Captain J.V. Young's Company G was on detached service collecting taxes from secessionists. Young wrote to his wife, Paulina, on August 22:

> I regret I was not in Camp when you came up, but was in the line of my duty. I was out three days but never saw the track of a Rebel but I understand that there is quite a force on the Middle Fork of Mud, stealing horses and plundering houses. But I did not have time to go after them. My business was to make the secessionists fork over their tax and they did fork, and when they didn't we forked their horses. Oh wonderful ! Six of Garret's negroes fol-

lowed us in, but nobody followed them — yet. I learned that that notorious Harvey Bowyer is in the neighborhood again. He was on the head of Chimer last Sunday night, and of course he is lurking about until he can steal some more horses. I think there are several in and about Teay's Valley, and the first thing the citizens know they will be minus their horses. [...][1093]

Young also wrote a long letter to Governor Pierpont (dated "Camp Defiance, Wednesday, August 12, 1863") in which he appealed to Pierpont to use his influence to permit his company to remain with the 13th Virginia Infantry rather than be returned to the 11th Virginia Infantry Regiment.

> I would respectfully request permission to make the following statements with regard to my company which has and is now doing duty in the 13th Regt Va. V.I and has been ever since October 1862. In the first place I was mustered in as 1st Lieut Nov 15th 1861 the comp then numbering 44 men. On the 6th of May following I was promoted to Captain and mustered with my men into the 11th Regt V.V.I. in October 1863 [I think he means 1862] I was ordered to report to the 13th Regt Vol Inft with some 40 or 59 more Recruits these were then mustered into the 13th by R.R. Crawford under instructions from Adj Genl Samuels with the understanding that the company should be consoledated and permanently attached to the 13th. But the thing has remained in this state ever since which you will perceive at once produces a very unpleasant and dissatisfactory state of things and I am exceedingly anxious to have the thing permanently settled one way or the other if I am to go into the 11th which I would very much regret indeed I want to know it or if the comp is to be consoledated and remain where it is which would be much more satisfactory I want to know that I would first state in this connection that there are not more than 25 or 30 men of those that were mustered into the 11th Remaining and they are exceeding unwilling to leave the 13th or to be separated from the balance of the company. I think that the good of the company and service both demand that the company be consolidated at once and remain where it is I hope that if the power is in your hands to do anything in the matter that you will do it at once I may further state that I have frequently refused to be mustered into the 11th knowing that it would result injuriously to the cause to divide the company and this would necessarily follow If you can do anything

for me in this matter you will confer a great favor and I would respectfully urge that it be done immediately.

N.B. all the now com'd officers are commissioned in the 13ᵗʰ and the name of the whole company are on our Regimental Descriptive Book which I think are very good reasens why the company ought to be consolidated and remain where it is.[1094]

Young was actively recruiting at Coalsmouth. Aug. 10, he enrolled Samuel Horton, Private, age 44; Aug. 15, John Bordette, Private, age 17; Aug. 19, Lewis I. Pursinger, Private, age 39; and on Aug. 23, Samuel I. Moses, Private, age 23 and William H. Moses, Private, age 17.[1095] These recruits were mustered in at Barboursville, Oct. 6, 1863, but did not remain with the 13ᵗʰ when at last the War Department decided that Young's Company was to return to the 11ᵗʰ Virginia.

Col. Brown also attempted to cement the connection between Young's Company G and the 13ᵗʰ by requesting commissions for officers. See Brown's letter to Gov. A.I. Boreman, of Saturday, August 22, 1863, in which he requested a 1ˢᵗ Lieutenant's commission for Clark Elkins.

13ᵗʰ Regt., Va. Vol. I. said commission to date back to June 6ᵗʰ 1863 at which time Governor Pierpont issued him a 2ⁿᵈ Lieut's commission in Co. (G) 11ᵗʰ Regt. V.V. Inft. There now being however a sufficient number of men claimed by the 11ᵗʰ Regt., in said company to justify his muster in I make this Request said applicant has been doing the duty of a 1ˢᵗ Lieut. since some time before June 6ᵗʰ. I can get him mustered into the 13ᵗʰ as there are men enough in said Co (G) which have already been mustered into this Regiment to entitle them to a 1ˢᵗ Lieut.[1096]

Reports for August submitted for Company H indicate that the company was stationed at Point Pleasant, on August 1. August 2, the company left Point Pleasant at "half past 6 P.M. for Charleston" and "[a]rrived at Charleston at 5 o'clock A.M. [on August 3]. Landed at the mouth of Elk River [and m]arched up Elk one mile and camped same day."[1097] August 3 to 5, the company remained at Charleston. It left Charleston on August 5 and arrived at Coals Mouth the same day. August 8, they marched one mile down Kanawha River and camped. (Another notation in morning reports records that the company was at Camp Defiance from August 6 through 31.)

Co. H had the following present and absent. One drummer was present each day of the month. One enlisted man was absent with leave from August 15 to 27. One enlisted man, Thomas M. Hacker, was absent without leave beginning August 2. He returned to the company on August 4. Francis M. Crofoot (Crowfoot) was absent without leave beginning August 10. Two enlisted men were absent without leave August 24 through 30. The command gained one recruit. Twenty-nine-year-old farmer of Perry County, Ohio, Thomas Faught/Fought/Fout, was enrolled in the regiment on August 7 at Point Pleasant by W.P. Mathers. Faught probably joined the company in camp on August 10. He was mustered in at Charleston on October 6, 1863.[1098] At the end of August, Co. H tallied present and absent three commissioned and seventy-seven enlisted men.

Two to eleven enlisted men were detailed for detached duty each day of the month. Eleven enlisted men were on detached service August 6 and 7. On August 25, Private Samuel Waran (Weaver?) was detailed "permanently" together with William Dunlap, Private of Co. C, "as Teamsters [to] report without delay to Lieut. S. Comstock, Regimental Quarter Master."[1099] Co. H had three to ten men present sick each day. This included Capt. W.I. Mathews, sick on August 1 and 2 and then absent sick beginning August 2. Ten to thirteen men were absent sick during the month. August 1 through 10, Private Jarrot F. Rigg was "absent in confinement" at Charleston. He was released and returned to the company on August 11. A notation entered for August 14 records that George W. Fulwiller had been in arrest at Charleston since May 1, 1863. Two enlisted men were reported absent in confinement from August 16 through 31.[1100]

Reports for Company I record that the company was stationed at the Mouth of Big Sandy on Elk River from August 1 through 8. They were at Camp Defiance August 9 through 31. 1st Lieutenant William Feazel joined for duty on August 5. He remained present with the company through the 31st. The company continued to be undersized with two to four sergeants, four corporals and between fifteen and thirty-seven privates present each day of the month. No one was absent with leave and two enlisted men were absent without leave from August 10 through 31.[1101] There was a gain to the company of three new recruits, who joined the regiment on August 28. At the end of August, Co. I had present and absent, one commissioned officer and fifty-five enlisted men. Ten to twelve enlisted men had detached duty for the period of August 10 through 31. These men were likely those detailed by Col. Brown in Special Orders No. 30, 31 and 34, issued respectively, on August 9, 10 and 21.

Special Order No. 30:

> Sergt. R.L. Young Co. I is hereby detailed to take charge of ten men and will proceed with two waggons to the mouth of Big Sandy on Elk River in Kanawha Co as a Guard. He will see that his command is Kept together and does its duty properly They will gather up whatever tents + Camp + Garrison equipage that belongs

to said Company and return to these Hd Qtrs without unnecessary delay[1102]

Special Order No. 31:

> Corpl Nathaniel M Jackson is hereby detailed to take a prisoner to Charleston W Va and deliver him to the Provost Marshal Capt. Williams — after which he will return to his Co without delay[1103]

Special Order No. 34:

> Sergeant Darnold [Darnell] Co (F) is hereby detailed to take charge of Twelve (12) men from Co (I) and will proceed up Elk River Kanawha Co. West Va on a recruiting expedition Sergeant Darnold will see that his men are kept together as fast as is possible and will report weekly either in person or by letter to these Head Quarters[1104]

Co. I reported five to eleven enlisted men and non-commissioned officers present sick for the month of August. Lieut. Feazel was absent sick August 1 through 4; five enlisted men were reported absent sick August 11 through 22; and three enlisted men were absent sick August 23 through 31. One musician was present in arrest or confinement from August 10 through 23. One enlisted man was reported absent in confinement August 2 through 31.[1105]

The members of Company K first drew clothing over a period of a week's time, beginning the last week of July until the very beginning of August (i.e., Elmore Taylor first drew clothes on July 25, 1863, while officer William P. Cunningham first drew clothes on Saturday, August 1.[1106])

For the 13th Regiment, August closed on a quiet note. David Burrows, Co. F, wrote home: "There is nothing going on up here. Everything is still. We expect to stay here some time yet. [...] Write the news about down there. We have good times up here. Send me some postage stamps."[1107] At the end of the month, on Friday, August 28, Col. Brown reported to Adjutant General F.H. Pierpont that at organization the 13th Regiment had 624 men and since then, the regiment had gained ninety-six recruits making seven hundred and twenty mustered into the regiment. Co. G (Young's), originally belonging to the 11th Virginia Infantry Regiment but since then ordered to duty in the 13th, had forty-one men making in all a total of seven hundred and sixty-one men. Brown continued in his report to Pierpont to explain the difficulties which had arisen as a result of the muster of men and officers of Co. G in two regiments:

At the organization of their Regt[1108] in Oct last the squad of Recruits that were mustered into the 11th Va Regt at Coalsmouth Va in May 1862 under Capt John V Young, were ordered to duty in the 13th Regt they numbered 41 — Subsequently 49 were mustered into the 13th Rgt making in all 90 men that are in that company. — The officers were commissioned in the 11th Va Regt — Since the organization one of the officers (Lt John S Cunningham) has been promoted to Adjutant of this (13th) Regt one (Lt Robert Brook) has resigned and his resignation has been accepted The Captain (John V Young) only remains — The Orderly Sergeant of the Co Clark Elkins has been commissioned but not mustered into the Service

Here the question arises about his discharge to receive promotion on account of the Company being mustered into the two Regiments whether the Company is entitled to have the vacancies filled —

There was some irregularity in the muster of the Squad + Capt John V Young May 23rd 1862 by Major Weed of General Fremonts Staff as Capt Youngs Muster is not recognized at the War Department Washington City and his pay is being withheld

I wrote to you some time since on the subject perhaps it did not come to hand as no communication has been recd from You.[1109]

At this time, field, staff and company officers of the 13th Regiment including Young's Co. G were: Colonel William R. Brown; Lieutenant Colonel James R. Hall; Surgeon Samuel G. Shaw; 1st Assistant Surgeon Abraham D. Williams; 1st Lieutenant and Adjutant John S. Cunningham; 1st Lieutenant and Quartermaster Stephen Comstock. The positions of major, 1st lieutenant and commissary, 2nd assistant surgeon and regimental chaplain remained unfilled. Company officers were, for Company A, Captain Greenbury Slack; 1st Lieutenant George Danner; 2nd Lieutenant Samuel S. Mathers. For Company B, Captain Milton Stewart; 1st Lieutenant Lovell C. Rayburn; 2nd Lieutenant Charles T. Latham. For Company C, Captain Van D. McDaniel; 1st Lieutenant Lemuel Harpold; no 2nd Lieutenant. For Company D, Captain Simon Williams; 1st Lieutenant James W. Hanna; 2nd Lieutenant George Snowdon. For Company E, Captain John D. Carter; 1st Lieutenant William N. Hawkins; 2nd Lieutenant John H. Rosler. For Company F, Captain Albert F. McCown; 1st Lieutenant Timothy Russel; 2nd Lieutenant Joseph Brumley. For Company G, Captain John V. Young; no 1st Lieutenant; 2nd Lieutenant Clark Elkins. For Company H, Captain William I. Mathews; 1st Lieutenant Oliver W. Griswold; 2nd Lieutenant William Perdue. For Company I, no captain; 1st Lieutenant William E. Feazel; no 2nd Lieutenant.[1110]

September 1863

Protecting the people by "driving out rebel scouts, horse thieves +c&c."

September found the 13th West Virginia settled once again into the routine of garrison duty. The soldiers continued to be encamped in the field. The colder weather brought more incidence of sickness such as camp dysentery and typhoid fever. The regiment attended to replenishing it's commissary and equipment. New clothing was issued. The problem of musters in relation to Company G was swatted about between 13th Headquarters and the State Government at Wheeling. Reductions and promotions of officers continued. Duty included considerable scouting in the counties of Cabell, Wayne and in Putnam guarding the Winfield Court House while the court was in session. The 13th arrested disloyal persons to be held as hostages pursuant to an act passed by the Loyal Virginia General Assembly on February 4, 1863, to secure the release of Union citizens held by the Confederates. Otherwise, the 13th continued to escort troublesome individuals (civilians, rebels and errant Union soldiers) to Charleston for confinement.

The lack of alarms and mundane duty resulted in a significant increase in the number of 13th soldiers in the guard house and civilian jail. Prospects that the war would soon end abounded. It was rumored that "Rosecrance and Burnside was giving them the devil."[1111] As overnight accommodations were readily obtainable in Barboursville, some soldiers had visits from family. Cos. I and K, the last companies to be filled before the regiment could be completed, were actively being recruited by 1st Lieutenant Henry Stump despite a number of frustrating obstacles.

On September 3, Brig. Gen. Scammon sent orders to 1st Brigade headquarters directing that the 13th Regiment and Lieut. Witcher's cavalry company be sent to Barboursville.[1112] Brown was unaware of this order. Col. Hayes ordered Brown to

> Start 3 Cos. of your command for Barboursville in the morn-
> ing with 3 days rations and follow with the remainder of your Regt.
> as soon as possible. Send all Baggage that you cannot hall by Steam-
> boat to Guyandotte. Send an officer to Gallipolis to make arrange-
> ment for your supplies to be shipped to Guyandotte.[1113]

Hayes' order was complied with. *Field History* confirms that the 13th was ordered away from Camp Defiance on the 3rd "to proceed to Barboursville without delay and camp at that place. The Regt. was in camp at Barboursville Sept. 4th."[1114] September 4, Col. Brown still at Coals Mouth, wrote to 1st Brigade Headquarters from "Head Qrs. 13th Regt. Va.V.I. Coals Mouth Va. Sept. 4th 1863" informing that he had "for-warded Adison Bias a Rebel Soldier to these 1st Brigade Head Quarters."[1115] Pursuant to orders, the 13th left for Barboursville, Cabell County, on Saturday, September 5. September 5 and 6, the 13th marched to Barboursville traveling a distance of twen-ty-nine miles.[1116]

February 10, 1863, when 13th Head Quarters moved from Point Pleasant to Hurricane Bridge until this latest move to Barboursville on September 5 and 6, the regiment had marched a total of seven hundred and forty-three miles.[1117] 13th head-quarters was located at Barboursville for the months of September and October.[1118] During this time, "Field and Staff and Companies" were stationed at Barboursville with the exception of Co. I, which was stationed at Guyandotte.[1119] Co. I had been ordered to proceed to Guyandotte on September 9 to "take Command of the Post at that Place" and "assist the Aid to the Revinue in Making the arrests he may deem necessary."[1120]

The regiment remained divided and stationed, at Barboursville and Guyandotte until November 12.[1121] Around September 4, a number of 13th soldiers were on detached duty at the Provost Marshal's Office (3d Div., 8th Army Corps) at Charles-ton.[1122] Col. Brown, not realizing that Scammon had ordered the 13th together with Witcher's Cavalry Company to Barboursville on September 3, wrote to 1st Brigade Headquarters on September 7 requesting "that Captain Witchers Co. 3rd Va Cavalry be sent down to ass't in cleaning out the horse thieves."[1123]

Col. Hayes wrote to Brown on September 8 confirming and instructing him as to the purposes of the regiment's presence in that locale:

> Your Regt was sent to Barboursville to protect the people of
> that region by driving out Rebel scouts, horse thieves +c&c. It is
> requested that you take such steps also as will protect Winfield
> during the session of the Putnam Court. It sits on Thursday the
> 10th, I believe. You will do what in your judgement is necessary. It
> was suggested that two or three Companies at Hurricane for two

or three days would answer the purpose — The General told me
that your absence from Coalsmouth would be temporary. [...] P.S.
Let me hear from you often if practicable while you are out on this
Campaign. Write by return Courier — Some say the Court is on
the 15th in Putnam — You will know —[1124]

Later that day (September 8), Hayes wrote again to Brown regarding the date the
Putnam County Court would convene correcting his earlier statement. Hayes wrote,
"Court meets at Winfield in the 15th inst. instead of the 11th as I informed you."[1125]
In compliance with Hayes' orders, Col. Brown issued Special Order No. 40 on September 14 in which he directed:

Lt. Col. James R. Hall

Sir:

You will take one hundred and ten (110) men of this Regt and
proceed to Winfield, West Va. and remain there as long as you may
deem proper to protect the court, the session of which commences
on Sept 15 1863 after which you will return and report with your
command to these Head Quarters without unnecessary delay.[1126]

September 10, Brown issued orders regarding the practice among the men of
freely absenting themselves without leave: General Order No. 23 provided as follows:

Still finding discipline trampled upon by men absenting themselves from their companies without leave — Commanders of
Companies are hereby directed to prefer charges for every violation
of this kind — A neglect for so doing will be considered a criminal
neglect of duty and disobedience of orders.[1127]

Captain Van D. McDaniel, commanding Company C, died on September 9
of acute dysentery. His death necessitated the issuance of orders regarding ordnance
and other unspecified public property which had been in his care. September 13,
Colonel Brown issued Special Order No. 40 appointing Captains McCown, Carter
and Mathers as "a Board of Survey to take an Inventory of the Public Property and
report upon the condition of the same which was in charge of Capt Van D. McDaniel
deceased, late of Co. 'C' of this Regt. 1st Lieut. Lemuel Harpold of Co. 'C' is the officer designated to take charge of the said property."[1128] September 14, Brown issued
Special Order No. 41 directing that Orderly Sergeant John P. Wood of Co. C be

detailed to proceed direct to Point Pleasant, to procure the last
Quarterly return of Ordinance + ordinance Stores of Capt Van D
McDaniel Deceased late of Co C and such other papers as the board

of Survey may deem necessary to enable them to take an inventory of the Public Property which was in charge of Capt McDaniel at the time of his decease after which [Sergt. Wood] will report to these Head Quarters without delay.[1129]

Unaware that the Captain Witcher's Co. G of the 3rd Virginia Cavalry was under command of Col. Brown and working in conjunction with the 13th to scout the area around Barboursville, 1st Brigade Quarter Master William McKinley wrote to 13th Quarter Master Stephen Comstock on September 11:

> Lt. Yours of the 10th inst. with Requisitions for approval received; amongst the latter I find one for C. C + G Equipage to clothe and equip Recruits from 3rd Va Cavalry. I have no knowledge of the existence of any such Company, if there is give me the necessary information. The two other Requisitions have been approved and forwarded.[1130]

On September 11, General B.F. Kelley, commanding the Department of West Virginia, received notice from the War Department at Washington regarding his endorsement referenced in correspondence from Captain John V. Young to Washington requesting that Young's Company be transferred to the 13th Virginia Volunteers. A copy of the letter resolving this issue was forwarded to Col. Brown. Adjutant General Thomas Vincent writing on behalf of the War Department responded as follows:

> In reply I am directed to inform you that the records of this Department show the Company to have been mustered into the 11th Regiment, and there is no evidence of its transfer to the 13th as Stated by Captain Young; Even if the Transfer was ordered by the Adjutant General of the State, it was invalid, and cannot be recognized. This Company is to all intents and purposes Company 'G' of the 11th Regiment, and must be so known.

This notwithstanding (or perhaps Col Brown had not yet received notice of the War Department's decision regarding Co. G), Col. Brown continued to appeal to West Virginia State Adjutant General Pierpont on behalf of Co. G officers, Young and Elkins, who wished to remain with the 13th. September 15, Brown wrote:

> There are 47 men of Co G mustered into the 13th Regt, and 38 men mustered into the 11th Regt who are doing duty in the same Co Which makes the whole No 85 men — Not having sufficient men in the 11th Regt I cannot get Clark Elkins mustered into

service on the commission issued to him in the 11th Regt June 6th 1863. —

He has been performing the duties of 1st Lieut ever since the date of his commission of 2nd Lt in the 11th Regt — There being a sufficient No of men in this Regt to enable him to be Mustered in as 1st Lt I respectfully request a commission to be issued to him as 1st Lt in the 13th Regt V.V.I. to fill vacancy to rank as such from Jan 6th 1863 — (the date of his former commission) This will enable the Co to have a <u>Lieut</u> which I deem absolutely necessary

Clark Elkins has been performing the duties of 2nd Lt since about the 6th of June last I therefore respectfully recommend that Capt John V Young + Clark Elkins both be commissioned into the 13th Regt with the squad of 41 men who were originally mustered by Major Weed in May 1862 in order to make the whole muster regular this proceeding would give satisfaction to the officers + men of that Co[1131]

Heavy scouting duty continued continued for the Kanawha military district. Scouting expeditions of several days duration were sent out from Camp Piatt to Fayetteville September 11 through the 13. September 12, there was a skirmish in Roane County. The 13th was likewise sent out by companies on scouting expeditions. The result of one of these sweeps was the case of George W. Hensley, who was brought in by Co. I. On September 10, Col. Brown wrote to Judge H.J. Samuels regarding George W. Hensley, then in Brown's custody, who was suspected of supplying Confederates in West Virginia with supplies such as salt, bacon, arms and ammunition. Brown wrote:

I have a man under Guard at this place by the name of George W. Hensley who is charged

I. For bringing 21 Bbls salt into the state and selling it without permission —

II. For purchasing quantities of bacon + bringing it into the State without permit —

III. For having in his possession arms and ammunition, such as Lead, powder, Guncaps, and guns, one a breech loading carbine without authority —

What had I better do with him? Turn him over to the Civil Authority? He has been violating the Revenue laws and is a fit subject

for your disposal — I think you had better make him give good
security for two or three thousand dollars for his future behavior —
Please write to me what I had better do with him.[1132]

On the back of Brown's letter is Samuels' response (dated "Guyandotte, Sept. 12th
[18]63"). It reads: "Col. Brown State Courts have no jurisdiction, for, violation of
N.I. laws + regulations, if you can prove he has sold liquor (without license) it will
give the circuit court jurisdiction. H.J. Samuels"

September 14, Col. Brown wrote to Rigdon Williams, Captain and Provost Marshall, at Charleston:

I have the honor to forward to your charge George W. Hensley
for trial + punishment he being guilty of several charges as follows

Charge 1st

of violation of the Revenue laws of the United States

Specifications In this that the said George W. Hensley did purchase
from Boats passing down the Ohio River and bring into this state
(West Va) a certain lot or quantity of salt say 21 Bbls more or less
and did sell the same to the citizens of this State without permission from the proper authority

Specification 2d In this that the said George W. Hensley did purchase a quantity of Bacon and brought it into this state in Cabell
County without permission

Charge 2d

Specification In this that the said George W Hensley was found in
possession of a Lot of arms and ammunition such as Guns Powder
Lead and persussion Caps + all this without written authority The
lead found in his possession is now at these Head Quarters The
Powder Arms + Caps are to be had at Guyandotte.

This man was arrested and brought here by Lt. Wm. E. Feazel
who now commands the Post at Guyandotte and he will be the
proper person to call upon for the witnesses against him

This man being guilty of violating the U S laws will of necessity be tried before the U.S. Court from all the information that I
can gain he seems to be a very bad man and deserves to be punished
severely for his actions.[1133]

On the back of Brown's letter is a notation made by Provost Marshal Rigdon Williams, dated "Charleston, September 14." It reads: "Send some person to swear out a warrant and also send witnesses, names + residence."

Thursday, September 17, Col. Hayes sent word to Col. Brown, to arrest and send to 1st Brigade Head Quarters three soldiers belonging to the 1st Virginia Cavalry (Union). Privates Benjamin D. Fortny, Miller Dunaway and John Shafer had all deserted from Camp White and were "supposed to be in your [Col. Brown's] neighborhood."[1134] September 19 through 20, the battle of Chickamauga, a great tactical victory to the Confederate arms, was fought in Chickamauga, Tennessee. On September 21, Brown issued a series of Special Orders. Among them was Special Order No. 42 directing that:

> Capt. Greenbury Slack of Company A will take the recruits of the regiment not mustered into service and proceed to Charleston W.Va. and have them mustered into the U.S. service. After which he will detail six (6) men of Said recruits and proceed to Walton Roan C[o]. for the purpose of having the recruits which are at that place[1135] mustered into the U.S. service. The remaining recruits at Charleston will be placed in Charge of Sergeant Hemmings who will report with them without delay to these Head Quarters. Capt Slack after accomplishing his mission will return Also without delay to these Head Quarters.[1136]

Outside of their military duties, the men found diversion in politics and vice. Captain Carter gave voice to his political opinion regarding Kellian V. Whaley, candidate. Carter spoke at a public meeting held at Point Pleasant Court House (the Court House at that time also in use as a U.S. Army Hospital) in the evening of September 19. An anonymous member of the audience, self-described as a convalescent soldier took note of Carter's remarks.

"Convalescent" indicated that Carter took issue with Congressman Kellian V. Whaley, who was at that time seeking re-election as Representative to the U.S. House of Representatives from the newly re-organized Western portion of the State. Specifically, he was seeking election as Representative for the Third District (formed in 1863 from the counties of Kanawha, Jackson, Mason, Putnam, Cabell, Clay, his home County of Wayne and also Logan, Boone, Braxton, Nicholas, Roane and McDowell counties). Whaley ran as member of what then was termed the Unconditional Unionist Party (precursor of the Republican Party in Western Virginia). No dispute, Whaley had force. He was a major player in Western Virginia, first as entrepreneur and mercantilist (trading chiefly in lumber), then as Congressman from Wayne County. Having lost his seat in Congress after Virginia's secession from the Union,

Whaley then served as a soldier in 9th Virginia Infantry with the rank of Major. He had command of forces at Guyandotte when that place was overrun by Confederate military and local secessionists in November of 1861. He was captured at Guyandotte with other Union soldiers and pro-Union civilians. Whaley with other prisoners were marched off to Richmond and starved for about six days, when he effected an escape from his captors. After his return to the Northwest, he served as aide to Governor Francis Pierpont, who had been appointed to act as Governor for the Restored Government of Virginia—restored upon a loyal footing to the U.S. national Union.

As aide to Pierpont, Whaley furthered the militarization of Western Virginia by recruiting and supervising the organization and outfitting of new regiments such as the 13th Regiment. Captain J.D. Carter's comments suggest that he disagreed with what was perceived by him and others as a tempering of Whaley's posture, a watering-down of his earlier, stronger pro-Union positions to placate local secessionists and slave-holders. A strong pro-Unionist stance and a full-on willingness to do what was necessary to squash the rebellion and force peace was what had been promised and what was expected—this was Carter's posture. This included raising fighting men by conscription; raising taxes; and confiscation of property as the reward for rebellion. Such issues were understandably of importance to men in hospital and in the field. They were enduring hardship, suffering and risking life and limb for the Union cause as both a national and local matter. "Convalescent" wrote (dated "U.S. Gen[eral] Hospital, Sept. 21"):

> *Mr. Editor:*— Capt. J.D. Carter, Co. E. 13th Reg. W.V.V.I addressed the soldiers and citizens of this place, Saturday evening the 19th inst[anter], at the Court House.
>
> The Captain started off on facts, as he introduced the record to prove his assertions—asking no one to believe him but to take the [Giebes?] and examine for themselves. He reviewed the history of Mr. Whaley—ex-Congressman from Wayne District—now before the people for their suffrage from the beginning of his political career up to this time. He scrutinized and sifted the acts of Mr. Whaley, in the last Congress very close, and introduced some stubborn facts, which we as friends of Mr. Whaley's ask him to disprove or explain to our satisfaction, if he expects our support in this election. They are as follows, viz:
>
> Voting against the Resolution, 'That it was no part of the duty of the Army of the United States, to catch and return fugitive slaves.' Dodging the revenue Bill; Voting against the Conscription Bill; Dodging Confiscation Bill & c. These bills we consider of vital importance, and we demand an explanation of your actions

in regard to them Mr. Whaley, for in this crisis the affairs of our country, we need men of nerve and backbone—men who will show their hand—men who have the moral courage to stand up to a thousand Vallandigham's [Carter refers to the Peace Democrats or "Copperheads"] and vote men and money to put down this unholy rebellion. These are the kind of men we want in the 88th Congress. Let us hear from you.[1137]

September 23, Col. Brown wrote to G.M. Bascom, Captain and A.A. General regarding the unauthorized sale of spirits to his soldiers by Ohio merchants. Seeking relief from the problems his command was experiencing as a result of these sales Brown wrote to the Adjutant General:

> This Regiment is stationed at this place [Barboursville] which is about Four (4) miles from Haskinsville a small village on the Ohio side of the River In that Town there are several whiskey Shops that make it a business to smuggle it across the River and sell it to my men which tends much to demoralize them and causes me a great deal of unnecessary trouble These same liquor mongers practiced this same thing when the 5th Va was here and had well nigh distroyed the Regiment I would respectfully request your permission to suppress this trafick either by destroying the Liquor or shutting up the shops This I think to be necessary for the benefit of the men + also the good of the service.[1138]

September 24, Adjutant General Martin P. Avery at 1st Brigade Headquarters wrote to Col. Brown notifying him that, he (Avery) had on this day, forwarded to Brown "one box containing a 3 months supply of books and blankets. Please acknowledge the receipt on the accompanying blank."[1139]

September 27, Col. Brown inquired of 1st Brigade Head Quarters as to whether there was any authority to issue rations to the destitute families in the locality of Coals Mouth. 1st Brigade's ledger of "Letters Received" records that Brown wrote, that "there are many families in vicinity of Coals Mouth in destitute circumstances. States their support is in the 8th and 1[3th] Va. I.: Requests to know if there is any order to issue rations."[1140] September 28, Brown issued Special Order No. 51 directing that Capt. John Witcher's 3rd Va Cavalry "direct a detail of 25 men under the Command of Lieut LeSage to accompany the Deputy Sheriff of Cabell Co. He will report to these Head Quarters with his Command on or by October 4 next."[1141]

On Tuesday, September 29, Col. Hayes wrote to Col. Brown requesting that when sending prisoners to Charleston – "soldiers or citizens" — the guard should report to 1st Brigade Head Quarters before delivering them to Charleston.[1142] September 30, the 13th Regiment drew clothing.[1143] Also on this date Col. Brown wrote a third letter to 1st Brigade Head Quarters. All three letters requested that "an inspector be appointed to condemn tents in possession of 13th Va.V.I."[1144]

Morning reports for the month of September 1863 provide the following information. Reports submitted for Co. A show that the it was stationed at Camp Defiance from September 1 to 4. On the 4th, the company left Coals Mouth and marched eighteen miles to Mud Bridge. The next day, the company left Mud Bridge and marched eleven miles to Barboursville. Co. A was stationed at Barboursville, Cabell County, for the month.

On September 2, Private R.H. Snodgrass (Co. A) died of disease at "Post Hospital Point Pleasant." September 3, one enlisted man, Elijah F. Newell, returned to the company from being absent sick. John Moore was reported "returned to his Company" on Sept. 4. George W. King was reported "left at Coals Mouth" and Nathan Snodgrass was for the same date (September 4) reported "on duty." September 5, George W. King, Nathan Snodgrass and Charles P. Quigley returned to the company. September 7, Corporals Edens and Page and Joseph Scott "returned to the Company and reported for duty." I. Hurel reported back to the company for duty on September 9. John M. Naylor and Francis M. Cobb were reported "returned for duty" on September 12. The number of present and absent reported for Co. A at the end of the month was noted as three commissioned officers and eighty-three enlisted men. One drummer was reported present for the month and one artificier was reported present with the company for the period of September 1 through the 15. After an absence, an artificer was reported present again with the company at Barboursvlle September 21 to 30.

No commissioned officers, nor enlisted men were absent with leave. Enlisted men absent without leave continued to be a problem everyday until September 20. Until that date from one to three men were absent at any one time. Abraham E. Chandler was reported absent without leave on September 2. Corporal James Tully and Samuel Midkiff returned to camp with Chandler in tow on September 20. Andrew J. Davis was also reported absent without leave "since Sept[ember] 4." Davis returned to the company on September 9. After September 20, no one was reported absent without leave.

No one was assigned to special or extra duty. One private had daily duty September 1 to 4 and September 21 to 30. For the most part, the number of men on detached duty was small. September 9, Corporal James A. Means was detached to Co. I at Guyandotte. September 14, Lieutenant G. Danner and four men were ordered to Charleston with prisoners. Among the prisoners escorted on that day was Richard George of Co. A, who was sent to Charleston, in arrest on the 14th. Lieut. Danner and "two men returned from Charleston, September 21st." Also on the 14th,

Sergeant Thomas Moore and wagoner Jabe [Job] Hall were detached to go to Winfield, Putnam County. They returned to the company from Winfield on September 20. Captain Greenbury Slack and twenty men were ordered on a scout on September 17. They returned to camp on September 18. September 21, Corporal James A. Means returned to camp from being on detached service. Capt. Slack (together with Sergeant Elisha Hennings and one other man, all ordered to Charleston) returned to the company on September 21. Sergeant Hennings returned to camp from Charleston on September 25.[1145] On September 24, Col. Brown issued Special Order No. 49, which directed that Robert Gray of Co. A be ordered back to his company and "report for duty immediately."[1146]

Company A had few men present sick in camp during the month. Just three men were present sick for twelve days out of the month. Six to sixteen enlisted men were absent sick over the course of the month. Just six men were absent sick during the last week of September. Corporals Edens and Page, Irdell Hurel, Ralph Pauley, Joseph Scott, John G. Moore, James Toothman and William Grey were all left behind at the "Hospital, at Coals Mouth" on September 4 when the company marched to Mud Bridge. Perry Gatewood had been "absent sick since Sept[ember]" and James Light had been absent sick at the "Hospital, at Point Pleasant" since September 5. John Moore and James Toothman reported to the company "unfit for duty" on September 7, and James Light and William Grey reported to the company unfit for duty on September 8. On September 20, Perry Gatewood reported back to the company for duty and the same day was placed on daily duty at the hospital. Corporal Ashar Ramsy reported to the company unfit for duty on September 21. September 23, William W. Riley reported to the company for duty. Riley had been discharged from "Post Hospital at Point Pleasant." One enlisted man was reported absent in confinement September 15 through 30.[1147]

On Wednesday, September 2, Robert H. Snodgrass, Private, Co. A, died at "General Hospital Point Pleasant." He had been admitted there the previous month on August 17 with "enteric fever" (typhoid fever); had been treated at Point Pleasant regimental hospital; and died of complications brought on by this sickness. Robert, was the eldest son (of eight children) of Martin Snodgrass, who also served in Co. A. Records at the National Archives indicate that at the time of Robert's death, Martin, the father, was very ill himself and was not expected to live[1148] – "taken down with fever which resulted in trouble of head and back + rheumatism."[1149]

Company B reports indicate that the command was stationed at Camp Defiance September 1 through 4. September 5 and 6, the company was marching. September 7 through 30, the company was at Barboursville. "Present and absent" for the month included a drummer present with the company September 1 and 2 and 7 through 24. One fifer was present September 7 through 30. Three enlisted men were absent with leave September 3 through 10, and two enlisted men were absent with leave September 22 through 28. One enlisted man was absent with leave September 29 and 30. Three enlisted men were absent without leave September 11 and 21 through 24. One enlisted man was absent without leave September 12 through the 20. Total present and absent reported for Co. B at the end of the month was three commissioned and ninety-eight enlisted men. Special Order No. 48 issued from Head Quarters Co. B "[r]educing Corpl A C. Waller [Anderson C. Wallar] to the ranks and of promoting private H.J. Sines to fill the Vacancy" was approved by Colonel Brown on September 24. Sines was to "be obeyed and respected accordingly."[1150]

No one was reported doing special or extra duty during the month. Two to four privates had daily duty each day. Three to sixty-five men were detailed for detached service each day with large squads detached September 15 through 20 (consisting of two commissioned officers and sixty-three enlisted men) and September 25 through 30 (comprised of one commissioned and seventeen enlisted men). "Remarks" following morning reports provide more details and some discrepancies. The detachment which departed on September 15 returned on the 20th with the notation that "all but three men absent without leave viz. M. Cullins, C.J. Eckard and R.B. Marshall, Anderson C. Waller Corporal reduced to the ranks for misdemeanor."[1151] Adam W. Roberts was detached for Provost Guard at Charleston on September 1. On September 14, Captain Milton Stewart, Lieutenant L.C. Rayburn and fifty-two enlisted men of Co. B went to Winfield. September 24, orders detailing detachments from Co. B and Co. F for recruiting duty were issued from regimental head quarters. Capt. Milton Stewart, one sergeant, three corporals and ten privates left camp on September 24 pursuant to this order to drum up enlistments for the regiment.[1152]

Co. B reported three to ten men present sick during the month and nine to twelve men absent sick. Three privates were reported present in confinement September 25 through 30. No one was reported absent in confinement.[1153]

Morning reports submitted for Company C indicate that the company was stationed at Camp Defiance September 1 through 6. On September 5, they received marching orders and left Camp Defiance at 7 o'clock in the morning of the 6th. The company was stationed at Barboursville, September 7 through 31. Co. C had one drummer and one fifer present each day. No one was absent with leave. One enlisted men, Asa S. Johnson, was absent without leave September 7 through 11. Johnson returned to the company for duty on September 12.

During the course of the month, Co. A gained two enlisted men. Hezekiah S. Patterson enlisted at Coals Mouth on September 3 and Hiram Nevill enlisted on September 5 at Coals Mouth. The company lost Capt. Van D. McDaniel, who died at Point Pleasant, of "Camp Dysentery" on September 10. On Sunday, September

13, commissioned officers of the 13th met at Thirteenth Regiment Head Quarters Barboursville

> for the purpose of giving suitable expression to their profound and deep felt regret and sorrow at the death of their much beloved and highly esteemed friend and companion in arms, Capt. V. McDaniel, of Co. C, of this Regiment.
>
> Lieut. Col. James R. Hall was called to the Chair, and Captain A.F. McCown appointed Secretary.
>
> The following resolutions were offered and adopted unanimously:
>
> *Resolved,* That in the death of Capt. V. McDaniel, the Regiment has lost one of its most worthy members, and an efficient officer; one, who was always to be found at the post of duty, and one who as a patriot and a soldier, was ever ready to sacrifice everything for his country.
>
> *Resolved,* That while we shed the tear of sorrow, over one so suddenly cut down in the prime and vigor of manhood, yet in humble submission we bow to the inscrutable decree of an overruling Providence "who doeth all things well."
>
> *Resolved,* That to alleviate and soften down those pangs of grief and sorrow that now fill the bosom of his bereaved and widowed wife, we wish to assure her, that the noblest instinct of our nature, are enlisted in her behalf and that his surviving companions in arms fully realize the loss that she sustains in this sudden bereavement.
>
> *Resolved,* That the officers of this Regiment, wear for thirty days the usual badge of mourning, as prescribed by Army Regulations, paragraph 299.
>
> *Resolved,* That the widow of the deceased, be furnished with a copy of these resolutions, and a copy be sent to the Point Pleasant *Register,* for publication.
>
> Lieut. Col. James R. Hall, Chairman.
Capt. A. F. McCown, Secretary.[1154]

The following obituary appeared in the *Weekly Register*:

> After a lingering illness of four weeks duration, our esteemed friend—Capt. McDaniel—peacefully breathed his last under his paternal roof and surrounded by the consoling presence of a loving

wife, a doting father, an affectionate mother and sympathizing sisters and brothers.

We knew the lamented deceased well. He possessed a combination of estimable qualities seldom found in one so young. He was entirely devoid of love for ostentation; choosing, rather, to act well than to simply appear well. He possessed a kind, genial, affectionate nature, that drew around him, as if by some magnetic influence, all deserving of his companionship or respect.

Previous to his entrance into the army, Captain McDaniel followed the quiet and honorable occupation of a farmer[1155]—living upon his father's farm, on Old Town Creek. As a young farmer, he was ever honest, frugal, industrious and enterprising.

Upon the breaking out of the war, the militia in this part of the old State, was re-organized, and Capt. McDaniel was elected first Major, which position he held 'till July, 1862, when he 'went into the service.'

When the 13th Regiment of V.V.I. was recruiting, Major McDaniel forsook his pleasant home, and raised from among his neighbors and acquaintances, a company of men, and was elected Captain; which position he filled at the time of his death.

As an officer, he bore an enviable reputation among his brother Officers—ever attentive, vigilant, kind and obliging, and while he possessed in an eminent degree, the confidence, respect and love of his superiors and equals in rank; he also, by his assiduous regard for their welfare, by his forbearing, conciliating treatment of faults, and at the same time by his rigid adherence to impartial justice, so won upon the good opinion of the patriotic soldiers, as to insure their unbounded respect for him as an officer, and their heartfelt affection as a man.

Capt. McDaniel acted well his part in life. Domestic in his habits, and retiring in his disposition, he dearly loved the secluded, quiet walks of life; yet, being a patriot, when his imperiled country called for his help, he hesitated not, to cast all these fond pursuits and endearments behind him, and to buckle on his sword and stand forth as one of his loved country's defenders, though to do this cost him peculiar struggles and sacrifices.

Doubtless his proud, though undemonstrative spirit longed to meet the great conqueror—Death—on the battlefield, and that his last lingering gaze might decry the discomfiture of the severed ranks of ingrate rebels against his cherished, his beneficent country; yet, in the inscrutable workings of a mysterious Providence, his ardent desires were never to be realized; yet death cheated him in but part; for he died the death of a soldier—died from a fever caused by the exposures of a soldier's life—died amid the surroundings and sympathies of loving kindred—yet he yielded up his life as a sacrifice upon his country's altar. *Requiem pace*, though true, gallant and brave soldier! Ah! how true is it, that 'Death loves a shining mark.'

Capt. V. McDaniel, of Company C, 13th Reg. V.V.I.—the pride of his father, the idol of his mother, the cherished companion of an estimable wife, the loving brother of younger brothers and sisters, the respected citizen and the honored soldier—died in the full fruition of vigorous manhood, but not 'till he had chiseled a nich[e] in the temple of fame, that will endure 'till time is no more.'

Though it was denied to our loved friend, to witness the complete triumph of the glorious cause he had so zealously espoused, yet he was permitted to see the sure evidence of speedy and ultimate success, and in fancy doubtless he rejoiced over a re-united country.

In tendering our warmest condolence to the grief stricken wife, the disconsolate parents, and the bereaved relatives of the gallant dead, let us, his friends and associates, honor and revere his memory, by emulating his intrinsic excellency of character, his noble, generous, magnanimous and *patriotic* nature.[1156]

September 10, Private Ira Williams, Co. C, was discharged for disability by order of General B.F. Kelley (commanding Department of West Virginia). At the end of the month, Co. C reported a total of one commissioned officer and eighty-four enlisted men present and absent. No one was assigned to special or extra duty during the month. Three privates had daily duty September 19 to 30. Six enlisted men were detailed for detached duty September 1 to 18, and three enlisted men were detached for the period of September 19 to 30. Company C had for the month of September, two to four men present sick, and one to seven men absent sick. More men were present sick and fewer men were absent sick as the month progressed. No one was reported present or absent in arrest or confinement.[1157]

Morning reports for Company D show that for the month of September the company was stationed at Camp Defiance September 1 to 3. September 4 and 5, the company was marching. September 4, they marched from Camp Defiance to Mud Bridge a distance of "eighteen miles," and on September 5, they marched from Camp Mud Bridge to Barboursville a distance of "twelve miles." They remained at

Barboursville September 6 through 30. Co. D had one fifer present on September 12. No one was absent on leave. One to three enlisted men were reported absent without leave September 9 and 10 and 12 through 17. On September 9, it was reported that William Wolf had not returned from leave of absence. Wolf had received a pass at Point Pleasant on August 11 for five days leave to go to his home in Jackson County, West Virginia, and had not yet returned.

Recruits were added to the company during the month. Four men enlisted in the regiment on September 9, and another man, James Swain, enlisted on September 30.[1158] One enlisted man, Lewis C. Barnes, died on September 5 at his home in Jackson County of typhoid fever. Two enlisted men were transferred. Samuel D. Hanna was transferred to the non-commissioned staff roll on September 1. John Piatte was transferred to Co. I 13th West Virginia on September 7. September 10, two soldiers of Co. D, Privates John Embleton and John Pickens were discharged from Point Pleasant "for disability" by order of Major General B.F. Kelley (cmdg. Dept. of West Virginia). September 20, Assistant Surgeon C.D. Dally wrote to Col. Brown from "U. S. General Hospital, at Point Pleasant" requesting "proper Descriptive Lists for John Pickins + John G Embleton of Co. D 13th V.V.I. so that I can make out their final discharge."[1159] The company finished the month with three commissioned officers and eighty enlisted men present and absent.

As for Private Lewis C. Barnes, who died of typhoid fever, his mother, Jane Barnes, applied for the soldier's pension as dependent parent. Her affadavit reveals much about the soldier, the family, the challenges and sacrifice of people living in West Virginia at this time. Lewis was described as an industrious young man, who, from an early age, worked to support his parents and sisters. Lewis had two brothers. He may have been the oldest as the lot of supporting the family fell to him. He did some work as a day laborer but in the main, he managed the family farm before enlisting in the 13th Virginia. He enrolled in the regiment on August 15, 1862, at Mason City and was mustered in as Private, in Co. D, on October 8, 1862. He was at this time about twenty-two years of age (born in Belmont County, Ohio). Descriptive rolls record that he was five feet and six and one half inches tall; of fair complexion; with grey eyes and brown hair. He was a farmer.

Lewis was reported on company rolls from enrollment until August 1863 when he was taken sick and sent to hospital on the 23rd. Hospital records for the regiment record that he received medical treatment on July 4; August 18; September 20; September 22; and September 23 of the year 1863. He was treated with "powders, camphor and morphine" among other things written in abbreviated form and which are fairly illegible now on the old ledger cards.

By September, if not before, Barnes was taken ill with typhoid fever. He returned to his home on the family farm in Jackson County a few days before his death. The family physician, Dr. Henry Lisle, stated in sworn affidavit that he lived half a mile from the Barnes farm, which was located on "the road leading from the river," and that he assisted the soldier's father in bringing Lewis home. Lewis' date of death is given uniformly in company records as "September 5, 1863," but this could hardly be, if he was yet to receive medical treatment on September 20, 22 and 23. It would not be the first time in these old records that a date was gotten wrong in the first instance and then was repeatedly copied into other records. Perhaps a death date of September [2]5 is actually the correct one.

Lewis' father was Dorsey H. Barnes. Beginning about the year 1863, Dorsey had become broken down in health and "deranged" and was incapable of supporting his family. Eventually, he was repossessed of the family farm. He died "on or about March 28, 1882."

The family was loyal to the national government. Three sons had joined the Union army. These were James M. Barnes in Co. F 4th West Virginia Infantry Regiment; Wesley Barnes, also Co. F 4th West Virginia; and Lewis in Co. D 13th Regiment. Lewis and Wesley both died in the service. Tragically, one of Lewis' sisters, while nursing the brother, herself contracted typhoid fever and she also died. The family was shattered by the war. The family farm was repossessed. Lewis' state medal issued after the war went unclaimed.[1160]

To return to the duty assignment section of regimental records for the month of September: one non-commissioned officer and four privates were detailed for daily duty September 1 through 3 and September 6 through 30. Four to six enlisted men had detached duty September 1 through 3 and September 6 through 30. Over the course of the month, Co. D had one to ten men present sick and ten to nineteen men absent sick. The company reported the highest numbers of sick men after the march to Barboursville. There was apparently some concern at regimental headquarters regarding the health of Captain Simon Williams and perhaps reluctance on his part to have himself seen to inasmuch as on September 28 Colonel Brown issued Special Order No. 52 directing that:

> Capt Simon Williams of Co. D. of this Regt will proceed to
> Charleston W Va and report without delay to Geo M. Kellogg —
> Surg. U.S.A. + Medical Director, for examination[1161]

Co. D had the following men present in arrest or confinement: seven privates on September 1; three privates on September 2, 3, 6 and 8; one private September 9 through 11. No one was absent in confinement.[1162]

On September 19, Gen. Scammon wrote to 1st Brigade Head Quarters complaining of Captain J.D. Carter, commanding Co. E 13th Regiment. Scammon informed that "Capt. Carter, of the 13th Va. V.I. has been reported for making political speeches and being absent from his command," and he ordered the commanding officer of 1ˢᵗ Brigade (Col. Hayes presumably) to "Direct the commanding officer of the 13ᵗʰ Regt. to forbid an officer or soldier to be absent from that regiment without authority from Division H'd Qrt. Except on Military duty."[1163] Indeed, we know that at least on one occasion Carter did hold forth regarding candidate Kellian V. Whaley. September 26, Brown wrote in reply to Scammon's complaint that Carter was at the time "on duty by order of Col. <u>Hayes</u> Comd'g Brig. & etc."[1164]

Clothing was drawn in September. Private George W. Smith, Co. E, for one, drew on September 10: 4 pair of socks at $1.28; 1 shirt @ $1.30; 1 pair of pants @ $3.55; 1 shirt at $1.30; 1 haversack @ .56; and 1 pair of shoes @ $ 1.48.[1165] Mark E. Robison, Private, Co. E, found occasion to write home from Barboursville. He penned the following letter on September 25:

Camp Barbers Ville Cabbille Co West Va

September the :25: 1863

The Cotage Home

> after a silence of a few days and being sper rated from Home i will wonce mear take my pen in hand to let you no that i am well and harty at this time and hope thes few lines may find you ingoying the same Blessing and the blessing of God this is the second leter i have riten you and i hope you all will not for get mea as i cant cume Home ner i cant tell you When but i hope soon fer i never nue the loss of a home be four all though i ame well sadisfied with my life a solgea life a greas with mea very Well and all wase hase and blesse the lord for hit for i hope and trust i may sea this war to a close soon for they is a great prospect of hat now for rose crance and Burnsides is giving them devil know they have had a prety hard fight at Chatnuga Witch lasted 2 days he repulst tham with great loss the loss one both sids was a bout thirty thousant and rose crance was Still giting reinforse ment and vancing one to tham so i think We Will soon no What We Will do for i think the gig Will Soon be upe with the Rebles i ame in hops so far hit cant go upe eny too soon for mea We ar still scouting hear the Boys is out one a scout know and i thik we will stay her for some time and perhaps all Winter the Boys all don't like to stay her be case hit tis sech out of a way place and they is nothing hear and Wilderness looking place so i will bring my few lines to a close

Joel is well and harty he has had a rasen one his foot but he is a
bout well know so no mour at present yours untell death so good
By Dear father and Mother un tell i rite a gane Mark E Robison to
Mr John Robison rite to mea soon[1166]

Columbus Shrewsbury, who together with Will Sherwood serviced the 13th Regiment as sutlers, wrote home to his wife from Barboursville. While the letter was not
dated, it seems to date to a time just after the regiment moved to Barboursville and
sometime before Shrewsbury sent another letter home on October 10. Shewsbury's
letter conveys the hope commonly expressed in letters home from the regiment at
this time—hope born of recent victories to Union arms—that the war would soon
be at an end.

> Dear Wife
>
> The Regiment has moved from Coals Mouth to this place
> probably for the winter I did not get to Coal before Sherwood left
> and had to walk about thirty miles it made me very sore and tiered
> My health is stil improveing we have got a good house for our shop
> and a good Boarding house
>
> Thier is no news of any kind to write about only I think the war is
> closeing very fast and probably will be over by next Spring Fanny
> you must be a good Girl until I come home and we will get some
> candy [Fanny was his daughter]
>
> Direct to Barbersville Cabell Co Va (in care of Suttler 13th Va Vol)
>
> C. Shrewsbury[1167]

Sixteen-year-old Jasper M. Hill, who would serve as a musician for Co. E, was
enrolled at Barboursville by Captain Milton Stewart on October 12. He was mustered into service next day on October 13, 1863, at Charleston. He was described as
five feet and three inches tall; of fair complexion; with black eyes and brown hair. He
was born in Mason County, Virginia, and prior to enlisting he had been a farmer.
He would die, a prisoner of war in Salisbury prison, North Carolina, on January 9,
1865.[1168]

Morning reports submitted for Company F for September 1863 indicate that
the company was stationed at Camp Defiance September 1 through 3. September
4 and 5, the company was marching. September 4, Cos. A, D and F left Camp
Defiance and marched to Mud Bridge and bivouacked. September 5, Co. F marched
from Mud Bridge to Barboursville. It arrived at the latter place at "12 o'clock m[id-
day] and went into camp." September 6 through 30, the company was stationed

at Barboursville. Co. F had a drummer present each day of the month. There were no buglers, fifers, farriers, blacksmiths, nor artificiers reported present. No one was absent with or with out leave. One recruit, Alexander McDonald, enlisted in the regiment on September 1 and joined the company "on the 5th inst[anter]." At the end of the month, Co. F numbered three commissioned officers and eight-one enlisted men. On September 7, Capt. A.F. McCown was on special duty as Provost Marshal. McCown returned to the company on September 27.

Soldiers of Co. F were detailed for detached duty each day of September. Over the course of the month, three to twenty-eight men had detached service each day. September 17, Thomas Buchanan returned to duty from "Post Hospital at Point Pleasant." September 14 through 20, a group consisting of one commissioned officer and twenty-seven enlisted men were detached. This detachment included "20 men of this company [Co. F] under Lt. Russell [who] marched 19 miles on their way to guard the <u>Court</u> [at Winfield]." They arrived at Winfield on September 15 at "12 o'clock M. having marched 30 miles from Barboursville." The detachment under Lt. Russell returned from Winfield on September 20. On September 25, "Lieut[enant] Russell with 5 men started to Mason Co[unty] on a recruiting tour."[1169] Captain Milton Stewart (cmdg. Co. B) with eight men from that company was likewise ordered out on September 24 to recruit in Mason County. From one to four men were reported present sick each day over the course of the month. Five to ten men were reported absent sick. Numbers of absent sick declined over the last three weeks of the month. No one was reported present or absent in arrest or confinement.[1170]

While at Barboursville, some soldiers had visits from family. Roush related in his book, *If Thou Wilt Remember,* that the two friends, Frank (Francis) Van Meter and David Burrows (both of Co. F), arranged to have their wives down for two days. Frank made arrangements for the women to stay with friends. Burrows wrote to his wife, Lovina, on September 7 to be ready to come with Frank's wife when she came down on the boat.[1171]

On Tuesday, September 15, Col. Brown wrote to F.H. Peirpont at Wheeling from 13th headquarters at Barboursville again pressing the need for a 1st lieutenant for Co. G:

> I have the honor to call your attention to the condition of Co G of this Regiment, and present you a statement of the same.
>
> There are 47 men of Co G mustered into the 13th Regt, and 38 men mustered into the 11th Regt who are doing duty in the same Co — which makes the whole No 85 men. — Not having sufficient men in the 11th Regt I cannot get Clark Elkins mustered into service on the commission issued to him in the 11th Regt June 6/63. —

He has been performing the duties of 1ˢᵗ Lt. Ever since the date of his commission of 2ⁿᵈ Lt. in the 11ᵗʰ Regt, — There being a sufficient number of men in the Regt to enable him to be mustered in as 1ˢᵗ Lt. I respectfully request a commission to be issued to him as 1ˢᵗ Lt. in the 13ᵗʰ Regt. V.V.I. to fill vacancy to rank as such from June 6ᵗʰ, 1863 (the date of his former commission). — This will enable the co. to have a 1ˢᵗ Lieut. which I deem absolutely necessary. — I wrote to you some time since on the subject — perhaps it did not come to hand as I have not received any communication from you.[1172]

There is a note on the back of this letter saying that Brown's letter was answered by the forwarding of a copy of an order from the War Department emphatically stating that Young's Company belonged to the 11ᵗʰ and not to the 13ᵗʰ Regiment.

J.V. Young wrote to his wife and children from Barboursville on September 10 giving details of the day-in-day-out of the company since transferring to Barboursville; the results of scouting; the obvious evidences of war in that locale; that secessionists were claiming "to be Union" but would not associate with "Union people;" and his own frustrations in leaving his own family unprotected.

I am well this morning and have just returned from a scout in Wayne Co. I was on the Beach Fork on Twelve Pole but did not find any Rebs. This is a good farming country and the people and have once lived well here but Secessionism, that foul friend, has destroyed the country. There are very few Union families, either in Cabell or Wayne; and they say they haven't any neighbors. But all the Secessionists down there are trying to be Union. In all my Scout I did not find one man or woman but who claimed to be Union. But they refuse to associate with the loyal citizens, which shows what they are. I want no better mark. But the time is coming when they will be glad to have the privilege of associating with Union families. The poor, contemptible devils have the impudence now to claim protection from us. But as far as Co. G. is concerned they won't get it. I tell them they cant get protection from me while my own dear family is unprotected.

I will tell you how one old rich fellow on Beach Fork got protection from us yesterday. We got there about three o'clock and the first thing the boys did was in his milk house, then his good things generally. Next they found an old fiddle; ordered supper, then took possession of the parlor and went to dancing and had a regular Ball. Danced until supper, making the Reb and his negroes wait on them. This old devil claimed protection just because he had been at Camp Chase. When we left the boys thanked the old traitor saying that they would call on him again. We had a good time on our scout.

Everything is pretty in this country. General Samuels [West Virginia State Adjutant General Samuels probably] is here and says we will winter here, and I think it will be a very good place. I am boarding at Mr. Blankenship's tavern — $2.50 per week. A very good place. [...] Company G is in better condition at this time that it has been since last Fall. It is next to Company B in numbers, and in fact is called the best company in the Regiment.[1173]

Young's Company was sent out on another scout on September 21. On this day, Col. Brown issued Special Order No. 43 directing:

Capt J.V. Young of Co G will take fifty (50) men and will proceed to the vicinity of Hamiline[1174] on Mud River in company with the sheriff of Cabell Co

You will obtain all the information possible of the enemy. Arrest all the disloyal persons in arms against the Government you may find. Then return with your command to these Head Quarters without unnecessary delay.[1175]

For the month of September, morning reports for Company H show that the company was stationed at Camp Defiance September 1 through 5. On September 5, they left Camp Defiance for Barboursville. September 6, they were at Mud Bridge and "[a]rrived at Barboursville," and September 7 through 30 they were stationed at Barboursville.

Co. H had the following present and absent: one drummer present each day of the month, and one enlisted man was absent with leave September 2 through 16. On September 3, "Andrew J. Allen rec'd furlough for 10 days."[1176] September 4, Privates Samuel Weaver and George W. Childers were promoted to corporal. Two enlisted men were reported absent without leave September 1 through 9. On September 7, James C. Blair was first reported absent without leave. September 14, Blair returned to camp and was that day placed in confinement. September 17, Andrew J. Allen was reported absent without leave. Monday, September 21, Assistant Surgeon C.D. Dally wrote to Col. Brown from "U.S. General Hospital Point Pleasant" regarding Francis M. Crofoot, absent without leave from Co. H:

> I apprehended F.M. Crowfoot yesterday upon his own state-
> ment that he had broken jail at Pomeroy Ohio. I telegraphed to
> Sheriff of Meigs County Sheriff came today and took him back
> to jail I understood that he was a deserter of your regiment if you
> want him I thought I would inform you where he was.[1177]

Co. H had no gain nor loss during the month. At the end of the month, the company had present and absent three commissioned officers and seventy-seven enlisted men. No one was detailed for special, extra or daily duties. Men were detailed for detached duty each day with three enlisted men detached September 1 through 15, 21, and 28 through 30. On September 1, it was reported that "Charles M. Laurence returned to Regimental Hospital from detached service at Charleston, Aug. 1863."[1178] Eighteen enlisted men had detached service on September 16 through 20. Company "Remarks" contain details regarding scouting duty. "Sergt. Wm. Shannon and 14 Privates [went] out scouting Sept. 16, 1863." Shannon "returned with his command from scout Sept[ember] 20, 1863." September 21, Lieutenant O.W. Griswold and fifteen men were ordered out on another scout. Col. Brown issued Special Order No. 44 on September 21 directing 1st Lieut. Griswold to "take fifteen (15) men and proceed to Wayne Co. and obtain all the information possible of the enemy and report to these Head Quarters with his command on the 28th inst."[1179] Morning reports note that Griswold and his squad returned from their scout on September 27.

The Capture of Charles and William Furgeson

On September 18, 1863, Governor Arthur I. Boreman wrote to Col. Brown authorizing him to arrest two disloyal citizens, Charles and William Furgeson. A warrant for the two men had been issued pursuant to an Act of the General Assembly of Virginia [West Virginia] passed February 4, 1863, entitled "An Act Authorizing the Governor to arrest disloyal persons as hostages." The warrant called for the arrest of Charles and William Furgeson as a necessary measure

> to effect the release and safe rendition [of Morgan Garrett] a loyal
> citizen of Wayne County, W. Va. [who] has been seized and came
> out of [Wayne] county by soldiers under command of Colonel Fer-
> guson, who is a Military Commander acting under authority of the
> so-called southern Confederacy and that the said Morgan Garrett
> is now held and confined by said authority [...][1180]

On September 25, pursuant to Gov. Boreman's directive, Brown issued Special Order No. 50 initiating another scout in Wayne County and directing that

> Lt. Lesage [of Captain Witcher's Cavalry company] will take 25 mounted men and proceed to the residences of Charles Furgeson and William Furgeson rebel citizens of Wayne Co West Va[1181] and arrest the said Charles + William Furgeson as hostages for Morgan Garrett a Loyal Citizen who was seized and carried out of said County by soldiers under command of Col. Furgeson who is a Military Commander acting under the authority of the so-called southern Confederacy — [Lt. Lesage to] report with the prisoners to [Head Quarters 13th Regt. Va. V.I., Barboursville] without unnecessary delay.[1182]

Tuesday, October 13, Brown wrote to Gov. Boreman from 13th Head Quarters at Barboursville wanting to know what to do with the hostages:

> I have the honor to state that in accordance with Orders issued from your Head Quarters dated Wheeling Sept. 18th 1863 I have arrested Charles Furgerson + William Furgerson both citizens of Wayne Co. State of West Virginia They are now under arrest at these Head Quarters The stage, of the River However is such as to render it impossible to forward them to Wheeling as directed in your order Shall I send them by Rail Road or detain them until I can forward them by the River I wait instructions.[1183]

On October 15, Brown reported that Charles and William Furgerson had been arrested and delivered to 13th Head Quarters at Barboursville on October 11 and that the prisoners had been forwarded to Charleston on October 15.[1184]

To return to regimental reports for September 1863, Co. H reported two to eight men present sick each day of the month. Captain W.I. Mathews was reported sick September 7 through 11. There were eleven to thirteen men absent sick each day over the course of the month. September 2, Preston C. Laurence was "sent to Post Hospital, P[oin]t Pleasant with final statement Aug[ust] 1, 1863." September 6, "George W. Irby and Charles M. Laurence were sent to Post Hospital Gallipolis with final statement." No one was reported present in arrest or confinement. Two enlisted men were absent in confinement September 1 through 14, and three enlisted men were absent in confinement September 15 through 30. Among those in confinement was James C. Blair returned from being absent without leave.[1185]

September morning reports submitted for Company I show that the company was stationed at Camp Defiance September 1 through 4. September 5, the company was on the march. September 6 through 8, they were at Barboursville. September 9 through 30, they were stationed at Guyandotte. Co. I had been ordered to Guyandotte on Wednesday the 9th.[1186]

The company had present each day her commanding officer, 1st Lieutenant William Feazel, from two to four sergeants and eight corporals. On September 1, Lewis A. Rader, John J. White, John Martin and Elihu Brown were promoted to the rank of corporal. The company tallied seventeen to forty-four privates present in camp over the course of the month. Three recruits were reported having joined the company on September 20. No one was reported absent with leave. They had two enlisted men return from desertion. Irvin Lowe was reported "deserted" September 1 through 11 but returned on September 12. Hiram G. Snyder, reported absent without authority September 12 through 14 returned on September 15. At the end of the September, Co. I tallied a total present and absent of one commissioned officer and sixty-one enlisted men.

One private had daily duty September 26 through 30. Fourteen enlisted men were detailed for detached service September 1 through 12, and ten enlisted men were detached September 13 through 21. From one to six privates were reported present sick each day during the month. At least three enlisted men (the numbers of enlisted and commissioned officers absent sick has disappeared into a fold in the manuscript) were absent sick during the month. Among those absent sick was Joseph M. Cobb sent to "Point Pleasant Post Hospital" on September 1. One enlisted man was absent in confinement September 1 through 21.[1187] This man, James Johnson, was reported "returned from Confinement" on September 22. The return of Johnson may be attributable to Lieut. Feazel's letter of September 14 to 1st Brigade Head Quarters in which he requested that the charges and specifications against "Mr. James Johnson of this company, be withdrawn and that he be restored to duty."[1188]

On September 27, Lieut. Feazel wrote to Adj. John Cunningham that he was sending to 13th headquarters

> a man who claims the name of John Webb and says he is a discharged Rebel Citizen Prisoner with the enclosed paper which he claims is a pass from fort Delaware
>
> Webb when arrested had on a Federal Uniform and as I believe him to be a hard Case at best I Send him to you for examination. [...] P.S. Webb is supposed to be a Deserter from Wheeling[1189]

1st Lieutenant Henry Stump and the recruiting of Company K

As indicated above, the 13th Regiment was still incomplete. Being unqualified to be mustered in as a fully-organized regiment, it was not entitled to its full complement of officers. Consequently, efforts to recruit the last company, Company I, went on and continued until at last, in January of 1864, the regiment was completely organized. Now three years into the war, recruiting officers had the inenviable task of seeking out recruits in the interior counties. The obstacles impeding recruiters in this rugged terrain largely populated by supporters of the Southern cause can well be imagined. Communication with the "outside world" was a challenge and obtaining supplies and equipment for new recruits another. A few manuscripts do survive, however, giving some idea of what amount of perseverance and effort was required. A letter rich in unembellished detail, written by 1st Lieutenant Henry Stump, 13th West Virginia, then in the process of recruiting Co. K in his home county of Roane conveys to the reader an idea of the difficulties born by recruiters and recruits at this time. On September 3, Stump reported to Col. Brown that:

> My Recruits Numbering only 28 are now Rendez vousing at Walton, Roane Cty. and have been with the aid of the Militia scouting that Region of County. Our immediate vicinity is now cleared of Rangers We have captured 3 rebel soldiers and forwarded them to the Provost Martial at this place. My Recruiting has been considerably retarded by not being able to receive supplies in due season but now find labeled to me two Boxes containing no doubt the necessary Clothing My men are not supplied with guns.
>
> I would be glad you could give me 20 Guns a Box of cartridges + 20 Cartridge Boxes + 40 gun slings.
>
> I have now as I think Overcome the Opposition to my Recruiting which was so ungenerously brot to bear against me and think I will and am determined to Recruit Co K, cost what it will
>
> I have paid for the larger portion of My transportation already and will use all honorable Means to complete my undertaking I leave in the Morning with my present sup[p]lies and will endeavor to get my guns +c as soon as you can furnish them I will Muster my Recruits between now and the next pay day.[1190]

On Monday, September 21, Lieut. Stump wrote to both West Virginia Gov. Arthur I. Boreman and Col. Brown regarding his efforts and adventures recruiting for the 13th. To Boreman, Stump related that he had been recruiting for Cos. I and K "with much difficulty. I have succeeded in getting 41 recruits for [Company I] and 10 [for] Co. K with a fair prospect of doubling that number in the course of a week or two." Stump continued describing what appears to be sharp competition among recruiters to enlist from the same small population of civilian men:

> The worst opposition I have met with in recruiting was from Capt. King [...] who claimed the Right under the authority of the Governor of West Virginia to control this entire Region of country, to wit a portion of Kanawha, Roane, Jackson and Calhoun. But from his

numerous acts of Outlawry the county was bound to complain to provost martial at this place, who has frequently arrested him. [...] I have, however, in the face of all the influence of secession, King and that turbulent creature Lyle Paxton [captain of a unit Roane Co. Home Guards?] succeeded without the aid of Co. I (as promised to me by Col. Brown) in getting 41 regularly enlisted and sworn men for 3 years service.[1191]

To Brown, Stump wrote that he now had

38 Enlisted Men for Co K 13th Reg V V Infy and am drawing their Rations but am informed by the Ordnance department that they have no guns fit for service I wrote to you when here last that I desired One other Box of 20 guns, to be forwarded to this place (They are not here) Neither have I Recd clothing for my men save shirts + Blouses. I am in nead of 20 guns + 2 boxes of cartridge I wish you could [provide] them to me at this place I have drawn my Muster Rolls from the Mustering Officer and expect to bring 41 or more men here to Muster the last of this week —

I would be glad to be visited by yourself or Col. Hall or some other member of your Regt. I have been the Recipient of — and have labored under Very discouraging circumstances during my labors in Recruiting —

I hope for the better in future[1192]

October 1863

Anxious for news and corralling horse thieves 'beyond God's country'

The 13th Regiment continued to garrison Cabell County. Field officers, staff and all companies continued to be stationed at Barboursville with the exception of Co. I which remained at Guyandotte.[1193] Colonel Brown continued intermittently, to be instructed about some detail of his monthly returns,[1194] and the 13th was visited by the paymaster and mustering officer during the first half of October. Mid-October President Lincoln called for 700,000 more volunteers.[1195] All men who enlisted between October 24, 1863, and March 1, 1864, were entitled to receive a "premium $2, + 300 $ bounty," a hefty inducement to volunteer. By comparison, those who had initially stepped forward and were mustered into the 13th Virginia October of the previous year (1862), had received government bounty only in the amount of one hundred dollars.[1196]

The military Department of West Virginia was reorganized in October 1863. Organization of the 3rd Division to which the 13th belonged remained unchanged. October 3, the name of the brigade in which the 13th was brigaded was changed from 1st Brigade 3rd Division 8th Army Corps to "1st Brigade 3rd Division Department of West Virginia."[1197] October 3, John Butler of the Allegheny Arsenal responded to Col. William Brown's letter of September 25:

> Your letter of the 25th ult. inclosing 36 main spring swivels,[1198] came to hand this morning. I inclose 36 as per sample sent. I will thank you to change my Invoice from 10 to 36 for the Enfield Rifled musket, and I will make the same correction on your receipts.[1199]

Gen. Scammon wrote to Col. R.B. Hayes on October 3 regarding some complaint made about what could only be the use of public buildings in Point Pleasant for the army hospital there. Scammon wrote:

Hd Qtrs 3rd Div. Dept. West Va.
Charleston W. Va. Oct. 3rd, 1863

Col. R. B. Hayes Comdg. Brig.

You will direct the troops at Point Pleasant to vacate the Public Buildings belonging to the State and County at Point Pleasant.[1200]

October 5, Col. Brown issued Special Order No. 54 providing that:

Capt John S. Witcher of the 3d Va Cavalry will take ten (10) men of his Company and take charge of three prisoners and proceed to Camp White, and report to Capt M.P. Avery, A.A.A.G. of the 1st Brigade. After the transaction of all necessary business, he will report to these Head Quarters, without unnecessary delay.[1201]

On the 5th, Brown issued Special Order No. 56 to Lieutenant LeSage of Captain Witcher's command directing him to "detail one Sergeant, and three (3) men as an escort for the Sheriff, a few miles into the country, and return. They will report to these Head Quarters for instructions."[1202] October 7, Brown issued Special Order No. 57 directing Lieut. Col. James R. Hall "to proceed to Charleston to transact business pertaining to the Reg't, thereafter to report to Capt. James L. Botsford A.A.G. at division Head Quarters" and then "to report to these [13th Regiment] Head Quarters without unnecessary delay."[1203] Also the 7th, Brig. Gen. Scammon issued Special Order No. 112 appointing Captain Albert F. McCown (cmdg. Co. F) an "inspector, to inspect Quartermasters stores in that Regiment for which Lt. Comstock R Q M is responsible. He will be governed by paragraph 1022 Revised Army Regulations."[1204] 13th sutler Columbus Shrewsbury wrote to his wife on Saturday, October 10:

Dear Wife

Yourse came to hand a few days ago I was glad to see you were geting along so well and did not want for anything for that is my only studdy to make you comfortable and see that you do not want anything I begin to want to come home but it is no use to put in all the time going and coming if I do I won't make much It seems to me like it has been a long time since I was at home but it has been but about one month yet Mr Sherwood [sutler to the 13th] has gone to Cincinnatti for goods and will not be back for a week or ten days So I cant think of being at home for some time after his return the worst thing we have to contend with heare is Fleas they bite me so I can't sleep at night Everything is going right for Phelps heare in the Regiment I think he will be Elected without a dout All is quiet out heare about all we heare of is Horse Steeling The Regiment may probably move from heare this fall it is not settled yet about it The Paymaster has been heare and payed the Troops If you have any chance to send my Over Coat I wish you would for I will need it before I come home probably you can send it by some one but don't go to much trouble about it if you see a chance do so if not let it alone I will try and be at home long enough to eat up your canned fruit before Spring Fanny [his daughter] you must be a good girl until [I] come home and I will bring you them coppers to buy candy with write often and try to improve in your writing spelling Nothing more at present but remain yours

> Very Respectfully
> C Shrewsbury[1205]

On October 14, by order of Maj. Gen. B.F. Kelley, a group of ailing 13th West Virginia soldiers were discharged from military service at U.S. General Hospital Gallipolis, Ohio, and from hospital at Barboursville:

> John F. Blackburn, Private, Co. D, was discharged on a Surgeon's Certificate of disability to take effect on October 16, 1863. Cause for disability "confirmed phthisis" (pulmonary consumption). John Blackburn had enlisted on August 15, 1862. He had been a shoemaker before the war, and although a young man (about 26 years old) when mustered into service, he was sickly for most of the time. He was detailed to serve as "nurse" on 13th regimental field staff. On April 10, 1863, he was detached from the regiment to serve as a hospital nurse at Gallispolis Hospital.[1206]

> Morrison Miller, Private, Co. E, was discharged on a Surgeon's Certificate of disability on October 14, 1863.[1207]

> Canon (Canaan) Pierce, Private, Co. E, was discharged due to disability on October 13 or 14, 1863.[1208]

> John Hayse, Private, Co. F, was discharged October 14, 1863, due to disability caused by "chronic nephritis."[1209]

John Smith, Private, Co. F, was discharged for disability October 14, 1863. Cause of disability was "partial blindness and old age" (he was about 43 years of age).[1210]

Harrison Frost, Private, Co. B, was discharged for disability, at Barboursville, October 15, 1863.[1211]

On October 15, Colonel Brown issued Special Order No. 64 ordering Captain John S. Witcher, 3rd Virginia Cavalry, to seize horses:

> Sir you will take a detail of fifteen (15) men from your Command and will proceed with them to the lower part of Cabell Co W Va below Guyandotte River, and will Seize all such Horses as are in the possession of Rebel Citizens and are likely to be used against the interests of the Government You will See that this Order is executed and Report to these Head Quarters without unnecessary delay[1212]

Friday, October 16, William F. Dusenberry wrote to Col. Brown from Guyandotte requesting that prisoner Perry Adkins not be released in a prisoner exchange:

> I learn you have Prisoners by the name of Perry Green Adkins, if this is the case, cannot arrangements be made to have him turned over to the civil authorities, he was the leader of the gang who plundered my House in August last, for which he is now indited, he is a very dangerous man, and ought not to get loose again. No exchanging such men as he and the only Parol an Ounce of Lead or six feet of Rope, The Traitors Jewels. I do hope he will be left for the civil law.[1213]

On October 18, 1st Brigade Adjutant General, Captain Martin P. Avery reminded Col. Brown again to "forward his Monthly Return to 1st Brigade headquarters by November 5th of each month."[1214] October 22, Brown wrote to 1st Brigade Head Quarters informing that "Rebel horse thieves infest that part of the country" and asking for instructions as how to obtain horses[1215] that these local thieves rebels might be pursued with result. Col. Hayes responded to Brown as follows:

> Hd. Qtrs. 1st Brig. 3rd Div. 8th AC
> Camp White Va. Oct. 24, 1863

> Col. Brown

> In taking horses from citizens in your part of the country for the use of Government you will be governed by instructions contained

in letter from this Hd. Qtrs. dated Oct. 6, 1863 addressed to Capt. Witcher Co. 3ʳᵈ Regt. Va. Cav. 'Horses of loyal citizens will not be taken others will receive a certificate for the value of the horses taken payable on proof of loyalty or on continued loyalty at the end of the war. By command of R.B. Hayes[1216]

In another missive to Brown, written on the October 24, Hayes directed Brown to "page 451 of the *Revised Army Regulations* in regard to addressing 'official' communications."[1217] On October 25, Colonel I.H. Duval, commanding the 9ᵗʰVirginia Infantry Volunteers, wrote to Col. Brown requesting his aid in apprehending two deserters, who were supposed to be in Brown's area:

> I send descriptive list of Flowers + Stephensen Deserters from this Regt You will please use every effort to arrest them. If you succeed, put them in <u>irons</u> and forward them to this post — Flowers has deserted twice before this — You will probably find them near Guyandotte.[1218]

On October 27, his father's trial for murder being imminent, Lieutenant Colonel James R. Hall wrote to James Botsford, Assistant Adjutant General requesting that

> Captain Milton Stewart and myself be permitted to visit Point Pleasant for five or six days. For the following reasons, My father's trial comes off on the 2d day of next month. Captain Stewart and myself are important witnesses in the case. Owing to the fact, that the case is to be tried by a Judge from another Circuit, and only learning that fact very recently, it has precluded the possibility of getting a formal furlough.[1219]

Furlough was granted.

While minor actions went on all around them (most recently at Charleston, Ravenswood, and near Elizabeth) the 13ᵗʰ did not seem to know of them. Indeed, the 13ᵗʰ West Virginia companies were all too isolated—posted as they were in what one soldier called 'beyond God's country'—to receive news, be it local or national, and the men were anxious for news. Lieutenant John S. Cunningham, Adjutant for the 13ᵗʰ, wrote to his wife on October 28 from 13ᵗʰ Head Quarters at Barboursville eager to hear news of the recent elections in West Virginia; news of the war; and for news from the home neighborhood.

> My Dear little wife
>
> I walked all the way to this place the same day I left home, and let one of the soldiers lead the horse and placed all the blankets on

the saddle to help the soldiers along I am in excellent health — will you please send me all the news in the neighborhood I have not had any definite news in regard to the result of the election — there is a rumor that Father is elected + that the race is between Brooks + Cunningham. I hope Father is elected You can tell father to write immediately and leave the letter with Park and the day after tomorrow Messenger will come down to this place from Charleston he can give the letter to them when they pass do not forget to write the whole result how Kanawha has gone for congress + the legislature We have no news from the seat of War if you have a very late paper please send it along—[1220]

Morning reports for Company A indicate that the company continued to be stationed at Barboursville, Cabell County, during the month of October. On October 11, Sergeant Miletus Grinstead, Sergeant Amos Brown and John Orth were reported absent with leave to go to their homes by order of Brig. Gen. E.P. Scammon. Grinstead and Orth returned to the company on October 20, and Brown returned on October 22. Co. A gained four recruits in October. On October 1, John Craig, a "boatsman" by profession enlisted. October 23, John C. Hacker, John Strickline and John Edens enlisted in the company. These additions brought the number of men in Co. A, present and absent, to three commissioned officers and eighty-seven enlisted men — the highest numbers for the company thus far. One enlisted man, Nathan Snodgrass, was reported absent without leave for four days (October 21 to 24). He returned to the company on October 24. A drummer was present October 1 through 30. One artificier was present every day of the month. One private was assigned to daily duty from October 1 to the 29; two privates had daily duty on October 30; and three were detailed for daily duty on October 31. Among those on daily duty was Milton Wilson detailed for "Hospital nurse" in "Regimental Hospital" on October 29 by Special Order No. 69 issued by Col. Brown:

Orderly Sergt Co (A)

Will detail one man to relieve Milton Wilson who will report immediately to the Reg't Hospital for duty in the dep't for a few days[1221]

October 30, Levi Hemmings was detailed for company drummer. With the exception of a large detachment which went out October 13 to 14, the company had very little detached service during the month. Two commissioned officers were detached for service on October 2. Otherwise, just one commissioned officer was detached every two days. On Friday, October 2, Lieutenant Samuel S. Mathers was on duty at Col. Hayes' (1ˢᵗ Brigade) headquarters, Camp White.[1222] October 13 and 14, thirty-three enlisted men and two commissioned officers had detached duty. Capt. Greenbury Slack and one enlisted man returned from Charleston on October 1. Richard George returned from Charleston on October 7. The health of Co. A seems to have improved in October. Just one to four men were reported present sick (the higher number of '4' at the end of the month coinciding perhaps with the onset of colder weather.) Just five to six men were absent sick (the higher number during the first two weeks of the month). One enlisted man was reported absent in confinement October 1 to 7.[1223]

October morning reports for Company B note that the company was stationed at Barboursville October 1 through 31. One drummer was present October 13 through 31. One fifer was present October 1 to the 31. No "Buglers, Farriers, Blacksmiths or Artificiers" were reported present during the month. One enlisted man was absent with leave October 8 through 16 and October 27 to 28. Five enlisted men were absent without leave October 14 to 25, and one enlisted man was absent without leave October 26 through 31. To deal with the absentees, Col. Brown issued Special Order No. 62 on October 14:

> A.H. Windham, Oscar Chapman, John S. Adkins, David Fisher Five (5) privates Co B are hereby detailed to proceed to Mason Co. and will arrest James McCoy, James Jeffries, Josiah Mattox William Nichols and Alfred Sinclair all deserters from Co B and will report with them to these Head Quarters without unnecessary delay.[1224]

The trip to Mason County to apprehend the deserters proved successful and the absentees were sentenced by Brown pursuant to his orders of October 26 in which he directed that

> Josiah Mattox, James McCoy, James Jeffries, Alfred Sinclair of Co B, W[esley W.] Crowfoot Co H having left their Companies without permission from proper authority which is a gros[s] violation of military law — It is hereby ordered that one months wages be withheld from each one of them and that they be put to hard labor for Twenty days[1225]

Company B lost one enlisted man, Harrison Frost, discharged for disability on October 17 by order of Gen. Kelley. Co. B had at the end of October a total of three commissioned officers and ninety-seven enlisted men present and absent. Four to six men were detailed for daily duty each day of the month. Details for detached service continued to be made involving from three to eighteen soldiers. Of note was the large detachment made October 1 through 12 of one commissioned officer and seventeen enlisted men. The company had three to six men present sick during the month. More men were sick in camp at the end of the month than at the beginning and six to ten men were reported absent sick. Three privates were present in confinement October 1 through 11. No one was absent in confinement.[1226]

Company C reports indicate that drummer and fifer were present with the company for the month. No one was absent with leave. Two enlisted men were absent without leave October 9 and 10. One enlisted man was absent without leave October 11 through 13. George Burris was reported deserted on October 8. October 13, Burris returned and was sentenced "to work Fatigue duty 25 days under a guard by order of Col. Wm. R. Brown." On October 18, Bazeleel Meek "returned from desertion and [was] restored to duty."[1227]

The vacancy caused by Capt. Van D. McDaniel's death resulted in a series of promotions. On October 7, 1st Lieutenant Lemuel Harpold was promoted to Captain. 1st (Orderly) Sergeant, John P. Wood, was promoted to 2nd Lieutenant. Private Francis W. Sisson was, after many months in the regimental commissary, transferred by promotion to be Regimental Quartermaster Sergeant on October 8. October 11 and 15, Col. Brown approved other promotions and changes in Co. C in Special Orders No. 64 and No. 63. These included: William McDaniel promoted to Orderly Sergeant vice John P. Wood promoted, and Private E.M. Roseberry promoted to 4th Sergeant vice William McDaniel promoted to Orderly.

There were other appointments as well. Corporal William H. Thomson was promoted to 5th Sergeant to fill the vacancy created by the reduction of Frank A. Windon for disability. Privates Alonzo T. Moristen/Morrison and John P. Harman were promoted to Corporal vice William H. Thompson promoted to Sergeant and Corporal William G. Davis reduced to the ranks for disability.[1228] On October 16, the promotions went into effect. October 31, Col. Brown approved the order issued from Headquarters Co. C "reducing Corporal Bowman to the ranks for drunkenness and Corporal Marcus D. Noble for disability and of promoting Madison H. Stone and George W. Sands vice Bowman and Noble reduced."[1229]

There were also new recruits to the company. Andrew K. Hoscher was reported enlisted in Co. C on October 1. October 4, in "Remarks" is recorded that Charles O. Willis and Daniel McCloud also enlisted in the company. On or by October 14, four men were reported as new recruits and an additional recruit was reported on October 17. At the end of the month, Co. C tallied present and absent two commissioned officers and eighty-eight enlisted men.

No one had special or extra duty. Three privates were assigned to daily duty October 1 through 19. Two privates had daily duty October 20 through 31. Three enlisted men were detailed for detached service October 1 through 19, and two enlisted men were detached October 20 through 31. October 27, Col. Brown ordered that "Sergt. James S. Kelly and Privates John S. McCulloch, Peter H. Meadows all of Co. 'C' of this Regt. are hereby ordered to proceed to Point Pleasant and vicinity and arrest Abraham Asbury, a deserter from this regiment and deliver him at these Head Quarters without unnecessary delay."[1230]

Co. C had one man present sick October 2 through 31. Seven were reported absent sick October 1 through 19, and eight were absent sick October 20 through 31. John Gorman (or Garman), Private, was discharged for disability on October 26 at "Gallipolis Hospital" by order of Brigadier General B.F. Kelley, cmdg. Dept. of W. Va. The cause of disability, as given in the company descriptive roll, was stated as "Ing[uin]ial Hernia and old age" (he was about age 44 at the time).[1231] One private was reported present in arrest or confinement from October 14 through 31.[1232]

Morning reports submitted for Company D show that the company had no drummer, no fifer and no farriers, blacksmiths and no artificers present during the month. No one was reported absent on leave. One enlisted men was reported absent without leave on October 31. The company gained a new recruit, Winfield S. Kimes, who joined the regiment on October 17.

Two more recruits, Amos Woodyard and Manley P. Woodyard, joined Co. D on October 31. On October 6, Nelson Flinn, James Swain and William T. Harshey were mustered into the United States service. October 9, James A. Cartmill was reduced to the ranks and Lorean Ohlinger promoted to Corporal. This change was effected pursuant to General Order No. 24 issued by Col. Brown on October 9 which provided that:

> Order No [left blank] issued from Head Quarters Co D reducing Corporal James Cartmill to the ranks for drunkenness and neglect of duty is hereby approved — Private Lorean Ohlinger is promoted to Corporal vice James Cartmill reduced he will be obeyed and respected accordingly[1233]

One enlisted man was reported "lost by desertion" on October 17, and another enlisted man deserted on October 29. October 17, James Burnett returned from desertion. One enlisted man was discharged for disability on October 19 and 20. The company finished the month with three commissioned officers and eighty-three enlisted men present and absent.

No one was ordered on special or daily duty. Four to five privates had extra duty each day of the month. Two fair-sized detachments went out from Co. D during the month. On October 7, twenty-eight enlisted men and on October 13 and 14 forty-two enlisted men had detached service. All other days, four to seven enlisted men had detached duty. October 1 through 3 and 24 through 30, one commissioned officer had detached service.

Co. D reported four to seven men present sick for the month. This number included Captain Simon D. Williams. In compliance with Col. Brown's order of September 28 (Special Order No. 52 supra) and perhaps a second directive given on October 1, Captain Williams proceeded to Charleston and reported to Surgeon and Medical Director George M. Kellogg for a medical examination. Private John McSherry, a native of Scotland, had his "arm and hand wounds dressed at 13th Regimental Hospital" on Friday, October 9, and then again, on October 20 he was admitted to "Regimental Hospital to have his wounds dressed."[1234] Some idea of how McSherry came to need medical attention may be discerned from Special Order No. 60 issued by Col. Brown on October 10 which directed that:

> Capts McCown, Mathews and Slack, are hereby appointed a Court of Enquiry to investigate the cause and difficulty which arose between Lt George Snowden and private John McCherry on the road from Guayandotte and Barboursville Oct 8/inst. The Court will convene 2 O.C. P.M. this day and after investigation report to the Col. Com'dg.[1235]

Six to eleven men of Co. D were reported absent sick over the course of the month. October 11, John M. Graham was discharged from "U.S. General Hospital, Point Pleasant." John M. Graham, Sergeant (ca., 32 years of age), was discharged by order of the departmental commander Major General B.F. Kelly, dated "Sept. 21, 1863, Headquarters Dept. W. Va." to take effect Oct. 11, 1863. Graham was

> Discharged by reason of Induration of the cellular tissues of the foot and compl[ete] anchylossis of ankle joint rendering him a permanent cripple caused on account of eresypelas which has settled in his foot he was first taken with typhoid fever which left him in this condition. Degree of disability one fourth.[1236]

Within the month, Graham expressed his thanks in the columns of the Point Pleasant *Weekly Register* for the care given him by Surgeon Charles D. Dally, in charge of the Point Pleasant Hospital. (Dr. Dally was appointed Surgeon in charge of the Hospital by Captain W.H. Zimmerman, commander of the consolidated Posts of Gallipolis and Point Pleasant.) Graham also expressed his gratitude to "William French, Steward and Mr. Seaton, Ward Master,"

> for their uniform kind treatment and constant careful attention and sympathy for me, while for four months I was under their charge as an invalid and convalescent. May all Post Hospitals be

favored with officers of equal skill, assiduity, urbanity of manners and kindness of heart is the sincere wish of John M. Graham, late of Co. D, 13th W.V.V.I.[1237]

On October 14, John F. Blackburn was discharged for disability from "U.S. General Hospital at Point Pleasant by order of General Kelly." October 12, Sheldon Gibbs returned from desertion. He had reported to "Hospital at Point Pleasant" but had not been reported on the face of the Report Book until Oct. 29, 1863. One private was reported present in arrest or confinement on October 9. No one was absent in confinement.[1238]

On October 3, Col. Brown issued Special Order No. 53 authorizing Capt. John D. Carter, cmdg. Co. E, "to attend the Civil Court which will Convene at Charleston West Va on Monday Oct. 5th 1863."[1239] Carter had been summoned to appear as a witness. After transacting "the Business upon which he is Summoned," Brown instructed Carter to "report to Capt James L. Botsford A.A.G. at Division Hd Quarters" and then "report without unnecessary delay to these [13th Regiment] Hd Quarters."[1240] Sergt. John W. Overshiner of Co. G had also been summoned to attend court at Charleston on October 5 to appear as a witness. On the 5th, Col. Brown issued Special Order No. 55 granting Overshiner permission "to attend civil court, which will convene at Charleston this day [...] He will report to Capt. M.P. Avery at Brigade Hd. Quarters, after the business upon which he is summoned is transacted, he will report to these Head Quarters without delay."[1241]

On October 9, Private Robert Cobbs, Private, Co. E, wrote a letter from "Camp at Barbours vill Cabell Co" and enclosed money from his pay: three dollars to his mother and five to his father. He closed with the request: "I want you to write to me as soon as you get this and let me no some. [...] Direct your leters to Barbours vill Cabell Co West Va in care of Capt J D Carter Co E 13 Va reg."[1242]

On October 10, Private George W. Smith, Co. E, drew clothes: 1 blouse at $3.14; 2 pair drawers @ $1.90; and 2 pair socks @ .64.[1243] Private Mark E. Robison, Co. E, also wrote home:

> Camp Barbers Ville Cabelle Co West Va
> October the "14" 1863
>
> Dear father and Mother after my best and kindes respect to you i will say to you i am Well and harty at this time and Wish thes few lines may find you ingoying the Same life blessing i receivd your kind leter one the 13 of this month i was truley glad to hear from you all But was truley Sorry to hear that you all had bin Sick and not well yet i hope you all will recover a gane in your health i wold like to Sea you all and to helpe you one the farme But my country

has cald one mea for aid and i thought hit my duty to go and help pute this rebellion doon and Wold rather go than to Sea the old men a goin and having their Wifs and thar famlis hit tis won of the most desperdis thing one earth all tho i hope this war wont last all wase for i think the last Strugel [struggle] is one hand now and i hope hit will soon be over and i can cume home Safe and un harmed father i don't want you ner mother to be uneasy a bout mea for i think we all will be at home a gane wonce moar in life and i don't think the time is far off so i have nothing much to rite to you this time We have bin payd a gane we had our Clothing bill to setle upe so i dident get very much this pay day So i dident have eny to send you but i will Send you the moar next pay day We are still taken a few rebls in hear our scouts is out every day hear and they keape all things hear qiat [quiet] i hant [haven't] got a leter from Moses for Some time but i ame looking for a leter evry mail So all the boys is Well in Camp So i have nothing mour to rite this time you must rite to mea Soon and often So Still remains your true and fecnent Son untell death So good by un tell rite a gane

Mark E. Robison to his father Mr John Robison[1244]

October morning reports for Co. F indicate that the company had drummer and one wagoner present each day of the month. One fifer was reported present with the company October 1 through 8 and 31. October 18, Col. Brown issued General Order No. 25 from regimental headquarters at Barboursville approving Special Order No. 4 issued by Captain McCown from headquarters Co. F, which directed that "Corporal James Edwards is hereby reduced to the ranks, at his own request, as he wishes to go into the Band, and Private John W. Fry is promoted to Corporal vice Corporal James Edwards reduced."[1245]

No one in Co. F was absent with leave. On October 13, Peter Donnelly returned from desertion after an absence of one year. October 16, Gustavus A. Andrews deserted. His clothing bill in the amount of $46.25 was charged against him on payroll. Eighteen-year-old carpenter Edgar D. Shank reported to Co. F for duty on October 1 "having enlisted September 28th, at West Columbia, W[est] V[irgini]a."[1246] On October 6, Alexander McDonald, Alexander Boles and Henry C. Casdorph were mustered into the service by Lieut. Colonel C.S. Roberts. October 7, Co. F was paid up to August 31 by Major Cowan. Clothing accounts were also settled up to the August 31 pay roll.

More recruits joined the company during the month. October 21, Samuel N. Proffitt enlisted. October 25, John Wallace enlisted, and October 28 Joseph F. Gibbs enlisted in the company. October 19, Privates John Smith and John Hays were discharged from the U.S. Service for disability. They had been discharged from "Hospital at Gallipolis" on September 14. October 10, Col. Brown issued Special Order No. 59 ordering Bradford N. Tilden of Co. F to "report to his company for duty immediately."[1247] October 12, Brown issued Special Order No. 61 directing Brafford N. Tilden

> to proceed to Paynesville and Lancaster Ohio. There to arrest Frank E. Barret and Gustavus A. Andrew, if they can be found (They being deserters from this Regt) and return without unnecessary delay.[1248]

At the end of the month, Co. F numbered three commissioned officers and eighty-eight enlisted men.

On extra duty were two privates October 1 through 16, and four privates October 17 through 30. Three privates were assigned to daily duty October 1 through 30, and seven privates had daily duty on October 31. Over the course of the month, four to forty-nine men had detached duty. On October 9, Lieutenant Joseph Brumley with forty-one men from Co. F started on a scout up Mud River. Col. Brown had issued Special Order No. 58 on October 9 detailing "Lt. Bromly Co. F and 86 men [...] for scouting purposes." They were to "proceed to the neighborhood of Mud River and Scout that section of Country" and then return and report to 13th Regiment Head Quarters "on or before the 15th in't [instanter]."[1249] Brumley and his men returned from their scout to Barboursville on October 14 and he submitted a written report of the scout to Lieutenant John S. Cunningham that day as follows:

Head Quarters Co F 13th Regt Va Vol Infty
Barboursville West Virginia Oct 14th 1863

Lieut J S Cunningham

Adgt

Sir

I have the Honor to report that in pursuance of Special order No — dated at Barboursville W. Va Oct 9th 1863 and directed to me. I proceeded to Mud River Country with one Lieut and 85 men to Scout on the different forks + creeks of Mud river, and from there to Coal River above the falls and from there to this place by way of Coals Mouth distance 90 miles which we accomplished in 5-1/2 days reaching this place on the 14th. — I did not ascertain that there were any armed Rebels in that Country Except Some local bushwhackers which fired on us from the brush on Sugar Creek. I sent out Skirmishers but they found nothing. The men are all in Good Condition

> Joseph Brumley
> 2d Lt Co F 13th V V I[1250]

On October 12, 1st Lieutenant Timothy Russell "returned from his recruiting expedition to Mason County, having obtained 5 recruits."[1251] Russell submitted a written report to Lieutenant Cunningham upon his return to the company on Monday, October 12. Russell informed that he had been ordered pursuant to Special Order No. 46 (dated "Barboursville, West Virginia, Sept. 24, 1863") to proceed to Mason County on a recruiting expedition. He reported that he had

> Succeeded in Enlisting four Recruits who have reported to this Company for duty I also arrested and brought Peter Donally with me a private of this Company who deserted about one year ago. I also arrested Sheldon Gibbs a private of Company D, who being sick was left at the Hospital at Point Pleasant.[1252]

Lieut. Russell's new recruits "Sam C. [or I.] Slack, Napoleon B. Russell, Timothy R. Davis and David W. Doolittle having enlisted previously reported for duty"[1253] on October 12. October 10 to 12, two commissioned officers and forty-seven enlisted men were detailed for detached service. October 13 to 14, a group consisting of one commissioned officer and forty-one enlisted men were detached.

Russell was ordered out again, this time to scout with thirty-one men. October 24, Col. Brown issued Special Orders from headquarters at Barboursville directing that Russell with thirty-one men was "to scout through the County of Wayne on the watters of Guyan and ____hale near Louisa and down to Trouts Hill and back to [Barbours]ville." Russell reported that he and his men "arived Oct 30th 8 oclock in the Evening found no [Re]bels, I was informed both by Union and Rebel citizens [tha]t four hundred Rebel soldiers camped at or near the _____ pigeon creek in Logan County W. Va."[1254] On October 26, Brown ordered

> Bradford N. Tilden of Co. (F) with two Privates Sylvester Keith and Richard Woodman [...] to Proceed to St Marys Ville and Pomeroy there to arrest Hazard Farley, John Twaddle and Louis Boys [Bays] Deserters from this Regt and Return them under arrest to these Hd Quarters without unnecessary Delay — Any Quarter Master of the U. S. A. will furnish necessary transportation[1255]

Two to four men of Co. F were reported present sick each day of October. Four to seven were absent sick each day. Numbers of sick declined during the last two weeks of the month. No one was present or absent in arrest or confinement.[1256]

Captain J.V. Young wrote to his wife, Paulina, from Barboursville on October 11 providing details of life in camp, duty and the condition and status of Co. G:

> I have just returned to Camp Mud, my field of labor. I am enjoying good health and my men are all well; and I think my company is in better condition now than it ever was. I was sent to Mud to move Mr. Curry out of that infernal den. I had Co. C. with me, and when I returned Lieut. Elkins was gone with Company G to move Mr. Holstein and Mrs. Griffith to Coals Mouth.
>
> I captured Emberson Chapman, the man who helped kill Mr. Gibson. He says he did not help. But he says he held their horses for them while they killed him.
>
> The paymaster has been here and paid the 13th and also the mustering officer, and says my affairs will soon be settled. He says there is an order from the War Department for him to muster me in the U.S. Service, and as soon as he gets the order he will send for me to come to Charleston and will muster me in service, and then send me to Wheeling for my money.
>
> Colonel Hall and Colonel Frost1257 have a great time about Company G. They are both at Charleston but who will win I can't tell. But one thing I can tell, I want my company straightened up. I don't care much to which Regiment we belong. At all events it will be settled in a few days.
>
> While I am writing prayer meeting is going on over my head by those Methodist soldiers and their wives of which there is quite a number.[1258]

Young wrote again to Paulina from Barboursville on October 13 regarding politics, candidates running for office while in the army and his concern about 'butternuts' in Putnam County, his home county, running for State offices:

> Colonel Frost [commander of the 11th Virginia Infantry] was here yesterday and made a rousing speech. The Colonel is all right and will be elected for Congress, and J.M. Phelps for Senate sure. The boys are voting today and Mr. Whaley [Kellian V. Whaley] has gotten 4 votes. I don't think he will get ten votes in the Regiment. Captain McKown [Captain Albert F. McCown, commander of

Company F 13th Virginia Infantry] has gotten 2 or 3 today. I think he had better give it up as a bad job.

What are they doing in Putnam? Who are the candidates? I will send Van up to see what you are all doing, and see if any butternuts are running in Putnam for office. Write by him everything you know about the candidates, and who they are.[1259]

October morning reports for Company H show that a drummer was present each day of the month. G.W. Turner was detailed as fifer on October 21. He continued as fifer through the end of October and perhaps longer. No one was absent with leave. Three enlisted men were reported absent without leave October 1 through 20; two enlisted men were reported absent without leave October 21; and one enlisted man was absent without leave October 22 through 30. On October 31, Doliver Workman was dropped from the company roll having been absent as a deserter since August 22, 1863.[1260] The company gained a new recruit, Stephen F. Fuller, who enlisted on October 21. One man was somehow wounded in action. At the end of the month, Co. H tallied present and absent three commissioned and seventy-seven enlisted men.

October 1, George W. Fulwiler and Stephen Fuller were restored to duty by order of Col. Brown. October 11, Thomas M. Hackett was ordered back to duty with the company. October 17, James C. Blair was restored to duty by order of Col. Brown. October 2, Capt. William I. Mathews was on special duty as Provost Marshal. Two privates were detailed for extra duty on October 31. Co. H had detached service each day of the month. Three to five enlisted men were detached October 1 through 19.

On October 20, Col. Brown issued Special Order No. 65 ordering Lieutenant William Purdue (Co. H) to "take Charge of nine (9) men and [he] will proceed with them to Wayne Co West Va. [and] Scout in that region. Recruiting all he can, and [he] will return and report to these Head Quarters on the 28th inst—."[1261] Lieut. Perdue returned from scouting on October 26. October 20 through 26, one to sixteen men were away on detached duty, and ten to eleven men had detached duty October 27 to 30. Among these was a squad composed of Lieutenant Russell, who with five men left to go out on a scout October 25. The scout returned on October 30. Four enlisted men had detached duty on October 31. Among these was Wesley W. Crofoot, who had been detailed for duty as "Mail Carrier from Barboursville to Guyandotte. [Crofoot to] report to Provost Marshall's office for instructions."[1262] Crofoot returned to the company from mail carrying on October 19. October 13, Peter Huchinson was detailed as teamster. Sergeant William Shannon was put on detached service by order of Col. Brown on October 16. October 28, Oliver Taylor was detailed for daily duty as cook for the teamsters.

On October 26, Col. Brown issued orders regarding deserters from the 13th Regiment, who had been brought back from Mason County. Among them was Wesley Crowfoot, who was named along with four other offenders as "having left their Companies without permission from proper authority which is a gros[s] violation of military law." Their punishment was to have "one months wages [...] withheld from each one of them and that they be put to hard labor for Twenty days."[1263]

Co. H had three to eight men present sick over the course of the month, and eleven to thirteen men absent sick with an improvement in the number of absent sick over the last twelve days of the month. Charles M. Laurence, who had been in "Hospital at Point Pleasant," returned to his company for duty on October 14. James Abbot returned to duty from "Post Hospital" on October 19. One private was present in arrest or confinement on October 31. One enlisted man was absent in confinement October 1 through 16. Andrew J. Allen returned from absence without authority and was reported absent in confinement on October 21. October 22, Francis M. Crofoot returned to the company and was placed in confinement. October 24, Crowfoot was released from confinement by order of Col. Brown. Two men were reported absent in confinement October 22 to 24. One man was reported absent in confinement October 25 through 30.[1264]

Morning reports submitted for Company I record that the company was stationed at Guyandotte for the month of October. 1st Lieuteanant William E. Feazel, four sergeants, eight corporals and forty-four to forty-seven privates were present in camp each day of the month. No one was absent with or without leave. The company gained six new recruits. One joined on October 7; another joined October 9; another on October 21; one on October 30 and October 31, respectively. At the end of October, Co. I reported one commissioned officer and sixty-seven enlisted men present and absent. One private had daily duty each day and one enlisted man had detached service on each day of the month. Two to three privates were present sick October 19 through 31, and three enlisted men were absent sick each day. No one was present or absent in arrest or confinement.[1265]

On October 26, P.H. McCullough wrote to Lieut. Feazel regarding a Confederate raid (rumored to be imminent) to rescue one of their own, a wounded comrade named Pine, who was being attended at McCullough's place.[1266] Feazel forwarded the letter to Col. Brown. McCullough wrote:

> Rheumor says there are Rebells in about the Falls of Guyandotte + on the Beech Fork + it has been intimated that part of their mission might be to carry out with them, their Comrade (Pine) whom they left with me I wish you to consult your Superiors today + have him taken to the Hospital immediately as I am determined not to be held Responsible for conquences over which I could

have no Controle were they to make a dash in at night He can be removed in an easy Ambulance — is improving Rapidly + will Recover in a short time with proper treatment. Do what you do, quickly [...] P. S. You need not let him know that I dictated this course as I might suffer by them hereafter McC.[1267]

October 26, Lieut. Feazel wrote to Adj. John S. Cunningham complaining of discipline problems he was having with men of the Co. A.

Sir I have the honor of Sending Robert Grey of Co. A 13th V.V.I. and [Philip] Wentz of same Co. to you under guard they sliped my guards and crossed the river yesterday and returned this evening each with a Canteen full of whiskey Concealed under their clothes and upon being detected by my patrol guards they refused to give it up and attempted to leave for the regiment with their whiskey I followed with a guard and arrested them and took possession of the whiskey and in taking the Canteen from Grey he acted very impertendtly and striped up his sleeves and spit in his hands to strike me and no doubt would have done so had I not been prepared to resist him Wentz did not resist neither did he misbehave after he was arrested information came to me that Grey swore that he could whip all of Co. I and that the damd Co could not take his whiskey as I have had a great deal of trouble with men from Co. A I would Respectfully make request that you inform Capt. Slack that his men are continuously abusing his Confidence in them and that it is a serious injury to the Service for him to give some of his men passes out of the lines

This I say for the good of the Service and especially our regiment which is of great importance to me and all of us I have possession of the whiskey please direct what I shall do with it

N.B. it was bought in violation of your orders and also of the revanue law[1268]

Saturday, October 31, the 13th Regiment drew clothing.[1269]

Lieutenant Feazel's reference to Englishman, Robert Grey (Gray), who absented himself from camp to acquire whiskey deserves some further amplification. Grey had worked as a miner before the war and must have been something of a loveable rogue for all his failings. He was a vigorous man. He liked to engage in sports and dancing but his weakness for strong drink would continue to get him into trouble with his superior officers, until finally, in 1865, he was brought up on charges of "Drunkenness while on duty;" "Disrespectful language towards a superior officer;"

and "Riotous conduct prejudicial to good order and military discipline."[1270] He was found guilty on all three counts and sentenced "to be confined at hard labor for the period of six months."

After the war, he relocated to Iowa. He applied for veteran's pension. This application generated considerable paperwork. Numerous statements were taken from his former comrades-in-arms from Co. A, including its officers. The question revolved around whether or not he was in fact deserving of pension. The quandary was not related to his habit for absenting himself from camp and getting drunk but grew out of a singular episode in the fall of 1864 at the battle of Cedar Creek, Virginia (fought October 19, 1864). The weight of consensus was that during the initial onslaught of that early foggy morning, Gray had advanced against the enemy with Co. A and then, when ordered to fall back, he had dropped down into a ravine instead of retreating with the rest of the company. There, presumably in the ravine, he surrendered or was captured by the enemy. He was sent to Salisbury Prison in North Carolina where he somehow ingratiated himself and became employed as cook to the Confederate officers manning the prison. This gave him considerable freedom as compared with the other prisoners. He improved his situation by escaping and walked back over the mountains to return to his company then encamped for the winter, at Cumberland, Maryland.

November 1863

So-called 'lines of occupation' and sweeping the counties of partisans, deserters and other felons

November and December of 1863, found the forces of the Department of West Virginia in a holding mode with considerable scouting duty. The lines of occupation had been drawn and the 13th West Virginia like other military units was posted for the winter. One correspondent to the Cincinnati *Commercial* calling himself "Thota" described in detail the Federal lines of "occupation" in West Virginia and the nature of military service and operations along these lines.

The lines in Western Virginia, as described by Thota, began at Louisa, Kentucky, on the Big Sandy River. From there they ran through Barboursville, Charleston, Fayette, Summersville and Beverly. From the latter place, they ran in a variable and indefinite line to the junction of the West Virginia line with the Potomac. At all the named points were garrisons "of varying strength, as the exposure of the place and the extent of its picket duty requires."[1271] The so-called 'lines' here were imaginary ones created by drawing a line between the garrisons and the posts. In the mountainous terrain, posts were strong points where roads passed "through narrow defiles which are easily commanded by a few pieces or a picket of determined men." Such strong points were selected and picketed on all roads passing between two garrisons which were passable for trains or artillery or regular forces of any size or even a single wagon. A line drawn from one garrison to another through these posts constituted the Federal lines in West Virginia in the winter of 1863.

Confederate and Federal forces operated along these lines with "similar strategy," remarked Thota and he continued. The Confederates, represented by a "single partisan horsemen or independent or by roving gangs lightly equipped, could pass between Federal posts, and penetrate many miles into the area thus occupied by the Federal military, where they committed their depredations and return with impunity." In this way, explained Thota, Confederates

> frequently made dashes on the Kanawha between Barboursville and Charleston, and despoil[ed] some luckless store of a part of its dry goods and shoes. On the other hand, the same course is always open to our boys, and they sometimes improve the occasion vigorously.[1272]

The rebel lines, continued Thota,

> extend along the Tenn & Virginia Railroad, though Parisburg, Union, and Hot Springs, toward Winchester. Thus in the south west there is left a great tract [the region about the Sandy and Guyandotte Rivers] which is neutral ground. It is not worth formal occupation by either party.[1273]

13th Head Quarters was reported to be at Barboursville for the months of November and December. Most 13th companies were also reported stationed at Barboursville for these months with the exception of Co. G, stationed at Hurricane Bridge and Co. I, stationed at Guyandotte.[1274] The 13th Regiment had been divided between Barboursville and Guyandotte since it had been ordered from Coals Mouth to Barboursville during the first week of September.[1275] Head Quarters and all companies but Co. I had been stationed at Barboursville. Co. I was ordered to Guyandotte on September 9 away from the concentration of companies at Barboursville soon after its arrival there. This arrangement of things persisted until November 12 when a further division of the regiment was ordered. On November 12, Young's Co. G, Company C and a squad detailed from Witcher's cavalry company were ordered to Hurricane Bridge. The 13th Regiment remained at these posts (Barboursville, Guyandotte and Hurricane Bridge) until December 31, 1863.[1276]

On Wednesday, November 4, Col. Brown wrote to State Adjutant General Pierpont informing him that the 13th was now "filled to the minimum" required. In consequence, Brown asked that Joseph E. McCoy be commissioned 1st Lieutenant and Henry Stump promoted to Captain.[1277] This month the answer to the question of whether Young's Company belonged to the 11th or 13th Virginia Infantry Regiment was finally settled on November 5 when T. Melvin, A.A.G., wrote to Lieut. Col. J.B. Frothingham, Commissioner of Musters, Department of W. Va. (A copy of Melvin's letter was sent to Adjutant John S. Cunningham 13th Regt. V.V.I.). Melvin wrote:

> [U]nder the decision and orders of the Secy of War, the Company commanded by Capt Young and known as Co G 11ᵗʰ Va. Infty, though attached to and serving with the 13ᵗʰ Regt Va Infty belongs to and is part of the 11ᵗʰ Va Regt —
>
> In view of this the Genl Comdg the Dept directs that you will please regard and treat the officers and enlisted men of this Company, in any case when mustering duty is to be performed, as belonging to the 11ᵗʰ Va. Infty — and this applies to any enlisted men of this Company who may have been through inadvertance or mistake mustered into the Service as members of the 13ᵗʰ Va Regt.[1278]

The foregoing

> order having been received from the War Department deciding that company 'G' commanded by Capt J.V. Young, was part of the 11ᵗʰ Regiment W Va Vol. Infty, and was 'to be considered and treated as such' measures were immediately taken to form a new company, to replace that of Capt Young's, his company having been reported as part of this Regiment from its organization untill the end of 1863.[1279]

New recruits for the tenth company of the regiment, Co. K, were also being actively sought out despite a slowdown in the recruiting business in the Kanawha Valley. At this point in the war, there were few recruits to be raised. In November, when it became apparent that a company to replace Co. G had to be recruited and while Co. K was being recruited, Gov. A. I. Boreman requested of Brig. Gen. B.F. Kelley, cmdg. Dept. of W. Va., that Private James W. Alexander of Holmes' Battery be authorized to recruit for the 13ᵗʰ Regiment. Kelley wrote to Boreman in response to this request from Cumberland, Maryland, on November 28:

> In reply to Governor's request for an order to empower James W. Alexander, private of Capt. Holmes' Battery to recruit a company for the 13ᵗʰ Infantry, I have the honor to state that a recruiting detail has been already made from that regiment and an additional one asked for with a view to the organization, principally, of the 10ᵗʰ Company until the result is known. It seems hardly advisable to detach an enlisted man from another command for the same purpose. Every effort, not incompatible with existing orders of the War Dept. will be made to secure for this Regt. the additional company.[1280]

On Monday, November 23, Paymaster Cowan made way to visit the 13th Regiment to pay the soldiers their wages. Col. R.B. Hayes wrote to Col. Brown informing him that Major B.R.Cowan "will start for your place 'via' the river to-day. Have your pay Rolls signed and send an ambulance and escort to meet him at Guyandotte. He will be at Guyandotte this evening or early tomorrow morning."[1281]

In November, the war sprang to life in the Virginias. In West Virginia, November 1 through 17, expeditions from Charleston—Averell's and Duffié's cavalry commands; 2nd Brigade, 2nd Infantry Division plus Independent Blazer Scouts, organized pursuant to Carr. B. White's counter-guerrilla initiative—went out against Lewisburg, West Virginia. One column left Charleston on November 1 moving toward Beverly and another column set out from Charleston on November 3 for Lewisburg.[1282] November 5 and 6, these columns "had engagements with 'Mudwall' Jackson driving his forces down through the valley east of Greenbriar mountain, through the town of Lewisburg."[1283] On November 6, Averell's force attacked and defeated the Confederates at Droop Mountain in Greenbrier County. In East Virginia on November 7, the Army of the Potomac, under command of General Meade, pushed across the Rappahannock River at Rappahannock Station and Kelley's Ford. Severe engagements flared at both of these points. At 1st Brigade headquarters, Charleston, a letter was received from Brig. Gen. Scammon, dated November 1 informing that "[a] regiment of rebel cavalry [had been] reported en route for Kanawha river" and the troops under Col. Hayes were ordered "to be on the 'que vie'."[1284] Col. Hayes wrote to Captain Botsford at 3rd Division Head Quarters at Charleston on November 3 regarding the cavalry referenced by Scammon in his letter. He wrote: "I am informed correctly I think that a rebel force perhaps three or four hundred strong is now encamped in Logan Co. and have instructed Col. Brown and Capt. Witcher to watch their movements."[1285]

On November 5, Scammon wrote to 1st Brigade Head Quarters complaining that "the 13th Va.V. I. [was] not scouting the country around Barboursville."[1286] That same day, Captain M.P. Avery, upon Hayes' order, informed Brown that:

> Col. Hayes is in receipt of the following communication from Divis Head Qrs which you will act in accordance with.
>
> Head Qrs 3d Divis + c
> Charleston Va Nov. 4th 63
>
> Col. R.B. Hayes
> Comdg 1st Brig
>
> Complaints received from Clarksburgh that the 13th Reg't Va.V.I. is reported not scouting the Country. Order Col. Brown

to scout thoroughly. Keeping parties of his command constantly in the woods and to render weekly reports of scouts made oftener if any thing of importance is learned.[1287]

Col. Brown, responded to the news that his regiment had been negligent with the following to Hayes, dated November 8:

> Head Quarters R 13 Regt Va Vol Infty
> Barboursville West Va Nov 8 1[8]6[3]
>
> My Dear Col
>
> It is with paine I learn that this Regt has been reported at Department Head Quarters as negligent in Scouting this part of the country I can assure you there has not been one Single day since the Regt has been Stationed at this Post that I have not had Scouts out in some direction
>
> At the time I received your letter their was one Company in Wayne Co Two Companies up Mud River and fifteen men in Mason Co Scouting
>
> I presume I am to blame for not Sending in reports of Scouting expeditions we have accomplished So little for the amount of labor performed, that I did not deem it necessary to report, the expeditions to you one good result is that the Sheriff of this County has colected by our assistance nearly $800 in taxes for the years 1861 & 1862 which he would not have colected, but for the assistence this Regt has rendered him
>
> Col visit us when you can Yours Respectfully

Col R B Hayes	Wm R Brown
Com'g 1ˢᵗ Brigade	Col 13 Regt[1288]

Extant paperwork for September, October and November (morning reports, orders, etc.) although providing a fragmentary mosaic of day to day military operations around Barboursville seems to support Brown's statement. His companies and Captain Witcher's cavalry had many scouting assignments. November's record of scouting is quite rich. In addition to the scouting assignments made by Col. Brown's infantry on November 6 Brown issued Special Order No. 75 ordering Capt. Witcher, cmdg. Company G 3ʳᵈ Virginia Cavalry, to "detail twelve men a guard for Sheriff, afterward to proceed up Guyandotte and Mud River." On November 9, Brown issued Special Order No. 77 directing that "Capt Witcher Co G 3d Va Cavalry will detail one sergt and Five men to accompany the Sheriff of Cabell Co for two days."[1289]

Other correspondence from this time includes the following that document scouting expeditions undertaken by Brown's command and other units that cooperated with the the 13th West Virginia to keep the peace in Cabell County. Partisans, Union deserters and other felons were gathered up on these sweeps. Judge H.J. Samuels wrote to Col. Brown: "Sir, I request you to detail a small force to protect my court at Ceredo commencing Monday the 20th day of November proximo And much oblige."[1290] Brown would oblige Judge Samuels. He ordered his companies to scout and report on the presence of rebel activity in surrounding Wayne County. On Sunday, November 1, Captain Isaac M. Rucker, commanding the 181st Regiment Mason County Militia, wrote to Col. Brown from Mason County. This correspondence gives some idea of the extent of cooperation between Rucker and Brown in prosecuting the sweep done there.

> I send under arrest Eli Chapman + William Chapman[1291] both Charged with being connected with the Rebel army one have just Returned these men are Both Bad Characters they Belonged to Keetings Co of Guerrileas as they acknowledge I would prefer not allowing them the benefit of the oath of allegiance if so they will certainly not Respect it pleas have them sent to Camp Chase or some other Prison Camp for I am confident they are nothing but Bushwhackers. [...] P.S. Pleas allow the Guard to fill their cartridge Boxes with ammunition as I am scarce[1292]

November 1, Lieut. Feazel, cmdg. Co. I, wrote to Col. Brown informing that he had apprehended the two deserters from the 9th Virginia Infantry wanted by Col. I.H. Duval pursuant to his letter of October 25 (see above).

> Sir I have the honor of Sending two Deserters to your [Mounted] Guard they belong to he 9th V.V.I. Henry Stephens and Washington Flowers
>
> I arrested them Flowers was at his Fathers and Stephens was at William Rays the families at both places denied their having been any Soldiers at their houses when at the Same time they were concealed in their houses they claimed to be sent as Scouts but had nothing to show.
>
> You will please attend to having thirty dollars each collected for me for their arrest as I wish my boys to have the money that made the arrests[1293]

John Slack, Jr., had written to Brown the month before on October 29 requesting aid in capturing one Henry Langhorn. Langhorn (or Lawhorn) was a soldier of the 1st Virginia Cavalry (Union), who had recently been convicted of a felony and who had escaped from the penitentiary and was presumed to be making for Barboursville where his mother lived. Slack wrote as follows:

> On Tuesday night last a man by the name of Henry Laughorn who was convicted + sentenced to the Penitentiary for two years for a felony Broke jail his mother lives near the town of Barboursville her name I believe is Fips he was a soldier who belonged to the 1st Va Cavalry he had on the uniform of a cavalry soldier he is about 24 years old heavy made man is a great rascal and should be punished his mother was married as I understand the 2nd time her name is either Fips or Nips I have no doubt but he will try to get to that place I wish you to keep a look out for him and arrest him if you can[1294]

November 12, Captain Dixon R. King, commanding forces at Ravenswood wrote that Lawhorn had made an appearance in his area:

> [...] a man by the name of Henry Lawhorn 'or so gave his name' took a horse from this place [Ravenswood] to go to Ripley as he said since which time he has not been heard from. [...]

> Lawhorn is 5 feet 9 inches hight Grey eyes, Black Hair, Dark complexion. He was also sworn into the U.S. Service on [October 31] by Lt. Uriah Lytle, he is supposed to be a Rebel Spy. [The mare Lawhorn took] was a large dun, about 15-1/2 hands high + under the knees of one or the others of the fore legs thare is a calious under the skin [Lawhorn said] he was born in Cabel County + was 24 years of age.

> Please make enquiries + if there is such a person within your reach have Apprehended + inform the Sheriff of Cabell County who will have a warrant for [Lawhorn].[1295]

How this episode played out is preserved in a notation appended at the bottom of Slack's letter written in Adjt. J.S. Cunningham's hand: "Arrested Nov. 15/63 + brought before Col. Wm R Brown for examination + released on condition that he report with papers proving his innocence the next day."[1296] At the top left hand side of Slack's letter was a note from Judge H.J. Samuels to Cabell County Sheriff, John Alford. Slack had apparently been notified by Brown that Lawhorn had been arrested by the 13th Regiment. Samuels in turn notified Alford and directed him to "take in charge the villain named Henry Laughorn and now under arrest and deliver him to the sheriff of Kanawha Co and for so doing this shall be your authority Given under my hand this 4th day of December 1863 H. J. Samuels Judge 8th C[ircui]t West Va."[1297]

As part of the drive to recruit for the 13th Regiment, a handsome looking notice calling for volunteers was placed in the Point Pleasant *Register* on November 12 to run for four weeks. At the top of the notice is an image of an American war eagle sitting on a shield which lies on the ground. Its head and neck are white, its body dark. Its tail feathers are spread in an aggressive manner. The image conveys a sense of strength and tension, of energy contained. Lines are kept simple. The eagle's body is low and stretched long, its wings arched above as if it has just landed or is about to take off in flight. The bird holds arrows in its talons and in its beak it holds a banner bearing the words: "A vigorous prosecution of the war, and no compromise with traitors." Beneath, we learn that Lieutenant W.N. Hawkins, Recruiting Officer was authorized to "raise a few more recruits for the 13th Regiment West Virginia Vols."

> Enlist now and avoid the degradation of being drafted. The volunteer will receive $13.00/month salary and $3.50/month for clothing and subsistence; bounty and premium making a total of $820.00 that the soldier will have received, by the end of his 3 year term of service. Pay will commence from the date of enlistment; clothing and subsistence will be furnished and transportation to the Regiment.[1298]

Other men besides members of the 13th were also authorized to recruit for the regiment. November 13, Col. Brown issued Special Order No. 82 authorizing Captain James C. McFadden "to recruit for the 13th Regt V.V.I. he will report weekly to these Head Quarters of his Success giving the names and No. of recruits enrol[l]ed."[1299]

At times, of course, in their zeal to clear the country of partisans, the 13th Regiment arrested the wrong people. Such was the case with Patrick Hensley and Daniel Langhorn (Lauhorn or Lawhorn), who had been arrested and jailed in Barboursville. On November 15, 186[3], Cassander Spurlock and Samuel A.G. Mguire [McGuire] wrote to Col. Brown on behalf of these men as follows:

> We understand by Mrs. Langhorn and Mrs. P. Hensley that their husbands are under arrest at barboursville upon what charge wee doe not know, but we deem it our duty to state to you that as far as our knowledge extend the s[ai]d Patrick Hensley + Daniel Langhorn [Lanhorn, Lauhorn or Lawhorn] have been at home as

> much or more than any two men that wee know in their neigh-
> bourhood wee the undersigned also state that neither of the men
> have had any thing to doe in the wor since its commensment I
> C. Spurlock lives in about 2 miles of one and about 3-1/2 of the
> other and I Samuel A.G. McGinis lives still further from them.
> Given under our hands this 15th Day of November 186[3 or 2]
> [...] Attest Wm. A. McGinnis.[1300]

Whether the two men were released to their wives is not known.

On Monday, November 16, Col. Brown wrote to State Adjutant General F.
Pierpont from 13th Headquarters at Barboursville recommending that Captain
John S. Witcher, 3rd West Virginia Cavalry Regiment, be commissioned Major to
command two companies of cavalry now full to the required amount and operating
in that section of the country:

> There are Two Companies of Cavalry now full in this section of
> the State one at Pt Pleasant + one at this place These Two Com-
> panies can be united into one Battalion under command of a
> Major and I would recommend Capt. John S Wicher as a young
> man of good moral character + an afficient + dutiful officer and
> would respectfully request that you commission him as such Capt
> Wicher is well acquainted with all that section of country lying
> between the great Kanawha + Big Sandy Rivers, and has pene-
> trated the country far into the interior, and I am satisfied that
> could this arrangement be made it would result in much good to
> the service in this part of our New State. These Two companies
> belong to the 3rd Virginia Cavalry which is now full and has but
> on[e] commissioned officer who ranks as Major and hoping that
> this request will meet your approbation[1301]

On November 18, Judge H.J. Samuels wrote to Col. Brown requesting that in
the absence of officers to oversee the poor, if it be within his power, could Brown
alleviate the severe poverty and suffering among the population in that section of
the country by issuing rations. Samuels wrote at length:

> It is come to my Knowledge that a number of persons in this
> county whose whole male support is in the U.S. Army are in great
> want and indigence and unable to procure food to sustain life,
> there are as yet no overseers of the poor or other officer of this
> county to give the public aid required And you are aware that our

people have suffered so much in property that private Charity will not meet the exigencies of the times.

I therefore request if your instructions permit, that you afford aid in the shape of food to such persons as come within the class above named and designate Mr John Alford Sheriff of this County as the Civil officer who best can inform you of worthy recipients of governmental bounty if you can comply with this request.[1302]

November 24, Col. Brown endorsed and forwarded Samuels' letter to 3rd Division headquarters with a note of his own saying:

Herewith inclosed please find Request sent me from H.J. Samuels Judge of this District asking that something be done for the relief of indigent + suffering families in this County (Cabell) What I wish to say is that the Statement set forth in the aforesaid Request are substantially correct And it would be a great relief to me and also to this suffering class of our community to have some aid furnished them in Rations by the Government I would therefore respectfully request that you give me whatever instructions you may deem necessary in the case.[1303]

Brig. General A.N. Duffié, who was in position to authorize Brown to issue rations to families needing relief, endorsed Samuel's request and authorized Brown to distribute rations providing he was fully satisfied as to the families' actual need.[1304]

Indeed, many families suffered for want of food. A myriad of factors contributed to this. Food was scarce. Money was scarce. People were afraid to leave their homes for fear of falling victim to partisans and renegade soldiers, who could be anywhere. Horses and other livestock had been stolen or sent away for safekeeping. Crops and gardens could not be properly sewn. Women, whose men were away in the army, worked in the fields by day and tended to their housework at night. Ever-present was the fear that the rebels would swoop down upon them. To buy meal or have their grain ground, people would travel at night to mills—what few still operated—with pack horses if they had them. Travel was done through the woods as prudence required that open roads be avoided. As a result of these conditions, many families became impoverished.

Part of the duties which fell to the 13th Regiment was to eliminate treasonable activities. These included 'treasonable' language as well as anything to which treasonable intent could be attributed. Questions as to what constituted treasonable language and actions were doubtless from time to time laid before Col. Brown as in the case of Milton Reese. On November 21, John Holroyde wrote to Brown:

[...] on or about [Nov. 18] I was on duty as Guard over John Smallridge + John Brown two Rebell soldiers who Came in this part of the Country to give themselves up to Federal Authorities [illeg. word] Said. Smallridge in a conversation with myself + Mr D Curry of Hurrican Bridge Stated that as he (Smallridge) passed the residence one M C colister he met Milton Reese who made Enquiries of him where he was going his Business what his intentions were +c. Smallridge told him that he was on his way to Give himself up and stop fighting on the rebel Side, Milton Reese replied to him in the following style, that they Smallridge + Brown were wrong deserting the rebels that it was on account of such men as they that the cause of the South was being weakened + injured, that they had better return to the rebels Army and Continue fighting on their side in fact doing all in his power to persuade them to return to the rebel Army they replied to him their families were in distress that they wanted to provide for their families he told them that he had helped to provide for them and would do so again if they would go Back to the Confederate Army

The above Statements I respectfully Submit to Your decision whether Mr Reece by saying the Language before Mentioned Commited Treason Or not.[1305]

On November 24, Col. Brown requested that he be granted permission to be absent from his command for ten days "to assist my wife in Pomeroy, Ohio who is in delicate state of health and to transact some private business which demands my immediate attention. Time to commence Dec. 15." Gen. Duffié denied Brown's request for furlough. Col. Hayes at Camp White, however, gave permission to Brown to be absent seven days justifying the move as within the bounds of his authority as Brown would be absent within the limits of his (Hayes') command.[1306]

By order of Brigadier General A.N. Duffié at Division headquarters, a day was set aside. "All business" to be "suspended [...] and [the day was to] be observed as a day of Thanksgiving and Prayer."[1307] On November 27, Col. Brown wrote to Captain Th[omas] Melvin requesting that he and Lieut. Col. James R. Hall be mustered in according to the Commissions held by them (i.e., Colonel and Lieutenant Colonel):

This Regiment was mustered into the U.S. service on the 10th day of Oct 1862. At that time it consisted of 8 companies making an aggregate of 657 men. Since that time we have recruited enough to raise the aggregate to over a thousand men provided there had been no deaths nor desertions. Myself + Lt Col Hall have had charge of this command ever since but have received pay only as Lt Col + Major though our commissions are for the Ranks of Col + Lt Col. The Regiment would now number nearly a maximum Regiment had Comp G (Three Fourths of which we recruited), been permitted to remain with us but it has now been ordered to the 11th Va Regiment by the War Department but is still in this command. The command now consists of 11 Companies Whichers Cavalry included each numbering considerably over the minimum number making in all an aggregate of about 975 men, and the Tenth company is now being recruited and there is every indication that the company will soon be filled up. Now what I am anxious to know is whether we could not under these considerations be mustered in on our Commissions proper. If this thing can be done consistently with the good of the service and military usages You would confer a great favor by granting permission to have it done at once.[1308]

Notations on the back of Brown's letter of request indicate its path up and down through the levels of command to the final denial of his request by the War Department in Washington. This denial came on January 5, 1864, from the Adjutant General's Office. The following reasons were given:

Respectfully returned to Lieut. Col. J.B. Frothingham, U.S.A. The Muster-in of these officers as Lieut. Col. and Major, respectively, will under the peculiar circumstances of the cases, be permitted to stand. The Regiment is below the minimum, so they cannot be mustered as Colonel and Lieut. Colonel. By order of Secretary of War (Signed) Thomas M. Vincent, Assistant Adjutant General. It will be perceived from the above endorsement, made in A.G.O., that the request of Lieut. Col. Brown cannot be granted. John R. Frothingham, Lieut. Col. A.D.C. U.S.A., Com. Of Musters, Dept. of West Va.[1309]

November 28, "Lt. James W. Johnson, A.A.I.C., 3d Brigade" wrote to Col. Brown regarding inaccuracies in his receipts for muskets received:

I have received your receipts for 74 Rifle Muskets I had 52 on hand last return. I am not able to account for the gain. To enable me to make a correct Return I ought to have your receipts for 52 Rifle Muskets + 1,760 Ball Cart. Doubtless you will be able to correct the mistake and send me the proper receipts and I will return the original ones[1310]

November 30, the 13th Regiment drew clothing.[1311]

Sometime during the month of November, John C. Bayler, writing on behalf of the Boyd County Kentucky Bible Society, wrote to Col. Brown regarding a donation of bibles to the regiment. Bayler wrote:

> It is proposed by the Boyd Co. (Ky) Bible Society to donate Testaments to the soldiers of your Regiment, if they have not been already supplied + with your approbation. You will please inform me of the facts, + state how many copies will be needed. It is hardly worth while to give to any who cannot read — yet if thought best, it will be done. We are expecting the testaments from New York by middle of December — + I would endeavor to superintend the distribution in person.[1312]

Morning reports November 1863

Morning reports submitted for Company A indicate that the company continued to be stationed at Barboursville for the month of November. The number of present and absent held firm at three commissioned officers and eighty-seven enlisted men: an aggregate total of ninety. No drummer or fifer was reported present with the company during the month. One artificier was present for the entire month. Co. A reported no farriers, no blacksmiths and no horses present. Two enlisted men, Daniel Snodgrass and Andrew Snodgrass, were reported absent without leave from November 5 to 8, and one enlisted man was reported absent without leave November 9 through 12. One enlisted man, George W. King, was absent with leave, by order of Brigadier General Scammon from November 26 to 30. On November 20, Samuel V. Flake was reported "absent at Charleston, W. Va. since Feb. 12, 1863."

Two privates from Co. A had daily duty November 1 to 7, and one private had daily duty November 8 to 30. One commissioned officer was out on detached service each day of the month. November 3, Captain Greenbury Slack was ordered on a scout with nineteen men. They returned on November 4. November 5, Samuel McCormac, James R. Spradling and Samuel Snodgrass were ordered to Charleston to guard a prisoner. McCormac returned from Charleston on November 10. Samuel Snodgrass and Andrew Snodgrass (who had previously been reported absent without leave beginning November 5) returned from Charleston on November 12. James R. Spradling returned to the company from Charleston on November 15. November 26, Joseph Scott and Abram E. Chandler were out on a scout to Coalsmouth. November 24 or 29, Corporal William H.G. Hoffman was ordered to Charleston to guard prisoners. November 29, six men were "ordered up Guian [Guyan] River on a scout." November 30, one commissioned officer and ten enlisted men were ordered out on detached duty. November 30, Capt. Greenbury Slack was reported "returned from Gallipolis 6 o'clock a.m." With the foregoing exceptions, the number of enlisted men on detached duty was very small—just one to four men—over the course of the month. November 1, Corporal James H. Tully was ordered to Charleston by Colonel Brown.

The number of present sick stood between two to five men for the first three weeks of the month. November 20 and 21, there were ten and then nine men reported present sick. On November 22, this number dropped to six and continued to decline until November 30 when only one enlisted man was reported present sick. Four to six men were tallied as absent sick with the higher number reported the last five days of the month. Martin Snodgrass returned to the company unfit for duty on November 1. November 7, Milton Wilson returned to duty from "Regimental Hospital." November 8, Daniel H. Snodgrass returned to the company, and on November 9 John G. Moore was reported absent sick at "General Hospital, Gallipolis, Ohio." On November 18, it was recorded that George W. Ramsey had been absent sick at "Point Pleasant Post Hospital" since August 26, 1863. November 19, Elijah F. Newil (Newell) was reported absent sick at "Post Hospital, Point Pleasant since April 22, 1863." November 21, John F. Teel was reported absent sick at "Post Hospital, Point Pleasant since July 15, 1863," and on November 26 James Light was noted as absent sick at Charleston. Young Private Abner P. (or T.) Nanley (or Nunley) was sent to Charleston in arrest on November 17. He was reported absent in confinement from the 18th to the end of the month.[1313] Court martial records (LL1439 #3 Private Abner Menley, Co. A 13th Va. V. I., NARA, Washington D.C.) record that he was tried by military tribunal at Charleston, West Virginia, on November 28, 1863.

Morning Reports for Company B provide the following information. The men continued to be stationed at Barboursville for the duration of November. Drummer and fifer were present with the company each day. No one was absent with leave. One enlisted men was absent without leave November 1 through 10 and November 13 through 20. The company gained one man, Jacob Sines, who enlisted in the regiment on November 14. He was duly sworn in on "the 3rd." One enlisted man was reported deserted on November 11. Co. B had at the end of November a total of three commissioned officers and ninety-seven enlisted men.

On November 26, Col. Brown issued General Order No. 28 appointing Captain Milton Stewart "one of the council of administration for this Regt vice Capt Van D McDaniel deceased."[1314] Four to five men had daily duty each day of the month. Over the course of the month, no less than three up to nineteen men had detached service each day. The first half of the month saw a particularly great number of large groups on detached service. On November 2, Col. Brown issued Special Order No. 72 ordering 1ˢᵗ Lieutenant Lovell C. Rayburn to "take 25 Men and proceed to Greenbottom farm [Confederate General Albert G. Jenkins' plantation; sale of his property in forfeiture?] for the purpose of Protecting the S[h]eriff at the sales of property advertised to be sold this day — after the Sale you [Rayburn and his men] will return to your Regt without unnecessary delay."[1315] November 29, Capt. Milton S. Stewart, cmdg. company, issued Special Order No. 88 directing:

> Sergeant [Robert O.] Boggess of Co (B) is hereby ordered to take Charge of five (5) Men of Company (B) Six (6) Men of Company (A) and four (4) Men of Co (G) 3d Va Cavaly and proceed with them to a point three (3) Miles above the Falls of Guyanotte river — in which vicinity he will Thouroughly Search the Country for a Gang of Bushwhackers and Horse Thieves reported to be Consealed there — he will report again to these Head Quarters on Tuesday December 1ˢᵗ 1863[1316]

As for the numbers of sick men in Co. B present and absent, the company had from one to eleven men present sick. More men were sick in camp as the month progressed. Five to seven men were absent sick for the month. No one was present or absent in arrest or confinement.[1317]

On November 9, Colonel Brown issued Special Order No. 78 appointing Capt. Lemuel Harpold, cmdg. Co. C, Capt. A.F. McCown and Lieut. Joseph Bromly, officers of Co. F, as "a board of survey to examine into and report upon the condition of the arms for which Capt John S. Witcher is responsible."[1318] November 26, Col. Brown issued Special Order No. 86 directing Capt. Harpold "to take 15 [or 50 , manuscript illeg.] men of his Co. and scout the country between this place and Coal river and up Coal to the Forks. [Harpold to report] with your Co. within 6 days at this Hd. Qtrs."[1319]

Morning reports submitted for Co. D record that the company was stationed at Barboursville for the month. The company had no drummer nor fifer present and no farriers, blacksmiths and no artificiers present. One commissioned officer was absent on leave November 28 and 29. One enlisted man was reported absent without leave on November 1, 2, 10 and 27. November 10, John Morarity was reported absent without leave. November 27, Louis F. Vincent was reported absent without leave beginning November 26. On November 28, a notation made by Lieutenant George Snowden (Co. D) in "Remarks" appended to morning reports indicated that "Louis Vincent and Nathaniel Birchfield will not be allowed to go outside of the lines for the term of 60 days. Lieut. Snowdon."

The company gained men during the month. On November 7, George H. Woods enlisted in the regiment and joined Company D. On November 16, four more recruits were gained: Jacob Whitehead, General L. Boso, George Proor and David H. Lisle. Amos Woodyard was rejected by Assistant Surgeon Abraham D. Williams on November 17. November 25, Jacob Whitehead, General H. Boso, George Proor, David H. Lisle, Manley Woodyard and George H. Woods were all mustered into the United States service. Hezekiah Mays enlisted in the Company on the 25th. He was mustered the same day with the other recruits. Co. D finished the month with present and absent three commissioned officers and eighty-eight enlisted men.

Each day one non-commissioned officer and five or seven privates had daily duty. There was detached service each day of the month. Three to nine enlisted men and one commissioned officer were detached November 2 and 3 and November 21 to 27. The company had from one to seven men present sick during the month. This number included the Captain Simon Williams, sick on November 1 and 15 to 26. Numbers of present sick declined toward the end of the month. There were five to eight men absent sick during the month. No one was present or absent in arrest or confinement.[1320] On November 28, "George S. Farley of Co. D, 13th Va. Vols., and Miss Miranda Ellis of this county [Mason], were this day united in the holy bonds of matrimony by the undersigned. [W.W. Harper]"[1321]

As for Company E for which no morning reports survive, on November 1 Col. Brown issued Special Order No. 71:

> 1st Lieut William N. Hawkins of Company E, has at his own request been detached on recruiting Service by Gen Scammon after Nov 1st he will take (10) ten Men and proceed to the Counties of Kanawha Mason Putnam + Jackson to Recruit for the 13th Regt. and report to the Col Cmdg — weekly"

November 3, Col. Brown issued Special Order No. 73 instructing Captain commanding Co. E, J.D. Carter, to "take one hundred and Five men and proceed to Hamiline or vicinity" where they were to "Scout the country thoughorly and assist the enrolling officer of Cabell Co." Carter was further directed to report back to 13th Head Quarters with his command on November 19. November 5, Brown issued Special Order No. 74 ordering Lieutenant Rosler, Co. E,

> to take fifteen men and [...] proceed to Charleston West Va in charge of six prisoners and deliver them to the Provost Marshall at that place He [Lieutenant Rosler] will report at Briga[d]e Head Quarters at Camp White On his return he will scout the country by Davis Creek across Coal River and thence down Mud River to this place and report to these Head Quarters as soon as possible.[1322]

November 9, Lieut. William Hawkins ordered Sergeant Vincent A. Hays, (both officers in Co. E) to "take Ten men and report to Col. Wm R Brown without unnecessary delay."[1323] Attached to this order was a note from Hawkins addressed to Brown:

> Col. Brown I have the men of reporting I have three men since I left the regt. which I have given leave to fix up their business I will bring them down with me I am still going for them you will please send me a new Squad up in as short a time as possible your obt Svt Wm. N. Hawkins.[1324]

On November 10, Private George W. Smith (Co. E) drew 1 pair of pants @ $3.55 and 1 pair of shoes @ $1.48. November 26 , he drew 1 blanket @ $3.61.[1325] November 10, Capt. J.D. Carter reported concerning a large scouting expedition undertaken by him and one hundred and five men under his command:

> Headquarters Co. E 13th Va Vols
> Barboursville W. Va. 10 Nov. 1863
>
> Wm R. Brown
> Col. 13 Va Vols
>
> Sir I have the honor herewith to State that in Accordance with Your Order I started for Hamline on the morning of the 3d of Nov with 105 men. Detachment of Cos. E. F. G + C of the 13th Va Vols. I arrived at Hamlin on the evening of the 4th On the 5th I sent out two Scouting Parties one up Main Mud into Boone County the other up the Middle fork of Mud. They returned + reported no Armed Rebels in the County. On the 6 + 7 We Scouted the County Generally through the upper end of Cabell + Lower end of Boone Counties. Finding no Syn of Rebels in all that Country, the County in that part, has nothing to subsist upon more than for their own necessities. There is no fat Cattle, no Hogs, and No Horses. On the morning of the 8th of Nov. I starting for Barboursville. Arriving on the 9th
>
> I am Sir
>
> Very Resptly J.D. Carter
>
> Capt 13 Va Vols

November 20, Lieut. Hawkins reported to Col. Brown from the Mouth of Sandy River regarding his success in recruiting:

I have got 7 recruits, Sargeant Hays Moriea had high time + am still going for the tadigomalions they think the bounty a big thing on Ice

I had to leave 2 of my men at their homes sick but I was lucky enough to have two recruits to take their arms and I still t[illeg 2 letters]prim my squad we live with <u>our Sothern</u> friends they treat us all right because thay cant help them We expect to leave here in a day or two for the blue creek and thence down Camels creek to Charleston at which place I will ship my spoils to the regmenttel Col Hall to fix the pay on those boots for I will need them by the time I get to the regment boys in high spirits[1326]

Morning reports submitted by Company F indicate that the company was stationed at Barboursville as it had been the month before. The company had drummer and fifer present November 1 to 3, and one waggoner present November 1 through 30. No one was absent with or without leave. The company gained three recruits. James R. McCoy and Robert Childers enlisted in the company on November 1. Francis Graham enlisted on November 11. At the end of the month, Co. F numbered three commissioned officers and ninety-one enlisted men.

Seven to ten men (including one non-commissioned officer detailed November 21 to 30) had extra duty each day. There was also detached service each day of the month involving from four to thirty-seven men on duty. November 4 to 8, a large group consisting of one commissioned officer and thirty-six enlisted men was detached. Then, two smaller groups were sent out. One commissioned officer with nine enlisted men was detailed November 17 through 25, and ten enlisted men were detached November 26 to 30. November 15, Colonel Brown issued Special Order No. 84 directing 2nd Lieutenant Joseph Brumley to "take three Men and proceed to Mason Co. W. Va. and arrest all absentees from this Regt and then to report without unnecessary delay to [13th Regiment] Head Quarters."[1327]

Co. F reported three to five present sick each day. Private Sylvester Keith, Co. F, was probably among those present sick. He wrote as much to his mother, Elizabeth McIlvan/McIlvain, on November 27:

Barboursville Va
Nov 27th 1863

Mother

I have once more seated my self to address you hoping you are well. I have bin unwell all day but I feele some better now I have bin looking ever day for a letter from you but none came so I thought I would toil some of [my] time away with writing to you I herd Bill was at work again over the river from here ef I get better by Monday I will go over and se him if he don't come over before then we have bin payed of since I came back for two months I have thirty Dollars for you if I had any chance to send it safe through to you if I se[e] Bill before he comes home I shall send it to you perhaps I will come home about new years day I think I can get the capt to let me come up about then give my love to Mary Ann and all that ask about me tell Mary Ann that my head is about to burst or was a few minutes ago. Tell all of my firends up about Coal run if they ever expect to here from me that I want them to write give them my address and if they don't write they may go to thunder my head paines me very oblige to close so [illeg.] you will excuse me for the present still remain your old son Sylvester Keith to his mother

Elizabeth Mc[Ilvan]

Direct your letters to Guiandott Cabell Co W Va and then will come sooner here or there will answer I don't care which place you send them just so they come to me. Here wo[uld] be the best I think

Sylvester Keith Co F 13th Regt Barboursvill Cabell Co West Va[1328]

Four enlisted men of Co. F were reported absent sick each day of the month. No one was present or absent in arrest or confinement. November 25, Company F was paid up to October 31, by Major Cowan.[1329]

On Tuesday, November 10, Clark Elkins, 2nd Lieut. (Young's Company G) issued Special Orders reducing Corporal Jacob C. May (Co. G) to the ranks "for using unbecoming disrespectfull + imprudent language to his Superior Officers. And David Stevenson is hereby appointed Corporal vice Jacob C. May in said Co to rank as such from the present date." Elkins asked in an appended note that Col. Brown "pleas approve and have read on dress Parade the above Order."[1330] Also on the10th, Hayes at 1st Brigade Head Quarters received orders from Brig. Genl Scammon, commanding at 3rd Division Head Quarters, to send "Capt. Young's Company [G] of the 13th Va. V.I. to Hurricane Bridge [...] with four or five cavalrymen, to remain there until further orders."[1331] That same day, (November 10), Hayes ordered Brown to order Young's Company and, if it could be spared, "a squad of Cavalrymen say 5 or 6 under a Sergt. to proceed to Hurricane Bridge. To remain there until further orders." The cavalrymen were to be detailed from Capt. Witcher's command.[1332] The transfer of Co. G to Hurricane Bridge was not immediately executed as it seems

Witcher's cavalry escort for the company was destined to be detached on November 11 to escort Federal deserters to Charleston. Brown issued Special Order No. 79 on November 11 regarding this assignment:

> Capt. John S. Witcher, will take Five men and proceed to Charleston, with two deserters, one from the 1st Va Cavalry and the other from the 9th Va In'fty, he will deliver the prisoners, to the Provost Marshall, at Charleston, fi[r]st reporting at Brigade Head Quarters. He will report at these Hd Qtrs, without unnecessary delay.

Special Orders No. 80 and 81 issued next day (November 12) authorized the detachment of Co. G to Hurricane Bridge in compliance with Scammon wishes. Special Order No. 80 directed that:

> The Commanding officer of Co G 3rd Va Cavalry will detail one Corporal and five Men under the Command of a Sergeant, who will accompany Company (G) of this Regiment to Hurricane Bridge W Va. he will Report for duty to the Commanding officer of that Co. untill further Orders[1333]

Special Order No. 81 issued November 12, directed

> Capt[ain] J.V. Young of Co. (G) of this Regt will proceed with his Company to Hurricane Bridge West Va. and take Command of the Post at that place — he will take Ten days rations with him Five Men one Corporal under the Command of a Sergeant from Co (G) 3rd Va Cavalry will report to you for duty. You will remain at that post untill further orders
>
> Send your Tri Monthly reports by Messanger[1334]

On Thursday, November 12, Captain Lemuel Harpold's Co. C was ordered to take post at Hurricane Bridge to augment the force there. These companies remained at Hurricane Bridge until December 31, 1863.

On November 14, Capt. Young wrote to Col. Brown reporting on guerrilla activity in his home neighborhood and requesting support. Much agitated, to judge from his spelling mistakes, Young wrote:

> Last night there was a bout 20 Rebels at my home and took my Horse and a number others in the neighborhood this morning I sent out the Cavalry and mounted 15 or 20 of my men in persuit but I lern by the Sergt Capt Carpenter is here with 40 men and says there is more in the neighborhood and expects to attact me to day

can you send the Cavaltry up here[1335]

Brown straight away ordered that:

> Lt Lesage of Company (G) 3rd Va Cavalry will take Forty Men of his Company and proceed to Hurricane Bridge and Scout the Country thoroughly as a Rebel force under the Command of Peter Carpenter is supposed to be in that vicinity —
>
> you will report at these Head Quarters — at your Convenience
>
> You will report to Capt J.V. Young Commanding Post at that Place[1336]

Guerrilla operations were escalating in the vicinity of Coals Mouth as well. S. Benedict, a resident of the Coal River Valley, wrote to Captain Bottsford at 3rd Division Head Quarters on November 15 describing the state of affairs there and requesting that some troops be stationed at "Camp Defiance."

> Capt A I Botsford   Coals Mouth November 15 1863
>
> Dear Sir I presume that you are now satisfied that my fears in regard to depredations of Guerillas in this part of our valley were not unfounded as you have no doubt been informed of the late raids + robberies and is now within two miles of us a party of from forty to fifty, they are takeing all the horses + other property they can lay their hands on will unless we have help be in our town tonight. Our union men some twelve in number are well armed and watching day + night, but we can not hold out much longer against these Banditties without assistance. Now I ask shall we have it or shall we succumb and leave what little we have left to be sacrificed to such out laws and our lives imperiled we have a few unconditional union [men] here who have been so from the commencement of this rebellion we have already lost the most of out property in trying to sustain in every possible way the government in this great and desperate struggle for our national existence to preserve our Union intact and our flag from disgrace, so that the blessing of a free government may be continued to future generations. But enough of this. We calculate to fight and fight on to the end which will erelong be a successfull.
>
> If there can be a few troops stationed at Camp Defiance say one good company if no more can be spared we can I think with assistance of our Union men repel any force that may make an attack

upon us. You and through you to the General is the proper authority to whom we should apply for protection. Shall we have it! Or must we look elsewhere!

Be kind enough to show this to Gen. Scammon and let me hear from by the bearer Mitchel C. Clay who is a perfectly reliable man. Respectfully your Obt. Servt

S. Benedict

Written on the back of Benedict's letter is this: "Hayes will send Compy of Infty Coalsmouth until Guerillas are driven out of the country [illeg.] send the company this eveg [...] By order Brig. Genl Scammon Jas L. Botsford AAG."[1337]

J.V. Young wrote to Paulina, his wife, from Hurricane Bridge on November 20 chaffing with annoyance over the stealing of Van, their horse. Young described with relish and biblical reference the effect his company's presence was having at Hurricane. Stolen horses were being rounded up by his men with assistance from very possibly, the 23rd Ohio Infantry:

> I would have sent some men up today but for the rain and warm weather. But if it turns cold this evening I will send up in the morning. I have sent Lt. Elkins to Coal this morning to inquire about our horse. He will return in the morning. And then I will send you the mare. [...]
>
> The scriptures say we shall place our feet on the necks of our enemies, and I believe it will hold good in this infernal Rebellion. It is really humiliating for the poor devils to have to come to Young for a pass to go to their neighbors. Mrs. Conner says if Young's company stays at the Hurricane Bridge she won't stay at home. I say, let her go to Dixie, if she wants to. [...]
>
> Brooks caught thirteen of them and some horses, I don't know how many, and yesterday the 23rd caught four more who had been back and stolen more horses on Brown's Creek. I hope among the captured we will find Van.
>
> We are looking for the Paymaster every hour. [...][1338]

November 29, Capt. Young enrolled nineteen-year-old Lewis Ellis at Winfield for a term of three years. The soldier was mustered in to date from November 29, 1863. The enrollment was filed with the 13th Virginia Infantry Volunteers on March 28, 1864, when Co. G was at last officially transferred back to the 11th Virginia Infantry.[1339]

Morning reports for Company H record that the company was stationed at Barboursville. A drummer and fifer were reported present November 1 through 30. No one was absent with leave. One enlisted man was reported absent without leave November 29 to 30. The company gained new recruits. John W. Swanson enlisted on November 6. He joined the company on November 8. James M. Owens and Edmond Cowan (Cowen) enlisted October 26. They joined the company for duty on November 9. November 15, Jacob Shumaker (Shoemaker) joined for duty.[1340] Joseph Blair and Charles A. Johnson enlisted in the regiment on November 22. Silas Ward enlisted on November 29. November 25, James M. Owens, Edmond Cowans (Cowen), William Clark, Joseph Blair, Charles Johnson, Stephen F. Fuller and John W. Swanson were mustered into service by Lieutenant and Mustering Officer, C.S. Roberts. At the end of the month, Co. H reported an aggregate present and absent of three commissioned and eighty-four enlisted men.

Captain W.I. Mathews, commanding Co. H, had special duty at departmental headquarters, on November 1 and November 3 to 9. November 7, Col. Brown issued Special Order No. 76:

> Capt. [I.] W. Mathews of Co H having been ordered to report to Capt Thayer Melvin A.A.G. at Head Quarters Dept of West Va for assignment to duty by Special Order No 127 issued from Hd Qtrs 3d Div'n will turn over all Ordnance and Ordnance Stores and Company property for which he is responsible to Lt. O.W. Griswold 13th Regt V.V.I.[1341]

November 9, Capt. Mathews' detached service at department headquarters was extended by order of Gen. B.F. Kelley, commanding Department of West Virginia.

Two privates of Co. H had extra duty November 1 to 5. One private had extra duty November 6 to 30. Overall, one to six men (including often times a commissioned officer) had detached duty each day of the month. November 1, Lieutenant O.W. Griswold and forty men left camp to go on scout pursuant to Special Order No. 70 from Col. Brown:

> Lt O.W. Griswold of Co (H) will take forty Men and proceed to Ceredo Wayn County for the purpose of Protecting the Civil Court which will Convene at Ceredo tomorrow Nov 2nd 1863. You will Remain there until the Court is adjourned and then return to your Regt without unnecessary delay[1342]

Griswold's scouting party returned to camp on November 8. Additional detached duty was recorded in reports. November 2 to 6, two commissioned officers and twelve enlisted men had detached duty. November 3 through 6, two commissioned officers and forty-four enlisted men were out on detached service. November 6, Lieut. Griswold returned to camp from Ceredo, Wayne County. He submitted the following report of the expedition to Adjutant John Cunningham:

> Agreeable to Special order No. 70 issued at your Head Quarters No[v]. 1ˢᵗ 1863. I immediately proceeded to Ceredo Wayne Co. Virginia with 40 Men for the purpose of protecting the Civil Court, arising there Nov. 1ˢᵗ and remained until Nov. 6ᵗʰ 1863, at which time court adjourned Returned to Regt. at Barboursville Nov. 6ᵗʰ, 1863. Received information while at Ceredo Nov. 3ʳᵈ that Rebel Jessee Dotson[1343] with Seven others had Stolen five horses on the night of 2ⁿᵈ near the Mouth of Whites Creek, which empties into Big Sandy River, nine miles from Catlettsburgh Ky. Was also informed on night 4ᵗʰ Nov. that a small squad of rebells were seen at the head of Walkers Branch in Wayne Co. W. Va. 5 miles from Ceredo W. Va.[1344]

November 9, Col. Brown was ordered by Hayes to "direct a detail of 1 Corpl to report to Prov[ost] Marshal Charleston W.Va. for duty immediately."[1345] Brown chose Corporal James P. Elkins (Co. H), for this assignment. Morning reports record that on November 12, Elkins returned to the company for duty from Provost Guard duty at Charleston. November 27, Sergeant John H. Raider and four enlisted men were ordered to "take Charge of David Shelton Rebel Prisoner and proceed with him to Charleston Ka Co. W. Va and [ ] deliver him to Brig Head Quarters Camp," thereafter to return and report to 13ᵗʰ Regiment Head Quarters at Barboursville "without unnecessary delay."[1346] Raider and the men departed on November 28. Brown notified Col. Hayes at 1ˢᵗ Brigade headquarters that "one rebel prisoner who gave himself up to the authorities" had been sent and requested "that the guards who accompanied the prisoner be permitted to visit their homes up Elk River."[1347]

One to six men of Company H were reported present sick each day over the course of the month. There was an improvement in these numbers during the last week of the month. Eight to eleven men were reported absent sick each day, and this number decreased as the month progressed. November 9, George W. Irby returned to camp from "Post Hospital at Point Pleasant." November 26, Irby was detailed as teamster. November 16, Rice King returned from "Post Hospital at Gallipolis, Ohio," for duty. November 19, Hiram Fuller returned for duty from hospital at Point Pleasant. No one was present in arrest or confinement. November 3, Private Andrew J. Allen returned to duty from confinement by order of Col. Brown.[1348]

Morning reports submitted for Company I show that the company was stationed at Guyandotte November 1 through 30. 1ˢᵗ Lieutenant W.E. Feazel was present each day. Four to five sergeants; seven to eight corporals; and from forty-nine to sixty-six privates were present each day of the month. No one was absent with leave. One enlisted man was reported absent without leave for the entire month. Co. I also gained in new recruits, who joined as follows: one joined on November 9; one joined November 10; two joined November 17; two joined November 19; one joined November 21; one November 24; James E. Lunsford and one other recruit joined on November 25; one more joined on November 26; and November 27 through 30, two joined each of those four days. At the end of the month, the company had present and absent one commissioned officer and eighty-six enlisted men.

One private had daily duty November 1 through 15. Two to four enlisted men had detached duty each day. Co. I had one private present sick November 1 through 15. Three enlisted men were absent sick each day. November 13, Sergeant John H. Davis returned to the company for duty. No one was present or absent in arrest or confinement.[1349]

Guyandotte had the reputation of being a virulently secessionist town and Lieut. Feazel took it upon himself to administer the oath of allegiance to people in and around Guyandotte. Several manuscripts preserved in collections today shed light on this rite of the times, at least as administered and enforced by Lieut. Feazel. On November 7, S. Bowden, a citizen, swore the following oath of allegiance to the United States of America before Feazel and Sergeant Ernest M. Ong. As will be readily seen, the oath of allegiance was a sweeping promise to also report secessionists; not to sell liquor to soldiers; and not to violate the U.S. Revenue laws.

> I of my own freewill do most solemnly promise and swear that I will support the Government and Constitution of the United States, + the Laws in pursuance there of as the Supreme law of the Land, also the Government + Constitution + Laws of the State of West Virginia, and will perform all my duties as a Citizen of the same honestly + faithfully.
>
> That I will not offer aid and comfort or information, directly or indirectly to any of the enemies of the same, but will give prompt information of the presence or design of any disloyal person who may reside in or visite my neighborhood, should the same come to my Knowledge, I also promise + swear that I will not sell, trade or give any Intoxicating Liquor of any Kind to any soldier of the United States or permit it to be done by others if in my power to prevent it, neither will I in any manner violate any of the U.S. Revenue Laws. To all of which I do solemnly promise + swear without

the least hesitation, mental reservation, or self evasion of mind to keep and perform the same.

Subscribed to + Sworn before Lt. Wm. E. Feazel in Guyandotte West Virginia this 7th day of November 1863 in presence of Ernest, M. Ong. S. Bowden.[1350]

On November 9, Lieut. Feazel wrote to regimental headquarters regarding his actions taken in the case of Larkin Noel (Newell/Nowel), who was accused of having violated his oath of allegiance:

Larken Nowel having twice taken the Oath of Allegiance to the U.S. and having unconditionally broken the same it is hereby ordered the said Larken Newel does leave the State of West Va. within fifteen days from this date and if found within ten miles of Guyandotte after said date the penalty will be death by order of Wm. E. Feazel, Lt. Com. Post.[1351]

On November 19, F.D. Beuhring, Noel's landlord, wrote to Col. Brown on behalf of his tenant:

I regret that I have not the honor of Your personal acquaintance, as I wish to appeal to your feelings of humanity, + sense of Justice in behalf of an old man, named Larkin Noel; from the common report of your willingness to address grievances, + dispense impartial justice, I am induced to hope that you will not consider the case as unworthy of your notice.

Larkin Noel, is a <u>simple</u>, honest, + industrious man, none more so in this country, — came from Eastern Va., some three years since, having been a tenant of mine for that length of time, he has but few acquaintance, but among them some bitter enemies who have several times had him arrested, — their false testimony has, until this time, failed to injure him, — but now he is ordered by Lieutenant Feazel (as you will see by order enclosed) to leave his sick wife, helpless family + the unsecured fruits of his year's hard work, to go, where, by the rigid force of the conscription he may be forced to serve against you — instead of quietly and harmlessly pursuing his duties as a <u>good</u> citizen at home. Of the charges brought against him, he says, the most heinous is the alleged feeding of 'Dodson + his men' — He is willing to swear that this is false, — he has not seen 'Dodson' since the war began, and he asks only that the charges against him may be investigated, having had no trial, witnesses confronted with him. He has always been a plain spoken man, + I have had no cause to doubt his word in this matter. His honesty is proverbial with all who knew him, — and his enemies are those who have never been even suspected of that virtue + therein find cause sufficient for their bad usage of his character.

I have seen him, + passed his house almost daily for the first year and am perfectly satisfied of his being guiltless of any serious offence. It is true he has two sons in the southern service,[1352] for which he is not responsible, — although this is one of his 'crimes,' — when the rebels came in on the Border his Sons came to his house, and he told them 'If you have come here for the purpose of disturbing the people, or their property I don't want you to come about me,' — and he has in no wise as far as I know, (+ he says he has told me every thing) deserved such summary treatment.

I do not know what discretion Lieut Feazel has in such cases, — nor what representations he was induced to thus punish the old man — but I conscairned [discerned?] the 'manner of his going' somewhat informal — and thought it my duty to endeavor to help Mr Noel to a fair hearing, or release from this order, by appealing to one who though a stranger to me, has drawn the favorable esteem of all who have the honor of his acquaintance. For my own character for '<u>veracity</u>', I would respectfully refer you to Mr Chas Qroffe Mr. Thos. Thornburg, or any <u>gentlemen</u> in this country.

If you desire to see Mr. Noel and hear his own relation of the matter at issue, — he is just without the lines awaiting your decision — not knowing whether I would be advisable to enter without express permission.—

Mr Noel is charged with a crime which he is willing to answer he has not committed, — his accusers he has not seen, — has had no trial, — is ordered beyond the lines under penalty of death, — leaving a sickly, + helpless family

He asks of you as a christian gentleman, and a superior officer, that you will give him a fair trial — confront Mr. Noel with his accusuers, — or that you will rescind the hard order of Lt Feazel + permit him to remain quietly at home as heretofore. [...] P.S. I came to the lines an hour or two since but was informed that I could not get out again, — therefore I have written this.[1353]

November 23, Col. Brown issued Special Order No. 85 to Lieut. Feazel who appears to have been remiss in attending to company discipline and paperwork:

> The attention of Lt Wm E Feazel is called to the Condition
> of his Company books He will, devote his exclusive attention to
> having his Company Books Completed or entered up by the 15th
> of Dec next — Also recruiting and Desipl[in]ing his Company —
> These will occupy all his attention[1354]

Company K was organized by Lieutenant Henry Stump at Barboursville in November 1863. It was mustered in by Lieutenant Cyrus S. Roberts, 150th Ohio Volunteer Infantry and Assistant Commissioner of Musters, 3th Division, Department of West Virginia, on November 26, 1863, at Barboursville. The recruits pledged to serve "from the dates of enrollment for the term of three years, unless sooner discharged." The new company was reported stationed at Barboursville for the months of November and December 1863.[1355] November 28, Henry Stump wrote to Col. Brown, from Charleston reporting on misplaced accouterments, marching, the optimal situation of the 23rd Ohio Volunteers, recruits and recruiting:

> I delivered your Order to the Ordnance Officer here and we have
> made search for the box of accoutrements but have failed in finding
> them the Ordnance Officer seems to have no recollection of the
> Box as described by you +c
>
> We arrived here after a fatiguing March of about 18 hours we
> turned your prisoner over to the provost Martial, I think we will
> get transportation for a few rations to Roane Cty tomorrow though
> teams are very scarce and Busily engaged — the 23d, seems to be
> nicely fitted Out, with their Winter Quarters The Order in which
> they have stocked their Tents presents every thing in the shape of
> Beauty + Comfort. I saw Lieut Hawkins this evening who informs
> me that he will forward some 7 Recruits to your Regt tomorrow, he
> says however that the Recruiting Business is very dull in Jackson +
> Roane — I have however some hope of Obtaining Recruits in this
> region of Country[1356]

December 1863

The "rebels are poking around"

— James M. Comly, Lieutenant Colonel Commanding 1ˢᵗ
Brigade, 3ʳᵈ Division to Capt. G.W. Gilmore, Gauley, manu.
dated "Head Quarters Charleston, Dec. 15, 1863," Letter Book
July 25, 1862 to 16 December 186[?], Headquarters 23ʳᵈ Regt
Ohio Vols., James Comly Diaries, micro. Roll 1, Ohio Historical
Center, Columbus, p. 344.

Campaigning should have been over for the year but operations were still being undertaken. December 8 through 25, a Federal raid was made on the Virginia and Tennessee Railroad and Federal demonstrations were made up the Shenandoah Valley and from the Great Kanawha Valley upon Lewisburg. Gen. Scammon attacked Gen. Echols at Lewisburg on December 12 driving him from there; killing and wounding a few; and capturing thirteen "including the ordnance officer on Genl Echols staff." "[A] lot of Q[uarter] M[aster] Stores" were taken and Scammon now held "the country to the Greenbrier river."[1357] Brigadier General A.N. Duffié was in pursuit of the enemy with his cavalry.[1358] In response, Confederates countered with an operation into West Virginia and Virginia in December, which despite the onset of severe weather, illicited considerable commotion and activity within country patrolled by the Kanawha Division.

Military life for the 13ᵗʰ West Virginia Regiment continued as before—same posts same duty and problems. 13ᵗʰ headquarters continued to be at Barboursville, Cabell County, for the month of December. All 13ᵗʰ Companies (including Company C) with the exception of Companies G and I, were stationed at Barboursville. Company G was still at Hurricane Bridge[1359] and Company I at Guyandotte.[1360]

Confederate partisan activities bubbled up in various places and to 'keep a lid on things' the regiment had to be divided. It was not easy. Around December 11, it was reported that there was a force of enemy cavalry in Logan and Boone counties. West Virginia was not as loyal as she had on occasion been portrayed "with the sacred Goddess of Liberty perching upon the huts of the sturdy mountaineers."[1361]

While Scammon attacked the Confederates at Lewisburg, there was a shift of command at 1st Brigade Headquarters at Charleston. General Order No. 1, issued from "Head Quarters Post of Charleston," on December 9, 1863, gave Lieutenant Colonel James Comly command of the "Post of Charleston; including Camp Piatt and all posts below Charleston except Barboursville and the posts garrisoned by the 13th Virginia."[1362] The military draft was also on again in the 13th soldiers' home counties. The Point Pleasant *Register* of December 3 carried: "Let the shattered ranks of the 2d, 4th, 5th, 8th, 9th, 10th, 13th and other West Virginia Regiments in which Mason is represented be filled up at once."[1363] The 13th was recruiting to fill up Cos. I and K and to create from whole cloth a new Co. G now that it was clear that Young's Company would be taken from them and be given to the 11th Virginia Infantry.

The newspaper article, "Volunteering," which appeared in the Point Pleasant *Weekly Register* of December 24 reported enthusiastically about the briskness of volunteering during the months of November and December. West Virginia Governor A.I. Boreman's inducements to volunteering had been published in the previous issue. It was hoped that these inducements would stimulate enlistments, and it would seem that they had. The number of men required of Mason County to satisfy the national quota was not known at this time but all who volunteered before the year was out would certainly receive the "liberal bounty now offered." Not that Mason was the only county that recruits came from. Nonetheless, with pride of place the writer maintained that replacement Co. G

> is being rapidly made up in Mason county, indeed, the glorious 13th will soon be entirely full, notwithstanding Captain Young's Company, was lately given to another Regiment. Our 13th has gained an enviable name in this part of our State, and promises to soon surpass all others in this district.[1364]

All men between the ages of twenty and forty-five were urged to volunteer to avoid the draft inasmuch as:

> Congress will soon—if they did not, the first of this week—abolish the distinction between classes, and the $300.00 commutation clause, and probably raise the pay to privates to $20.00 a month and do away with all bounties for future volunteers. The great need just at this time, has induced the Government to offer such large bounties, but the necessity passing, they will probably be denied those deferring to volunteer 'till after the 5th of January.

Such bounties are necessarily a very heavy drain on the treasury and will be stopped as soon as possible. Let all desiring to make sure of the present bounties, join at once or it may be too late.[1365]

William W. Harper, correspondent for the 13th West Virginia Regiment, wrote another letter to the Point Pleasant *Register* for the benefit of the people at home. He wrote from Barboursville on December 2:

Editor, Register:— It has now been a long time since we have given an account of ourselves. We want it to be understood, however, that we are still about and trying to obey our superior officers as best we can, though some of the boys at times become a little self-willed, and the consequence is the next place they find themselves is in the Guard House, which by the way, they find is not very pleasant after all. This I am happy to say, happens less frequently with our Regiment, I believe, than most any other in the field. For taking if from Alpha to Omega, I think that this Regiment will stand as bright for morals and upright conduct both among commissioned officers and privates as [any] other Regiment in the service: for drunkenness, we have comparatively speaking, but few of that class, and they only get drunk when they can get the whiskey. The boys will, once in a while, overstep the bounds of propriety, and will indulge a little too much in the use of profane language. One of the causes that gives rise to this perhaps, is that they unfortunately suffer themselves to forget that the broad eyes of Deity are constantly scanning their every thought and every word. This uncautious state of mind, is perhaps the cause of nine-tenths of all our crimes.— We don't steal much; and indeed, I believe that it has never been decided yet whether the appropriating of such things as chickens, ducks, geese, turkeys, pigs, sheep, and calves to the benefit of the soldier, will properly come under the head of stealing, and especially if taken from rebels. I believe that the boys have decided that it is an impressment into the service, and generally put it down under the head of a military necessity. What your Courts of Justice may decide it to be, has never yet been taken into the account. The truth is, we don't know much about courts of justice out in this land of desolation, treason and [secession]. The Judges have all migrated as Parson Brownlow says "h-ll-ward," and as for lawyers we have succeeded [with] our own cases and care but little whether we ever see one again until the war is over.

And by that time, if it don't close up too soon, and we continue to improve as rapidly in the future as in the past, we will have learned all the tricks of the profession, and will be able to attend to our own business. Our business out here is scouting, of which we have done and are now doing a good share, and also aiding in collecting the revenue, a good amount of which has been collected since we have been here. The people think the military must be obeyed and consequently shell out, without much complaining. Rebels are coming into our lines almost every day and delivering themselves up. Their statement in regard to the affairs on the Confederacy all agree as representing that destitution, starvation and ruination are staring them in the face, and that a general despondency is taking hold of the minds of the people, and they are becoming anxious for peace. So mote it be.

W.W.H. [1366]

Although Harper's letter makes no mention of the weather—severe freezing cold moved into the area towards the end of the month—a letter from William McKinley, Quarter Master of the 3rd Division to 13th Regiment Quarter Master Stephen Comstock suggests that the Kanawha River may well had been iced over in December 1863 hampering communications and the supplying of military detachments. McKinley wrote to Comstock as follows:

> Lieut. I send herewith a second set of Invoices and receipts for Q.M. Stores and Clothing C and G Equipage for Dec. 1863. My first set of transfers may not have reached you on account of the River being closed up, if not they are at Gallipolis, Ohio. Don't fail to send Receipts by return courier. [1367]

On December 1, Lieut. Feazel wrote to Adjutant J.S. Cunningham informing him

> that information had been received at these Head Quarters by Capt. Carrall of the 9th Va Vol. Inft that Bill Smith was at the falls of twelve pole with about 300 men and had captured Capt. Wm. Turner and Sare [?] and the Capt. thinks that he may attack me

> I feel perfectly safe against 300 but I think it would be advisable for you to send a heavy Scout towards Buffalo Shoals at once as they may find Smith and clean him out.

N.B. Capt. Carroll Started home from Guyandotte and upon hearing the news of Smith he turned back to inform me of the fact and is here with me now[1368]

Certainly the great number of detachments detailed by Col. Brown during the month of December despite the inclement weather seems to indicate that neither side suffered for lack of incentive and stamina. Alarms continued to be raised and scouting made necessary although it was too cold for men to be out in the field or on the march for any length of time.

Col. Brown's concerns over lack of qualified staff for his regimental hospital(s) were brought into focus with the inclement weather. December 7, Brown wrote to Gov. A.I. Boreman concerning the 13th Hospital which needed "immediate attention."

> I would respectfully beg leave to make the following statements relating to the affairs of our Hospital hoping that they may meet with your favorable consideration.
>
> You are doubtless aware ere this time that Doc Shaw the former Surgeon of this Regt. has resigned his position thereby leaving the whole charge of the Hospital in the hands of the Asst Surgeon A.D. Williams. I recommend to your honor Doc Dalley as a fit person for the position of Surgeon with the Request that you commission him as such this was done because it was the wish of a large majority of the officers of the Regiment. Since this recommendation has been sent forward I learn that there is an effort being made to counteract it which seems to have caused some delay on your part in issuing the commission.
>
> Now what I wish to say is that Dr Williams is a very good man so far as his abilities are concerned. He is somewhat dilitary. His qualifications otherways I doubt not are good and I presume would make a good Surgeon with a little experience. Dr [Jacob] Lallance the Hospital Steward is a very good and attentive man to his post and deserves promotion and I would like very much to have him commissioned as Assistant Surgeon in case Dr Williams should be promoted to the position of Surgeon. I understand also that the Assistant Surgeon of the 23d Ohio has been spoken of for the position of Surgeon. He I believe is a very good man and I would have no objections to that. Either of the three men named will suit me for the positions named. I am anxious to have the matter settled some way and hope that you will give it your immediate attention.[1369]

The paper trail left behind by Colonel W.R. Brown, preserved in a variety of letters official and unofficial, suggest that Brown was a compassionate and charitable man. This is revealed in a couple of letters written by him on behalf of several of his men to Capt. James L. Botsford, Adjutant at 3rd Division headquarters. The first letter concerns two men, who had been deserters but since rejoining the regiment had been doing consistent duty and for this reason, Brown argued, deserved to have their salary reinstated. Bazalleel Meek, Private, Co. C, had

> left his company without leave about the 20th day of January and was absent untill the 11th of October ensuing when he voluntarily returned and reported for duty + George W. Fulwiler [of Co. H] left, his Company about the 1st of May 1863 without leave and was shortly after arrested at Barboursville Cabell Co. West Va and sent to Charleston and left in confinement untill the 1st of October last Both these men have been doing duty well since their Return but only by my Order which is not authority enough to ennable them to get their pay. I would respectfully request that they be ordered to duty from Division Head Quarters which I believe is the only proper authority in cases of this kind thereby enabling them to be paid. Fulwiler is a poor man and has a large family and has received no pay since Dec. 31st 1862.[1370]

The second letter asks for leniency for two underage boys who, if restored to duty, "might yet make good soldiers." Brown wrote again to Assistant Adjutant General Botsford. He requested that

> an order be issued restoring to duty Peter Donelly and Frank E. Barrett, privates of Company F of this Regiment. They have been reported as deserters, and the Paymaster refused to pay them unless they were restored to duty either by General Court Martial, or an order from Division Head Quarters. In this case I am of the opinion that a Court Martial would not be for the good of the Service, And am satisfied that these two boys, if restored to duty, will make good Soldiers. They are both very young boys, (about 15 years of age) and I think that a forfeiture of all pay and allowances, during the time they were absent, would be, with the punishment I have already put on them, sufficient.
>
> I therefore respectfully request, that if it can be done according to law, and Regulations, that these boys be restored to duty, they forfeiting all pay and allowances during the time they were absent from their Regiment.[1371]

Casual absenteeism continued to undermine military discipline in December. Always a problem in the 13th, absenteeism may have been further aggravated by the uniformity of duty, the onset of cold weather and the fact that some men had not seen their families in many months and had not been able to get money to them. On December 12, Col. Brown wrote to 1st Brigade headquarters requesting "that 1000 copies of revised regulations be sent him,"[1372] presumably for distribution to the men. On December 13, exceedingly annoyed with men absenting themselves without leave, Col. Brown issued General Order No. 29 which set out:

> In view of the continued violation of military Law on the part of the Soldiers of this command, by absenting themselves from their respective companies, without leave, it is hereby ordered that any Non-commissioned Officer or Soldier who shall after due notice here given be guilty of so offending, he or they shall be dealt with according to the 21st Article of War.[1373]

An alternate strategy to curb absenteeism may have been to permit the soldiers to go home. Captains of 13th companies were liberally granting furloughs to their men at this time. In fact, more furloughs were granted than official records attest to. Leave was granted that the men might return home for the Christmas holiday. This gave them opportunity to take home their pay to their dependent family members, who after so many months of separation from their support were badly in need of money.

It may have also been a little of what Francis Lord called "no easy matter to discipline soldiers from the West." (Lord, "Comments and Conclusions," *They Fought For the Union*). Soldiers from frontier areas, offered Lord, possessed too "keen a sense of equality" and consequently they freely "exercised their individual judgement as to the immediate need for service and the relative calls of home and Army."[1374] With or without leave soldiers went home. Granting of leaves and punishment—the carrot and the stick—may have been Brown's calculated two-prong approach to alleviating the problem. Col. Brown may have felt duty-bound to enforce military discipline with the 21st Article of War (giving complete discretion as to the nature of punishment to the superior officer) for incidents of absenteeism but this posture was tempered by company captains, who for humanitarian reasons, seem to have decided to accommodate their men rather than undermine morale.

On December 4, Deputy Provost Marshal William F. Dusenberry wrote to Col. Brown from Guyandotte proposing a joint venture to curb illicit trade in Kanawha Valley salt — to strangle the flow of this all important commodity to the Confederate armies and the South. The smuggling out of salt from the Kanawha Valley was now a thriving rebel enterprise. Dusenberry wrote:

I am informed that salt is sold in almost any quantity in Bar-boursville and taken up the Guyandotte River and on the Beach fork, by which means, the rebels + their infernal <u>tools</u> get all their supplies, as the country is all open above Barboursville, don't you think it would be well to put some restriction on it and not allow any one person to take out more than he will actually need for 60 days. I have adopted this plan here. I think the rebels will <u>keep</u> as long as I want them without salt.[1375]

Col. Brown took steps to squelch the illegal trade in salt referred to by Dusen-berry by sending Capt. John Witcher's cavalry company (Co. G 3rd West Virginia Cavalry Regiment) to scout. On December 10, Witcher reported regarding the scout undertaken by him and fifty men of his company which began on Monday, Decem-ber 7, from camp at Charleston. Witcher scouted through "Wayne, Logan and Boone Counties."[1376] He and his men left camp at about noon on the 7th and proceeded to cross the Guyandotte river at Dusenberry's Mill. They continued on until they struck "Beech Fork at Jeff Rowen's." Two rebel soldiers belonging to the 8th Virginia Cavalry were captured at the Mouth of Raccoon Fork of Beechfork and then, that night, Witcher crossed "on Raccoon." On Tuesday morning, they

> passed out of the head of Beech fork: thence by way of Four mile to the mouth of Nine-mile: thence up Guyandotte to the mouth of 'Big Ugly,' and camped there Tuesday night. Wednesday went up Big Ugly, thence on the Guyandotte up the same to mouth of Big Creek.
>
> About one mile up Big Creek killed George Adkins who escaped from the Guard House at Charleston, and captured a rebel Lieut. A-mile-and-a-half further up. Killed a rebel orderly sergeant and captured seven (7) rebel soldiers and their Captain. Thence up North fork of Big Creek, and captured a rebel Lieut. and three (3) men, and still further on captured four (4) men. Thence to the head of Turtle Creek, down the same striking Coal river at Boone Court House. Thence to Peytona on Big Coal river, marching that day fifty five (55) miles. Thence to-day twenty-five (25) miles to this place, bringing all prisoners here without the loss of a man or one getting wounded, although we had some skirmishing.[1377]

Witcher reported that he and his men marched "altogether one hundred and fifty (150) miles, in three (3) days" and that he sustained the loss of "three horses with equip-ments." Witcher stated that the rebel loss was two killed and three wounded. In all, during the course of the expedition, eighteen rebel prisoners were captured and "twenty (20) horses and equipments, thirty (30) guns of different kinds."[1378] The prisoners captured represented the following ranks and commands: one captain; one 1st lieutenant, one 3rd lieutenant; one quarter master sergeant and eleven privates from Beckley's Battalion; men belonging to the 8th, 14th, 22d, 26th, 34th and 36th Virginia Cavalry Regiments; Moreheads Battalion Kentucky Cavalry; and the 1st Virginia State Line.[1379]

Upon reporting at 1st Brigade headquarters at Charleston, Witcher conveyed other intelligence. There was "[a]ny amount of subsistence in Logan County. Plenty of cattle, Hogs + Grain, and roads in good condition [...] and an army of 10,000 men [could be taken and subsisted] through Logan County."[1380] Witcher also supplied considerable evidence regarding another Confederate incursion in the near future. Witcher stated that "a considerable force [was] organizing in the region of Abington for a raid into Kanawha Valley on Barboursville."[1381] "Beckley," stated Witcher

> has a Lieut. Colonel's Commission, & is organizing a Regiment — has about 350 to 400 men on Island Creek. Lieut. Allen one of the prisoners to Capt. W [said] that Col. Fergeson of the 16th Va. Cavalry is organizing for a raid into the Kanawha Valley or Barboursville. He has about 1100 men somewhere in the region of Abingdon.[1382]

Upon questioning his prisoners, Capt. Witcher also heard from them of the Federal advance on Lewisburg and about what counter measures were being taken by the Confederates. Prisoners informed Witcher that "Jones' command and a large rebel force who had been on the way to meet Longstreet, were now returning to reinforce the troops at Lewisburg [and] Bill Smith's command about 150 men is in Logan."[1383]

According Columbus Shrewsberry, sometimes sutler to the 13th Virginia, then recuperating from an ailment, things were rather quiet around Barboursville. Shrewsbury wrote to his wife from Barbousville on December 8:

> Dear Wife
> I am still geting better have got a good appetite and think I will get well soon have got a good Bed to sleep in all is quiet heare I had a talk with Marshall Smith in regard to my mare he says that it does not look at all like Mr Roushes and that he will sware that it is not his you need not say anything about it I am safe
>
> Nothing more at present Keep the <u>Children</u> in the House out of the cold let Fanny go in to Mr Slaygles and learn how to talk the Gearman language
>
> Yourse
> C. Shrewsbury[1384]

December 15, Col. Brown issued Special Order No. 100 directing Lieut. Col. James R. Hall "to Proceed to Mason County West Va for the purpose of Collecting together Some Recruits, that have been enlisted for this Regt. He will return and report with them to these Head Quarters without unnecessary delay."[1385] December 18, Brig. Gen. E.P. Scammon issued General Order No. 33 from 3rd Division Headquarters, Charleston, authorizing the appointment of Lieut. Frank Millnard, 34th Regiment Ohio Volunteer Infantry to the position of aide de camp and Acting Assistant Engineer for "this Division. All information of wades, papes etc. — in fact all topographical information obtained by commanders of parties scouting through the Territory embrac[ed] in this command, will be promptly reported to him to be used in the correction of existing maps."[1386]

December 22, "[t]wo Rebel prisoners with description Lists" were forwarded from 13th headquarters at Barboursville to 1st Brigade headquarters at Charleston.[1387] December 23, Col. Brown issued Special Order No. 104 to John Likins granting him permission to operate his "Mill for the exclusive use of the neighborhood untill further Orders — But in no Case will it be used to the benefit of disloyal persons unless Compelled to do so."[1388] December 23, the Assistant Adjutant General at the War Department, Washington, sent orders and cover letter to Col. Brown reminding him that his October reports were delinquent and

> I am directed by the Provost Marshal General to call your attention to enclosed General Orders, and to inform you that the reports therein called for, for the month of October, have not been received at this office.
>
> The Secretary of War directs that you forward such reports without delay, with a letter of explanation why you failed to do so at the proper time.
>
> The necessary Blanks, if any are needed will be promptly furnished upon requisition being made.[1389]

Convalescent soldiers at the Point Pleasant Post Hospital, which included many soldiers belonging to the 13th, were given "a Christmas dinner" by the "friends of our convalescent soldiers." The friends also assisted "those in charge of our Post Hospital here — Dr. C.D. Dally, Dr. Strickland, Steward, William French and Ward Master Warren Seaton" with their duties.[1390]

Another request to put a stop to the illegal sale of goods in West Virginia came at the end of the month, when on December 26, Special Agent Heaton wrote to Colonel Brown saying:

I have rec'd from J.P. McGlaughlin Esquire Local Special Agent of the Treasury Department at Guyandotte W.Va of which this is an Extract:

A.W. Brown a merchant of Catlettsburg says there is no restriction at Catlettsburg on the sale of goods coming into Va. and it appears that my recommendation is not a pre-requisite to the purchase of supplies there: and the Soldiers at this place (Guyandotte) say they cannot guard the roads through Wayne County with their present force. If you would send me a request on Col. Brown of the 13ᵗʰ Va. to see to the guarding of those points much good would be done, by cutting short the suplies to Wayne and Logan counties, unless recommended.

I send you herewith a copy of the Regulations of the Secretary of the Treasury and of the Local Rules of the Supervising Special Agent, and I will feel obliged if you will co-operate with the officers of the Department in seeing to it that the Regulations are observed and Enforced.[1391]

December 27, an order from the Provost Marshal's office, 3ʳᵈ District, Charleston, to "cause the arrest" of James E. Smith of Ceredo[1392] was referred to Col. R.B. Hayes, who forwarded it to Col. Brown.

Monday December 28, Mrs. Mary Eggars wrote to Gen. Scammon complaining of the bad behavior of the soldiers of the 13ᵗʰ Regiment and particularly of their Colonel:

I beg permission to ask you one favor And I hope wou will grant it. I am a soldier wife my husband is out in eastern Va and I Am here alone the 13 Va is here and there A grate meney of them get their wives here and I am living in alarge hous here + I have let them have two rooms of my hous and have Accommodated them in every way I could and that wonte do they want to take the rest of my Hous and turn me out of doors because I Wouldant give up to let them have the last room

They air throwing rocks through my windows at night and the col. don't pretend to mind it For he is so bizeley ingaged in Spreeing around With the secesh that he lets the rebels come Around here and go off unharmed men I would Be happy to have abill of portectien from you if you pleas[1393]

It is not known what became of Mrs. Eggars complaint.

On Wednesday, December 30, Gen. Scammon issued orders directing that Lieut. Col. James R. Hall be appointed "an Inspector to examine and report upon the condition of public property in Quartermasters Dept. of the 13th Va. Regt. Inf. and for which Lt. S. Comstock R.Q.M. 13th Va. Inf. is responsible. He will make the Inspection immediately on receipt of this order."[1394]

Morning reports December 1863

Morning reports for Company A indicate that the company gained in recruits during the month. Among these were Samuel Field, Theophilus Gillespie, Edward Hammons and James Coleman, who were reported as having enlisted in the company on December 15. This brought the number of present and absent in Co. A to three commissioned officers and ninety-one enlisted men—the highest number for the company to date. Already on December 24, however, Col. Hayes wrote to Col. Brown regarding two of the recruits, Samuel Field and Edward Hammons/Hammonds informing Brown that:

> Samuel Field aged 17 next March and Edward Hammon aged 15 next February are claimed by their mothers to have been enlisted under military age without consent of parents. If you find this is so you will discharged them <u>unless they have been mustered into service by the mustering officer</u>.[1395]

On December 30, Hayes wrote again to Col. Brown:

> Your note as to Fields and Hammond the boys enlisted under age is received. — The bearer tells me the boys have not been mustered. [illeg.] as you will discharge the boys <u>unless you are satisfied that they are eighteen years of age</u> — It would be unlawful and a disobedience of orders to muster them if under age against the wishes of their parents —[1396]

Apparently Col. Brown was satisfied that the boys qualified for service inasmuch as both Field and Hammonds were recorded as mustered into the U.S. Service on February 27, 1864, at Barboursville. Both young men served until mustered out with the regiment at Wheeling on June 22, 1865. Company Records indicate that both Fields and Hammonds were enlisted by Captain Greenbury Slack on December 15, 1863. Field was recorded as nineteen years old, and Hammonds as eighteen.[1397]

Company A reported no drummer, fifer, farriers, blacksmiths and no horses present with the company during the month. One artificier was present each day. One enlisted man was absent with leave December 1 to 6. December 4, Captain Slack informed 13th Adjutant J.S. Cunningham that he was forwarding a furlough for "Cop

[Cornelius] Page and one for Samuel Teel and Agrippa Samples asking eight days absence you will approve and send them on by Order of G. Slack"[1398] John Tully and another man whose name appears in the ledger as "Jassof Teath" (Joseph Scott?) —two enlisted men—were reported absent without leave December 7 to 12. They returned to camp on the 12th. Enlisted men reported absent without leave were: one man December 13 to 26, and two men on December 29. These latter were probably William Grey and Baron D. Kalb Wintz, who were reported absent without leave December 26 to 29. The two men returned to the company on the 29th.

On December 3, Capt. Greenbury Slack, cmdg. Co. A, was appointed to a Board of Survey together with Capt. Lemuel Harpold (cmdg. Co. C) and Lieut. James W. Hanna (Co. D) to inquire into the loss of "Sertain arms for which Capt A.F. McCown" was responsible; Harpold to meet together with the other members of the board at two o'clock in the afternoon at Company F headquarters.[1399] December 5, Col. Brown issued Special Order No. 94 ordering Capt. Greenbery Slack "on the Board of the Council of Administration Vice Capt Van D. McDaniel Deceased."[1400] December 8, Col. Brown issued Special Order No. 97 ordering Perry Gatewood, Co. A, then on duty as Warden Master in Regimental Hospital together with Lieutenant John H. Rossler, Co. E

> to proceed to Charleston W. Va as witnesses before the Court now in Session immediately upon arrival at that Place — You [Gatewood and Rossler] will report ay Brigade Head Quarters — at Camp White —- after the Court has received the evidence required from you. —- You will report to your regiment without unnecessary delay.[1401]

No one belonging to Co. A was reported detailed for special duty. One private had extra duty December 1 to 6, and one private had daily duty December 7 to 31. On December 1, "six men returned from on a scout up Guian [Guyan] river." The number of details made from Co. A for detached service was heavy this month. Detached details included one commissioned officer and ten enlisted men on December 1, and one commissioned officer and four enlisted men on December 2. Also on the 2nd, "Sergeant Miletus Grinstad, Sergt. R.H. Davis, Richard George, William George, William Gray and Philip Wintz were ordered out to recruit for the 13th in Kanawha County."[1402] "Abraham Chandler returned from on a scout Dec. 2, 1863 [in the] P.M."[1403] Sergt. Grinstead and four men returned from Charleston to camp on December 23 as directed. December 3 to 7, one commissioned officer and ten enlisted men were out on detached duty. December 8 to 11, one commissioned officer and nine enlisted men were out. Among those on detached service at this time was Sergeant R.H. Davis, who was recorded as having returned to camp on December 11.

Further details for detached duty were made: on December 13, one commissioned officer and seven enlisted men, and December 14, three commissioned officers and fifty enlisted men were detached. Captain Greenbury Slack and Lieutenant George Danner with forty-three men were ordered on a scout up Mud River on the 14th. They returned to camp from scout in the afternoon of the 14th. December 15 through 20, one commissioned officer and seven enlisted men were detached. December 21 to 23, two commissioned and seventeen enlisted men were detached. Special Order No. 101 issued December 16 by Col. Brown directed the detailing of a detachment under command of Lieutenant George Snowden. Special Order No. 101 directed Lieut. Snowden "with 18 Men [...] to Proceed to Charleston W. Va. he [Snowden] will take in Charge one Prisoner of War and one deserter from the 2nd Va Cavalry." Snowden was to "report at Brig-Hd Quarters at Camp White — With the Prisoners." He was thereafter to return with his "Guard to these Head Quarters without unnecessary delay."[1404] December 20, Col. Brown issued Special Order No. 102 detailing

> Lt George Danner + Ten Men of Company (A) [...] as a Scouting party — they will take Charge of G.W. Hensley a Citizen Prisoner and will proceed with him to Charleston W. Va. and deliver him to the provost Marshal at that place — you [Lieutenant Danner and men] will first report at Brig-Head Quarters[1405]

Also on the 20th, Brown wrote to 1st Brigade headquarters requesting "permission to send a Lt. and 20 men to Buffalo in Putnam Co., W.Va. to remain there for 15 or 20 days for Recruiting purposes." Col. Hayes granted permission.[1406] Morning reports indicate that Lieut. Danner was ordered out with ten men of Co. A to go into Kanawha County in the afternoon of December 20 for recruiting service. December 24, two commissioned and twelve enlisted men were detailed for detached duty. These included three men who were ordered on recruiting service on December 24. December 25 through 30, two commissioned and fifteen enlisted men were ordered on unspecified detached service. Among those ordered out during this period were Capt. Greenbury Slack, who, with forty-one men was ordered on a scout on December 30. December 31, three commissioned officers and fifty enlisted men were detached for duty away from camp.

Over the course of the month, Co. A reported from three to nine men present sick. The number of present sick hovered at from five to seven men. Six men were absent sick. One enlisted man was reported absent in confinement December 1 through 3. December 4 to 8, three men were absent in confinement, and December 9 through 31, two men were absent in confinement. Joseph Scott, who was absent from the company for some reason, was noted in "Remarks," as having "returned to the Company" on December 26.[1407]

Company B morning reports record that a drummer and fifer were present each day of the month of December. No buglers, farriers, blacksmiths and no artificiers were reported present. One to two enlisted men were absent with leave December 3 to 18. John R. Gaskins was reported absent without leave December 4 to 13. December 14, Gaskins was arrested by Third Sergeant John McDaniel and placed in confinement, and December 17 the prisoner was reported in confinement at Charleston.[1408] A new recruit, William L. Harris, enlisted in the regiment and was duly sworn in as a member of Co. B on July 4. At the end of the month Co. B tallied present and absent three commissioned and ninety-eight enlisted men.

No one had special or extra duty. Four privates were assigned to daily duty. Over the course of the month, a minimum of four to as many as sixty-one men were detailed for detached service each day. December 3, Lieut. Col. James R. Hall ordered Sergt. John McDaniel, Co. B, "to Mason Co West Va to recruit for this Regt — he will report to A.C. Mason who is now recruiting in said County."[1409] Large detachments were set in motion on December 10 (consisting of one commissioned officer and seventeen enlisted men); on December 14 (consisting of two commissioned and fifty-nine enlisted men); and a group of two commissioned officers and forty-five enlisted men detached on December 31.

Co. B still enjoying its characteristically good health, had over the course of December reported just three to six men present sick and six men absent sick. Taylor Hogg was admitted to General Hospital Point Pleasant on December 7 with a contusion to the right leg. He returned to duty on February 12, 1864.[1410] One to two privates were present in camp in arrest or confinement December 15 to 18, and one enlisted man was absent in confinement December 17 to 31.[1411]

On December 1, Lieut. Col. J.R. Hall issued Special Order No. 90 ordering Sergeant William McDaniel, Co. C, to proceed "to Charleston Ka Co West Va to give evidence before a Court Marshal now in Session at that place. He [McDaniel] will report to Capt M.P. Avery Judge Advocate."[1412] December 3, Capt. Lemuel Harpold, cmdg. Co. C, was appointed to a Board of Survey to together with Captain Greenbury Slack and Lieutenant James W. Hanna, Co. D, to inquire into the loss of "Sertain arms for which Capt A.F. McCown" was responsible. Harpold was to meet with the other members of the Board at two o'clock in the afternoon at Company F Head Quarters.[1413] December 21, Col. Brown ordered Capt. Harpold to take five privates of his company and "proceed to Mason Co., W.Va., for the purpose of gathering up some absentees from this Regiment. He [Harpold] will report at these Hd. Qtrs. on the 29th inst."[1414] December 24, Col. Brown ordered Corporal Madison H. Stone (Co. C) and Hamilton Johnson (Co. F) to proceed to Buffalo, Putnam County, where they were to join a party under command of Lieut. Joseph Brumley (Co. F) then recruiting in that area. Stone and Johnson were to go by way of Gallipolis, Ohio, "where they will take charge of Some Commissary Stores for said recruiting party and see that they are taking care of and properly delivered to the persons for whom they are intended."[1415]

December morning reports for Company D preserve that one enlisted men—probably Jacob Messick—was reported absent without leave December 5 to 11. Julies H. Hatcher deserted the company from "U.S. Army Hospital at Point Pleasant" on November 10. Word of this did not reach company headquarters until December 21 when it was at last entered upon company books. Co. D gained two enlisted men, Frederick Olinger (Ohlinger) and William B. Cherrey (Cherry), who enlisted on December 1. Another recruit, David H. Harper, joined the company on December 18. Several men deserted during the month. The company finished the month with an aggregate of three commissioned officers and ninety enlisted men present and absent.

Over the course of the month, one non-commissioned officer and five to eight privates had daily duty. Some members of Co. D had detached duty each day. Among these details were William Jackson, originally from Hartford City, who together with Leonard or Thomas Oliver was out recruiting for the 13th Regiment in the vicinity of Hartford City. On Thursday, December 17, William Jackson wrote the following to Col. Brown, reporting his progress:

> I write you a few lines to let you know how we are getting recruiting we have got 4 more making 8 in all since we come home they are all stout men not boyes we travel every day a good deal and it is hard work for the roads is very muddy now for we have no horse to ride please tell Dr williams[1416] how i am getting along and not to get mad at me for i am doing the best i can Cornal pleas rite back and tell me what to do no news at present from yours truley william Jackson and oliver[1417]

December 17 and 18, one commissioned officer and thirteen enlisted men, all of Co. D, had detached service. December 25 through 31, one commissioned officer and six enlisted men were away on duty. On all other days from three to six enlisted men had detached duty. December 9, John Jones (5th Orderly Sergeant) went to Jackson County, West Virginia, on detached duty. December 10, "5th Sergeant Roseberry and two men went in pursuit of Jacob Messie [Messick]."[1418] On December 22, Sergeant (Arthur) Darnel went to Mason County to recruit. December 25, Col. Brown issued orders directing that Lieutenant George Snowden, Co. D,

> take Charge of Sergeant Scott Co (D). Sergeant William McDaniel Co (C) and Sergeant Boggess Co (B) and will Proceed with them to Charleston W Va to give evidence in Certain Cases now

before Court Marshal now in Session then he will report to Capt M.P. Avery Judge Advocate for instructions. He [Snowden] will return and report to these Head Quarters Without unnecessary delay[1419]

December 30, "Co. D left Barboursville on a scout at one o'clock P.M. in command of Lieut. J.W. Hanna." December 31, all or some part of Co. D, went "[o]n a scout in Way[n]e Co. on Twelve Pole creek and waters."[1420]

Co D had few sick men in camp. One non-commissioned officer was reported present sick December 13 to 15, and one private was present sick December 26 to 31. Just four to six men were absent sick. In the middle of the month, three men returned from home presumably from sick leave. These were William Swain returned from his home in Jackson County on December 11; 6th Sergeant George W. Maggi returned from his home in Pittsburgh, Pennsylvania, on December 12; and Lorean Olinger returned from his home in Mason County on December 13.

Co. D reported a number of men in arrest or confinement. There were two privates so reported on December 5 to 6; three privates on December 6 to 7; one December 12 to 13; three December 14 to 15; and four were in the guard house on December 16. Three enlisted men were absent in confinement December 17 through 31. Among those arrested were John Morarity and Henry Schlosser confined for drunkenness on December 6. Nathaniel Birchfield was reported in confinement for ten days beginning December 7. On December 8, John Morarity, Henry Schlosser and Nathaniel Birchfield were "relieved from confinement," and on December 16, Louis Vincent, Nathaniel Birchfield and Jacob Messie or Messick were sent to Charleston for absenting themselves from the company—these men had been first reported absent on December 14.[1421]

December 2, Gen. Scammon directed that "Capt. Carter 13th Va. V. I. be called to Charleston to give evidence in the case of James Ruffman and Capt. Ford."[1422] December 3, Private George W. Smith, Co. E, drew "1 pair pants @ $3.55; 2 pair socks @ .64."[1423] December 4, Col. Brown again ordered Capt. Carter to Charleston as embodied in Special Order No. 93:

> John D. Carter
> Capt Co (E)
>
> Sir
>
> In pursuance to Order from Brigade Head Quarters dated at Camp White Dec 3rd 1863. You will report in Person immediately to Charleston Ka Co W. Va. To give evidence in a Suit now

pending in a Court of Common Pleas for Said County — between James Ruffner + Captain Ford —

You will report to Brigade Head Quarters upon your arrival.[1424]

December 8, Brown issued Special Order No. 97 ordering Lieutenant John H. Rossler, Co. E, together with Warden Master Perry Gatewood, Co. A,

> to proceed to Charleston W. Va as witnesses before the Court now in Session immediately upon arrival at that Place — You [Gatewood and Rossler] will report ay Brigade Head Quarters — at Camp White — after the Court has received the evidence required from you. — You will report to your regiment without unnecessary delay.[1425]

December 12, Col. W.R. Brown issued Special Order No. 99 ordering Augustus McDaniel, George W. Lacy [Legg], William Sigman and Charles W. Young—all of Co. E—"to Charleston Ka Co West Va. Without delay — They will report in Person to Lt W.N. Hawkins for instructions."[1426] On Wednesday, December 23, Lieutenant William N. Hawkins, Co. E, who had been detached on recruiting service while he recuperated from a severe wound received at the battle of Point Pleasant, wrote to Col. Brown regarding his success in recruiting:

> I have enlisted in all 17 men one of which was the sone of Benjamin Burns of my company and when I went down I took the enlistment paper to him for to get his signature at which time he told me that if he could get home a short time to fix up his business he wood consent and I brought him up with me and sent him home and he has remained home ever since I sent one of my men up after the young man and his father refused to let him come up after sining [signing] the enlistment papers I wish you to instruct me what to do in the premicies Burns hasent done me any service since I left the regment tell me whether to send him and his son down or not I go to Malden there is a good many refugees come in the last raid one of which I have enlisted and got mustered and clothed and is doing good service in recruiting
>
> I have one of the worst cough that I ever had my woond [wound] is running a little but I will still try to recruit on yet I wish to now whether the bounty will be paid after January or not.[1427]

Morning reports for Company F record that the company had one waggoner present each day of December. No one was reported absent with leave, clearly however, some men had been granted furlough. David Burrows informed his wife that he would start for home on December 18. While at home, he learned that an "old friend Alex Gorby" had also joined the 13th. Gorby had enlisted December 22, 1863, at Hartford City and was mustered February 28, 1864, at Guyandotte for the new Co. G. Burrows also learned from his wife, Lovina, they were expecting a child in July. When it came time for Burrows to return to camp, he overstayed his leave at home by "several days," to stay with Lovina, who was suffering from morning sickness. When he did return to camp, he took back with him "a box of cookies from his mother, his father's gold watch and two red flannel shirts Lovina had made him."[1428] Also, on December 5, Captain McCown wrote to Adjutant John S. Cunningham transmitting "a furlough for Robert Darnel and Henry Miller. They have not been at home for nine or ten months, and their families need their money badly. Please approve + forward."[1429]

Beginning about December 18, two enlisted men of Co. F were reported absent without leave until December 29 when from then until December 31 just one man was reported absent without authority. On December 1, Sergeant Peter Darnel was discharged to be promoted 1st Lieutenant of Co. I by order of General B.F. Kelley. His discharge "came to hand Jan. 4, 1864."[1430] December 2, Corporal J. Hamilton Johnson was promoted to Sergeant to replace Peter Darnel. Two recruits were added to the rolls. On December 24, John Daugherty enlisted in the company but did not report for duty until January 6, 1864. December 25, John W. Lanham enlisted. He also did not report for duty until January 6. At the end of the month, the company's numbers held again at three commissioned officers and ninety-one enlisted men.

No one had special or daily duty. Seven to five privates had extra duty each day. There was detached service each day as follows: December 1 to 4, eleven enlisted men; December 5 to 7, ten enlisted men; December 8 to 9, nine enlisted men; and December 10 to 13, ten enlisted men detached. On December 13, two commissioned officers "Started at 9 o'clock p.m. to Mud Bridge with 40 men of this company in pursuit of rebels." The detachment arrived at Mud Bridge at 4 o'clock a.m., on December 14. They "slept an hour and followed the retreating rebels up Mud River. Dec. 15 [the detachment] [a]rrived safe at Barboursville without the loss of a man, captured one rebel and several horses."[1431] December 15, morning reports preserve that ten (presumably ten more) enlisted men were detailed for detached duty; December 16 to 17, one commissioned and eighteen enlisted men were detached; December 18 to 21, one commissioned and sixteen enlisted men; December 22 to 24, one commissioned officer and twenty-eight enlisted men; and December 25 to 26, one commissioned officer and sixteen enlisted men were out on detached service. The latter group was probably the one ordered on out on December 24 by Col. Brown in Special Order No. 105, in which Lieut. Joseph Brumley was directed to proceed to Bufalow in Putnam Co W. Va. with 20 Men and remain there 20 days or untill further Orders, for the purpose of recruiting for this Regt and Scouting the Country in that Vicinity — You [Brumley] will report weekly to this Regt — and at the expiration of the 20 days Report in person with your Command and recruits to these Hd Quarters[1432]

Special Order No. 106, issued on December 24 by Col. Brown, ordered Corporal Stone of Co. C and Hamilton Johnson of Company F

to proceed to Bufalow Putnam Co W Va where they will join Recruiting party under Command of Lt Joseph Brumly of this Regt. they will go by way of Gallipolis Ohio where they will take charge of Some Commissary Stores for said recruiting party and see that they are taking care of and properly delivered to the persons for whom they are intended[1433]

More details for detached service followed. On December 27 to 28, one commissioned and eighteen enlisted were detached from Co. F for duty; and December 29 to 31, one commissioned and thirteen enlisted men were away on detached service. Co. F reported four to ten men present sick over the course of the month, and three to four enlisted men absent sick each day. No one was reported present or absent in arrest or confinement.[1434]

December 8, Colonel Brown issued Special Order No. 98 directing that Private Jeremiah Webb Co (G) of this Regt is hereby Ordered to proceed to Charleston W Va. as a Witness before the Court now in Session. — You [Webb] will report at Brigade Head Quarters at Camp White — on the 15th of this Month you will report to your Company in Person.[1435]

December 8, Capt. J.V. Young wrote to Col. Brown from his post at Hurricane Bridge reporting on the result of his scouts and other issues:

Sir we are all right here yet, my men is in good helth and spirit. Sergt Morris returned last night from a <u>scout</u>, on big Creek Turky Creek and Climer finds no rebs, nor hears of non but Holly Kirk Bragg and a few others of the same <u>Class</u>.

William Henson who is now in Dixie with his negros + Horses has (100) Bushels rent corn here in the Valley. John Moses who is a union man owes this corn to Henson and wants me to get it. But I don't like to do it without orders from you. If the Q.M. will send me some nails I will Commence on my winter quarters[1436]

December 9, Brigadier General Scammon apprehending some trouble in the vicinity of Coals Mouth ordered Capt. Young, cmdg. detachment at Hurricane Bridge, to "direct twenty-five (25) of your men under a Lieutenant to proceed to Coalsmouth to remain there untill further Orders."[1437] On Thursday, December 10, Lieutenant Colonel J.W. Comly wrote to Young confirming the need for this further division of Young's command and directing him to notify his "Regimental Commander immediately of the change."[1438] December 11, Capt. Young, now thoroughly annoyed with what he perceived as a cavalier disregard for the safety of his men, wrote to Col. Brown enclosing

> Copys of orders received today and you may imagine my surprise on there receipt. To think that I have been placed here on this rebel thoroughfare with my Company where we are in danger day + night and then by an other representation to have the Company divided so as to be sure that some of us my be Captured or Kiled (it makes me indignant.) You know that I never flinched from duty since I have been in the service but this trys my pacients and I think the request bey on precedent I suppose that you have not learned that Provost Marshal has arrested a great number of rebel women and men in Putnam, this week and taken them to Charleston and they are now in the Guard Hous and the rebels here says they will have revenge in a few days I understand to day that they are threatening that they will have union women taken as hostages and that in a few days
>
> I learned yesterday that that ntorious Horse theif Wake Dudding is in, I had Sergt McDaniel and 12 men after him last night, and but for this order we would have got him to night.[1439]

In the afternoon of December 13, Young's detachment at Hurricane Bridge was attacked by a Confederate force. Captain John S. Witcher's information gained from prisoners taken on his scout through Wayne, Logan and Boone counties proved correct. The news of the Confederate advance into the Kanawha Valley came "simultaneously" from Barboursville, Gallispolis and Camp Piatt and the news and fresh alarms spread like wildfire.[1440] Captain Isaac M. Rucker, commanding "Mason County Independent Scouts," who continued to work in conjunction with Col. Brown wrote to Brown on December 13 with the message that he had learned that morning that

> a considerable force of Rebels passed Mud Bridge this morning in this V[i]scenty they passed within two miles of me Capturing all of the Horses men +c they could get hold of [on] their cou[r]se I have

not fully Learned but Suppose they wil either pass out by Winfield + Huricane Bridge or down the Kanawha. I have sent men to watch their movements + will inform you pleas Give me any information that you have [...] Rolen Balleger and William Rousey bearers of dispatch[1441]

Captain W.S. Rice wrote from Camp Piatt on December 13, that he had "just received positive information that Jenkins' old command of three hundred men under command of Col. Smith left Logan C[ourt] H[ouse], Friday [December 11] with the intention of making a raid into Guyandotte or some point on this river in the absence of most of our troops."[1442] Col. Brown directed Lieut. Feazel at Guyandotte (no telegraph service apparently in operation at Barboursville) to have the telegraph operator at the Guyandotte telegraph office wire Lieut. Comly at Brigade Head Quarters. Feazel sent the following from Guyandotte on December 13:

> I am directed by Col. W.R. Brown Com'd'g 13th Va at Barboursville to notify you that three hundred (300) Cavalry crossed from Guyandotte to Mud Bridge this morning and passed down the Gallipolis road and, it is supposed they will go to Point Pleasant. He wants you to telegraph this to Gallipolis at once and you will notify all the Steamboats of the fact that pass by, as they may intend to try and capture a boat.[1443]

Captain W.H. Zimmerman, commanding the military post at Gallispolis, also wrote to Comly on December 13 saying that he had likewise "received reliable information that there was a rebel force about twenty (20) miles beyond Gallipolis on the Va side. Report says they number five hundred (500). They are supposed to be making for Point Pleasant or some point on the Kanawha."[1444]

Comly sent off immediately to notify Col. Brown and department headquarters that a rebel force of from "three to five hundred" was "reported 20 miles from Gallipolis approaching the Ohio on the Kanawha."[1445] To Zimmerman, in command of the military post at Gallipolis, Comly wrote directing him to take one of the boats he had there at hand and "extemporize a gunboat" and put "Hicks" in charge of it if he was there at Gallipolis and "to have everything at Pt. Pleasant so that it can be removed, if it becomes absolutely certain that the Rebs will attack Pt. Pleasant."[1446] On December 14, Zimmerman wrote to Comly notifying him that he had "armed fifty (50) men from Hospital here [Gallipolis] and will send them to Point Pleasant. Will send arms to Hospital at Point Pleasant. Have ordered Lieut. Hicks to patrol with Victor No. 2 to Red House."[1447] Comly also improvised a gun boat on his side of things at Charleston and on December 14 sent the boat down from Charleston to patrol and rendez-vous with Zimmerman's boat at Red House.[1448] The two boats (the

Victor and the *Watson* or the *Levi*?) patrolled together—one moving down the river the other moving up river to prevent the rebels from crossing the Kanawha river and attacking Point Pleasant. Just one small party of about "twenty-eight rebels crossed the Kanawha river three miles below Coals Mouth"[1449] during the night of December 14-15 and "cut the wires near Red House. No other damage has been done." By the afternoon of December 15, the lines had been put up again.[1450]

December 13, Lieutenant Colonel J.R. Hall left camp at "10 P.M." with a detachment of two hundred men from the 13th West Virginia bound for Mud Bridge "to intercept if possible the Rebels (then between the Charleston Pike and Kanawha River)."[1451] Hall reached Mud Bridge and found that the Rebels "had passed out about the time we left Barboursville." Hall continued:

> From Mud Bridge I sent a squad of Cavalry to communicate with Capt Young — they assertained that Capt. Young had left some time during the night — We then started in persute and followed as far of Guyan River — thinking further persute useless as they had 4 hours advance I returned to this Post — Think there object was to cross Kanawha River at Buffalo pass out through Jackson + Braxten Counties — There was 7 Cos of men belonging to 17th Va Rebel Regt + number about 300[1452]

December 14, Col. Brown, concerned for the safety of Capt. Young's men at Hurricane Bridge and also for Lieut. Col. Hall at Mud Bridge, wrote to Captain Isaac Rucker:

> There is one detachment at Hurricane Bridge, you had better place your self in communication with them if you have any information or can do anything.

> Spare no means to inform Col. Hall as he marched from here last night at about 10 o.c. to Mud Bridge and keep him advised.[1453]

Brown's official report of Lieutenant Colonel Hall's expedition to intercept the rebels reportedly making for Mud Bridge was as follows:

> On the evening of the 13th of this month information deemed reliable was received at these Hd Qrts that a considerable force of mounted Rebels had come down Guyandotte River within Ten miles of this place and were moving in the direction of Mud Bridge in Putnam Co upon which I immediately dispatched 4 companies of Infantry and Co G 3rd Va Cavaly Capt Witcher all under command of Lt. Col. J.R. Hall This force left this Post at 10 oclock

same evening + marched direct to Mud Bridge distance 11 miles here intelligence was received that the enemy had returned and gone out in the same direction they came and in compliance with my instructions a squad of Cavalry was dispatched immediately to Hurricane Bridge to gain some information of Capt Young who when they returned reported the Post evacuated My forced then started in pursuit following as far as the falls of Guyandotte and learning there that the enemy were four hours in the advance and our men returned and arrived in camp all safe on the evening of the 14th

This force of Rebels was the 17th Va Rebel Vol Infantry mounted were under command of Major Smith of Jackson Co. It consisted of 7 Companies numbering about 300 or 350 men. They were recruited principally in Jackson + Braxton Counties. From all the information that I can gather their intention was to cross the Kanawha River at Buffalo and pass out through Jackson + Braxton counties. In their retreat they left a considerable number of worn down horses along the road. They were of such an inferior quality however that it did not justify any efforts to take care of them. I am sir Respectfully Wm. R. Brown Col Com'd'g[1454]

The Confederates had attacked the detachment of Capt. Young's company in the afternoon of the 13th and had driven it from Hurricane Bridge. Young fell back towards Coals Mouth where the other detachment of his company was posted under Lieut. Clark Elkins. Upon Young's arrival at Coals Mouth on December 14, Young conveyed to Elkins that the enemy "450 strong,"[1455] had "completely surrounded him" at Hurricane Bridge, and that when night and darkness set in he made his escape through the woods to Coals Mouth.[1456] Young stated further, that the enemy left Hurricane Bridge in the morning of December 14 and headed towards Barboursville "abandoning everything in the shape of Camp + Garrison Equipage."[1457] Young lost just two men captured. Elkins wrote again to Comly, probably later in the day on the 14th, that "Report says that after Capt. Young had made his escape the enemy became alarmed and left in such a hurry as would not admit of them taking any of our property with them. Whether this is so I cannot say. We have sent men down to learn the facts, and save the property if possible."[1458] Comly responded to Elkins writing "[k]eep a good look out and keep the enemy from crossing Coal if possible. There are two Gunboats patrolling the river to prevent the enemy from crossing Kanawha."[1459] To division headquarters Comly reported that the enemy probably departed about the same time as Capt. Young and they, "in great haste, going towards

Barboursville, taking nothing with them. It is doubtful whether they discovered that [Young's] force had left."[1460]

Elkins reported to Comly regarding the salvaging effort writing that on the morning of December 15, Capt. Young went back to Hurricane Bridge

> with thirty (30) or forty (40) men [...] to try and save as much of the Company property as he can. He thinks he will not lose much property. He has already succeeded in saving the wagon and team; the most valuable property he had.
>
> If anything transpires we will let you know. If they attempt to cross Coal we will resist them to the last: but we do not think they will attack us here in so strong a position as this.[1461]

Young returned to Coals Mouth with whatever he could collect of company property left behind at Hurricane Bridge and then remained at Coals Mouth waiting for further orders.[1462]

Comly claimed that he had no force at Charleston with which to intercept the enemy and cut them off. He was in hopes that the 13th West Virginia at Barboursville might intercept the Confederates' retreat. The 13th Regiment, however, was not under Comly's command, and Comly had "no telegraphic communication with Barboursville, Col. Brown is in command there."[1463] All boats had been halted at the first sign of danger. On December 15, boats were running to Loup Creek[1464] but it was thought that perhaps there were small parties of rebels yet hanging out along the Kanawha River, who might attempt to commandeer a boat,[1465] and so it was not until December 16 that all boats were again running "accompanied by sufficient guards,"[1466] who could perhaps render assistance in countering further mischief.

Ten days later, the Point Pleasant *Register* carried its own account of the late rebel movements under command of Major (formerly Captain) Bill Smith of Logan and Boone counties. Smith's incursion, informed the *Register*, was met by a joint effort involving the Point Pleasant militia and three other nearby militia companies; the convalescents of the Gallipolis and Point Pleasant Post Hospitals; independent citizens; the government gunboat, the *Victor No. 2*; and the 13th West Virginia at Barboursville. The article reported that about 9 p.m. on December 13, a telegram was received at Point Pleasant from Captain William Zimmerman, commander of the consolidated posts of Point Pleasant and Gallipolis. The telegram informed that five hundred rebels were twenty miles away moving towards Gallipolis or Point Pleasant.

> [...] boats of every description were collected from the lower side of the Kanawha and guarded, the medical stores at our Post Hospital were properly disposed of — the ferry boat was ordered in read-

iness to cross the Ohio, if necessary with the Government property — while the convalescents armed themselves as best they could awaited an attack. — Only part of our citizens knew of the alarm 'till morning, those, however, who were apprized of it, were gun in hand, on the quevive. Half hoping half dreading their approach. The night, however, wore away and no rebels came. The *Victor No. 2,* which had lain during the latter part of the night at the wharf with a strong guard and those dreaded howitzers on board ready for action, at about daylight steamed up the Kanawha in search of the 'thieving rebs.'

Col. J.P.R.B. Smith, on Monday morning, called out the Point Pleasant and three other companies of militia nearest here, who rallied with commendable alacrity and in conjunction with the convalescents from the hospital here and at Gallipolis and a few others, Capt. Zimmerman promptly sent here, guarded the different roads, and held themselves ready to promptly repel attack. The rebels not appearing the militia were drilled, and after the arrival of the *Victor No. 2,* with information that the rebels were retreating into Logan, were dismissed — carrying with them the thanks of Col. S. and the Union people of the Point for their prompt discharge of duty. Thus ended the late "Rebel scare." Captain Young quietly withdrew from his position at Hurricane, without encountering the rebels and retreated to a place of safety.

The 13th at Barboursville, unsuccessfully attempted to capture the free-booters, who, after stealing all the good horses and impressing all the serviceable militia in that part of Putnam county, made good their escape. The rebels were commanded by Maj. Smith — long Capt. Bill Smith of Logan and Boone counties, whose thorough knowledge of the country has hitherto enabled him to defy the Union forces by eluding their pursuing forces. Gen. Duffie was absent during this raid with his cavalry brigade on an expedition to Lewisburg. We can't forbear to remark in this connection, the marked contrast in the late action of Captain Zimmerman and that under similar circumstances, last Spring of Capt. E.P. Fitch, and are forced to the conclusion that had Capt. F. [acted] as Capt. Z. did, Jenkins and his band never could have recrossed the Kanawha.[1467]

On December 19, pursuant to orders Capt. Young returned with his company to Hurricane Bridge. After several days, he wrote to Col. Brown saying that information received from people in the vicinity of Hurricane Bridge prompted him to believe that the appearance of a rebel force could be expected in that neighborhood. Young wrote on December 23:

> Col Brown Sir on the 19th I received verbal orders from Col Hays to return with my company to the Hurricane Bridge on the 22d we was all here, and I suppose by the rebel citizens reports that we my soon look for a nother general skedaddle or fight in the next 10 or twenty they say that Joe Morris says that will come in here at the sacrifise of his whole Command Your Friend J.V. Young[1468]

December 23, Captain Isaac Rucker, cmdg. Independent Mason County Scouts, also wrote to Col. Brown about the possibility of another raid and his own critical need for ammunition.

> I send enclosed two blank Receips for cartridge if it is in your Power pleas have the Recepts filled as I am near out of ammunition + cannot get ammunition from Whe[e]ling for probably two weeks as I have ordered it it may be delayed in transportation I would wish if you are not scarce yourself to have Two Thousand Rounds if not two thousand one thousand by al[l] means as I may have use for it before I can get from Whe[e]ling as I learn the Rebels will attempt another Raid between this + new year but if I find out I will inform you immediately.[1469]

As anticipated, another attack came. On December 26, P.H. Shaw, Sheriff of Putnam County was taken prisoner,[1470] and on the 27th Hurricane Bridge was attacked. December 27, Capt. Young, then posted at Hurricane, wrote to Col. Brown:

> this morning at 2 o clock a number of rebels drove in our pickets, and captured our team I sent out men in pursuit but they was soon driven back. I sent Corp May + William Enick, to Winfield for the purpose of dispaching to Charleston But they met the rebels Enick is wounded in thigh, I suppose Corp May is in the hands of the rebels, Sergt Overshiner has just returned from Mud and has brought with him Jud Kirk the notorious rober. [...] P.S. I understand that thare is a Regt of thoes rebels they robed Coals Mouth last night and mad[e] a clean deal with Horses, there was 20 rebs crossed the turnpike at Cal McCalisters 10 this morning[1471]

Towards the end of December and into January, a 'gang' of rebels was operating in the Coal River area—an area which included Hurricane Bridge. Lieut. Col. James R. Hall, commanding 13th Regiment, was sent out with two hundred men to intercept the rebels on December 30. He reported to Col. Brown concerning his movements:

> In obedience to your order I left this place on the 30th of Dec. with two Hundred men to ascertain if possible the whereabouts of the Rebels which were lately in the vicinity of Hurricane Bridge. We passed through the County between this and Wayne Court House and found that they were camping in the neighborhood of Wayne C.H. I found it impossible to force them to Fight as they were well mounted and appeared to be only disposed to interupt us by harrassing our advance and rear Guards — I ascertained the Rebel force to be composed of the 16th Regt Va Vols under Col. Furgeson numbering from three to five Hundred. I would have remained out longer but for the want of Rations and the sudden change in and inclemency of the weather which rendered it impossible for the men to march.[1472]

According to Andrew Stiarwalt, musician, 23rd Ohio, the Confederate gang operating in the Coal River area captured soldiers and Union citizens and "carried off about 10000 worth of property which was their only object." Stiarwalt went on to describe operations undertaken by Kanawha Valley forces, specifically the 13th West Virginia and 2nd West Virginia Cavalry, to apprehend the gang despite the freezing weather, which continued into January. Stiarwalt wrote in his diary that a Federal force

> went out in persuit of them returned without finding any of them on the 29th Dec the 2d Va Cavely have scowered the Country in the vicinity of the mouth of Coal River but did not find any of the thieving gang on the 3d Jan 1864 a Scouting party out of 13th VAI captured a Rebble major a Lieutenant and three privates on Mud River and brought them to Charleston the 4th of Jan on the 6th of Jan it had been reported that a considerable force of Rebbles are posted on the south side of Mud River who intend to move upon the 13th VAI who are posted at Barbersville as soon as the high water will fall and they can cross the mud river but it is now reported that they have left for fear they might [freeze] and account of the need of supplies the oldest citizens say this is the couldest weather from the first of Jan to the 16th of Jan 1864 that there has been for six or seven year in West Virginia [...].[1473]

An article from *Confederate Veteran* sheds light on the Confederate operations in Wayne and Cabell counties during the winter of 1863-64 which required the services of the 13th West Virginia and other forces of the Kanawha Division. To judge from this article more than just Bill Smith's men were in the area. Charles A. Lattin originally from Barboursville and a member of J.H. Nounnan's company (later Co. K 16th Regiment Virginia Cavalry) wrote that:

> During the winter of 1863-64, Col. M.J. Ferguson, of the 16th Regiment Va. Cav., with about one hundred and fifty men, went on an expedition to Wayne and Cabell Counties, Va., within the enemy's lines, to give the boys a chance to go home an recruit their wardrobes, which at that time were very much depleted.[1474]

On December 31, according to Lattin, Ferguson's men were at the mouth of the Big Sandy River near Round Bottom, West Virginia, and on the Kentucky side of the river. They crossed the river on ice to harrass a company of what they had been told was the 19th Kentucky Infantry. Lattin wrote that "[t]he cold was so intense that we could hardly load our guns, and it was known for several years as the 'cold New Year morning.' "[1475]

On December 28, Col. Brown wrote to 1st Brigade headquarters stating that Capt. Young's company had been "dropped from the Rolls of the 13th V.V.I. and ask[ed] if it will not be best for Capt. Young to report direct to Brig. Head Quarters."[1476] As indicated above, a second Company G to replace Young's command was being recruited at this time. Among those who joined the ranks of this new company was an unusual individual, Clarington D. Saunders. He was at the time of his enlistment working as a farmer; was described as five feet and ten inches tall; of fair complexion with grey eyes and dark hair. Nothing about him would have suggested that he was the biological oddity subsequent newspaper reports claimed him to be. Saunders entered the U.S. service on December 28, 1863, enlisted by Lieutenant Joseph Brumley, and he was appointed Orderly Sergeant of the new company on January 15, 1864.[1477]

Saunders was born in Botetourt County, Virginia. He was forty-three years old at the time of his enlistment at Buffalo, West Virginia. He had a wife and children. During the war, they resided in Putnam County. Saunders was described as "an intelligent man," who had "the benefit of a moderate education," and who "led a temperate life." With a few exceptions, he had enjoyed

> generally excellent health during his life. In 1847 he was attacked with cholera and since that period with lung fever on two occasions. In the summer of 1850 sleep forsook him, and since that time he has never felt the least drowsy.

Indeed, he had the singular distinction of not "having slept a single moment for fourteen years." In 1864, when he was admitted to Chesnut Hill Military Hospital in Philadelphia, Pennsylvania, from the campaign in the Shenandoah Valley, no less that four newspapers reported his story.

Saunders was brought into the lime-light when he was sent to Chesnut Hill Hospital (according to the *Philadelphia Press*) by order of the field surgeon. He was admitted into the hospital on November 17, 1864, suffering from chronic diarrhea and rheumatism. He remained there convalescing for several months, and at the time his story was published he had nearly recovered from his ailments. "[H]is appetite was good" but, it was reported, the man could not sleep. As claimed in Wheeling *Daily Intelligencer*, Saunders "retire[d] to bed the same as other soldiers, but he cannot sleep. He simply receives physical rest. This brief narrative of a most wonderful phenomena may seem fabulous, but the reader is assured that it is the truth." The *Philadelphia Press* (the source for all subsequent articles about Saunders) stated in its edition (probably February 1865) that in the time since Saunders had entered the army and then been admitted to hospital he had

> been on seven raids and on four charges, during which time he informs us that he never felt tired or sleepy.— He was in the four charges made beyond Harper's Ferry, Virginia, on the 17th, 18th, 19th, and 20th of last August, [Morgantown *Weekly Post* says 17th, 29th, and 30th of last August] and yet he did not feel sleepy. Why it is that he can not or does not sleep, is as much a mystery to him as it is to many scientific gentlemen, who having had their attention called to him, have been astounded in their attempts to investigate the cause.
>
> Upon one occasion, at his request, a number of curiously inclined gentlemen watched him for forty-two days and nights consecutively, in order, if possible, to arrive at the cause of the wonderful phenomena.— These gentlemen took turns with each other in the progress of watching, so that if he should chance to sleep it would be observed. Some of the watchers became drowsy, and it was as much as he could do to awaken them.[1478]

This being as it may, Saunders returned to duty from hospital and was honorably discharged and mustered out at Wheeling, West Virginia, with the regiment, on June 22, 1865.

Two more recruits for Co. G would come to camp on the next boat. December 29, William Jackson (Co. D?) wrote to Col. Brown, from Hartford City:

These two Recruits did Not Come in in time to get on the Ohio, So I send them on the first Boat that Come their Names are Alexander Gorby + James Roush Lieut. Allen Mason has got their Papers.[1479]

December 30, Captain William Feazel wrote to 13th headquarters notifying that he was "sending you fifteen Enlistment papers for Co. G. and I have no doubt that I will have enough by Sunday to make twenty so I will expect a 2nd Lts commission for Sergt. Darnel." Feazel added that he sent the papers forward at this time so that the new recruits might qualify to receive pay from the paymasters, who were scheduled to visit the 13th West Virginia on December 31. In postscript Feazel added that he had "98 enlisted left after this fifteen. My aggregate with those is 116 all is quiet. W.E. Feazel." Feazel provided the following list of men for whom enlistment papers were being forwarded. They may have been recruited from Captain Isaac M. Rucker's Mason County Militia Company of Independent Scouts:[1480]

1	William J Smith	10	Martin B Malory
2	Abraham Johnson	11	James J Forbush
3	William W Soward	12	William Birks
4	Merritt Adkins	13	Peter Nagley
5	John Wilson	14	Jesse Birks
6	William J Dillon	15	Ridger Gillis
7	William F Burchan	16	Franklin Elkins
8	John W Sneed	17	Thomas W Davis
9	Isaac M. Soward	18	Strother H Clendennin

On December 31, Capt. Feazel again wrote to Col. Brown notifying as to recruitment of men for the new Company G and of the status of things generally at the Guyandotte Post. Feazel wrote that he was sending

> three more enlistments for Co. G.[1481] by Sergt. Darnel making 18 I am expecting two or three more to day you will please inform Orderly Mason that I wish him to hold on to his Muster Roll as long as he can, so he can put on all the men that has enlisted up to this day in order that they can get paid the first pay day

> You will pleas send me two boxes of Cartridges by my waggon to day all is quiet this morning [in margin Feazel lists the names of three men:] Thomas W. Davis, Franklin Elkins, and Floyd Turley.[1482]

Morning reports for Company H record that a drummer was present with the company December 1 to 9. A fifer was present December 1 through 31. No one was reported absent with leave. One enlisted man was reported absent without leave on December 1. Archibald Tolbert was absent without leave December 13 to 19. On December 1, Sergeant William Shannon was discharged for promotion to the rank of 2nd Lieutenant, Co. I, by order of Gen. B.F. Kelley. Co. H gained new recruits. Two recruits were reported for December 11; one was reported December 18; one reported on the 19; and one was reported for December 31. Among those reported as new enlistments was Julius R. Mays and John T. Mays, who enlisted on December 10, and Calvin C. Shank enlisted December 11 and joined the company December 18. Alexander Elkins enlisted on December 27. At the end of the month, Co. H tallied present and absent three commissioned officers and eighty-eight enlisted men.

One private had extra duty each day of the month. Co. H had many details made upon them for detached duty. The following is the schedule for detached duty. December 1, a detachment was ordered to consist of one commissioned officer and five enlisted men to go out on detail until December 5. December 3, Lieutenant William Perdue and four men went out scouting and returned to camp on December 4. December 6 to 7, one commissioned officer and three enlisted men were detached; December 8, one commissioned officer and four enlisted men were detached; December 9, two commissioned and four enlisted men were detached. On December 8, Thomas M. Hackett and on December 9, Lieutenant William Perdue were on detached service recruiting by order of Col. Brown. December 10 to 13, two commissioned officers and three enlisted men were detached. December 14, three commissioned officers and thirty-eight enlisted men were detailed for detached duty. Among these was Lieutenant O.W. Griswold, who with thirty-five men, was sent out scouting on December 14. December 15 to 16, two commissioned officers and three enlisted men were detached; December 17 to 24, one commissioned officer and three enlisted men were detached. December 25 to 27, two commissioned officers and eighteen enlisted men were detailed for detachment. These included Lieut. O.W. Griswold and fifteen men sent out scouting on December 24. The schedual of detached duty continued without relief: December 28, two commissioned officers and nine enlisted men were detailed; December 29, two commissioned officers and six enlisted men went out; December 30, one commissioned officer and four enlisted men were detailed; and December 31, three commissioned officers and thirty-six enlisted men were ordered from camp. December 30, two lieutenants and thirty-five men were sent out scouting under the command of Lieut. Col. J.R. Hall.

Co. H had over the course of the month, one to seven men present sick each day, and at least eight enlisted men were reported absent sick each day. December 2 to 3, one private was present in arrest or confinement. This may have been Nathaniel

Simons, who on December 19 was noted as having been "returned to duty by order of Col. W.R. Brown." No one was reported absent in confinement.[1483]

On Sunday, December 27, Levett Perdue, Private, Co. H, wrote to his parents, who lived in Ceredo:

> Dec. 27, 1863
> Barboursville, W. Va.
>
> Dear Father and Mother
>
> I again seat myself to inform you that i am well and trully hope this may reach you and find you in the saim good health i sent you ten dollars by mail and it has been about a week since i started it to you and it has had time for you to get it you must write and tell me wether you got it or not you must come up and see me as soon as you can and if you cant come i will try and com and see you and stay with you a while so i hav nothing strange to write this time i did want to com down when sam was ther but the Lieutenant would not let me but i did not care much but I think i will be their in a few days so no mor this time but still remain your son untill death
>
> Levett Perdue to
> Lewis Perdue[1484]

Reports for Company I indicate that the company was stationed at Guyandotte, and that in terms of numbers of officers and enlisted men the company had finally 'arrived' beginning December 25. A full complement of officers was reported for the first time: captain; 1st and 2nd lieutenant; three to five sergeants; six to eight corporals and seventy-two privates. Recruits had been added as follows: December 2, one; December 3, one; December 4, one; December 6, two; December 7, two; December 8, one; December 9, two; December 12, two; December 13, perhaps one more; and December 14, two men joined. No one was absent with leave. One enlisted man was absent without leave December 1 to 8. At the end of the month, Co. I reported present and absent three commissioned officers and ninety-eight enlisted men.[1485]

No one in Co. I was reported detailed for special or extra duty. One private had daily duty December 1 to 10 and December 24 to 31. Two to seven enlisted men had detached service each day with the number of men detached decreasing as the month progressed. No one was reported present sick. Two to four enlisted men were absent sick each day. No one was present or absent in arrest or confinement.[1486]

On Wednesday, December 9, Laban T. M[oor]e wrote to Colonel W.R. Brown complaining about the foraging habits of the troops (probably Co. I) at Guyandotte. They had

> been and are now cutting off the wood from my land just above Guyandotte, there is but little timber attached to the land I own there and if it is cut off it renders far less valuable the arable or tillable lands. Now sir if it is absolutely necessary for the comfort of the soldiers I will not complain but it seems to me there is other lands where they can get wood that would not so injure the property, at least they ought to account for what is taken I believe the Regulations require this. I had the pleasure of once being associated with you when I commanded the 14th KY — and I hope for <u>Lang Syens</u> sake that you will see my rights protected My land lays immediately below the bridge above Guyandotte in the Guyandott River back of where Frank Hite lives please let me here from you[1487]

December 12, Lieutenant Feazel wrote to Adjutant J.S. Cunningham informing him that he was sending Henry Spicer, a deserter from Co. I 2nd Virginia Cavalry Mounted Guard. "Spicer was found at Wm Morrises 1 mile from this place he acknowledged to have been absent from his Co. four or five months or probably longer was his Statement, he did not know exactly how long."[1488] December 16, another citizen found cause to complain about the 13th Regiment. This time about the treatment his mother had received at the hands of Feazel's Company. I.H. Buffington wrote to Col. Brown the following:

> I feel it my duty to appeal to you on behalf of a widowed Mother for that protection which I as a private citizen cannot give her — against the unlawful en croachment of the soldiery under the command at this point of Lieutnt W.E. Feasel, to whom application has been made without any success, knowing that there is an authority, and a willingness in the government to redress the wrongs of her private Citizens. I am now taking the proper steps to obtain redress for one who has experienced forbearance till it has ceased to be virtue, But Sir permit me to say that I have no idea it will be necessary to appeal to any higher authority than Col. Brown. Believing that I have only to state the facts in the case to secure your prompt interference. A citizen asked permission to put two waggonloads of loose tobacco in her barn, which was already full of loose tobacco, with some Hey and corn, for this reason the refund, I at the same time telling him that I had a building a few hundred yards distant that

he was welcome to use. In a short time I saw two citizens putting tobacco in my mothers barn. I went to them and protested against it. They said a soldier gave them Liberty to do so. I then asked the soldier what right he had to forcible enter and occupy that building. He replied that he would show his right, and walked to his gun took [words blocked out by NARA stamp] it and said any interference on my part would result in my death or in other words he threatened to shoot me. I left him and applied to Feasel — only to receive additional insult — My Mother has invariably treated soldiers kindly and this is only one of many instances in which she has been treated rudly. Her barn is still occupied against her consent. I appeal to you as a Military Officer and as a gentleman for immediate redress and future protection for my Mother, during your sojourn to this section of the county.[1489]

Morning reports record that the Company K was stationed at Barboursville and that it had the following officers present: one captain on December 27; a 1st lieutenant present December 27 through 31; three to four sergeants; and two to five corporals were present each day of December. There were thirty-one to forty-nine privates reported present in camp each day of the month. One drummer was present December 27 to 31, and one fifer was present December 1 to 31. The following soldiers were reported "returned" to the company: Elias Jividen returned on December 1; December 2, Eli S. McGuire returned; December 14, Charles Hill and Jacob Searls returned; December 16, Allen Russell, Jonathan Jividen and John A. Boggess returned; December 17, James A. Hill returned; and December 21, William P. Cunningham returned. New recruits were added to the company as follows: on December 2, Abraham Kessel enlisted at Belleville; December 14, Parris Gray enlisted at Barboursville; December 17, Eliot Barnett enlisted at Leon; December 28, David P.S. Slaughter and Clark Westfall were enrolled respectively, at Ripley and Walton; and December 31, George W. Stover enlisted at Barboursville. At the end of the month, Co. K reported a total present at Barboursville of one commissioned officer and at most sixty-eight privates and with an aggregate present and absent of one commissioned officer and seventy-eight enlisted men, still below the minimum strength and under-staffed but with prospects. From one to three enlisted men were absent with leave December 9 through 27. No one was reported absent without leave.

No one in Co. K was detailed for special or extra duty. One private had daily duty each day of the month. On December 1, Henry Stephenson was detailed for "Hospital nurse." December 5, Daniel L. Walker was detailed for daily duty as teamster. He was relieved from this duty and returned to the company on December 14 at which time Parris Gray was detailed for teamster to serve until December 31.

Detached duty was assigned as follows: one commissioned officer, December 1 to 26; December 27 to 31, eight to ten enlisted men; and otherwise, groups of enlisted men numbering from twenty-two to thirty were detailed during the month. On December 27 and 28, squads of men, probably concluding their detached duty assignments, were reported returned to camp. "Remarks" following morning reports provide the following information:

> On December 27, "Sergt. Henry C. Hunter, Corporals Benjamin Hawkins, O.B. Hunt. John Harden and Abraham Wolf, Jonathan H. Casto returned to company" and "John H. Thompson, W.P. Herall, Epperson Harper, Nathan Westfall, Joseph Rhodes, Daniel M. Walker, Abraham Danham, Isaac M. Glaze, Denard Green, Jonathan H. Casto, Thomas Asbury, Edward Casto, Joseph Raines, Elmore Ranes, David Casto all returned to the Company, on Dec. 27, 1863."
>
> On December 28, "A.J. Harper, John Thornton, Moses Martain, Abraham Hesse, and Eliot Barnett returned to the Co."

One to five privates of Company K were reported present sick December 1 to 15, and one to three non-commissioned officers were present sick December 22 to 29. Two enlisted men were reported absent sick for each day of December. One of these men was Jordan Harper, who was reported at home sick on December 27. No one was present or absent in arrest or confinement.[1490]

On December 17, Col. W.R. Brown wrote to F. Pierpont, Adjutant General for the State of West Virginia, from 13th Headquarters at Barboursville recommending promotions for William E. Feazel, Peter Darnell, William Shannon (officers of Co. I) and Henry Stump, Joseph E. McCoy and Thomas Hill (Lieutenant T.C. Hill, recruiting officer for Company K?), officers of Co. K. Brown wrote:

> I have the honor to request that commissions be issued to 1st Lt William E Feazel to be captain of Co I 13th Regt V. V. I. to fill original vacancy — Peter Darnel to be 1st Lt Co I vice William E Feazel promoted —
>
> William Shannon to be 2nd Lt. of Co I to fill vacancy — That Company now number[s] 101 men and has only one commissioned officer.
>
> Also 1st Lt Henry Stump to be commissioned Capt of Co K 13th Regt V.V.I. to fill vacancy — Joseph E McCoy to be 1st Lt vice Henry Stump promoted — Thomas C Hill to be 2nd Lt to

fill vacancy — That co has now more than the minimum no of Enlisted men.[1491]

For the period of October 10 to December 31, 1863, the 13th had marched a total distance of "743 miles" and had sustained the following casualties: four men killed in action and seven wounded in action.[1492] West Virginia State *Adjutant General's Report for 1863* supplied other details as to the status of the 13th West Virginia Regiment for the year 1863. By December 31, 1863, the regiment was still undersized numbering just eight hundred and sixteen (816). Just eight companies had been reported and of course, there was the "difficulty" in regard to Co. G (Young's Company G) which had "been attached to this Regt [the 13th] since its organization."[1493]

Both the 11th and 13th Regiments claimed Captain Young's company inasmuch as a portion of the men had been mustered into the 11th and a portion into the 13th regiment. An order from the War Department settled this question, and it was assigned to the 11th Infantry Regiment. A replacement Co. G was recruited. The prospects for filling it and for completing the other two companies (Cos. I and K), that the regiment might be deemed "full" were reported to Adjutant General Pierpont as "good" by Col. Brown and indeed, in February 1864, the regiment was at last fully organized.

Unfilled positions of field officers and staff were reported as follows for December 31, 1863: the rank of major had never been filled; one of two positions for assistant surgeon remained unfilled, as was the position of chaplain; 1st lieutenant of Co. I; 1st lieutenant of Co. K; 2nd lieutenant for Co. C; 2nd lieutenant for Co. I; and 2nd lieutenant for Co. K—all were vacant at this time.

By December 31 of this year, thirty-one commissioned officers of the 13th West Virginia had been appointed from civil life. The 13th had one field officer; four captains; ten subalterns, for a total of fifteen commissioned officers. One field officer had resigned; two captains had resigned; two subalterns had resigned, for a total of five commissioned officers resigned. One captain had died. One subaltern had been dismissed.[1494]

Two hundred and ninety "Non-Commissioned officers, Privates &c." had joined the 13th by enlistment. Fifteen of these had been discharged from the regiment for disability; eight had been discharged to be promoted within the regiment; and one minor among the enlisted men, had been discharged from the regiment by civil authority, for a total of twenty four "Non-Commissioned Officers, Privates &c." discharged from the Regiment. A total of thirty-six "Non-Commissioned Officers, Privates &c.," had died since October 10, 1862, when the companies then in existence had been mustered into service. There had been thirty-one "Ordinary Deaths." Four had been killed in action, and one man had died accidentally. Seven enlisted men had been wounded in action, and eight had been wounded accidentally. Since October

10, 1862, eighty-one enlisted men had deserted. Sixty of these had been apprehended or returned voluntarily from desertion. Twenty men had been tried for desertion and twenty convicted for desertion. Five had been pardoned after sentencing. Two men had been tried by general courts martial, and a total number of twenty-one had been tried by regimental courts Martial. None had been tried by civil authority.[1495]

At the various posts or stations at which the 13th had been located since October 10, 1862, the following information was tallied:

> At Point Pleasant, West Virginia: one member of the 13th was wounded; twenty-four died; sixteen deserted; and eight were apprehended.
>
> At Winfield, West Virginia: two were wounded; six died; eleven deserted; and eight were apprehended.
>
> At Hurricane Bridge, West Virginia, six died; six deserted; and three were apprehended.
>
> At Coalsmouth, West Virginia, one died; a whopping thirty-two had deserted; and twenty-nine were apprehended.
>
> At Fayetteville, Camp White, and Guyandotte, West Virginia, no one died; deserted or was apprehended.
>
> At Charleston, West Virginia, one deserted; and one was apprehended.
>
> At Barboursville, West Virginia, ten deserted; and five were apprehended.
>
> At Pomeroy, Ohio, one deserted; and one was apprehended.
>
> At Mud Bridge, West Virginia, four deserted; and three were apprehended.

Taken together, these numbers comprised a total of thirty-seven deaths; eighty-one desertions; and sixty men who were apprehended.[1496]

> From October 10, 1862 to Dec. 31, 1863, the 13th sustained the following casualties. On March 28, 1863, at Hurricane Bridge the following men were killed in action: Private Henry Sands, Co. A; Private Jesse Hart, Co. B; Private Ultimus Young, Co. B; and Private Emory J. Bridgeman, Co. F. The following men were wounded in the action at Hurricane Bridge: Corporal James A. Rayburn, Co.

B; Private Henry Hoffman, Co. D; and Corporal Leroy Newman, Co. H.

On September 8, 1862, Private James Davis, Co. G, was wounded; September 6, 1862, [Private?] James Paul, Co. G, was wounded; December 27, 1863, Private William H. Enocks, Co. G, was wounded. March 30, 1863, at the battle of Point Pleasant, 1st Lieutenant William N. Hawkins, Co. E, was wounded.

From October 10, 1862, to December 31, 1863, Captain Greenbury Slack's Co. A had five deaths; six desertions; six apprehensions. Captain Milton Stewart's Co. B had five deaths; two desertions; one apprehension. Captain Lemuel Harpold's Co. C had three deaths; eleven desertions; eight apprehensions. Captain Simon Williams' Co. D had one death; seventeen desertions; twelve apprehensions. Captain John D. Carter's Co. E had eight deaths; one desertion; no apprehensions. Captain Albert F. McCown's Co. F had four deaths; five desertions; two apprehensions. Captain John V. Young's Co. G had nine deaths; thirty-five desertions; thirty apprehensions. Captain William I. Mathews' Co. H had two deaths; four desertions; one apprehension. Lieutenant William E. Feazel's Co. I had no deaths; no desertions; no apprehensions. Lieutenant Henry Stump's Co. K had no deaths; no desertions; no apprehensions. Total number of deaths, desertions, and apprehensions of the regiment for this period was: thirty-seven deaths; eighty-one desertions; and sixty apprehensions.[1497]

Regimental Flags

Sometime during the year of 1863, the West Virginia State legislature passed joint resolutions authorizing the governor to present State flags to all West Virginia regiments. The 13th Regiment was presented with its first State flag. This flag was in its general form and character the same as all the other West Virginia State flags. The flags were made by the Horstmann Brothers Company of Philadelphia, a company which by 1864 had become one of the largest manufacturers of military goods in the country. The Horstmann flags were described by West Virginia historian, Virgil A. Lewis, in his report to the West Virginia Department of Archives and History published in 1906. After the war, all flags (and their remnants) carried by West Virginia regiments in the struggle were turned back to the State. Due to their poor state of preservation, the flags remain locked away in the basement of the Charleston Capitol Complex where none are permitted to view them. This being the case, Lewis' century-old descriptions of the flags remain today one of the most valuable resources for historians. Lewis wrote that the State flags were

> six feet square of deep blue silk embroidered with long golden fringe. In the center, [...] was painted in colors an oval, as back-

ground, the transverse diameter being 32 inches and the conjugate 33 inches. In this was the Great Seal or Coat-of-Arms of the State painted also in colors, at the base of which a scroll bore the motto of the State 'Montani Semper Liberi.' Beneath all was a long reddish brown floating scroll on which appeared the number of the regiment and the arm of service to which it belonged.

Once the regiments had participated in battle, the names and dates of their engagements were added in gold letters surrounding the oval. On the reverse side, a spread eagle was painted "in colors." It measured forty-four inches from wingtip to wingtip. Protecting the bird's breast was a "barred shield" in the national colors—the red, white and blue; a sheaf of arrows in its right talon; and the olive branch of peace in its left. A floating scroll held in its beak bore the legend 'E Pluribus Unum.'

Beneath the eagle, in gilt letters on a red scroll was "Dulce et decorum est pro patria mori," a line from the third canto of <u>Odes</u>, written by Horace, a lyrical poet of ancient Rome. The line became a popular motto and was often quoted in literary and architectural context. (See, the motto displayed at the Arlington National Cemetery at the front entrance of the amphitheater there.) Translated from the latin into english the phrase means: "It is sweet and fitting to die for one's country." Near the flag's lower margin, still on the reverse side, appeared on a horizontal line of green, "Thirteenth Virginia Infantry." Above the eagle arranged in curved rows were thirty-four stars in gold representing the number of States admitted into the Union. A star for West Virginia, the 35th State, was not represented in these flags but perhaps was included in the reissue of State flags made in 1865. The flag staff was fashioned from "walnut or ash ten or eleven feet in length." The upper end was ornamented with "metal tips and brass ornaments" and "long silk cords of blue and white with long tassels of the same materials at the ends."

The 1863 State flag of the 13th was damaged to the extent that only two-thirds of it remained. It is possible that this State flag was rent at the same time as the regiment's national flag of which Lewis reported that "little" remained. Lewis wrote:

> The blue Canton with its thirty four stars is entirely gone, save a small remnant around the staff. Not a fourth part of the stripes remain and the part left of them is badly rent. Members of this Regiment now surviving say this Flag was torn to pieces by the bursting of a shell in battle. The staff of ash is ten feet in length and on it is roughly carved the number '13.'[1498]

For the men of the 13th West Virginia Regiment the campaign of 1863 had not been distinguished by any great events. Their duties of scouting, marching and fighting in the hills and mountainous country in back of the rivers offered up a steady diet of thankless, tiresome duty. These brought little glory no matter how zealously the tasks were performed.

The guerrilla element—guerrilla fighters such as partisans or "local service out-fits," as they were sometimes called and loose-canon regulars—was always something of an unknown quantity in West Virginia. These soldiers, who 'fought in the shad-ows,' tended to be overlooked by historians in assessing the history of the area and the contribution of her Union soldiery, who had to be compelled to contend with them. The task of keeping them in check, provided unique opportunities for Federal soldiers such as those of the 13th Regiment. They themselves become adept at guer-rilla warfare and in the course of so doing, they developed certain specialized skills and toughness of mind and body which only warfare in such mountainous terrain, against such an enemy, done in all kinds of weather could give.

This training in the field at the edge of Union army lines—'in the Wild and beyond God's country'—prepared the men of the 13th for the campaigns of 1864, when at last, they were cut loose from their base and were vaulted onto the front lines and spared no duty. In the Shenandoah Valley in 1864, their contribution would be of immeasureable worth. There was no give back to them. There, in the course of that remarkable campaign, on the great battlefields of that year, they would repeatedly show their uncommon speed on the march; their ability to move in secret; their skill and their powers of endurance on the flank attack, on the charge; and their coolness, aim and sticking power on the firing line, under the most trying circumstances.

Appendix

J. P. R. B. Smith

J.P.R.B. Smith (also called "Alphabet Smith") was short for John Peter Roman Bureau. Alphabet Smith had been named for a distinguished French emigrant to Gallipolis—one of the 'French 500' company of minor aristocracy and professionals, who emigrated to America to flee "The Terror" set in motion by the French Revolution of 1790. J.P.R.B. was born in Point Pleasant, March 17, 1838, to Nathan Smith and Ann (Roseberry) Smith. During "the Civil struggle he was made Colonel of the 106th Regiment Volunteer Militia, which became part of the West Virginia Home Guard."[1499] In summer of 1862, Colonel Smith, a young man "of popular manners," became the proprietor of The Virginia House, a popular hotel, which he reopened for business on August 11, 1862.[1500] Virginia House was located "on the corner of 1st Street and Front Street at the head of the wharf and ferry landing" in Point Pleasant.

The hotel was built of "hand manufactured brick, with wide halls, a great dining room, large parlor and spacious bed rooms."[1501] Bearing in mind that a number of the major and minor "movers and shakers" of the West Virginia New State movement were from Mason County (John Hall, Sr., Daniel Polsley, Lewis Wetzel and Kellian Whaley to name a few), the newly opened and "spacious" Virginia House centrally located on the high ground above the Ohio River surely provided a forum for the Union cause in Mason County and for the surrounding area. The Virginia House was a place for loyal community leaders to assemble to conduct the business of completing the re-organization of the new State and county government. This sweeping overhaul and re-delegating of power begun in 1861 with secession from Virginia and fully-realized in 1863 with West Virginia's admittance as the 35th State of the U.S. Union was a huge task in itself, but in reality, it was the easy part. The task of maintaining and asserting governance in midst of civil war was the big challenge and it took grit and stamina to do it.

Maintaining public safety and public confidence were critical challenges. Loyal community leaders didn't want citizens to be driven from the State. The challenge for local Committees of Safety, militia units, such as, the 106th Virginia Militia commanded by Colonel J.P.R.B. Smith and for U.S. volunteer regiments such as the 5th and 13th Virginia; the 23rd and 36th Ohio Infantry Regiments; and Colonel John Witcher's 3rd West Virginia Cavalry company was to suffieniently 'quiet' the secessionist element in the northwest counties that agriculture, mercantile business and all forms of livelihood be permitted to continue to hold body and soul together. To whatever extent possible a semblance of government 'in control' had to be put out in front of the people. To this end, the courts had to re-open and stay open; public referenda had to be held without molestation; and collection of revenue, collected from loyal and pro-South citizens alike, had to successfully proceed. The 13th West Virginia with all of the other organized military units herein named, individually and in collaboration, saw this done as best they might.

The re-opening of the Virginia House was a optimistic statement and symbol for the New State movement; a reference to a bright future for the State separated from the weight of "the old Dominion." After the shutting down of business in Mason County, which descended like a pall at the outbreak of the war, the re-opening of this inn was a positive indication, an encouraging sign of future posterity and goodwill in spite of the war. It goes almost without saying that the re-furbishing and re-opening of the inn was not done without Smith assuming a fair amount of personal risk of reprisal from pro-South individuals and groups nor was it done without financial risk.

As for Smith's command, the 106th Virginia Militia, it acquited itself admirably during the war. It saw active duty around Point Pleasant and served Mason County during the raids of 1862 and '63 during Morgan's raid through Ohio; Jenkins' raid in the Kanawha Valley; and in numberless details of guard and scouting duty. Further, the 106th supplied regiments recruited from this section (such as the 4th and 13th West Virginia Infantry Regiments) with men and officers. The 106th Militia seems to have been well organized and its members held to discipline. Not infrequently, the 106th convened its own military court at Point Pleasant where it tried and dispensed military justice upon convicted malfeasants. By 1863, Smith and Colonel Jacob Hornbrook served as agents for the State of West Virginia.[1502] On July 28, 1864, Smith was appointed to the rank of Major in the U.S. Veteran Army and was assigned to the Department of West Virginia and ordered to Wheeling. In post-war years, Smith continued to figure in local government serving a number of terms (at least 3 terms at 6 years to the term) as Mason County Clerk.

Dr. Elkanah Williams

Dr. Elkanah Williams was born in Lawrence County, Indiana, December 19, 1822. He graduated from Asbury University (today DePauw University, Greencastle, Indiana) in 1847, and maintained a general medical practice in Bedford, Indiana, until 1851, when after the death of his first wife, he closed his Bedford office and took graduate studies in surgery at the Louisville (Kentucky) Institute of Medicine (founded in the 1830s, later the University of Louisville School of Medicine).

Williams attended clinical teachings given at the Louisville City Hospital and at the Institute of Medicine. The institute's medical library was one of the finest in the country. Here, he had access to many volumes from Europe where scientific progress was preparing the way for cutting edge advances in the practice of medicine. Of great advantage to Williams was undoubtedly the fine teaching faculty of the Louisville Institute. Among this faculty were the foremost American scientists and medical doctors of the day. Notable among these were:

> Daniel Drake, M.D., 1816 from the University of Pennsylvania, he taught at Jefferson College of Medicine in 1830; organized the Cincinnati Medical College in 1835; came to Louisville in 1839 as Professor of Clinical Medicine and Pathological Anatomy; and founded a School for the Blind in Cincinnati and Louisville.

> J. Lawrence Smith, civil engineer, M.D., toxicologist, forensic scientist, chemist, inventor and mineralogist. Smith was professor of Medical Chemistry and Toxicology at Louisville.

> Benjamin Silliman, Jr., M.D., 1849 from the University of South Carolina, held the Chair of Medical Chemistry and Toxicology at University of Louisville from 1849 to 1854 and was editor of such notable scientific journals as *Journal of Science, World of Science, Art and Industry* and *Progress of Science and Mechanism*.

> Samuel David Gross, M.D., was Chair of Surgery at University of Louisville from 1841 to 1856. Gross was North America's most respected and influential surgeon in the 19th century. He was immortalized by painter (and anatomy student) Thomas Eakins in his enormous oil painting (8 by 6-1/2 feet; painted 1875) of the great surgeon rendered in the best American Realist tradition of man in the context of his profession—Dr. Gross in black frock coat, scalpel in ungloved bloodied hand, fiercely intent upon the task at hand, operating and lecturing to the group of students in the great surgical theater at Jefferson College, Pennsylvania.

Upon completion of his graduate studies in Louisville, Dr. E. Williams relocated to Cincinnati in 1852 and opened a practice there. Herman von Helmholtz (1821-1894), the great German physician and physicist revolutionized the field of ophthalmology with his version of the ophthalmoscope (first invented in 1847 by English mathematician Charles Babbage 1792-1871 but redesigned by Helmoltz in 1851). The ophthalmoscope permitted doctors to examine the inside of the human eye and understanding and treatment of the diseases of the eye were given enormous impetus. In November of 1852, to advance his knowledge of diseases of the eyes (ophthalmology) and ears (otology), E. Williams went for study to Europe. He remained in Europe for more than two years. He was supported during this time by his brothers, who were successful farmers and stock-breeders. In Paris, he remained 18 months studying with three important 19th century surgeons:

> Dr. Louis-Auguste Desmarres (1810-1882), one of the better known ophthalmic surgeons of the 19th century; also known for writing an important textbook on the diseases of the eye called Traite théoré que des maladies des yeux (pub. 1847).

> Dr. Auguste Nélation (1807-1863), famous French surgeon, professor at the Paris Academy of Medicine and very likely at work at this time on his five volume work about surgical pathology published 1854-1860.

> Dr. Phillibert Roux (1780-1854), pioneer in the field of plastic surgery; credited with performing one of the earliest surgical procedures to repair a cleft palette.

Williams had opportunity, and he studied widely. He spent several months at the famous Moorfields Eye Hospital in London—the oldest and largest eye hospital in the world.[1503] His mentors at Moorfields were Sir William Bowman (1816-1892), English surgeon and ophthalmologist and early user of the ophthalscope; and James Dixon, surgeon. Family historians also record that E. Williams went to Prague, Vienna and Berlin. Such was his talent as a student and physician that during his stay in Europe, Dr. Williams became so thoroughly familiar with the new invention of the ophthalmoscope that he published a two-part article on the revolutionary diagnostic tool in the *London Medical Times*, entitled "The Ophthalmoscope," published July 1 and 8, 1854.

In 1855, Williams returned to Cincinnati and resumed his practice but limited it to ophthalmology and otology. The same year he opened an ophthalmology and otology clinic at the Miami Medical College, Cincinnati, Ohio. This was a charity clinic for which he suffered some ridicule from general medical practitioners. Specialty medical practices and hospitals were regarded as something of an elitist adventure. Unperturbed, he maintained his office on "Eighth Street, between Vine and Walnut for more than thirty years." In 1860, he was given the Ophthalmology Chair (the first of its kind the United States) at Miami Medical College. In this position he taught for the next 20 years and trained many students in this specialty. Among these was his nephew, Abram D. Williams, who became Assistant Surgeon in the 13th West Virginia Regiment.

Dr. E. Williams was a recognized authority in his fields and published widely. He was co-editor of the Cincinnati *Lancet and Observer* from 1867-1873; President of the Ohio Medical Society, 1875; and in 1876, he was President of the American Ophthalmological Society and the New York Ophthalmic Congress. In 1862, he attended the Paris Medical Congress to present a paper (in French). During the Civil War, he served as Assistant Surgeon at the U.S. Marine Hospital (probably at Louisville). Dr. E. Williams died October 5, 1888.

Photo Gallery

Maps

Map of West Virginia.

Map of the Kanawha Valley in 1863.

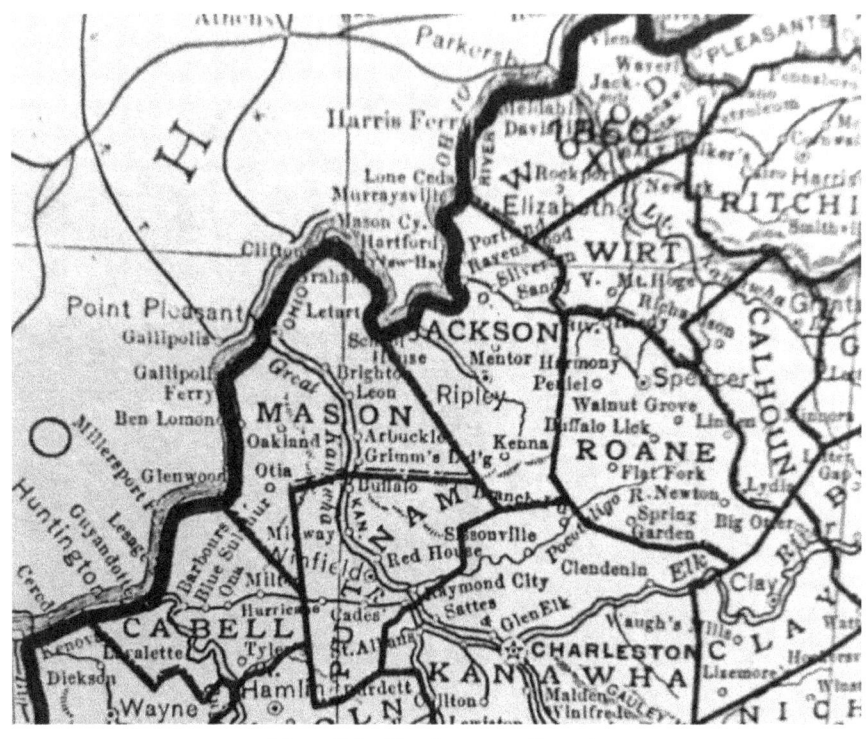

Map of the Northwestern Counties.

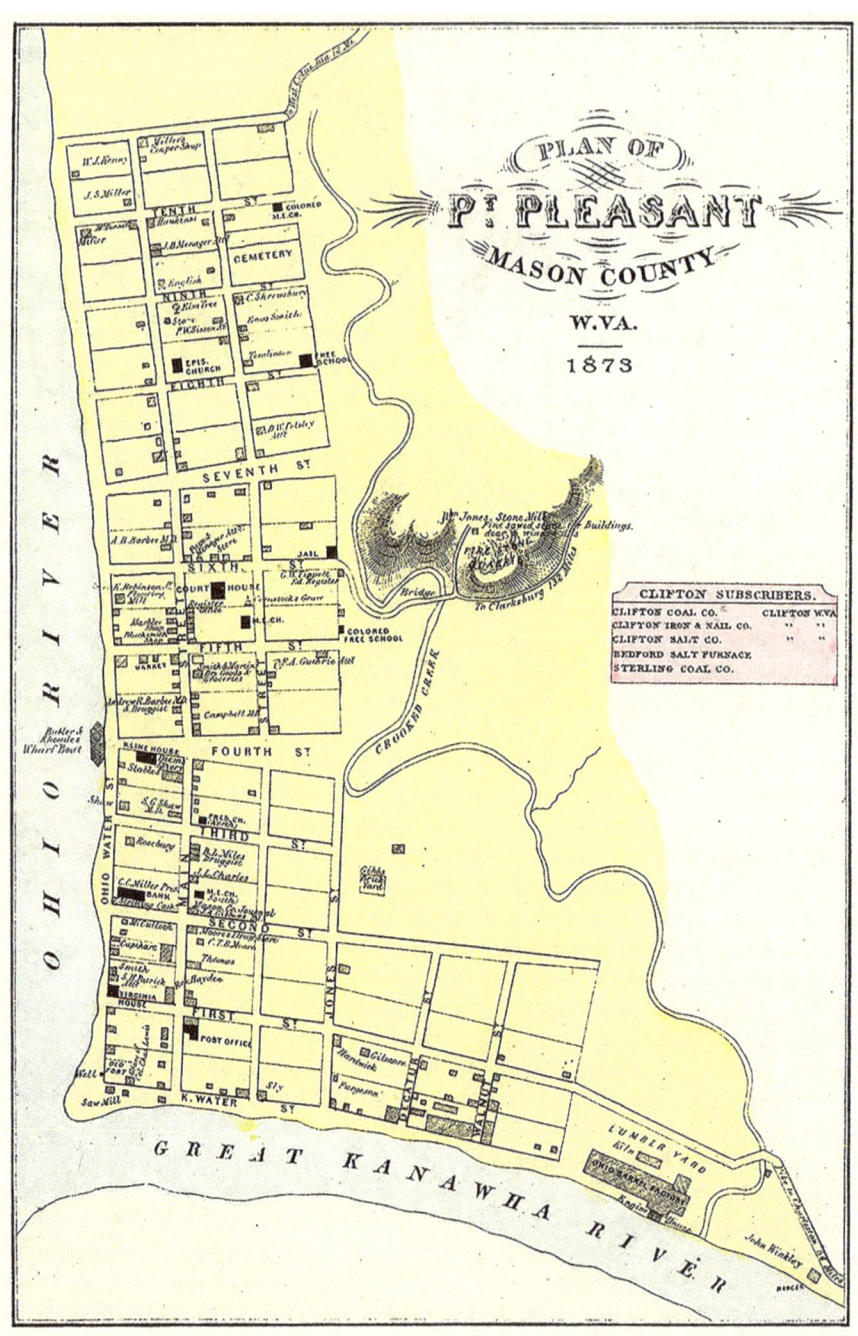

1873 Map of Point Pleasant, Mason County, West Virginia.

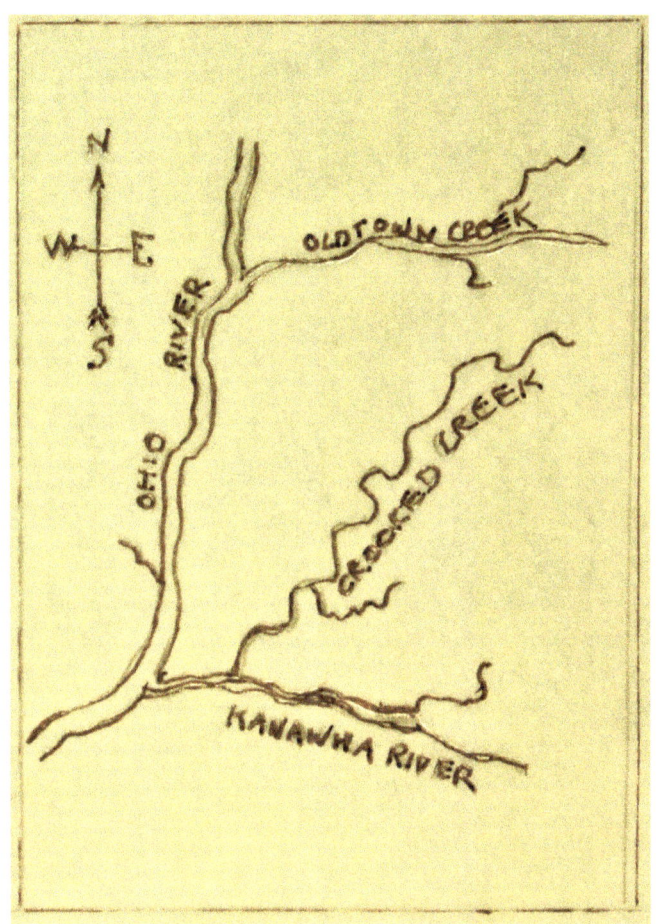

Map of the confluence of the Ohio and Kanawha Rivers and their respective tributaries, Crooked Creek and Old Town Creek, which frame the town of Point Pleasant.

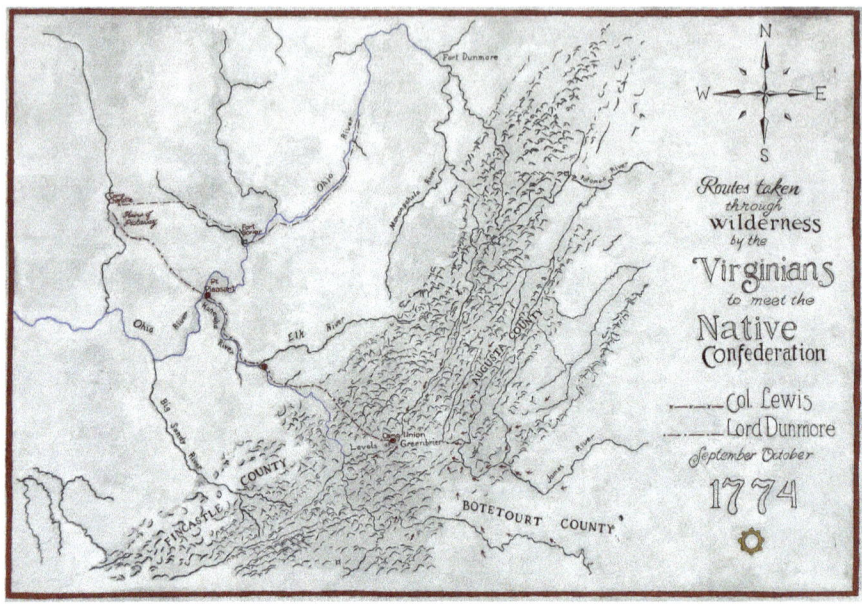

Map of the important rivers of the Northwestern part of the State and their tributaries at their sources in the Allegheny Mountains to the south and east. Source: "Map of the Routes taken through the Wilderness by the Virginians to meet the Native Confederation September October 1774." On display on Wharf Street, Point Pleasant, together with the 500 large-scale murals painted by renowned artist Robert Dafford illustrating the founding and settlement of Point Pleasant.

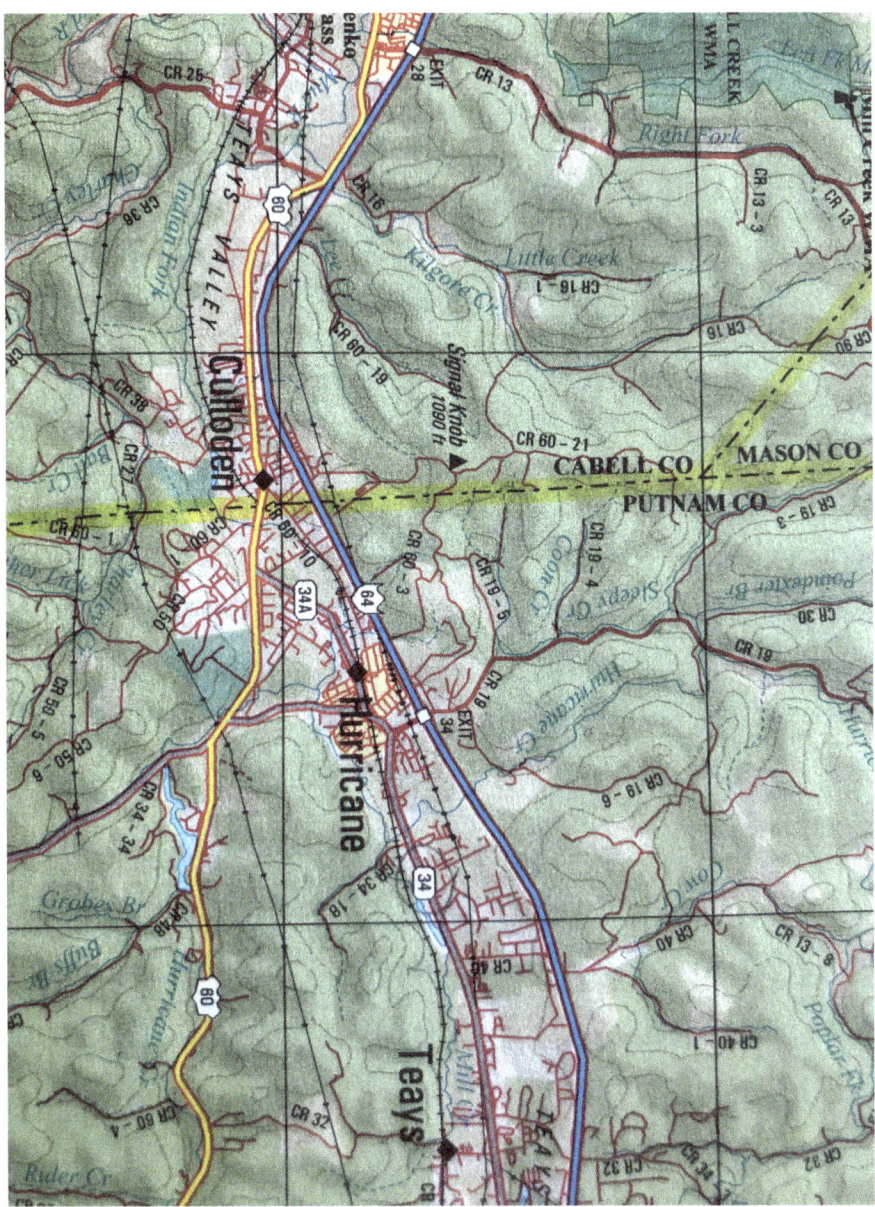

Map of the Hurricane River Valley, Putnam County, West Virginia, showing the town, the railroad, Hurricane Creek and the "circuitous ridge" to the northwest which wrapped around the town. Courtesy of DeLorme, West Virginia Atlas and Gazetteer, *Yarmouth, Maine:DeLorme, 1997, p. 42.*

Composite photograph of Hurricane Creek Valley. A segment of the "circuitous ridge" can be seen in the far distance above the trees. The battle of Hurricane Bridge, fought by a detachment from the 13th Regiment West Virginia Infantry under command of Captain James W. Johnson, took place between two heights at the crossing of 60 West and 34 North. The troops were stationed south of the town of Hurricane in the valley where they were pummeled from the adjacent heights. They were posted here to protect the all-important railroad (today CSX), its connection to the Kanawha River to the east at Scary and to hold in check partisan activity.

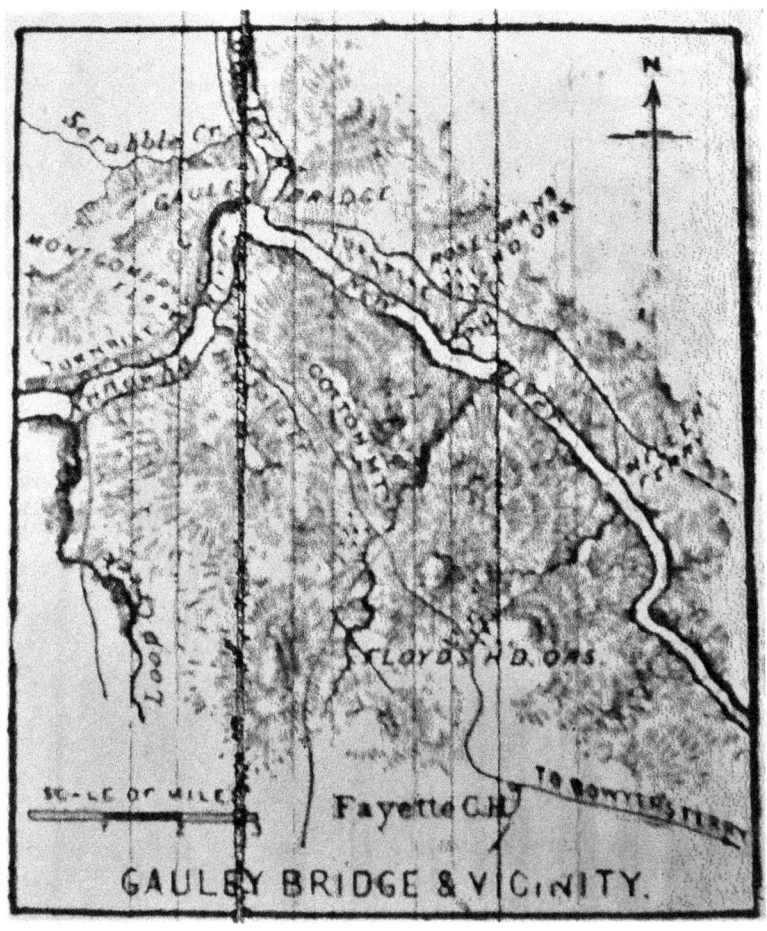

Map of Gauley Bridge and Vicinity from R.B. Wilson's letter to The National Tribune detailing his service with the Kanawha Division in the Civil War.

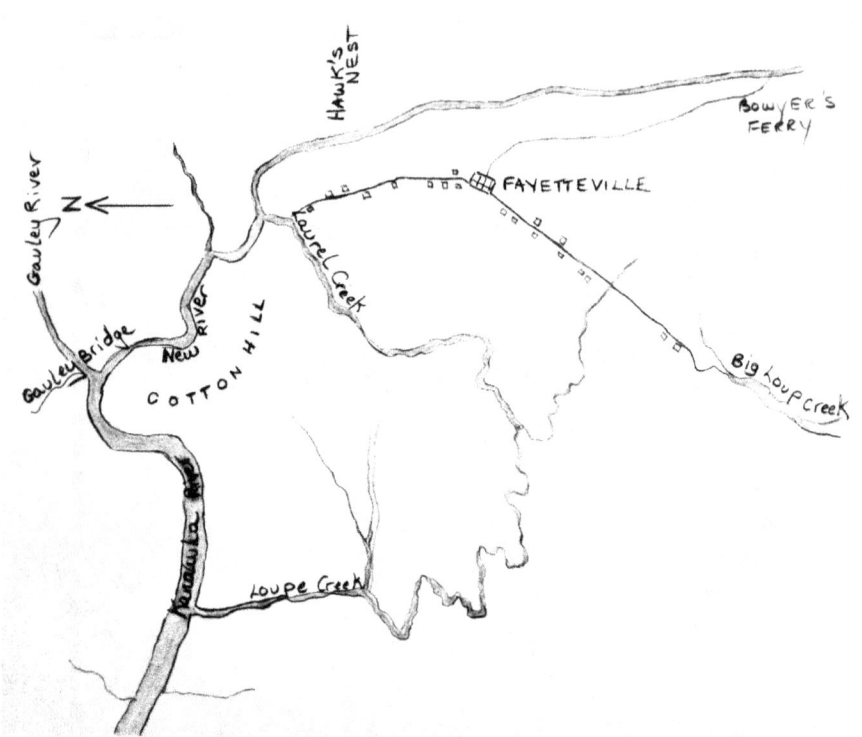

Map of Fayetteville showing waterways and a few landmarks.

General John McCausland attacked Fayetteville in this, the last attack made on Fayetteville on May 21, 1863. His command consisted of the 36th Virginia Infantry, six companies of the 60th Virginia Infantry and four guns of Bryan's Battery. The 13th West Virginia took position on Cotton Mountain with a detachment of forces sent by General Scammon to Fayetteville to repel McCausland's attack.

Source: Map of Fayetteville, Accession No. 895, Civil War Scrapbook, Regional History Collection, courtesy of West Virginia University, Morgantown. See also "Fayetteville, West Virginia During the Civil War," West Virginia History (Imprint: Charleston, W. Va., State Dept. of Archives and History) July 1963, which published a rendering of the battle (from an old print) with key identifying the location of landmarks and where the various regiments were located in the action. The 13th West Virginia is identified at 'H'.

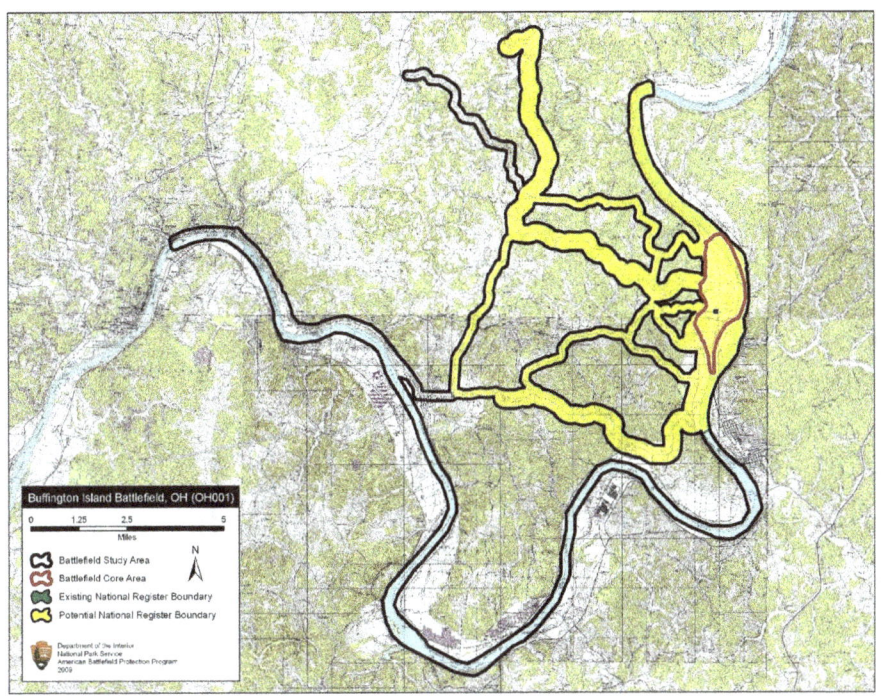

Map of the Battle of Buffington Island, West Virginia, fought July 19, 1863. Buffington Island lies three miles up the Ohio River from Ravenswood, West Virginia.

Places

Looking north up the Ohio River towards Gallipolis and the bridge crossing the river into the town and the view looking south down river. The bridge connecting 'the Point' to Henderson, Ben Lomond and to the counties below down the Ohio can be seen in the distance.

Mason County Courthouse of 1833. Linoleum block print from The Observer *(December 1933), p.6.*

Mason County Courthouse, Point Pleasant, West Virginia.
View of bas relief of the old 1933 courthouse on the 1937 courthouse wall.

The Courthouse was located upon 6th Street between Main and Viand, Point Pleasant. It was originally built in 1837. A separate jail stood adjacent on the east side of the courthouse. The old courthouse was replaced in 1933. Captain J.D. Carter and Company E were holding the Point Pleasant military post when they were attacked by Jenkins' men (the battle of Point Pleasant fought March 30, 1863). According to Myer's History of West Virginia, (Vol. I, p. 473), Company E was encamped "between Main and Viand streets, two blocks from the courthouse." Livia Simpson-Poffenbarger in her notes for the "Battle of Point Pleasant" (Poffenbarger Papers, Volume 9, p. 27) stated that Company E was encamped on a commons above the courthouse just north of 8th Street. From the open ground, Carter and Company E fell back fighting in the streets and took refuge in some brick buildings. Finally, Carter and the men of Company E established themselves inside the Mason County Courthouse where they had access to ammunition stored there.

People

William R. Brown, Colonel, commanding 13th Regiment West Virginia Volunteer Infantry, as he looked in 1862-63.

Source: From Massachusetts Commandery Military Order of the Loyal Legion of the United States, Army Military History Institute, Carlisle Barracks, Carlisle, Pennsylvania.

William R. Brown, Colonel, commanding 13th Regiment West Virginia Volunteer Infantry. Portrait cartes. Imprint Wheeling, 1865.

James R. Hall, Lieutenant Colonel, 13th Regiment West Virginia Volunteer Infantry.

Milton Stewart, Captain, Company B 13th Regiment West Virginia Volunteer Infantry.

John D. Carter, Captain, Co. E 13th Regiment West Virginia Volunteer Infantry.

William N. Hawkins, 1st Lieutenant, Company E 13th Regiment West Virginia Volunteer Infantry.

Allen C. Mason, 1st Sergeant, Company F 13th Regiment West Virginia Volunteer Infantry. Portrait cartes of Mason wearing the uniform of captain. Mason was promoted in 1864 to command the new Company G, recruited to take the place of Captain John V. Young's Company G, which was taken from the 13th West Virginia and returned to the 11th West Virginia Infantry.

John Ellsworth Alexander Patterson, postwar photograph. Patterson enrolled as Private in Company C 13th Virginia (West Virginia) on March 9, 1863, at Point Pleasant.

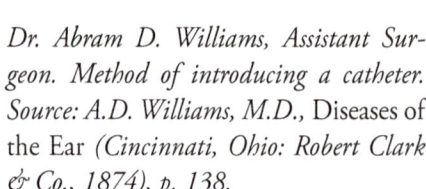

Dr. Abram D. Williams, Assistant Surgeon. Method of introducing a catheter. Source: A.D. Williams, M.D., Diseases of the Ear *(Cincinnati, Ohio: Robert Clark & Co., 1874), p. 138.*

Andrew Clark Hill

A.C. Hill enlisted on December 24, 1863, at Barboursville. He served as Private in Company G 13th West Virginia together with his brother, George W. Hill.

When the 106th Regiment Virginia (West Virginia) Militia was re-organized in late 1862, J.P.R.B. Smith became its Colonel. During the Civil War, the 106th became part of the Home Guard of West Virginia. In 1863, Smith coordinated with Colonel W. R. Brown, commanding the 13th Regiment West Virginia to effectively quiet the partisan element in the Northwestern counties to secure the safety of the communities that the newly established municipal government, operation of courts, mails, lines of transit and collection of taxes might operate without interference.

J.P.R.B Smith.
Source: Point Pleasant State Gazette.

VIRGINIA HOUSE.
Front Street, Point Pleasant, Va.,
J. P. R. B. SMITH, - - Proprietor.
Takes pleasure informing his friends and the travelling public that this popular hotel has just opened for the reception of visitors.
aug 14—1y.

Notice publishing the re-opening of the Virginia House, Point Pleasant, West Virginia. J.P.R.B Smith, proprietor. From the Point Pleasant Weekly Register.

The Virginia House served meals and offered accommodation for guests. Virginia House was located at First and Water Streets, Point Pleasant.

Other

ieutenant Colonel James R. Hall's letter to Mrs. Rebecca A. Barnett.
Letter dated January 20, 1863.

James. R. Hall wrote to Mrs. Rebecca A. Barnett at the request of her husband to inform her that her son, Columbus C. Barnett, was still very sick. The son being too ill to move, her husband had left their home to care for their son, Columbus. Hall reassured Mrs. Barnett that "whatever can be will be done for him and we will keep you informed of his condition." C.C. Barnett, Private Co. B 13th West Virginia Infantry Vols., Military Pension Records, NARA.

Christopher Columbus Barnett (called "C.C." or just "Columbus") to his father and mother, dated "Mud Bridge, June the 8th, 1863." C.C. Barnett, Private Co. B 13th West Virginia Infantry Vols., Military Pension Records, NARA.

"The Weekly Register Book and Job Office," advertisement of services available at the Point Pleasant Register office, George W. Tippett, proprietor. Tippett advertised that his office could do "Job printing plain and fancy." His type were "all good, and of the new, est. styles." He had "a large lot CUTS, ORNAMENS, BORDERS, &c., and guarantees satisfaction."

Source: Point Pleasant Weekly Register, *December 17, 1863.*

George W. Tippett, Sr. was a native of Maryland. He moved to Point Pleasant in 1855 and was actively engaged in the newspaper business for the next half century. The Register *office was located on the corner of Main and between Third and Fifth Streets in Point Pleasant. To support his large family and the* Weekly Register *begun in February 1862, Tippett offered printing services as outlined in this advertisement.*

The Weekly Register.

Advertisement and terms for subscription to the Point Pleasant Weekly Register.

RECRUITS WANTED
FOR 13TH VIRGINIA REGT.

This Regiment is, by a late order from the Secretary of War to be mounted.— The regiment is not yet quite full, and as it is now fully determined that it shall be mounted in time for the Spring campaign, we know of no better opportunity for those who have any desire to go voluntarily, into the service of their country than to come forward and immediately enlist in the Regiment. Congress has now passed a conscription law for the purpose of reaching all those who are able for the service, but have refused hitherto from going into it as volunteers. All, therefore, who wish to escape the odium which will necessarily result from being forced into the ranks to defend their country and liberties from treason and rebellion, can do so by coming forward and enlisting in the 13th, Va Vols. A better set of officers from the Colonel down, we venture to say, is not found in the service, kind, patient, and brave. Come forward then, and do your duty for your country. We must have more men, and if they cannot be had by volunteering they must be had by conscripting. This government must be saved, if it takes the last man and the last dollar to do it. Permit then the noble instincts of patriotism to prompt you to do your duty, and come out from your hiding places and help us save our country.   Lieut. WM. E. FEAZEL.
                              Recruiting Officer.

"Recruits Wanted for 13th Virginia Regt." Call for recruits taken out by Lieutenant Wm. E. Feazel published in the Point Pleasant Weekly Register, *March 4, 1863.*

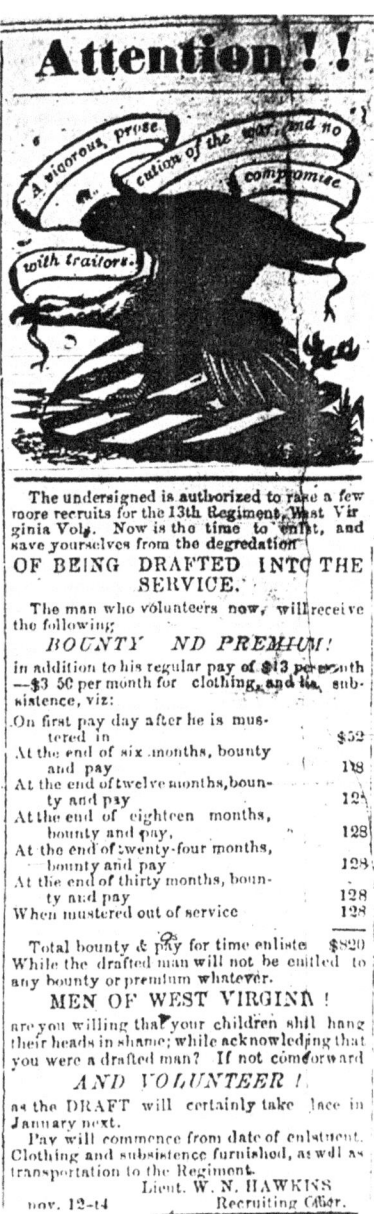

"Attention," notice placed by Lieutenant W.N. Hawkins, Recruiting Officer, Point Pleasant Weekly *Register, Dec. 17, 1863. This notice informed the readership that the undersigned, Lieutenant W.N. Hawkins, Recruiting Officer, was authorized to take enlistments for the 13th Regiment West Virginia Volunteers.*

Patriotic stationery header with invocation. Flying eagle image carrying a Union flag set against a shield surrounded by 34 stars. West Virginia, the 35th State to enter the Union, was not yet represented.

Orders, Colonel William R. Brown, commanding 13th Regiment Virginia (West) Virginia Infantry Volunteers, by Adjutants William I. Mathews and John S. Cunningham.

Order No. 7, dated January 9, 1863, Point Pleasant. Orders to Company C promoting Corporal William B. McCloud to sergeant; Private James S. Kelly to sergeant; Private James C. Davis to corporal; and Private John B. Patton to corporal.

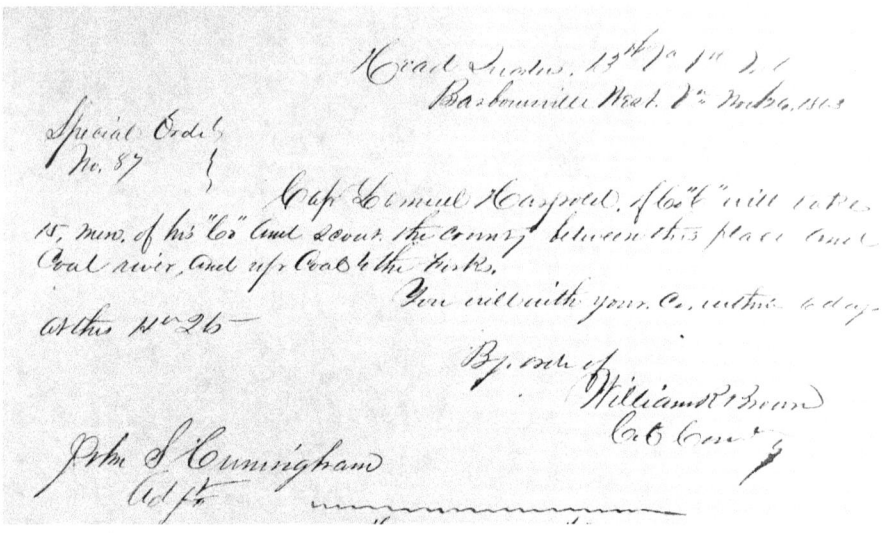

Special Orders No. 87, dated Head Quarters, 13th Virginia Vol. Infantry Barboursville, West Virginia, Nov. 26, 1863.

Captain Lemuel Harpold, commanding Company C, is ordered to take 15 men of his Company and "scout the country between Barboursville and Coal River and up Coal to the Fisks."

General Order No. 29, dated Head Quarters, 13th Virginia Volunteer Infantry,
Barboursville, December 13, 1863

This order was issued by Colonel W.R. Brown to curb absenteeism. "Due to continued
violations of military law incurred by soldiers absenting themselves from their respective
companies without leave, in future any non-commissioned or Soldier so absenting themselves shall be subject to punishment authorized by the 21st article of war." That is, "If any
officers or soldier shall leave his post or colors and go in search of plunder, he shall upon
being convicted thereof before a general court-martial, suffer death, or such other punishment as by a court-martial shall be inflicted."

Regular Gallipolis and Kanawha Packet

Vic 🚢 tor.

W. L. MADDY Captain,

C T! USLOW, Clerk,

LEAVES Gallipolis every Tuesday, Thursday and Saturday. Leaves Kanawha
every Monday, Wednesday and Friday. Every
attention paid to passengers and freight. Being a very light draught boat, (the lightest in
the trade,) she will run at all times when not
detained by ice.

The Staunch and Good Steamer

R. C. M. LOVELL,

Will ply between Syracuse and
Gallipolis daily, leaving Syracuse at
7 a m, Pomeroy at 8, and arriving at Gallipolis
at 12 a m. Returning, leaves Gallipolis at 1
P. M., Point Pleasant, at 2 p. m. All orders
promptly attended to. For freight or passage
apply on board.
W. D. PENNINGTON, Captain.

March, 6, 1862.—3m.

Steamboats travelling on the Ohio and Kanawha Rivers moved the army and supplies efficiently, far exceeding that of conveyance overland. Notices of the schedule of The Victor and the R.C.M. Lovell. Published in the Point Pleasant Weekly Register, March 6, 1862. Although it would require more research to confirm, it would seem that a fleet of steamboats was owned by one R.C.M. Lovell of Mason City, Mason County. The Victor No.2, named in reports about the battle of Point Pleasant, surely was one of this line of steamboats.

Endnotes

1   Major Geneneral Robert C. Schenk to A. Lincoln (dated Baltimore, Jan. 5, 1863), *Official Records of the Union and Confederate Armies*, Ser. I Vol. XXI (Washington: Government Printing Office, 1888), p. 947.

2   The manumission issue to be voted upon this day was also called the Wiley Amendment.

3   Returns for the military District of West Virginia (dated January 10, 1863) list the 2nd Kanawha Division, commanded by Brig. Gen. E.P. Scammon, with a total of 3,855 present for duty and 4,935 down from an aggregate present and absent in the last return of 7,562. The division had 10 pieces of field artillery and headquarters at Fayette Court House, W. Va. ("Abstract from return of the District of Western Va., Maj. Gen. Jacob D. Cox cmdg., for January 10, 1863, Headquarters Marietta, Ohio," O.R. Vol. XXI, p. 964.)

4   G.M. Bascom, Major and Assistant Adjutant General to Brig. Gen. E.P. Scammon (dated Hd. Qtrs., Marietta, Jan. 8, 1863), O.R. XXI, p. 960. The Giles, Fayette and Kanawha turnpike ran through Fayetteville south to the Virginia & Tennessee Railroad, an all-important lifeline connecting Richmond with the rest of Eastern Virginia and the South.

5   The *Journal* (Gallipolis, Ohio), dated January 16, 1863, p. 2, noted the departure of these "old regiments" adding that "[t]heir destination was a secret not to be ascertained, but from what we know of the officers and men, they will be sent where there is work for veterans like them to do. […] They all seemed glad to get out of the Kanawha Valley, and delighted at the prospect of having something more to do in the 'fighting line' than from the present status of affairs on the Kanawha, they are likely to have this winter."

6   Wright to Halleck, correspondences dated January 2-4, 1863, O.R. XXI, pp. 942-44.

7   Cox to McLean (dated Hd Qtrs. District of Western Va., Marietta, Ohio, Jan. 21, 1863), O.R. XXI, p. 992.

8   Cox, by Capt. S.L. Christie, Aide de Camp to Generals Cox and Scammon, at Fayette [W.] Va. (dated Marietta, Ohio, Jan. 22, 1863), O.R. XII, p. 997. Coals Mouth/Coalsmouth is known today as St. Albans. Camp Piatt was located on the bank of the Kanawha River about "twelve miles above Charleston [that is to say south of Charleston on the Kanawha] and opposite the old village of Brownstown." See Major General W.H. Powell, "Sinking Creek Valley Raid. In the Mountains of Va.," *The American Tribune* (pub. Indianapolis, Indiana), Friday, January 9, 1891, Vol. X, p. 3. When Crook was transferred with part of his command to Tennessee the "[o]ther Ohio and West Virginia regiments [such as the 5th and 13th] that had taken the place of the [Kanawha] division when it went East in 1862 took their places, and the command [of the division] devolved on Gen. Scammon." (R.B. Wilson, "Kanawha Division: Its Campaigns. [Part] I-cont.," *The National Tribune*, February 18, 1897, p. 2.)

9   Rutherford Birchard Hayes was born in Delaware, Ohio, on October 4, 1822. He was a graduate of Harvard Law School and before the outbreak of the rebellion practiced law in Marietta, Fremont and Cincinnati, Ohio. He was described by a fellow officer as being a "man of medium height, strong body, sandy hair, sanguine temperament, and was always self-possessed, and gentle in his intercourse with others. He was a most efficient officer and had the power to inspire his men to heroic deeds." See Joseph Warren

Keifer, *Slavery And Four Years of War A Political History of Slavery in the United States*, Vol. II 1863-1865 (New York & London: G.P. Putnam's Sons The Knickerbocker Press, 1900), p. 106. He had no military education prior to the war and seems to have been largely self-taught in military maneuver and strategy. Because he nearly always had too few troops with which to oppose the enemy, he came to appreciate the use of deception to surprise and outwit the enemy and minimize his own losses. In this latter respect, he learned much from his superior officer, General George Crook, whom he greatly admired for his considerable talents in this art. Hayes was wounded twice in service. After the war, he served a term in the U.S. Congress, three terms as Governor of Ohio, and he was elected 19th President of the United States serving from 1877 to 1881. During his term as president, he ended military occupation of the South, a decision which went far to promote reconciliation of the Southern people.

10   Tim McKinney, *The Civil War in Fayette County West Virginia* (Charleston, West Virginia: Pictorial Histories Publishing Company, 1988), pp. 180-81. Hayes wrote in his journal for Sunday, March 15, 1863: "My brigade is 23rd Ohio, 5th Va. Col. Ziegler, 13th Va. Col. Brown. [...] 13th at Coalsmouth and Hurricane Bridge." See Rutherford B. Hayes, Diaries 1834-May 1, 1865 (micro. Series 2/2), Rutherford B. Hayes Presidential Library, p. 22.

11   Maj. Gen. H.G. Wright to Brig. Gen. Julius White, cmdg. District of Eastern Kentucky (dated Hdqtrs. Dept. of the Ohio, Cincinnati, Ohio, Feb. 14, 1863), O. R. XXIII, Part II, p. 70.

12   After the formation of West Virginia, 44 companies of Home Guards were organized—in all about 2300 men. Home Guard units were more heavily concentrated in interior counties where guerrilla activities were more prevalent. Home Guard units, with more and less success and no small amount of irritation to local inhabitants, served a necessary role as an internal defense force.

13   Grotius, letter to the Gallispolis *Journal* (dated "Fort Pierpont, Ceredo, Wayne County, Va., Jan. 11, '63"), "Army Correspondence," *Journal*, January 22, 1863, [p.1].

14   Major General Sam Jones to A.T. Caperton (dated Hd. Qtrs. Dept. of Western Va., Dublin Depot, Jan. 17, 1863), O.R. XXI, p. 1095 and Jones to Brig. Gen. John S. Williams (dated Hd. Qtrs. Dept. of Western Va., Dublin Depot, Jan. 20, 1863), O.R. XXI, p. 1102.

15   Lee to Imboden (dated Hd. Qtrs. Army of Northern Va., Jan. 20, 1863), O.R. XXI, p. 1102.

16   "A Small Raid," Point Pleasant *Weekly Register*, January 15, 1863, [p. 2].

17   Joe Geiger, Jr., *Civil War in Cabell County, West Virginia 1861-1865* (Charleston, West Virginia: Pictorial Histories Publishing Co., 1991), p. 79. Guyandotte had for decades enjoyed strong commercial ties with Richmond and the South due to her overland connections. These included the extension of the James River and Kanawha turnpike to Guyandotte thirty years before and side-roads built connecting to it in the intervening years. The town's proximity to the Ohio River enabled the town to have a brisk trade by boat with the Southern States. In contrast to other towns in this section, and not surprisingly, Guyandotte's commercial and political ties came to be with the South. Surrounded by pro-Union "enemies," Secessionist sentiment and activity was so virulent there that after the "Guyandotte Massacre" of November 10, 1861, in which rebel forces and citizens attacked a Union recruiting force and citizens of the town, a sufficient number of Union troops were stationed there as an occupying force until the war was over.

18   Cox to Scammon (dated Hd Qtrs. District of Western Va., Marietta, Ohio, Jan. 22, 1863), O.R. XXI, pp. 997-98.

19   Ibid.

20   R.B. Wilson, "Kanawha Division: Its Campaigns. [Part] IV-continued," *The National Tribune*, February 25, 1897, [p.1]. As indicated by Wilson, the Kanawha Division operated in particularly difficult and broken terrain. The rigor of such challenging duty lent itself to turning out soldiers who tended to fight as individuals. As the great battles of 1864 would confirm, the duties performed in Western Virginia in 1863 were the cause of the extraordinary military discipline, stamina and and daring displayed in the mountains and on the battlefields of Eastern Virginia in 1864. In the din and roar of battle, officers and soldiers of the Kanawha Division time and time again held their lines to the last or fought themselves in the absence of orders in squads or by companies. Of particular

note in this regard is the 13th West Virginia, which on two documented occasions, amid the most tremendous firing, was the last regiment of the division to leave the field.

21   Francis Pierpont, "Adjutant General's Report for 1863" (dated State of West Va, Adjt. General's Office, Wheeling, January 18th, 1864), [n.p.]

22   Ibid., p. 5.

23   John H. Oley to H.J. Samuels, Adjutant General of Virginia (manu. dated "Wheeling 1863, 9 P.M."), Adjutant General's Papers—Miscellaneous, W.Va. State Archives, Charleston, courtesy of Terry Lowery. Oley had history with Keenan (or Kennan). On January 10, 1862, during the time that the 8th Va. Infantry Regiment was organized, Kennan was recommended for major and Oley was recommended for lieutenant colonel. See telegram (dated "Virginia, January 10th, 1862"), Registry #393 10W2 Part I, MTN Dept. E3124 Telegrams Sent Aug. 1861-July 1862 No. 2 Mtn., NARA.

24   George W. Tippett, publisher of the Point Pleasant *Register*, wrote to H.J. Samuels on January 30, 1863, regarding reimbursement for printing "Volunteer Enlistment" blanks, ordered by Lieutenant Colonel James R. Hall and Lieutenant William E. Feazel, recruiting officer; also "some Volunteer Posters for Lieutenant Feazel—amounting in all to about $31.00." These orders were placed with Tippett the previous September. Wrote Tippet to Adjutant Samuels: "Now I write to you for information as to how I shall proceed to get the above amount due me. Col. Hall told me that he was authorized to certify my account but since the killing of Lewis Wetzel by his father and his armed assault upon myself, he has obstinately refused to do so, I have written him twice but have received no answer in return. Lieut. Feazel is here. He says he will certify the account if you authorize him to do so." See George W. Tippett to H.J. Samuels, Adj. Gen. (manu. dated "Register Office, Point Pleasant, Va., Jan. 30th, 1863"), Letters to Samuels and Pierpont, WVSA. Tippett's letter to Samuels dated March 23, 1863, seems to indicate that the issue over reimbursement had been resolved. Tippett wrote to Samuels as follows: "Yours of the 1st of March has been received and in reply would say your action in respect to my acco[u]nt is approved by me. I am glad you did it for me, it is true the acco[u]nt was made about that time (September.) I hope you will try and process the amount of said bill for me as soon as possible, for it is hard to make collections at this time and the $29.50 would come in 'good play' at present. Any time you conveniently throw work into my hands, would like to have you do so cant you persuade some of your city Merchants to Advertise down this way. Anything that you may do in that way will not be forgotton at the proper time by your friend + well wisher Geo W Tippett publisher Register." George W. Tippett, publisher of the Point Pleasant *Weekly Register* to H.J. Samuels, Virginia State Adjutant General (manu. dated "March 23, 1863"), Letters to Samuels and Pierpont, WVSA. Another claim was made by Hutchinson McDaniel, operator of an inn at Point Pleasant, who wrote to Samuels requesting a reimbursement check for subsistence furnished to the 13th Virginia. See Hutchinson McDaniel to H.J. Samuels (manu. dated "Point Pleasant, Virginia May 18, 1863"), Uncatalogued 13th West Virginia Infantry Regiment Box, WVSA.

25   Field and Staff Muster Roll for January and February 1863, Record of Events Field and Staff 13th Va. Vol. Infantry and Regimental Return 13th Va. Vol. Infantry for January 1863 (micro. #594 Roll 196), NARA.

26   Company B, D and H Muster Rolls for Jan. and Feb. 1863, Record of Events Field and Staff (micro. #594 Roll 196), NARA and John Smoot, Co. B, Military Pension Records, NARA.

27   For this paragraph, see RG94 Morning Reports of James W. Johnson's Co. A 13th Regiment of Virginia Volunteers Army of the U.S., NARA, [n.p.]

28   RG94 Morning Reports Company A 13th Regt. Virginia Vols. for January 1863, [n.p.]

29   RG94 Morning Reports of Captain Milton Stewart's Co. B 13th Regiment Virginia Vols. for January 1863, NARA, [n.p.]

30   RG94 Morning Reports of Captain Van D. McDaniels' Co. C 13th Regiment Virginia Vols. for January 1863, NARA, [n.p.]

31   Private Nathaniel Burchfield had enlisted in the 13th Regiment on August 15, 1862, at Hartford City, Mason County. He was at enlistment, aged 34 years; "[f]ive feet and nine and one half inches tall" and working as a farmer. See Original Muster Roll of

Company D 13ᵗʰ V.V.I., Records of Mason County Soldiers #1541, W.V.U., Morgantown and RG94 Miscellaneous Records Descriptive Roll Co. D, NARA, [n.p.] For the charge of desertion (dated to January 10, 1863), see his case tried at Court Martial convened at Charleston, West Virginia, on November 28, 1863 (or Jan. 1864?). LL1439 #31 Private Nathaniel Birchfield, Co. D 13ᵗʰ V.V.I., NARA. The charge of desertion against Nathaniel Birchfield was removed. He was absent without leave from January 10 to April 1, 1863, because he had "substituted." See correspondence from the Adjutant General, Washington to W.Va. State Adjutant General (manu. dated "March 23, 1886"), Uncat. 13ᵗʰ W.Va. Inftry. Regt. Box, WVSA.

32   RG94 Morning Reports for Captain Simon Williams' Co. D of the 13ᵗʰ Regiment Virginia Vols. for January 1863, NARA, [n.p.]

33   RG94 Morning Reports for Captain A.F. McCown's Co. F of the 13ᵗʰ Regiment Virginia Vols. for January 1863, NARA, [n.p.]

34   Private Benjamin F. Kimberling was reported enlisted on January 5, 1863, at Point Pleasant, by J.B. Bumgardner and mustered in next day at the same place. Kimberling was a resident of Mason County; a farmer by occupation; and his age noted as 16 years. He seems to have deserted on March 6, 1863, from Hurricane Bridge. See RG94 Regimental Books Miscellaneous Records Descriptive Roll Co. H 13ᵗʰ Regiment V.V.V.I., Register of Deserters, NARA, [n.p.]

35   RG94 Regimental Books Morning Reports for Captain T.W. Hampton's Co. H of the 13ᵗʰ Regiment Virginia Vols. for January 1863, NARA, [n.p.]

36   RG94 Regimental Books Morning Reports for Lieutenant William Feazle's Co. I of the 13ᵗʰ Regiment Virginia Vols. for January 1863, NARA, [n.p.]

37   A memorandum from the Adjutant General's Office, Washington (dated "Dec. 24, 1870"), stated that there was no evidence on Company Rolls of John Young's enlistment muster in or service. M508 Compiled Military Service Records (microfilm Roll 211), NARA. Co. I Descriptive Rolls note only that Young "died at Post Hospital Point Pleasant on January 25ᵗʰ, 1863 of convulsions."

38   RG94 Regimental Books "Remarks" following Morning Reports of Co. I 13ᵗʰ Regiment Virginia Volunteer Army of the U.S. for January 1863, NARA, [n.p.]

39   RG94 "Remarks" following Morning Reports of Captain William Feazel's Co. I 13ᵗʰ Regt. Va. Vols. for March 1863, NARA, [n.p.]

40   W.W.H., dated "Headquarters, 13ᵗʰ Reg. V.V.I., January 27, 1863," Point Pleasant *Weekly Register*, January 29, 1863, [n.p.]

41   Col. W.R. Brown, by W.I. Mathews, Adjutant, Gen. Order No. 4 (manu. dated "Head Quarters 13ᵗʰ Reg[imen]t V[irgini]a Vol[unteer] I[nfantry], Point Pleasant, [Virginia], Jan[uary] 4ᵗʰ, 1863"), RG94 Regimental Books 13ᵗʰ Virginia Regiment Infantry Volunteers Order Book Co. C, NARA, [n.p.]

42   Col. W.R. Brown, by W.I. Mathews, Adj., Gen. Order No. 5 (manu. dated "Head Quarters 13ᵗʰ Reg[imen]t V[irgini]a V[olunteer] I[nfantry], Point Pleasant, [Virginia], Jan[uary] 5ᵗʰ, 1863"), RG94 Order Book Co. C, [n.p.]

43   Colonel W.R. Brown, by W.I. Mathews, Adj., General Order No. 6 (manu. dated "Head Quarters 13ᵗʰ Reg[imen]t V[irgini]a Vol[unteer] Inf[an]try, Point Pleasant, [Virginia], Jan[uary] 8ᵗʰ, 1863"), RG94 Order Book Co. C, [n.p.]

44   See CIRCULAR FROM THE PROVOST MARSHAL GENERAL OF THE STATE (dated "Office Provost Marshal General State of Virginia, Wheeling, Jan. 15. 1863"), Gallipolis *Journal*, Feb. 12, 1863, p. 4.

45   Colonel W.R. Brown, by W.I. Mathews, Adj., General Order No. 7 (manu. dated "Head Quarters 13ᵗʰ Reg[imen]t V[irgini]a V[olunteer] I[nfantry], Point Pleasant, [Virginia], Jan[uary] 9ᵗʰ, 1863"), RG94 Order Book Co. C, [n.p.]

46   Captain J.V. Young to Col. William Brown (manu. dated "Hurricane Bridge Putnam Co W.Va. 22 [J] 63"), RG94, NARA, courtesy of Darl Stephenson.

47   R. Brooks, Lieut., Co. G [11th/13th V.V.I.], to Lt. Col. J.R. Hall, cmdg. "Post Winfield, Putnam Co. Va." (manu. dated "Jan. 6, 1863"), RG94 Box 4839 Miscellaneous unbound regimental papers 13th West Virginia Letters Adjutant General's Office, Muster Rolls, Returns Regimental Papers, NARA.

48   Young was a Methodist, a form of Protestant Christianity, which was widespread in the Northwestern (loyal) counties of Virginia. A number of scholars feel that Methodism tipped the scales among its followers, encouraging them to remain loyal to the national government against the slave-holding South. Secessionists derogatorily termed the Methodist Episcopal Church, "the abolition church."

49   J.V. Young to his wife, Paulina Marshall Franklin Young (dated "Winfield, W. Va., Jan. 25, 1863"), Young Civil War Papers (Vol. 31 Box 10), trans. WVU, Morgantown.

50   A.W. Campbell, editor of the *Wheeling Daily Intelligencer*, was one of the great supporters of the Union and the new State movement in Western Virginia, and his daily paper was a forum for all news and issues related to these two causes.

51   W.W. Harper to the Editors of the Wheeling *Daily Intelligencer* (dated "Headq'rs 13th Reg't Va. Vols., Point Pleasant, Jan. 1, 1863"), *Intelligencer*, January 10, 1863, p. 2.

52   These measures included relief of the heavy tax debt on the people by prohibiting the "log-rolling bargain and sale" way of running up the State debt; inducing professionals to emigrate and remain in the area by radical reduction of the tax levied on professional people when purchasing a professional license; and eradication of East Virginia's system of land law, which had resulted in a hopeless tangle of property claims that effectively deterred outside entrepreneurs from investing in Western Virginia and defrauded purchasers of land west of the Alleghenies. The only beneficiary of the latter outdated system was the State treasury at Richmond. For eighty years prior to the Civil War, Richmond had termed the lands west of the Alleghenies as "waste and unappropriated," and record of claims and patents/titles issued for these lands was done with wanton disregard of individual property rights. These lands were sold to more than one person at a time (in some cases one parcel of land had five or six different patents or claims) with the result that property-ownership of the whole country became a snarl of claims. Two, three and often more owners, paid taxes on the same land at the same time, which enriched the State treasury but defrauded the people of the west. To add insult to injury, there were two different statute of limitations for the nonpayment of taxes on land. There was one for land west of the Alleghenies and another (a longer period in which to bring claims) for the lands east of the mountains, before the parcel of land was sold out from under purchaser to pay the State's tax lien. See Granville Parker, *The Formation of the State of West Virginia and other Incidents of the late Civil War with Remarks on Subjects of Public Interest Arising Since the War Closed* (Wellsburg, W.Va.: Glass and Son, Book and Job Printers, 1875), p. 125.

53   F.H. Pierpont to the Wheeling *Daily Intelligencer*, trans. of the orig., 9 pp. in length, WVU. See also below, W.W. Harper, Sergeant, Co. K 13th W.V.V.I., to the Wheeling *Intelligencer*, dated "Point Pleasant, Va., Feb. 7, '63."

54   William R. Brown, Col. to H.J. Samuels, Adj. Gen. of Virginia (manu. dated "Head Quarters 13th Va. Vol. Infty., Point Pleasant, Jan. 16, 1863"), Samuels-Pierpont Papers, WVSA.

55   Mark E. Robinson, Private, Co. E 13th Regiment Virginia Volunteers to his father, John Robinson (manu. dated "Point Pleasant, Virginia, January 16, 1863"), Mark E. Robinson, Military Pension Record, NARA. Mark was 18 years old when he enlisted, his father's youngest son. At the time of his enlistment, Mark's father, John, had "2 good horses + harness. These the rebels took and pretty much everything else he had, which left the family destitute." John sent three sons to the Federal army. Sons Joel G. and Moses both died in 1864. Private Joel G. Robison served together with Mark in Co. E. Joel died of typhoid fever on December 27, 1864, at Winchester, in the Shenandoah Valley. He was buried at Winchester, Virginia. Mark E., our correspondent, died on February 20, 1865, at Cumberland, Maryland, also of typhoid fever. He was buried at Cumberland. See RG94 Regimental Books Miscellaneous Records Descriptive Roll Co. E 13th Regiment V.V.V.I., Register of Deaths, NARA, [n.p.] and "Father's Affadavit" (dated "Kanawha Co. Mar. 28, 1881"), Robinson, Military Pension File, NARA, [n.p.]

56   James R. Hall, Lieut. Col., 13ᵗʰ Va. Infantry to H.J. Samuels, Adj. Gen. of Va. (manu. dated "Head Quarters Detachment 13ᵗʰ Va. Winfield, Va. January 17ᵗʰ, 1863"), Uncat. 13ᵗʰ W.Va. Inftry. Regt. Box, WVSA.

57   RG94 Misc. Records Descriptive Roll Co. E, Register of Deaths, NARA, [n.p.]

58   West Virginia Clothing Books Company C 13ᵗʰ Infantry, manu., WVSA, Charleston, [n.p.]

59   Brig. Genl. Crook, by R.R. Kennedy, A.A.G. (manu. dated "Head Quarters 1ˢᵗ Kanawha Div., Charleston, Va., Jan. 23, 1863"), RG94 Box 4838 Miscellaneous unbound regimental papers 13ᵗʰ West Virginia Special Orders and General Orders, NARA.

60   A.F. McCown, Captain, Co. F 13ᵗʰ Reg. V.V.I. to H.J. Samuels, Adj. General of Va. (manu.dated "Point Pleasant, Va. January 23, 1863"), Uncat. 13ᵗʰ W.Va. Inftry. Regt. Box WVSA.

61   W.J. Nelson, H[ospital] S[teward], U.S.V. to Adjutant W.[I.] Mathews, 13ᵗʰ Va. Vol. Inft. (manu. dated "Gen. Hospital Point Pleasant Va., Jan. 23, 1863"), RG94 Box 4839 Misc. unbound regimental papers 13ᵗʰ West Virginia, NARA.

62   Isaac West, Military Pension File, NARA.

63   John V. Young to his wife, Paulina (dated "Winfield, W.Va., Jan. 24, 1863"), Young Civil War Papers (Vol. 31 Box 10), trans. p. 55.

64   C. Mollohan, attorney to the Adjutant General State of W.Va. (manu. dated "Gallipolis, Ohio, Jan. 10, 1865"), Uncat. 13ᵗʰ W.Va. Inftry. Regt. Box, WVSA.

65   Memorandum from A.G.O. (dated "Washington, Dec. 24, 1870"), M508 Compiled Military Service Records (micro. Roll 211), NARA.

66   J.D.C. [John Deriah Carter] "to a member of the Legislature" (dated "Point Pleasant, West Virginia, Jan. 23 [26 or 28], 1863"), Wheeling *Daily Intelligencer*, January 31, 1863, p. 1.

67   Young to Paulina (dated "Winfield, W. Va., Jan. 28, 1863"), Young Civil War Papers (Vol. 31 Box 10), trans. p. 57.

68   Field History of the 13ᵗʰ Regiment West Virginia Infantry, addressed to F. Pierpont, Adj. Genl. W.Va. (manu. dated "Head Quarters 13ᵗʰ Regt. W.Va. Vol. Inft., Barboursville, W.Va., Jan'y 13ᵗʰ, 1864"), Uncat. 13ᵗʰ W.Va. Inftry. Regt. Box, WVSA, [n.p.]

69   See Evelyn Abraham Benson, comp., *With the Army of West Virginia. Reminiscences & Letters of Lt. James Abraham Pennsylvania Dragoons Company A First Regiment, Virginia Cavalry. Abraham Archives No. 1* (Evelyn A. Benson: Lancaster, Pa., 1974), p. 40.

70   Young to Paulina (dated "Winfield, Jan. 29, 1863"), Young Civil War Papers, trans. p. 58.

71   George W. Tippett to Jeff [H.] Samuels, Adj. Gen. (manu. "dated Register Office, Point Pleasant, Va., Jan. 30ᵗʰ, 1863"), Samuels-Pierpont Papers, WVSA.

72   Manu., Uncat. 13ᵗʰ W.Va. Inftry. Regt. Box, WVSA.

73   George W. Tippett, ed., "The Twenty-Ninth Year of the Register," *Weekly Register*, June 11, 1890, [p. 2].

74   Demarcus Ward to his father, William R. Ward (manu. dated "Point Pleasant Jan the 31ˢᵗ 1863"), Ward, Military Pension File, NARA.

75   Regimental Return 13ᵗʰ Va. Vol. Infantry for February 1863 (micro. #594 Roll 196), NARA.

76   Col. Wm. R. Brown to Gov. F.H. Pierpont (manu. dated "Hd. Qtrs. 13ᵗʰ Barboursville, W. Va. Feb. 2, 1864"), Uncat. 13ᵗʰ W.Va. Infry. Regt. Box.

77   Brig. Gen. E.P. Scammon, by Jas. L. Botsford, Capt. and Asst. Adjt Genl. to Col. W.R. Brown, 13th Reg. Va. Vol. Inf. (manu. dated "Head Quarters Dist. Kanawha, Charleston, W. Va. Feby. 6, 186[3]"), RG94 Box 4839 Misc. unbound regimental papers 13th W.Va.

78   John [M.?] Butler [Butter?], P.M. & M.S.K. to Col. Wm. R. Brown, 13th Va. Vol. Infty., Point Pleasant, Va., in care of Lieut. A.R. Buffington, Wheeling, Va. (manu. dated "Allegheny Arsenal, Pa. Pittsburgh P. O., February 7, 1863"), RG94 Box 4839 Misc. unbound regimental papers 13th W.Va. The Allegheny Arsenal was built in 1814 in Lawrenceville, near Pittsburg.

79   Maj. Gen. J.D. Cox, by G.M. Bascom, Maj. + A.A.G to Col. Wm. R. Brown, Cmdg. 13th Va. Vol. I., comdg post P[oin]t Pleasant ,Va. (manu. dated "Head Quarters Dist of W Va Marietta O. Feb 20, 1863"), RG94 Box 4839 Misc. unbound regimental papers 13th W.Va. and W.H. Tomlinson, Commonwealth [of Virginia], Attorney to Maj. Gen. Cox (manu. dated "Point Pleasant Mason Co Virginia Feb 7th 1863"), RG94 Miscellaneous Department of W.Va. Papers, NARA.

80   [Editorial], "Thirteenth," Point Pleasant *Weekly Register*, Thursday, February 12, 1863, [n.p.]

81   Ibid.

82   "The Paymaster," Point Pleasant *Weekly Register*, February 19, 1863, Vol. 1 No. 49, [n.p.]

83   Mark E. Robinson, Military Pension File, NARA.

84   "Dry Up," Point Pleasant *Weekly Register*, February 19, 1863, Vol. 1 No. 49, [n.p.]

85   "Something Wrong," *Weekly Register*, February 19, 1863, [p. 1].

86   Harper's casual reference to these powerful political leaders from the South calls for brief explanation. Jeff Davis is, of course, Jefferson F. Davis, graduate of West Point; veteran of the Mexican-American War; politician from Mississippi; and only President of the Confederate States of America. Robert Barnwell Rhett was a prominent politician from South Carolina, who advocated for State's rights and secession from the United States Union after Abraham Lincoln was elected President. Rhett, Yancy, Keit, Toombs, Wigfall and Edmund Ruffin, a Virginian, who, for some reason Harper did not name, became known as the group, the 'Fire-Eaters' because of their aggressive and outspoken support of States' rights, 'chattel ownership' and secession. After the war was lost to the North, Ruffin committed suicide. William L. Yancey was born August 10, 1814, and died July 27, 1863. He served as Senator in the Confederate States Senate. Perhaps Harper had not heard that Yancey had 'passed beyond the veil.' Howell Cobb was lawyer and politician from Georgia, who served as Speaker of the U.S. House of Representatives 1849-51. He advocated for the immediate secession of his State upon Lincoln's election to the U.S. presidency and supported the extension of slavery into new territories. Cobb served as President of the Provisional Confederate Congress, which convened in Montgomery, Alabama, 1861-62, and as major general in the Confederate States Army. Laurence M. Keit served as Representative from South Carolina in the U.S. Congress. He then served with the Provisional Congress of the Confederacy convened in Montgomery and in the Confederate States Army. He was wounded at the battle of Cold Harbor, fought near Richmond, Virginia, and died of his wounds the next day on June 4, 1864. Henry Wise served as Representative from Virginia in the U.S. Congress from 1833-1844 and as 33rd Governor of Virginia 1856-1860. He was instrumental in Virginia's move to secede and served with the rank of brigadier general in the Army of Northern Virginia. John Letcher was a lawyer, newspaper editor and politician from Lexington, Virginia, who served as Representative from Virginia to the U.S. Congress 1851-1859 and during the Civil War as the 34th Governor of Virginia. Letcher, unlike the others in the group named by Harper, took a more middle of the road stance to slavery advocating for gradual manumission of all slaves. Robert A. Toombs was a politician from Georgia, who served as a key figure in the formation of the Confederacy. He served as the first Secretary of State of the Confederate States and during the Civil War as brigadier general. He was wounded at Antietam. Roger A. Pryor was another Virginian, a newspaper editor, lawyer and politician, who was elected to office in U.S. and Confederate States governments. He served in the U.S. Congress resigning his seat upon Lincoln's election to the presidency. He became known for his fiery speeches in favor of secession. He served in the Confederate Army as colonel and brigadier general. After the war, he moved to New York with his wife, where he was able to reestablish himself and pull himself out of poverty. He died in New York in 1919. Louis T. Wigfall was a powerful speaker and political leader from Texas. He served as Senator from Texas 1862-1865 in the Confederate State Congress and

briefly as a brigadier general. He was uncompromising in his stance upon preserving and expanding aristocratic agrarian society with its reliance upon slave labor.

87   W.W.H. to the Wheeling *Daily Intelligencer* (dated "Point Pleasant, Va., Feb. 7, '63"), published as "What Voting Against the New State Will Do!" in the Point Pleasant *Weekly Register*, Vol. 2 No. 50, March 4, 1863 [this dated marked out by hand and "February 21" written in], [p.1].

88   Col. W.R. Brown et al. to H.J. Samuels, Adj. Genl. State of Va. (dated "Head Quarters 13th Reg. V.V.I., Point Pleasant, Feb. 15th, 1863"), Uncat. 13th W.Va. Inftry. Regt. Box.

89   F.H. Pierpont to Col. [W. R.] Brown, 13th Va., Point Pleasant (manu. dated "The Commonwealth of Virginia. Executive Department, Wheeling, Feb. 19, 1863"), RG94 Box 4839 Misc. unbound regimental papers 13th W.Va.

90   Jos. Darr, Maj[or] + P[rovost] M[arshal] G. to Comd'g Officer, Point Pleasant, Va. (manu. dated "Head Quarters, Office Prov. Mar. Genl of Va., Wheeling, Feb'y 21st, 1863"), RG94 Box 4839 Misc. unbound regimental papers 13th W.Va.

91   Col. W.R. Brown, by W.I. Mathews, Adjutant, Special Orders No. 3 (manu. dated "Head Quarters 13th V[irgini]a Vol[unteer] Inf[an]t[r]y, Point Pleasant, V[irgini]a, Feb[ruary] 26th, 1863"), RG94 13th Regt. Inftry. Vols. Order Book Co. C, NARA, [n.p.]

92   Col. C.B. White, by James L. Botsford, Captain and A.A.A.G., General Orders No. 3 (manu. dated "Head Quarters District Kanawha, Charleson, W.Va., Feb. 26, 1863"), RG94 Box 4838 Misc. unbound regimental papers 13th W.Va., Spec. Orders and Gen. Orders, NARA.

93   Samuel Preck, Asst. Adjutant General, War Dept. to Commanding Officer 13th Virginia Vols., [Orders] (manu. dated "Adjutant General's Office, Washington D. C., February 27, 1863"), RG94 Box 4838 Misc. unbound regimental papers 13th W.Va., Circulars, NARA.

94   Theophilus Maher had been serving as 2nd Lieutenant of Co. A since his appointment on July 28, 1862. His resignation was accepted in Special Order No. 54 Headquarters Department of the Ohio, by order of Major General Horatio G. Wright, to take effect on the first day of February 1863. See RG94 Misc. Records, List of Commissioned Officers of Co. A 13th Regt. W.V.V.I., NARA, [p 1].

95   RG94 Morning Reports of James W. Johnson's Co. A 13th Regiment of Virginia Volunteers Army of the U.S. for February 1863, NARA, [n.p.] and entry for February 28, 1863, in Record of Events Field and Staff, Muster Roll Co. A 13th Va. Vol. Infantry (micro. #594 Roll 196), NARA.

96   Jesse Hart, Private, Company B 13th West Virginia Infantry Regiment to his father (manu. dated "herican brig Putnam county Virginia February the 24, 1863") Jesse Hart, Military Pension File, NARA.

97   RG94 Morning Reports of Captain Milton Stewart's Co. B 13th Regiment Virginia Vols. for February 1863, NARA, [n.p.] and RG94 Regimental Books Miscellaneous Records Descriptive Roll Co. B. 13th Regiment V.V.V.I., Register of Men Discharged, NARA, [n.p.]

98   RG94 Records of the Adjutant General's Office Carded Medical Records Vols. Mexican and Civil War 1846-1865 Entry 534 Box 3849, NARA.

99   RG94 Misc. Records Descriptive Roll Co. C. 13th Regiment V.V.V.I., Register of Deaths, [n.p.]

100   RG94 Morning Reports of Captain Van D. McDaniels's Co. C 13th Regiment Virginia Vols. for February 1863, [n.p.]; RG94 Misc. Records Descriptive Roll Co. C 13th Regiment V.V.V.I., Register of Deaths, [n.p.]; and James R. Dicken, Military Pension File, NARA.

101   RG94 Morning Reports of Captain Simon Williams' Co. D 13th Regiment Virginia Vols. for February 1863, NARA, [n.p.]

102   John Pauley, Military Pension File, NARA; RG94 Regimental Books Misc. Records Descriptive Roll Co. E 13th Regiment V.V.V.I., Register of Deaths, NARA, [n.p.]; and RG94 Owen B. Cark, Hospital Cards 13th Virginia (West) Volunteer Infantry Regiment.

103   RG94 Morning Reports of Captain A.F. McCown's Co. F 13th Regiment Virginia Vols. for February 1863, NARA, [n.p.]

104   Herbert L. Roush, Sr., *If Thou Wilt Remember* (Lowell, Michigan: Modern Printing, 1981), p. 26.

105   Colonel W.R. Brown, by W.I. Mathews, Adj., Special Orders No. 2 (manu. dated "Head Quarters 13th V[irgini]a Vol[unteer] Inf[an]t[r]ly, Point Pleasant, V[irgini]a, Feb[ruary] 10th, 1863"), RG94 Company H 13th Virginia Regiment Infantry Vols. Order Book, NARA, [n.p.]

106   Letter to the Adjutant General of W.Va. (manu. dated "August 10th, 1889"), Uncat. 13th W.Va. Inftry. Regt., [n.p.]

107   Allen returned from desertion on April 1, 1863. See RG94 Morning Reports Co. H 13th Regiment Virginia Volunteers for April 1863, NARA, [n.p.]

108   RG94 Morning Reports Co. H 13th Regiment Virginia Vols. for February 1863, [n.p.]

109   The Virginia House was a popular hotel located on the corner of 1st and Front Street at the head of the wharf and ferry landing in Point Pleasant. It had been built of "hand manufactured brick, with wide halls, a great dining room, large parlor and spacious bed rooms." ("Looking Backward Eighty Years. Reminiscences As Gleaned from Mr. Hiram H. Swallow and others Who Remembers the Best of this Locality Article IV," *The State Gazette*, Point Pleasant, W. Va., August 19, 1909, in *History of Mason County, W. Va.* Accession #203668, Call #RL 975.433H Huntington Public Library, [p. 1].) The Virginia House had been closed but was reopened on August 11, 1862, by Colonel J.P.R.B. Smith (John or Jonas Peter Roman Bureau Smith, named for the distinguished French emigrant to Gallipolis by that name), the new proprietor of the inn and Colonel of the reorganized 106th Mason County Militia. ("Virginia House," Point Pleasant *Weekly Register*, August 14, 1862, [p. 2].)

110   T.W. Hampton to H.J. Samuels (manu. dated "Head Quarters 13th Regt. V.V.I., Point Pleasant Feb. 7, 1863"), Uncat. 13th W.Va. Inftry. Regt. Box.

111   RG94 Misc. Records Descriptive Roll Co. H. 13th Regiment V.V.V.I., [n.p.]

112   RG94 List of Commissioned Officers, [n.p.]

113   Col. Wm. R. Brown to Lieut[enant] and Recruiting Officer W.E. Feazel (manu. dated "Hd. Qtrs. 13th Pt. Pleasant Feb. 1, 1863"), Uncat. 13th W.Va. Inftry. Regt. Box.

114   Colonel W.R. Brown, by W.I. Mathews, Adj. Special Orders No. 1 (manu. dated "Head Quarters 13th V[irgini]a Vol[unteer] Inf[an]t[r]ly, Point Pleasant, V[irgini]a, Feb[ruary] 1st, 1863"), RG94 13th Regt. Inftry. Vols. Order Book Co. C, [n.p.]

115   Lieut. Wm. E. Feazel, Recruiting Officer, "Recruits Wanted for 13th Regiment," Point Pleasant *Weekly Register*, Thursday, February 26, 1863, Vol. 1 No. 50, [n.p.] It should be added here that at no time during the war was it necessary to conscript men in Mason County for the 13th Regiment.

116   RG94 Morning Reports for Lieutenant William Feazels Company I of the 13th Regiment of Virginia Volunteers Army of the United States, Colonel William R. Brown for the Month for the month of February 1863, NARA, [n.p.]

117   Col. William R. Brown to Adj. Gen. H.J. Samuels (manu. dated "Head Quarters, Point Pleasant, Feb. 9, 1863"), Uncat. 13th W.Va. Inftry. Regt. Box.

118   There is, however, no official record to document that Butler was in fact granted a recruiting commission for the 13th or that he was enrolled in the 13th Regiment.

119   Austin D. Butler, *Butler Generation in America* (undated manu.), pp. 6-7, courtesy of Terry Lowery.

120   Annual Return of Casualties + c. 13th W. Va. Infy. Oct. 10/62 to Dec. 31/63 (manu. "Return completed Jan. 13, 1864"), RG94 Box 4838 Misc. unbound regimental papers 13th W.Va., NARA.

121   Field History 13th Regiment West Virginia Infantry (manu. addressed to F. Pierpont, Adj. Genl. W. Va. and dated "Head Quarters 13th Regt. W.Va. Vol. Inft., Barboursville, W. Va., Jan'y 13th, 1864"), Uncat. 13th W.Va. Inftry. Regt. Box, WVSA, [n.p.]

122   James L. Botsford, Captain and A.A.G. to Captain J.V. Young, cmdg. at Coalsmouth (dated "Headquarters District of Kanawha, Charleston W.Va. Feb. 10, 1863"), Civil War Scrapbook. Accession #859, Regional History Library, WVU.

123   "The 13th West Va. Infantry List of Engagements," *The Journal* (imprint: Charleston, W.Va.), July 5, 1865, p. [2] and "The Thirteenth West Va. Infantry," Wheeling *Daily Intelligencer*, June 24, 1865, [p. 3].

124   Brig. Genl. Scammon to Col. Brown, cmdg. Post Pt. Pleasant (manu. dated "Hd. Qtrs. Dist. Kanawha, Charleston, Va., Feb. 10, 1863"), RG94 Box 4839 Misc. unbound regimental papers 13th W.Va..

125   J.V. Young, com'dg Post to Paulina, (dated "Coals Mouth, Feb. 21, 1863"), Young Civil War Papers, Vol. 31 Box 10, trans. p. 8.

126   J.M.R. to the Editor of the Point Pleasant *Weekly Register* (dated "Union Ridge, Mason county, Va., February 11th, 1863"), *Weekly Register*, March 4, 1863, [p. 4].

127   Ibid.

128   Letter to the Editors of the Wheeling *Daily Intelligencer* (dated "Headquarters 13th Regt. V.V.I., Point Pleasant [Virginia] Feb. 17, 1863"), reprinted as "From Point Pleasant. Headquarters 13th Regt. V.V.I." in the Point Pleasant *Weekly Register*, February 24, 1863, [p. 1].

129   "Guerrillas are again becoming troublesome in West Virginia," Gallipolis *Journal*, March 5, 1863, p. 2.

130   Ibid.

131   Milton Stewart was the brother of Robert L. Stewart, who wrote for the Gallipolis *Journal* and he may well have received the detailed information of Col. Brown's actions in this instance from his brother, Capt.Milton Stewart, commanding Co. B 13th West Virginia..

132   [Untitled], reprinted from the Gallipolis, Ohio, *Dispatch* in the Point Pleasant *Weekly Register*, March 5, 1863, [p. 2].

133   [Editorial], Point Pleasant *Weekly Register*, February 12, 1863, p. 2.

134   Gallipolis *Journal*, March 5, 1863, p. 2.

135   [Untitled], Point Pleasant *Weekly Register*, March 5, 1863, [p. 2].

136   "Southern Chivalry," Point Pleasant *Weekly Register*, February 12, 1863, [n.p.].

137   "One of the Fifth," letter from a soldier in the 5th Virginia (Union) Regiment, "Two Days Scout on Guyan River" (no place or date given but probably written from Ceredo, where the regiment was stationed), *Ironton Register*, March 26, 1863, p. 2 and reprinted as "Capture of Nine Murderers of Gibson Letter From the 5th Virginia Two Days Scout on Guyan River," *Gallipolis Journal*, April 2, 1863, p. 3.

138   Company H was also stationed at Hurricane Bridge at this time but no detachments seem to have been made from this undersized company.

139   Maj. Gen. H.G. Wright to Brig. Gen. Julius White, cmdg. District of Eastern Ky. (dated Hdqtrs. Dept. of the Ohio, Cincinnati, Ohio, Feb. 14, 1863), O.R. XXIII, Part II, p. 70. The town of Ceredo had just recently come into existence having been founded largely by settlers from New England as a social engineering experiment to undermine the slavery base in Western Virginia. It consisted of only several hundred people as many had by this time 'pulled up stakes' and moved after Virginia's adoption of the Secession Ordinance in 1861. The 5th W.Va. Infantry was largely recruited from Ceredo.

140   Grotius, a soldier of the 5th Virginia Regiment stationed at Camp Pierpont, Ceredo, letter to the Editor of the *Journal* (dated "Fort Pierpont, Ceredo, Wayne County, Va., Jan. 11, 1863"), "Army Correspondence," Gallipolis *Journal*, January 22, 1863, p. 1.

141   Ibid.

142   This reference may give some indication as to the time of day. In this section of the country, 'dinner' most usually referred to the midday meal, while 'supper' was the evening meal. Dinner was customarily the more substantial meal, consisting of cooked meats, vegetables in season, starches of all kinds, beans, pickles stewed fruit, etc. Supper consisted of lighter fare such as cold meats, bread and butter and stewed fruit.

143   The Marion Adkins mentioned here may be Private Marion Adkins of the 34th Battalion of Virginia Cavalry from Lincoln County. See Jack L. Dickinson, *Tattered Uniforms and Bright Bayonets West Virginia's Confederate Soldiers* (Huntington, West Virginia: Marshall University Library Associates, 1995), p. 70.

144   At that point in time, dry goods of almost every description were reportedly bringing very high prices. A calico dress cost "as much of an object as a silken one used to be." Point Pleasant *Weekly Register*, February 12, 1863, [n.p.]

145   "One of the Fifth," "Two Days Scout on Guyan River," *Ironton Register*, March 26, 1863, p. 2 and reprinted as "Capture of Nine Murderers of Gibson," in the *Gallipolis Journal*, April 2, 1863 p. 3.

146   A.I. Boreman to F.H. Peirpont (dated "Parkersburg, February 27, 1863"), trans. photostat copy no. 38, *Pierpont Papers* A + M9 F.H. Pierpont Correspondence 1863 -1865, WVU, pp. 2-3.

147   Ibid., [p.1].

148   Ibid., pp. 1-2.

149   Ibid., pp. 2-3.

150   Maj. Gen. Ambrose E. Burnside, U.S. Army, Cmdg. Department of the Ohio, Report of operations March 25—August 10, 1863, O.R., p. 12.

151   An arrest of one key figure was Clement L. Vallandigham, arrested May 6, 1863, at Dayton, Ohio. Honorable Vallandigham, of Ohio, had headed up a secret anti-war organization known as the Sons of Liberty. They were part of a larger group known as Copperheads as they were termed by their detractors. They were so named for the copper pennies they wore as their emblem and the venomous snake of the same name lying in wait to strike. They termed themselves Butternuts or "Peace Democrats." The Peace Democrats were a small but motivated minority. They were more conciliatory towards the South; maintained that the South had been provoked into secession by abolitionist Republicans; and they called for immediate end to the war, whatever the price. Their influence waxed and waned with the fortunes of war. They were annoyingly most conspicuous when the Union had suffered some defeat or string of setbacks, when they called for peace and argued that it was again made obvious that there could be no military resolution to the conflict. There was a greater concentration of them in the Midwestern States such as Indiana, Ohio and Pennsylvania—Northern States whose residents had both more extensive dealings with the South and who harbored a natural distrust of the rich and powerful States of the Northeast. Copperheads were particularly hated by pro-Union West Virginians (many of whom were Democrats themseves), who wanted no slur of a connection to the abolitionist movement and they were especially loathed by her volunteer soldiers in harm's way and fighting in their home counties. Their fight was not for racial equality but for equal opportunity, equal rights and equal application of law. West Virginia civilians and military were highly critical of their neighbor States across the Ohio River for condoning Copperhead activities and permitting a proliferation of their platform.

152   "Petition of the Members of the Convention at Wheeling Recommending William E. Feazel for Major 13th Regt. Va. Vol. Inftry." to Governor of Virginia F.H. Pierpont (manu. dated "Feb. 1863"), Uncat. 13th W.Va. Inftry. Regt. Box, WVSA.

153   Col. William R. Brown to H.J. Samuels (manu. dated "Head Quarters 13th Va. Point Pleasant, Va Feb. 23, 1863"), Uncat. 13th W. Va. Inftry. Box, WVSA.

154   James R. Hall, Lieut. Col., 13th Va. Infantry to H.J. Samuels, Adj. Gen. of Va. (manu. dated "Head Quarters 13th Va. Infantry, Point Pleasant, Va., March 2, 1863"), Uncat. 13th W. Va. Inftry. Box.

155   Lt. Col. James R. Hall to Lieut. W.E. Feazel (manu. dated "Head Quarters 13th Va. Infty., Point Pleasant, Va., March 6, 1863"), Uncat. 13th W.Va. Inftry. Box.

156   J.B. Ackley to Hon. D[aniel] Polsley (manu. dated "Racine [Ohio,] Feb[ruar]y 15, 1863"), Uncat. 13th W.Va. Inftry. Box. This appears to be the first of a series of recommendations and endorsements written on behalf of Captain Carson.

157   Ibid.

158   Wm. E. Feazel, 2nd Lieutenant, 13th Virginia Volunteer Infantry, Rec[rui]ting officer to F.H. Peirpoint, Governor of Virginia (manu. dated "P[oin]t Pleasant, V[irgini]a, March 2nd, 1863"), Uncat. 13th W.Va. Inftry. Regt. Box.

159   Jas R. Hall, L[ieutenan]t Col[onel] 13th V[irgini]a Inf[an]t[r]y to F.H. Peirpoint, Gov[ernor] of V[irgini]a (manu. dated "Head Quarters 13th Va. Infty. Pt. Pleasant, Va. March 2d, 1863"), Uncat. 13th W.Va. Inftry. Regt. Box.

160   W.R. Brown, Col., 13th Vol. Infantry to [illeg.] Samuels, Adjutant General, State of Virginia, at Wheeling (dated "Head Quarters 13th Regt. Va. Vol. Inftry., Point Pleasant, Va., March 9, 1863"), Letters to Samuels and Pierpont, WVSA.

161   Three brothers with surname Fry served in the 13th Regiment. They were Marion and Peter Fry of Co. D and John W. Fry of Co. F. Company descriptive rolls indicate that they were each about six feet tall or more, and all were born in Mason County and worked as farmers before their enlistment in 1862. They were the sons of Peter and Hannah (Knapp) Fry, who had settled in Mason County in 1817.

162   Washington Davis, comp., "Camp-Fire XVII," *Camp-Fire Chats of the Civil War; Being the Incident, Adventure & Wayside Exploit of the Bivouac & Battlefield, as related by Veteran Soldiers Themselves* (Mansfield, Ohio: Union Publishing Co., 1887), pp. 204-10.

163   The Conscription Act passed March 3, 1863, was much criticized because it made distinction between rich and poor. The bill designated that a drafted man could sidestep his obligation to serve if he could find 'a substitute,' who would fight in his place and to whom he could pay up to but not more than $300.00. The expression "rich man's war, poor man's fight" came into usage with the passage of this act. The old laws regarding mandatory military service that exempted "members of Congress, custom house officers and clerks, postmasters and their clerks, professors and students at college, clergymen, judges, and many other officials" were superceded. Under the new law all classes and persons were liable to serve. Few exemptions were permitted. Those exempted were "only sons" or "only brothers," who supported their widowed mothers, aged or infirm parents, or children under twelve years without father or mother." See Samuel M. Schmucker, *A History of the Civil War in the United States*,Vol 2 (Philadelphia: Bradley & Co.66 North Fourth Street and Baltimore: Jones Brothers & Co. 71 West Fayette St., 1863), pp. 294-95.

164   This county meeting on the upcoming anti-slavery amendment and statement of resolutions may have been necessary for other reasons as well. Non-slave holders were sensitive that their intentions not be misunderstood on several issues in the community at large. One concern was that passage of the amendment would take money—i.e., the value of emancipated slaves—out of their neighbors' pockets and another was that New State supporters were loath to be perceived as being 'abolitionist.'

165   "Soldiers' Meeting," Point Pleasant *Weekly Register*, March 5, 1863, Vol. 1 No. 51, [n.p.]

166   The offensive newspaper may well have been penned and published by Reuben Spicer, an editor resident in Shelbyville, Indiana, who persisted in publishing his shockingly hardline articles in support of slavery, secession and the Southern Confederacy in Shelby County, Indiana, until 1864. Special thanks to Donna Dennisom, Indiana History Librarian, Shelby County Public Library, for this information.

167   "How the Officers and Men of the 13th Virginia Reg't Feel About the Butternuts of the Free States," Point Pleasant *Weekly Register*, March 12, 1863, [p. 1].

168   Editor's preface to an anonymous poem about the 13th Regiment entitled "The Gallant 13th Virginia Volunteers," Point Pleasant *Weekly Register*, March 5, 1863, [p. 1].

169   [Anonymous], "Poetical. The Gallant 13th Virginia Volunteers," Point Pleasant *Weekly Register*, March 5, 1863, [p. 1].

170   1st Lieut. Timothy Russell was about thirty-five years of age at this time. Prior to his enlistment in the 13th Regiment, he had served as Captain of the Mason County 106th Virginia Militia (reorganized November 1861) and been engaged in the milling and lumber business at West Columbia, Mason County. The town of West Columbia was located at the mouth of Ice Creek, Waggoner District, about twelve miles above Point Pleasant.

171   E.M. Fitzgerald (also written Fitz Gerald) wore many hats. He was an attorney-at-law. His office was located opposite the the the Point Pleasant Courthouse. He also served as secretary at the Mason and Putnam county public meetings held in 1862 and 1863. Upon the death of Lewis Wetzel in October 1862, Fitzgerald took over as editor of the Point Pleasant *Weekly Register*.

172   A New State meeting was also held at the Methodist Episcopal Church at Hartford City, Mason County, on March 16, 1863. ("New State Meeting," Point Pleasant *Weekly Register*, March 19, 1863, [n.p.]) This particular church was where Reverend W.W. Harper, now Sergeant Major Harper of the 13th West Virginia, served as minister.

173   Point Pleasant *Weekly Register*, March 12, 1863, Vol. 1 No 52, [n.p.]

174   The phrase "the negro humbug exposed" probably referred to public concern related to the issue of immediate emancipation as required by the U.S. Congress as a condition of West Virginia statehood. Put bluntly, West Virginians worried about what would happen when the Black American slave was suddenly free and loose in the public domain. Ideas in circulation to counter such concerns were that there really weren't that many enslaved negroes in West Virginia to begin with, and these once freed, would probably go north to join their already emancipated friends and family.

175   Point Pleasant *Weekly Register*, March 12, 1863, Vol. 1 No. 52, [n.p.]

176   T. Harry Williams, *Hayes of the Twenty-Third The Civil War Volunteer Officer* (New York: Alfred A. Knopf, 1965), p. 148.

177   On March 17, Brig. Gen. Eliakim P. Scammon issued Special Order No. 36, directing that "The 5th and 13th Regiments Virginia Volunteer Infantry, Lieut. Witcher's Co. 3rd Va. Cavalry are hereby attached to the first Brigade District of Kanawha and will immediately report to Col. R.B. Hayes commanding." See Brig. Gen. E.P. Scammon, by Jas. L. Botsford, Asst Adjt. Gen'l. to Col. W.R. Brown, cmdg. 13th Va. Vols., Special Order No. 36 (manu. dated "Head Quarters District Kanawha, Charleston, Va., March 17, 1863"), RG94 Box 4838 Misc. unbound regimental papers 13th W.Va., Spec. Orders and Gen. Orders, NARA.

178   Rutherford B. Hayes, Diaries 1834-May 1, 1865 (microfilm Series 2/2), Rutherford B. Hayes Presidential Library, p. 22. See Charles R. Williams' compilation of Hayes' diary and letters in *Diary and Letters of Rutherford Birchard Hayes*, Vol. II 1861-1865 (Columbus, Ohio: F.J. Heek Printing Company, 1922), p. 395. Williams lists Hayes' troops and their locations at this time. At Camp White Charleston was the 23rd Ohio Infantry, Lieut. Col. James Comly, cmdg.; at Ceredo, the 5th Virginia Infantry, Col. John Ziegler, cmdg.; at Coal's Mouth and Hurricane Bridge, the 13th Virginia, Col. W.R. Brown, cmdg.; at Charleston, Captain George W. Gilmore's cavalry; at "Loup and Tompkins Farm," Lieut. Gonseman [with cavalry?]; and at Gauley Bridge, Captain Seth J. Simmond's battery.

179   Both quotations are from Hayes' letter to his mother, Sophia Hayes (dated "Camp White, near Charleston, March 22, 1863"), reprinted in Williams, *Diary and Letters of R.B. Hayes*, Vol. II, p. 396.

180   Ibid., Williams, pp. 395-96.

181   Colonel R.B. Hayes, by M.P. Avery, Capt. and A.A.A.G. to Colonel W.R. Brown (manu. dated "Hd. Qtrs. 1st Brig. Dist. Kanawha, Charleston, Va., March 17, 1863"), RG94 Box 4839 Misc. unbound regimental papers 13th W.Va., NARA.

182   Theodore F. Lang, Major, 6th W.Va. Cavalry and Brevet Colonel, *Loyal West Virginia From 1861 to 1865* (Baltimore, Md.: The Deutsch Publishing Co., 1895), pp.289-90.

183   "S. Comstock, Proprietor of Eagle Mills," Point Pleasant *Weekly Register*, February 26, 1863, [n.p.] and "Eagle Mills — A Change," Point Pleasant *Weekly Register*, March 12, 1863, Vol. II No. 52, [p. 2].

184   W.W. Holmes, Medical Director, District Western Virginia to the Governor of Virginia (dated "Head Quarters District Western Virginia, Marietta, Ohio, March 26, 1863"), Samuels-Pierpont Papers, WVSA.

185   RG94 Morning Reports of James W. Johnson's Co. A 13th Regiment of Virginia Volunteers Army of the U.S. for the month of March 1863, NARA, [n.p.]

186   RG94 Morning Reports of Capt. Milton Stewart's Co. B 13th Regiment Virginia Vols. for March 1863, [n.p.]

187   RG94 Morning Reports of Capt. Van D. McDaniels' Co. C 13th Regiment of Virginia Vols. for March 1863, [n.p.] and Record of Events Field and Staff, Muster Roll Co. C 13th Va. Vol. Infantry dated April 30, 1863 (micro. #594 Roll 196), NARA.

188   Marshall Statton was mustered in August 5, 1863. There are no further records for this recruit. State records preserve that he was throught to have died at Kanawha Falls, W.Va., time and date unknown or unrecorded. The company would have several occasions to pass through Kanawha Falls as the regiment became deployed later that summer to the Gauley Bridge area and in 1864 to the Shenandoah Valley.

189   RG94 Morning Reports of Capt Simon Williams' Co. D 13th Regiment Virginia Vols. for March 1863, [n.p.] and RG94 Regimental Books Misc. Records Descriptive Roll Co. C. 13th Regiment V.V.V.I., NARA [n.p.]

190   RG94 Morning Reports of Capt. A.F. McCown's Co. F 13th Regiment Virginia Vols. for March 1863, [n.p.]

191   Col. W.R. Brown to Capt. J.V. Young, commanding at Coalsmouth (manu. dated "Head Quarters Detachment 13th Va Vols, Hurricane Bridge, March 6th [1863]"), Civil War Scrapbook Accession #859, Regional History Library, WVU.

192   J.V. Young to Emma Young (dated "Headquarters Company G 11th Va., Barboursville, Cabell Co., Mar. 9"), Young Civil War Papers, Vol. 31 Box 10, WVU, trans. p. 8.

193   RG94 Regimental Books Misc. Records Descriptive Roll Co. H 13th Regiment V.V.V.I., Register of Deserters, [n.p.]

194   RG94 Morning Reports for Capt. T.W. Hampton's Co. H of the 13th Regiment Virginia Vols. for March 1863, [n.p.] and RG94 Misc. Records Descriptive Roll Co. H, [n.p.]

195   RG94 Morning Reports for Lieut. William Feazel's Co. I of the 13th Regiment Virginia Vols. for March 1863, [n.p.]

196   James Comly, Diaries, dated "Thursday, March 10, 1863" (micro. Roll 1), Ohio Historical Center, Columbus, [n.p.]

197   Comly, Diaries dated "Thursday, March 10, 1863" (micro. Roll 1), OHC, [n.p.]

198   See Maj. Gen. H.G. Wright to Brig. Gen. Julius White (dated Hdqtrs. Dept. of Ohio, Cincinnati, Oh., Mar. 13, 1863), O.R. XXIII, Part II, p. 140 and Maj. Gen. H.G. Wright to Brig. Gen. Scammon at Charleston, W.Va. (dated Hdqtrs. Dept. of the Ohio, Cincinnati, Oh., Mar. 14, 1863), O.R. XXIII Part II, p. 141.

199   "West Virginia," Point Pleasant *Weekly Register*, March 26, 1863, [p. 2].

200   Charles H. Ambler, *A History of West Virginia* (New York: Prentice Hall Inc., 1933), p. 333.

201   "Appointed," Point Pleasant *Weekly Register*, March 12, 1863, Vol. II No. 52, [p. 2].

202   "West Virginia," Point Pleasant *Weekly Register*, March 26, 1863, [p. 2]. In reporting this round number, the editor added that the number of soldiers voting against the amendment "hardly amount[ed] to a 'baker's dozen.'"

203   Ambler, *History of West Virginia*, p. 392.

204   Theodore F. Lang, Major, 6th W. Virginia Cavalry, wrote in reference to this important vote that the soldiers in the field were permitted to vote on the new State question in an election held in their camps on March 12. "[A]ll the qualified voters availed themselves of this opportunity to attest their loyalty to the new state." (Lang, *Loyal West Virginia*, p.106.)

205   "New State – Soldier's Vote," Point Pleasant *Weekly Register*, May 26, 1863, [p. 2].

206   "Cabell," Wheeling *Daily Intelligencer*, March 31, 1863, [n.p.]

207   "Cabell," Wheeling *Daily Intelligencer*, March 31, 1863, reprinted in Geiger, *Civil War in Cabell County*, p. 81, cited in Ch. 7 fn. 2, p. 139. For a view of the town of Barboursville, see "Barboursville Nov. 1861," a print in the collection of Marshall University, Huntington, West Virginia. This image depicts the 34th Ohio Zouaves marching on dress parade through the town center. This print has often been been published. The more serious student of history might appreciate seeing it in Frances B. Gunter's pamphlet about the old town entitled *Barboursville*.

208   "New State — Official," Point Pleasant *Weekly Register*, April 16, 1863, [p. 2]. April 20, Lincoln issued the proclamation declaring that the State of West Virginia had the approval of Congress and would officially join the Union on June 20, 1863. From first to last, however, it was touch and go. Had Lincoln vetoed admission of West Virginia as a new State, West Virginians could not have hoped to consolidate the two-thirds votes in Congress to have over-ridden the President's veto.

209   "Battle of Point Pleasant," Point Pleasant *Weekly Register*, April 6, 1863, [n.p.]

210   Paraphrase of Livia Simpson-Poffenbarger, "Battle of Point Pleasant," Poffenbarger Papers, Vol. 9, p. 10, Mason County Public Library, Point Pleasant.

211   See Sam Jones, Maj. Gen. to A.G. Jenkins, Brig. Gen. (dated Hdqtrs. Dept. of Western Va., Dublin, Va., Dec. 7, 1862); Sam Jones, Maj. Gen. to James A. Seddon, Sec. of War (dated Hdqtrs. Dept. of Western Va., Dublin, Va., March 8, 1863); and Sam Jones, Maj. Gen. to J.B. Floyd, Maj. Gen. (dated Hdqtrs. Dept. of Western Va., Dublin, Va., March 8, 1863), all references in O.R. XXV Part II, as quoted by Poffenbarger in her notes for "Battle of Point Pleasant," Poffenbarger Papers, Vol. 9, pp. 29-30.

212   Colonel Hayes, comdg. brigade at Charleston, reported the raid somewhat differently in a letter to his wife, writing that "General Jenkins and about eight hundred men left the railroad at Marion, Smith County, southwestern Virginia, and crossed the mountains to the head waters of Sandy River and so across towards the mouth of Kanawha. They reached [the 13th] outpost [at Hurricane Bridge] twenty-four miles from here and demanded a surrender." R.B. Hayes to his wife, Lucy (dated "Camp White, April 10, 1863"), reprinted in Williams, *Diary and Letters*, p. 404.

213   [Untitled], reprinted from the Gallipolis *Journal* in the Point Pleasant *Weekly Register*, May 14, 1863, [p.1].

214   Information learned from questioning officers Samuels, Timms and Holderby, summarized and published in "The Late Raid Upon Point Pleasant—What Jenkins Intended to Do," Wheeling *Daily Intelligencer*, April 7, 1863, p. 3.

215   B.J. Redmond, attorney-at-law with the firm Hoge & Redmond, Point Pleasant, West Virginia to F[rancis] H. Pierpont, Governor of West Virginia (dated "Point Pleasant, April 2, 1863"), Wheeling *Daily Intelligencer*, April 9, 1863, p. 1.

216   Paraphrase of a letter received from Point Pleasant on April 4, 1863, at the offices of the Wheeling *Daily Intelligencer* providing "particulars" of Jenkins' raid on that town, published as "The Jenkins' Raid Upon Point Pleasant" in the Wheeling *Daily Intelligencer*, April 4, 1863, p. 3.

217   For an insightful biography of the very human Captain Johnson and well-researched account of this engagement see Philip Hatfield, PhD., *The Battle of Hurricane Bridge,* Charleston, West Virginia: 35th Star Publishing, 2019.

218   J.W. Johnson, Capt. Comd'g Detachment 13th V V I at Hurricane Bridge, Virginia to Col. W.R. Brown, comdg 13th Reg Va Vol Infty, "Report of the engagement fought at Hurricane Bridge, March 28, 1863" (manu. dated "Hurricane Bridge Va April 3, 1863"), RG94 Box 4839 Misc. unbound regimental papers 13th W.Va., NARA.

219   Jim Comstock, ed., *The West Virginia Heritage Encyclopedia. Supplemental Series Hardesty's,* Vol. 1 "Putnam County" (Richwood, W.Va.: Jim Comstock, 1973), p. 131.

220   Interview of Samuels et.al., "The Late Raid Upon Point Pleasant—What Jenkins Intended to Do," Wheeling *Daily Intelligencer,* April 7, 1863, p. 3.

221   J.W. Johnson, Capt. "Report of the engagement fought at Hurricane Bridge, March 28, 1863," (manu. dated "Hurricane Bridge Va April 3, 1863"), RG 94 Box 4839, NARA.

222   B.J. Redmond to F[rancis] H. Pierpont, Governor of West Virginia (dated "Point Pleasant, April 2, 1863"), Wheeling *Daily Intelligencer,* April 9, 1863, p. 1.

223   Major James H. Nounnan enrolled in the Confederate States' service in Putnam County. Jack L. Dickinson, *Tattered Uniforms and Bright Bayonets West Virginia's Confederate Soldiers,* p. 287.

224   B.J. Redmond, a Point Pleasant attorney, wrote to Governor Pierpont in response to misreporting provided in four installments by a certain "gentleman in Gallipolis" to the newspaper of West Virginia's seat of government, the Wheeling *Daily Intelligencer.* Redmond proposed "as an act of justice to this gallant regiment to give [Pierpont] the facts as well as connection with the battle of Point Pleasant [and] Hurricane Bridge." See B.J. Redmond to F[ranci]s H. Pierpont, Gov. of West Virginia (dated "Point Pleasant, April 2, 1863"), Wheeling *Daily Intelligencer,* April 9, 1863, p. 1.

225   J.W. Johnson, Capt., "Report of the engagement fought at Hurricane Bridge, March 28, 1863" (manu. dated "Hurricane Bridge Va April 3, 1863"), RG94 Box 4839, NARA.

226   Ibid.

227   See Col. R.B. Hayes who noted in his diary for Saturday, March 28, 1863, that the fight at Hurricane Bridge was "between four hundred Jenkins' or Floyd's men and two hundred and seventy-five Thirteen Virginia [men] at Hurricane Bridge." Williams, *Diary and Letters,* p. 398.

228   Morning reports submitted for Company B 13th W. Va. Regiment, indicate that the attacking Confederates numbered five hundred. Other estimates vary. R.B. Hayes' diary entry for March 29 contained that "[r]umors of the fight at Hurricane Bridge represent the Rebels as Jenkins' men, four hundred to seven hundred strong." Williams, *Diary and Letters,* p. 399.

229   Comstock, *West Virginia Heritage Encyclopedia. Supp. Ser. Hardesty's,* Vol. 1 "Putnam County," p. 131.

230   Field History 13th Regiment W.Va. Infantry (manu. dated "Head Quarters 13th Regt. W.Va. Vol. Inft., Barboursville, W.Va., Jan'y 13th, 1864"), Uncat. 13th W.Va. Inftry. Regt. Box, [n.p.]

231   B.J. Redmond to F[rancis] H. Pierpont, Governor of West Virginia (dated "Point Pleasant, April 2, 1863"), Wheeling *Daily Intelligencer,* April 9, 1863, p. 1.

232   Ibid.

233   Information learned from questioning officers Samuels, Timms and Holderby, summarized and published in "The Late Raid Upon Point Pleasant—What Jenkins Intended to Do," Wheeling *Daily Intelligencer,* April 7, 1863, p. 3.

234   J.W. Johnson, Capt., "Report of the engagement fought at Hurricane Bridge, March 28, 1863" (manu. dated "Hurricane Bridge Va April 3, 1863"), RG94 Box 4839 Misc. unbound regimental papers 13ᵗʰ W.Va.

235   Redmond to Pierpont (dated "Point Pleasant, April 2, 1863"), Wheeling *Daily Intelligencer*, April 9, 1863, p. 1.

236   Alexander Samuels, Captain, Co. E 8ᵗʰ Virginia Cavalry Regiment and brother to Judge H.J. Samuels, Adjutant General of the Restored Government of West Virginia. Alexander Samuels (Alex) had enlisted in the Border Rangers towards the end of May 1861, at Greenbottom, Cabell County. The Border Rangers was formed by Albert G. Jenkins as a militia company on December 10, 1860, at Guyandotte to steady the locale in wake of Lincoln's election to the U.S. presidency and specifically—so the story goes—to guard a Virginia State flag with its famous motto "Sic Semper Tyannis" ("Thus always to Tyrants") and image of Lady Liberty standing sword in hand over a vanquished tyrant, whose crown—symbol of sovereignty—has been unceremoniously struck from his head. The flag had been raised on the banks of the Ohio as a declaration and challenge, where it might be seen by all, who traveled this major throughfare. The militia company guarded the flag that it might not be taken down and indeed, reportedly, it remained aloft until April 20, 1861, when word that the convention at Richmond had voted in favor of the secession ordinance was received. On May 29, 1861, the Border Rangers were sworn in as Co. E 8ᵗʰ Virginia Cavalry, Confederate States Army. Paraphrase of George Selden Wallace, *Cabell County Annals & Families* (Richmond: Garrett & Massie, Publishers, 1935), pp. 77-79.

237   This likely refers to the Trent Affair. James M. Mason of Virginia and John Slidell of Lousianna, both appointed to the Confederate Foreign Servce, were sent, respectively, to England and France in late 1861 to secure 1) recognition of the Confederacy as an independent country and 2) to curry policies favorable to the Confederacy. On Nov. 8, 1861, while aboard the steamer *The Trent*, a British steamer then passing through the Bermuda Channel in the Caribbean on their return trip, the two diplomats were arrested.

238   M.V.B. Edens, Co. A 13ᵗʰ Virginia Volunteers, "Battle of Hurricane Bridge!," Point Pleasant *Weekly Register*, April 30, 1863, [p. 1]. The poem may have been set to some rhythm of drum or fife tune or perhaps was fitted to the melody of some popular song of the time.

239   J.W. Johnson, Capt., "Report of the engagement fought at Hurricane Bridge, March 28, 1863" (manu. dated "Hurricane Bridge Va April 3, 1863"), RG94 Box 4839.

240   Field History 13ᵗʰ Regiment W. Va. Infantry (manu. dated "Head Quarters 13ᵗʰ Regt. W. Va. Vol. Inft., Barboursville, W.Va., Jan'y 13ᵗʰ, 1864"), Uncat. 13ᵗʰ W.Va. Inftry. Regt. Box, [n.p.]

241   RG94 Morning Reports of Captain A.F. McCown's Co. F 13ᵗʰ Regiment Virginia Vols. for March 1863, [n.p.]

242   See "History of 23ʳᵈ Ohio Regiment," Comly Papers (micro. Reel 1), OHC, pp. 388-89.

243   B.J. Redmond to F[rancis] H. Pierpont, Governor of West Virginia (dated "Point Pleasant, April 2, 1863") Wheeling *Daily Intelligencer*, April 9, 1863, p. 1.

244   RG94 Morning Reports of James W. Johnson's Co. A 13ᵗʰ Regiment Virginia Vols. for March 1863, [n.p.]

245   RG94 Morning Reports of Captain Milton Stewart's Co. B 13ᵗʰ Regiment Virginia Vols. for March 1863, [n.p.]

246   Comstock, *West Virginia Heritage Encyclopedia. Supp. Hardesty's*, Vol. 1 "Putnam County," p. 131.

247   Applications for Headstones Mason County, West Virginia, NARA.

248   "Robinson District. James A. Rayburn," H.H. Hardesty, *Hardesty's Historical and Geographical Encyclopedia*, "Mason County Vol. 5" (orig. Chicago: H.H. Hardesty, 1883), reprint Jim Comstock, *Hardesty's West Virginia Counties*, Richwood, W.Va., 1973, p. 93.

249   RG94 Morning Reports of Capt. Simon Williams's Co. D 13ᵗʰ Regiment Virginia Volunteers for March 1863, [n.p.]

250   RG94 Morning Reports of Co. H 13th Regiment Va. Vols. for March 1863, NARA, [n.p.] and RG94 Misc. Records Descriptive Roll Co. H. 13th Regiment V.V.V.I., List of Non-Commissioned Officers, [n.p.] Corporal Fullwiler returned to his company after his capture. Inasmuch as orders were received "from Brigade Head Quarters that no paroles given by Jenkins during his raid should be respected [Fulwiler] was ordered to duty." He "[w]ent with his Co. to Point Pleasant and then deserted" and charges were preferred against him. See W.I. Mathews, Capt. Co. H 13th Regt. V.V.I. to Joseph Dan, Maj. + P.M.G. (dated "Head Quarters C.H. 13th Regt. V.V.I. Coalsmouth, Va. May 31, 1863"), George W. Fullwiler, M508 Compiled Service Records (micro. Roll 260), NARA.

251   RG94 Morning Reports of Capt. Milton Stewart's Co. B 13th Regiment Virginia Vols. for March 1863, [n.p.]

252   B.J. Redmond to F[rancis] H. Pierpont, Governor of West Virginia (dated "Point Pleasant, April 2, 1863"), Wheeling *Daily Intelligencer*, April 9, 1863, p. 1.

253   Emory J. Bridgeman, Military Pension File, NARA.

254   Field History 13th Regiment W.Va. Infantry (manu. dated "Head Quarters 13th Regt. W.Va. Vol. Inft., Barboursville, W.Va., Jan'y 13th, 1864"), Uncat. 13th W.Va. Inftry. Regt., [n.p.]

255   Redmond to Pierpont (dated "Point Pleasant, April 2, 1863"), Wheeling *Daily Intelligencer*, April 9, 1863, p. 1.

256   See Morning Reports for March 1863 for Cos. C and F. Company C reports state that the detachment arrived at 5 o'clock. Company F reports record that they arrived at 4 o'clock a.m. *13th West Virginia Field History* records that C and F marched to the relief of the companies engaged at Hurricane Bridge and reached "the scene of action at 4 O.C. A.M. March 29."

257   Redmond to Pierpont, (dated "Point Pleasant, April 2, 1863"), *Daily Intelligencer*, April 9, 1863, p. 1. 13th regimental records preserve that "after abandoning the attack" at Hurricane Bridge, the Confederates "took a circuitous route and came out on the Kanawha river, a short distance below Winfield, and proceeded down the Kanawha River to Point Pleasant." See Record of Events Field and Staff in Company Muster Rolls March and April 1863 (micro. #594 Roll 196); Compiled Service Records of Volunteers (Union) West Virginia (Station: Winchester March and April 1865 micro. #M508 Roll 204), NARA; and Record of Events for the Thirteenth West Virginia Infantry, October 1862–June 1865 Field and Staff, *Supplement to the Official Records of the Union and Confederate Armies*, Vol. 74 (Series 86) Part II, p. 545. The Confederates may well have followed Hurricane Creek in whole or in part to come out on the Kanawha below Winfield and continued from there on their way to Point Pleasant. It is not known how the 13th Companies proceeded from Point Pleasant to Hurricane but a march overland seems likely as it might have afforded them opportunity to also scout the country as they moved into "Dixie."

258   Roush, *If Thou Wilt Remember*, p. 29.

259   Capt. Fitch, A.Q.M. to Governor Pierpoint. Dispatch dated Gallipolis, [Ohio, March 30, 1863]. Paraphrased and published under title "REBEL RAID ON THE KANAWHA." Wheeling *Daily Intelligencer*, March 31, 1863, p. 3.

260   The author of a history of the 23rd Ohio Volunteers stated that the boats were about twenty miles from Point Pleasant "when they were fired on by Rebels drawn up in line on either side of the river, and were ordered to heave to" and "the boats were riddled with shot." See "History of 23rd Ohio Regiment," Comly Papers (micro. Reel 1), OHC, pp. 388-89.

261   [Account of the attack on the Victor No. 2 as given by Paymaster Major B.R. Cowan], *Wheeling Daily Intelligencer*, April 2, 1863, p. 3.

262   Captain F[red] Ford, commandg. U.S. Steamer Victor No. 2 to the Officers of Victor No. 2 (dated "U.S. Steamer, Victor No. 2, April 14th, 1863"), Point Pleasant *Weekly Register*, April 16, 1863, Vol. II No. 5, [p. 2]. Paymaster Cowan in his account, however, stated that the officers of the *Victor* had been warned "by some negroes" "some considerable distance above" Hall's Landing of danger "but as such warnings had been very frequent, no attention was paid to it." [Account of the attack on the Victor No. 2 as given by Paymaster Major Cowan], Wheeling *Daily Intelligencer*, April 2, 1863, p. 3.

263   Ibid., [Cowan], *Daily Intelligencer*, April 2, 1863, p. 3.

264   The steamer seems to have been attacked just as it came near Frazier's Bottom or Frazier's Landing on the Kanawha. See the Cincinnati *Daily Commercial* which reported that the *Victor No. 2* was fired into "4 miles above Buffalo." This latter would place the attack, according to a DeLorme map of today, just above Frazier's Landing. ("Particulars of the Rebel Raid on Point Pleasant — Government Steamboats Fired Into, etc.," Cincinnati *Daily Commercial*, April 2, 1863, p. 2.)

265   Capt. Fitch, A.Q.M. to Governor Pierpoint (dispatch dated "Gallipolis, [Ohio, March 30, 1863]") paraphrased and published under title "REBEL RAID ON THE KANAWHA,"Wheeling *Daily Intelligencer*, March 31, 1863, p. 3. Col. R.B. Hayes noted in his diary that "ten companies of Jenkins' men" fired into the *Victor* when it was "nearly opposite Buffalo." They were mostly on foot "a few had horses." (Hayes, diary entry for "Monday, March 30,"Williams, *Diary and Letters,* pp. 399-400.)

266   Capt. F[red] Ford, cmdg. U.S. Steamer Victor No. 2 to the Officers of Victor No. 2 (dated "U.S. Steamer, Victor No. 2, April 14th, 1863"), Point Pleasant *Weekly Register*, April 16, 1863, [p. 2].

267   [Account of the attack on the Victor No. 2 as given by Paymaster Major Cowan], *Wheeling Daily Intelligencer*, April 2, 1863, p. 3.

268   Account given by 'Mr. Brammer' to the *Ironton Register*, published as "Jenkins Turned Up Again," *Ironton Register*, April 2, 1863, p. 2.

269   [Cowan], *Wheeling Daily Intelligencer*, April 2, 1863, p. 3.

270   [Untitled], reprint from the Gallipolis *Journal* of April 9, 1863, in the Point Pleasant *Weekly Register*, April 30, 1863, [n.p.]

271   [Cowan], *Wheeling Daily Intelligencer*, April 2, 1863, p. 3.

272   Field History 13th Regiment W.Va. Infantry, manu. dated "Head Quarters 13th Regt. W. Va. Vol. Inft., Barboursville, W.Va., Jan'y 13th, 1864," [n.p.]

273   [Cowan], *Wheeling Daily Intelligencer*, April 2, 1863, p. 3.

274   "ONE WHO KNOWS THE FACTS" for the Point Pleasant *Weekly Register*, published as "What the Government Makes by Having the Right Man in the Right Place," *Weekly Register*, Thursday Morning, April 16, 1863, Vol. II No. 5, [p. 1].

275   "Gallia County in the War of the Rebellion," *History of Gallia County* (Chicago and Toledo: H.H. Hardesty & Co., Publishers, 1882), pp. 29-30.

276   John Hall, Esq., had been arrested and arraigned the previous autumn for the shooting death of Lewis Wetzel, another Union man and editor of the Point Pleasant *Weekly Register*. John Hall, also the father of the Lieutenant Colonel James R. Hall, 13th West Virginia Regiment, was a staunch Unionist and influential figure in his community. Hall was an attorney with an office on Main Street, Point Pleasant; a representative from Mason County in both the old and new State government of Virginia; and a longtime entrepreneur and developer of industry.

277   Simpson-Poffenbarger, notes for the "Battle of Point Pleasant," Poffenbarger Papers, Vol. 9, p. 49.

278   Mr. Brammer to the *Ironton Register*, published as "Jenkins Turned Up Again," *Ironton Register*, April 2, 1863, p. 2.

279   Ibid. The earnestness of the attack, as evidenced by the number of bullet holes made considerable impression. A week later the Gallipolis *Journal* carried the following: "No loyal citizen can look at the effect of the murderous fire poured into the Government steamboats Gen. Meigs and Victor No. 2, by Jenkins guerrillas, on the Kanawha, last Sunday week, without feeling how narrowly our town escaped destruction, and how many happy homes might now be the scene of desolation and mourning. We say LOYAL citizens because a goodly number of the 'butternuts' that in case of an attack they would be safe from injury. This may be true, for Jenkins doubtless knows his friends here as elsewhere. But to the loyal, law abiding, peaceful lover of his country more than party,

every bullet hole in those boats conveys an idea of terrible import." ([Untitled], reprint from the Gallipolis *Journal* of April 9, 1863, in the Point Pleasant *Weekly Register*, April 30, 1863, [n.p.])

280   [Account of the attack on the Victor No. 2 as given by Paymaster Major Cowan], *Wheeling Daily Intelligencer*, April 2, 1863, p. 3.

281   "From the Point Pleasant *Register* April 2d. Battle of Point Pleasant," reprinted in the *Ironton Register* April 9, 1863, p. 2.

282   "Battle of Point Pleasant," Point Pleasant *Weekly Register*, reprinted in the Wheeling *Daily Intelligencer*, April 6, 1863, p. 2.

283   "Rebel Prisoners from Point Pleasant—More Particulars of the Raid of Jenkins,"Wheeling *Daily Intelligencer*, April 6, 1863 p. 3.

284   [Anonymous correspondent], dated "KANAWHA VALLEY, Charleston, W.Va. 1863,""Albert G. Jenkins," Point Pleasant *Weekly Register*, April 16, 1863, Vol. II No. 5, [p. 1].

285   Capt[ain] J.D. Carter, Co[mpany] E 13th V[irgini]a V[olunteer] I[nfantry], commanding Post [at Point Pleasant] to the Editor of the *Gallipolis Dispatch* (dated "April 6, 1862"), "POINT PLEASANT BATTLE. Letter from Capt. Carter Commanding Post," Point Pleasant *Weekly Register*, Thursday Morning, April 16, 1863, Vol. II No. 5, [p. 1].

286   Hayes to his wife, Lucy (dated "Camp White, April 10, 1863"), Williams, *Diary and Letters,* p. 404.

287   "Unpub. "History of 23rd Ohio Regiment," Comly Papers (micro. Reel 1), OHC, pp. 388-89.

288   A few sources, including Captain J.D. Carter, stated that the battle commenced at 11:00 a.m. The Point Pleasant *Register* reported that about 10 o'clock a.m., the Point Pleasant military Post "was surprised by Jenkins' scurvy, thieving, ragamuffins." ("Battle of Point Pleasant," reprinted from the Point Pleasant *Weekly Register* in the Wheeling *Daily Intelligencer*, April 6, 1863, p. 2.) The Cincinnati *Daily Commercial* also reported that Jenkins men "made a dash into Point Pleasant, at 10 o'clock, Monday morning last." ("Particulars of the Rebels Raid on Point Pleasant — Government Steamboats Fired Into &c.," Cincinnati *Daily Commercial*, April 2, 1863, p. 2.) Captain Ford stated that the Confederates attacked the boats at 10 a.m. (Capt. F[red] Ford, cmdg. U.S. Steamer Victor No. 2 to the Officers of Victor No. 2, dated "U.S. Steamer, Victor No. 2, April 14th, 1863," in the Point Pleasant *Weekly Register*, April 16, 1863, [p. 2].) Hardesty and Livia Poffenbarger both recorded that the battle commenced at ten o'clock a.m. See "Mason County," *Hardesty's History and Encyclopedia* (1833), p. 10 and Poffenbarger, notes for "Battle of Point Pleasant," Poffenbarger Papers, Vol. 9, p. 27.

289   Those defending Point Pleasant were apparently accused of being taken by surprise. Many years after the war, the edited reissue of Hardesty's: *The West Virginia Heritage Encyclopedia* (1973), carried under "Putnam County" that the although the officers of the *Victor* notified Captain Carter that there was a sizeable Confederate force in the valley and that the *Victor No. 2* had been attacked, "[Carter] made no preparations for defense, and permitted himself to be taken completely by surprise, the next day when the force of General Jenkins entered [Point Pleasant]." See Jim Comstock, ed., *The West Virginia Heritage Encyclopedia. Supplemental Series Hardesty's*, Vol. 1 "Putnam County" (Richwood, W.Va.: Jim Comstock, 1973), pp. 131-32.

290   Redmond to Pierpont (dated "Point Pleasant, April 2, 1863"),Wheeling *Daily Intelligencer*, April 9, 1863, p. 1.

291   Paraphrase of a letter received from Point Pleasant on April 4, 1863, at the offices of the Wheeling *Daily Intelligencer* providing "particulars" of Jenkins' raid on that town. ("The Jenkins' Raid Upon Point Pleasant," Wheeling *Daily Intelligencer*, April 4, 1863, p. 3.)

292   John D. Carter, Capt[ain] Co[mpany] E, 13th V[irgini]a Vol[unteer] Inf[antr]y, Commanding Post to the Editors of the Wheeling *Daily Intelligencer* concerning misinformation provided by a dispatch from E.P. Fitch and published in the *Intelligencer* on March 30, 1863. Carter's letter published under the title "Letter from Capt. Carter about the Point Pleasant Fight," Wheeling *Daily Intelligencer*, April 9, 1863, p. 2. Co. E was an undersized company from the get-go and at the time of these events had been reduced further. It

was mustered in with just 82 men (instead of the 100 plus it should have had) when the regiment was organized October 9, 1862, and at the time of the Point Pleasant fight numbered just 60 men.

293   Sylvester Myers, *Myers' History of West Virginia,* Vol. I. (Wheeling, W.Va.: Wheeling News Lithograph Co., 1915), p. 473 and Simpson-Poffenbarger, her notes for "Battle of Point Pleasant" in Poffenbarger Papers, Vol. 9, p. 27, Mason Co. Pub. Lib., Point Pleasant.

294   Co. E 13th Va. Vol. Infantry, Company Muster Roll March and April 1863 (micro. #594 Roll 196), NARA.

295   Ibid.

296   Carter to the Editor of the *Gallipolis Dispatch* (dated "April 6, 1862"), reprinted as "POINT PLEASANT BATTLE. Letter from Capt. Carter Commanding Post," Point Pleasant *Weekly Register,* April 16, 1863, Vol. II No. 5, [p. 1].

297   Paraphrase of a letter received from Point Pleasant on April 4, 1863, at the offices of the Wheeling *Daily Intelligencer* providing "particulars" of Jenkins' raid on that town. "The Jenkins' Raid Upon Point Pleasant," Wheeling *Daily Intelligencer,* April 4, 1863 p. 3.

298   "Battle of Point Pleasant," reprint from Point Pleasant *Weekly Register* in the Wheeling *Daily Intelligencer,* April 6, 1863, p. 2.

299   Co. E 13th Va. Vol. Infantry, Company Muster Roll March and April 1863 (micro. # 594 Roll 196), NARA.

300   Ibid.

301   No casualties were sustained by the defenders from Cos. C or F at the Point Pleasant fight. See Company Muster Rolls for Cos. C and F for March and April 1863 (micro. #594 Roll 196) and RG94 Morning Reports Capt. A.F. McCown's Co. F 13th Regiment of Virginia Volunteers Army of the U.S. for the month of March 1863, NARA, [n.p.]

302   R.B. Hayes to his wife, Lucy (dated "Camp White, April 10, 1863"), *Diary and Letters,* p. 404.

303   Account provided by 'Mr. Brammer' to the *Ironton Register,* "Jenkins Turned Up Again," *Ironton Register,* April 2, 1863, p. 2.

304   Hayes to Lucy (dated "Camp White, April 10, 1863"), Williams, *Diary and Letters,* p. 404.

305   "Particulars of the Rebel Raid on Point Pleasant — Government Steamboats Fired Into &c.," Cincinnati *Daily Commercial,* April 2, 1863, p. 2.

306   "Col. J.P.R.B. Smith Dead," Poffenbarger Papers, Vol. 5, Mason Co. Pub. Lib., p. 66.

307   "Battle of Point Pleasant," reprinted from Point Pleasant *Weekly Register* in the Wheeling *Daily Intelligencer,* April 6, 1863, p. 2.

308   Ford to the Officers of Victor No. 2 (dated "U.S. Steamer, Victor No. 2, April 14th, 1863"), Point Pleasant *Weekly Register,* April 16, 1863, [p. 2].

309   "Gallia County in the War of the Rebellion," *History of Gallia County,* pp. 29-30.

310   Ford to the Officers of Victor No. 2, Point Pleasant *Weekly Register,* April 16, 1863, [p. 2].

311   Ibid.

312   "An Incident at Point Pleasant," reprinted from the *Kanawha Republican* in the Wheeling *Intelligencer,* April 16, 1863, p. 2.

313   In the columns of the Point Pleasant *Register,* it was related that when the Confederates ordered Captain Carter to surrender the town, he told them to "go to hell," "whither 20 that we know of unwillingly went." ("Battle of Point Pleasant," reprinted from the Point Pleasant *Weekly Register,* Wheeling *Daily Intelligencer,* April 6, 1863, p. 2.)

314   Redmond to Pierpont (dated "Point Pleasant, April 2, 1863"), Wheeling *Daily Intelligencer*, April 9, 1863, p. 1.

315   Gallipolis *Journal*, Thursday, April 2, 1863, p. 2.

316   "Rebel Prisoners from Point Pleasant—More Particulars of the Raid of Jenkins," Wheeling *Intelligencer*, April 6, 1863, p. 3.

317   "From the Point Pleasant *Register* April 2d. Battle of Point Pleasant," reprinted in the *Ironton Register*, April 9, 1863, p. 2.

318   Gallipolis *Journal*, April 2, 1863, p. 2.

319   Robert H. Ferguson, *History of Mason Co., W.Va.* (Point Pleasant, W.Va." Col. Charles Lewis Chapter N.S.D.A.R., 1961).

320   Poffenbarger, "Battle of Point Pleasant," Poffenbarger Papers, Vol. 9, pp. 20-21 and "Rebel Prisoners from Point Pleasant—More Particulars of the Raid of Jenkins," Wheeling *Intelligencer*, April 6, 1863 p. 3.

321   Gallipolis *Journal*, April 2, 1863, p. 2. The Gallispolis General Army Hospital was located near where the militia likely embarked to cross the river to Point Pleasant. The hospital was located on the high ground at the northern end of Camp Carrington. The camp came into being on May 29, 1861, when the 21st Ohio Infantry, a unit enrolled for three months service, (Colonel Jesse S. Norton, cmdg.) went into camp "[a]bout a mile" northeast from Gallipolis in a "wheatfield on the Barlow farm." (E., letter dated "Sept. 15, 1862, General U.S.A. Hospital at Gallipolis, Ohio," pub. in the Cincinnati *Commercial*, reprinted in *The Guerrilla. Devoted to Southern Rights and Institutions*, Vol. 1 No. 2, p. 2; and *Gallipolis 1861-1865*, Ohio State Museum, Cols., [1961, n.p.]) The southernmost corner of the property fronted on the Ohio River giving access to river transport. ("Gallia County in the War of the Rebellion" from *History of Gallia County*, pp. 29-30, as reprinted in Michael L. Trowbridge, comp., *Camp Carrington Gallispolis Ohio*, 2003, [n.p.], courtesy of Michael L. Trowbridge.) Today the Gallipolis Developmental Center is located upon the site of the old encampment.

322   Gallipolis *Journal*, April 2, 1863, p. 2.

323   Mr. Brammer stated in his account that the impromptu battery mounted on the *Victor* consisted of three howitzers. (Brammer to the *Ironton Register*, "Jenkins Turned Up Again," *Ironton Register*, April 2, 1863, p. 2.)

324   Gallipolis *Journal*, April 2, 1863, p. 2. Although the hierarchy of command as given in newspaper accounts for this day is somewhat confusing, the Trumble Guards seem to have been under the command of Lieutenant Gilmer and the Gallia Guards under command of one Lieutenant Freer. Overall command of the militia seems to have been led by Captain James Harper, although one Captain Baggs was also recorded as having general command.

325   [Brammer account], "Jenkins Turned Up Again," *Ironton Register*, April 2, 1863, p. 2 and Gallipolis *Journal*, April 2, 1863, p. 2.

326   Gallipolis *Journal*, April 2, 1863, p. 2.

327   Carter to the Editor of the *Gallipolis Dispatch*, "POINT PLEASANT BATTLE. Letter from Capt. Carter Commanding Post" (dated "April 6, 1862"), Point Pleasant *Weekly Register*, April 16, 1863, [p. 1].

328   Captain Fitch to Kanawha Division Headquarters at Charleston, West Virginia, dispatch (dated "6 P.M."), as referenced by Col. R.B. Hayes in his diary entry for Monday, March 30, *Diary and Letters*. Vol. II, pp. 399-400.

329   Carter, "POINT PLEASANT BATTLE. Letter from Capt. Carter Commanding Post" (dated "April 6, 1862"), Point Pleasant *Weekly Register*, April 16, 1863, Vol. II No. 5, [p. 1].

330   Ibid.

331   "The Point Pleasant Fight," *Marietta Register* (Marietta, Ohio: R.M. Stimson, 1862-1906), April 10, 1863, p. 3.

332   The foregoing details in quotation were supplied by Captain Carter in a letter he wrote refuting statements made by the editor of the *Gallipolis Dispatch*. Carter's full statement was the following: "As to the shells from the artillery which caused so much consternation among the enemy it was not owing to the destruction they caused, for they fell short, (as some of your citizens can

testify) the noise may have scared the Rebs some." Carter, "POINT PLEASANT BATTLE. Letter from Capt. Carter Commanding Post" (dated "April 6, 1862"), Point Pleasant *Weekly Register*, April 16, 1863, Vol. II No. 5, [p. 1]. Livia Poffenbarger, owner-editor of the Point Pleasant *Gazette* (1888-1913) indicated a time for the arrival in her otherwise summary research, writing that "About 3 o'clock re-enforcements from Gallipolis, Ohio, arrived on the opposite side of the Ohio, across from Point Pleasant. They brought with them a battery of Artillery, which was at once made up (a floating battery) to shell the town, inasmuch as the reinforcements mistakenly believed that it was the Confederates (instead of Federals) who occupied the Mason County Court House. Fortunately, however, before the firing could begin this error was corrected and the town and Carter's Co. E, who had taken cover there in the Court-house were saved." (Livia Simpson-Poffenbarger, "Battle of Point Pleasant," Poffenbarger Papers, Vol. 9, pp. 27-28, Mason Co. Pub. Lib., Point Pleasant.)

333   Capt. Fitch, A.Q.M. to Governor Pierpoint (dispatch dated "Gallipolis, [Ohio, March 30, 1863]") paraphrased and published under title "REBEL RAID ON THE KANAWHA," in the Wheeling *Daily Intelligencer*, March 31, 1863, p. 3.

334   Hayes, diary entries for March 30 and 31, *Diary and Letters*, pp. 399-400.

335   John Hall to Adj. Gen. H.J. Samuels (manu. dated "Point Pleasant, [W. Va.], April 6, 1863"), Adjutant General Papers, Miscellaneous 1861-65, W. Va. State Archives, courtesy of Terry Lowery.

336   Point Pleasant *Weekly Register*, April 16, 1863, [n.p.]

337   Gallipolis *Journal*, April 2, 1863, p. 2.

338   The foregoing quotations are from John D. Carter, Capt[ain] Co[mpany] E, 13ᵗʰ V[irgini]a Vol[unteer] Inf[antr]y, Commanding Post to the Editors of the Wheeling *Daily Intelligencer* concerning misinformation provided by a dispatch from E.P. Fitch (pub. in the *Intelligencer* March 30, 1863). Carter's letter was published under the title "Letter from Capt. Carter about the Point Pleasant Fight," Wheeling *Daily Intelligencer*, April 9, 1863, p. 2.

339   The foregoing quotations are taken from E.M. FitzGerald, editor, May 7, 1863, *Weekly Register*, [p. 2]. Some prisoners were loaded on the boat as indicated by FitzGerald and from Point Pleasant they were conveyed to Gallipolis where they were jailed. Some prisoners seem also have been jailed at Point Pleasant.

340   Capt[ain] J.D. Carter, Co[mpany] E 13ᵗʰ V[irgini]a V[olunteer] I[nfantry], commanding Post [at Point Pleasant] to the Editor of the *Gallipolis Dispatch*, "POINT PLEASANT BATTLE. Letter from Capt. Carter Commanding Post" (dated "April 6, 1862"), Point Pleasant *Weekly Register*, Thursday Morning, April 16, 1863, Vol. II No. 5, [p. 1].

341   R.B. Hayes to his wife, Lucy (dated "Camp White, April 10, 1863"), Williams, *Diary and Letters*, p. 404.

342   Account provided by 'Mr. Brammer' to the *Ironton Register*, published as "Jenkins Turned Up Again," *Ironton Register*, April 2, 1863, p. 2.

343   Hayes to Lucy (dated "Camp White, April 10, 1863"), *Diary and Letters*, p. 404.

344   "Particulars of the Rebels Raid on Point Pleasant — Government Steamboats Fired Into &c.," Cincinnati *Daily Commercial*, April 2, 1863, p. 2.

345   Record of Events Field and Staff, Company Muster Rolls March and April 1863 (micro. #594 Roll 196); M508 Compiled Service Records of Volunteers (Union) West Virginia Station: Winchester March and April 1865 (micro. # M508 Roll 204), NARA; and Record of Events, Thirteenth West Virginia Infantry October 1862-June 1865 Field and Staff, Supp. O.R. Vol. LXXIV, p. 545.

346   See *Myer's History of West Virginia*, Vol. 1, p. 473 and Poffenbarger, "Battle of Point Pleasant," Poffenbarger Papers, Vol. 9, pp. 27-28.

347   Ibid. Point Pleasant was bordered by the Ohio River on the west and the Kanawha River to the south and by Crooked Creek and rising hills on the east. The northern perimeter of the town at that time probably extended as far north as the Pioneer Cemetery (no farther north than 10th Street), and it was here, presumably, in the flat open area that the government horses were stabled and forage was stored.

348   The *Marietta Register* (Marietta, Ohio) reported that government property destroyed or taken by the rebels consisted of two cribs of corn, about "7000 bushels" of corn burned and about "40 horses taken." ("The Point Pleasant Fight," *Marietta Register*, April 10, 1863, p. 3.) Brammer reported to the *Ironton Register* that corn stored at a "government stable" valued in the amount of "$10,000" was "fired and destroyed." ("Jenkins Turned Up Again," *Ironton Register*, April 2, 1863, p. 2.) Accurate or not, this was the only monetary assessment made as to damage of any kind at Point Pleasant that came to be offered up in reports found by me.

349   Letter from Point Pleasant to Wheeling *Daily Intelligencer* providing "particulars of Jenkins' raid on that town" (dated "April 4, 1863"), "The Jenkins' Raid Upon Point Pleasant," Wheeling *Daily Intelligencer*, April 4, 1863 p. 3.

350   Confederates reportedly stole "a quantity of goods from two Union stores part of which was recovered." The mercantile establishments were those of R [or B.] Gilmore and Leonard & Gates). ("Battle of Point Pleasant," reprinted from the Point Pleasant *Weekly Register* in the Wheeling *Daily Intelligencer*, April 6, 1863, p. 2.)

351   Carter, "POINT PLEASANT BATTLE. Letter from Capt. Carter Commanding Post" (dated "April 6, 1862"), Point Pleasant *Weekly Register*, April 16, 1863, [p. 1]. See also an earlier local account in which it was stated that while in Point Pleasant, the Confederates "burned two cribs of corn, and one stable, and stole a small amount of government clothing." ("Battle of Point Pleasant," reprinted from the Point Pleasant *Weekly Register* in the Wheeling *Daily Intelligencer*, April 6, 1863, p. 2.)

352   John Hall to Adj. Gen. H.J. Samuels (manu. dated "Point Pleasant, [W. Va.], April 6, 1863"), Adjutant General Papers, Miscellaneous 1861-65, WVSA, courtesy of Terry Lowery. Capt J.V. Young at Coals Mouth (or Coalsmouth) at the time of the raid, wrote to a friend that Jenkins' soldiers "stole about one hundred head of horses and a great deal of other property." (Young to Captain [Edgar B. Blundon?, Young's friend and Captain in the 8th Loyal Va.] (dated "Coals Mouth, April 5, 1863"), Young Civil War Papers (Vol. 31 Box 10), WVU, trans. p. 3. Livia Poffenbarger, editor and publisher of the Point Pleasant *Gazette* (a postwar newspaper which frequently presented historical articles), wrote that the horses Jenkins' men obtained at Point Pleasant were acquired from their own supporters. ("Battle of Point Pleasant," Poffenbarger Papers, Vol. 9, p. 38.)

353   Gallipolis *Journal*, Thursday April 2, 1863, p. 2.

354   "Rebel Prisoners from Point Pleasant — More Particulars of the Raid of Jenkins," Wheeling *Daily Intelligencer*, April 6, 1863, p. 3.

355   "Particulars of the Rebels Raid on Point Pleasant — Government Steamboats Fired Into &c.," Cincinnati *Daily Commercial*, April 2, 1863, p. 2.

356   Letter requesting a duplicate copy of Lieutenant Joseph Bromley's commission "as his commission was destroyed by Rebels in the fight at Point Pleasant." Manuscript dated "Head Quarters 13th V.V.I. Regiment, Hurricane Bridge, West Virginia, June 3, 1863," Uncat. 13th W.Va. Infantry Regiment Box, WVSA.

357   "Rebel Prisoners from Point Pleasant—More Particulars of the Raid of Jenkins," Wheeling *Intelligencer*, April 6, 1863 p. 3.

358   "ONE WHO KNOWS THE FACTS," "What the Government Makes by Having the Right Man in the Right Place," Point Pleasant *Weekly Register*, Thursday Morning, April 16, 1863, Vol. II No. 5, [p. 1].

359   "Rebel Prisoners from Point Pleasant—More Particulars," *Wheeling Daily Intelligencer*, April 6, 1863, p. 3.

360   [Bammer account], "Jenkins Turned Up Again," *Ironton Register*, April 2, 1863, p. 2 and Gallipolis *Journal*, April 2, 1863, p. 2. Col. R.B. Hayes made a similar remark in his journal, writing that the rebels had retreated up Kanawha "starving out of shoes

and ammunition." (Hayes, diary entry for "10 P.M., March 30," Williams, *Diary and Letters,* pp. 399-400.) Poffenbarger also noted writing in her "War Notes" that during the Point Pleasant attack "[s]ome of the rebels appeared in our streets bareheaded, others barefooted. Some skulked in wood sheds and back yards. Some searched for horses. Others gave attention to firing of buildings. [...] Incredible as it may seem, a gun and cartridge box was taken from a wounded rebel, the box containing a piece of soap, some blue vitrol [vitriol or sulphuric acid] and minnie balls covered with that deadly poison." (Poffenbarger, "War Notes" in Poffenbarger Papers, Vol. 9, p. 57.)

361   "ONE WHO KNOWS THE FACTS" for the *Weekly Register,* pub. as "What the Government Makes by Having the Right Man in the Right Place," Point Pleasant *Weekly Register,* Thursday Morning, April 16, 1863, Vol. II No. 5, [p. 1].

362   Information learned from questioning officers Samuels, Timms and Holderby, summarized and published in "The Late Raid Upon Point Pleasant—What Jenkins Intended to Do," Wheeling *Daily Intelligencer,* April 7, 1863, p. 3.

363   Ibid., all quotes in this paragraph.

364   Letter received from Point Pleasant on April 4, 1863, at the offices of the Wheeling *Daily Intelligencer* providing "particulars of Jenkins' raid on that town," "The Jenkins' Raid Upon Point Pleasant," Wheeling *Daily Intelligencer,* April 4, 1863 p. 3.

365   "The Point Pleasant Fight," *Marietta Register,* April 10, 1863, p. 3.

366   *Gallipolis Journal,* July 2, 1863, p. 2.

367   Mark E. Robison to his father, John Robison (dated "[Point Pleasant] April the 6 1863"), Mark E. Robinson, Military Pension File, NARA.

368   Brig. Gen. E.P. Scammon, by Jas. L. Botsford, Assistant Adjutant General to Col. W.R. Brown 13th Regt. Va. Vol. Inf., Special Order No. 51 (dated "Head Quarters District Kanawha, Charleston, Va., April 4, 1863"), RG94 Box 4838 Misc. unbound regimental papers 13th W.Va., Spec. Orders and Gen. Orders, NARA.

369   Hayes, diary entry for "Friday, April 3, 1863," Williams, *Diary and Letters,* p. 481.

370   Williams, *Hayes of the Twenty-Third,* p. 150.

371   R.B. Hayes to his wife, Lucy (dated "Camp White, Sunday April 5, 1863"), Williams, *Diary and Letters,* p. 402. The saltworks referred to by Hayes would have been the Charleston saltworks. These were scattered along the Kanawha River road between Charleston and Camp Piatt (a distance of about twelve miles). The saltworks had been low on man-power since the start of the war and especially since Loring's occupation of Charleston the previous year, when Black laborers employed in the saltworks had departed for Ohio and the Northern States. The majority of saltworks were owned by men openly in sympathy with the South, and they would have done all in their power to provide that precious commodity to the Confederacy in the absence of U.S. military, thus the wisdom of destroying the saltworks in the event that Union forces were driven out of the Kanawha Valley.

372   "From the Point Pleasant *Register* April 2d. Battle of Point Pleasant," reprinted in the *Ironton Register,* April 9, 1863, p. 2.

373   Field History 13th Regiment West Virginia Infantry, manu. dated "Head Quarters 13th Regt. W. Va. Vol. Inft., Barboursville, W. Va., Jan'y 13th, 1864," [n.p.] Lieutenant William N. Hawkins of Kanawha Co. received a severe and painful wound which in time healed sufficiently to permit his return to duty.

374   Record of Events Field and Staff, Muster Roll Co. E 13th Va. Vol. Infantry for March and April 1863 (micro. #594 Roll 196), NARA.

375   Poffenbarger, "Battle of Point Pleasant," Poffenbarger Papers, Vol. 9, pp. 27-28.

376   Ibid.

377   B.J. Redmond to F.H. Pierpont, Governor of West Virginia (dated "Point Pleasant, April 2, 1863"), Wheeling *Daily Intelligencer*, April 9, 1863, p. 1.

378   Col. R.B. Hayes to Capt. James L. Botsford, Assist. A.G., Dist. Kanawha, Charleston, Va. (dated "Head Qtrs. 1st Brigade District of Kanawha, Camp White, Va., April 14, 1863") Registry #393 Pt. 2 Entry 1219 Vol. 139, Press Copies of Letters Sent and Special Orders 1863, NARA.

379   Poffenbarger, "Battle of Point Pleasant," Poffenbarger Papers, Vol. 9, pp. 27-28.

380   Capt. Fitch at Gallipolis to Kanawha Division Headquarters at Charleston, West Virginia (dispatch dated "9 P.M."), as referenced by Hayes, diary entry for "Monday, March 30," Williams, *Diary and Letters*, pp. 399-400.

381   *The Civil War Diary of Mrs. Henrietta Fitzhugh Barr [Barre] 1862-1863* (Marietta, Ohio: Hyde Brothers Printing Co., 1963), p. 21.

382   Colonel W.R. Brown, by W.I. Mathews, Adj., Special Order No. 5 (manu. dated "Head Quarters 13th Regt Va Vol Inftry Hurricane Bridge April 18th 1863"), RG94 13th Regt. Inftry. Vols. Order Book, NARA, [n.p.]

383   Field History 13th Regiment W.Va. Infantry (manu. addressed to F. Pierpont, Adj. Genl. W.Va. and dated "Head Quarters 13th Regt. W. Va. Vol. Inft., Barboursville, W. Va., Jan'y 13th, 1864"), Uncat. 13th W.Va. Infantry Regiment Box, WVSA, [n.p.] Alexander H. Samuels was originally from Barboursville, Cabell County, brother of Judge H.J. Samuels, State Adjutant General of Virginia, Restored (or Provisional) Virginia State Government. "Alex" enlisted in the "Border Rangers," a company raised by Jenkins from lower Mason and Cabell Counties. This company later became Company E 8th Virginia Cavalry Regiment, Jenkins' command, at Greenbottom, Cabell County, May 29, 1861. Samuel was elected 2nd Lieutenant upon organization of Co. E. He was captured at Point Pleasant during the course of the fight there, on March 30, 1863, and imprisoned. He was soon released and re-enlisted in the Confederate army on April 30, when he was promoted to the rank of 1st Lieutenant. (Geiger, *Civil War in Cabell County*, p. 96.)

384   Record of Events Field and Staff, Muster Roll Co. E for March and April 1863 (micro. #594 Roll 196), NARA.

385   Redmond to F.H. Pierpont, Gov. of W. Va., dated "Point Pleasant, April 2, 1863," reprinted in the Wheeling *Daily Intelligencer*, April 9, 1863, p. 1.

386   "The Late Raid Upon Point Pleasant—What Jenkins Intended to Do," Wheeling *Daily Intelligencer*, April 7, 1863, p. 3.

387   Hayes, diary entry for "Friday, April 3, 1863," Williams, *Diary and Letters,* p. 481.

388   Poffenbarger, "Battle of Point Pleasant" in Poffenbarger Papers, Vol. 9, pp. 27-28. Both men were Privates in the 8th Virginia Cavalry Regiment. Albert Neale in Co. D and Edward S. Guthrie in Co. E. See Dickinson, *Tattered Uniforms and Bright Bayonets,* respectively pp. 283 and 183.

389   Gallipolis *Journal*, Thursday, April 2, 1863, p. 2.

390   Ibid.

391   "Jenkins Turned Up Again," *Ironton Register*, April 2, 1863, p. 2.

392   "Battle of Point Pleasant," reprinted from the Point Pleasant *Weekly Register* in the Wheeling *Daily Intelligencer*, April 6, 1863, p. 2.

393   John Hall to [State Adjutant] General H.J. Samuels, dated "Point Pleasant, [W. Va.], April 6, 1863," Adjutant General Papers Miscellaneous 1861-65, WVSA, courtesy of Terry Lowery.

394   Ibid.

395   Ibid.

396   RG94 Morning Reports of Capt. Simon Williams' Co. D 13th Regiment Virginia Vols. for April 1863, NARA, [n.p.] April morning reports for Co. D indicate that among those detailed to pursue Jenkins was 1st Lieutenant James Hanna and fifty-two enlisted men who "went in pursuit of General Jenkins forces towards Barboursville" on April 1 or 2.

397   Ibid., Hall to Samuels (dated "Point Pleasant, [W. Va.], April 6, 1863"), Adj. Gen. Papers Misc. 1861-65, WVSA. The surgery Hall referred to was performed by Dr. Abraham D. Williams, who would join the 13th Regiment as Assistant Surgeon on May 10, 1863.

398   Information learned from questioning officers Samuels, Timms and Holderby, summarized and published in "The Late Raid Upon Point Pleasant—What Jenkins Intended to Do," Wheeling *Daily Intelligencer*, April 7, 1863, p. 3.

399   J.V. Young to Captain [Edgar B. Blundon?] (dated "Coals Mouth, April 5, 1863"), Young Civil War Papers, trans. p. 62. Capt. Young and Capt. Rucker had been comrades-in-arms in the Putnam County 181st Virginia Militia before the war.

400   R.B. Hayes, diary entry for "Friday, April 3, 1863," Williams, *Diary and Letters*, p. 481.

401   Mrs. Walter Mitchell, "History of Cabell Creek Community" (1925), as cited by Geiger, *Civil War in Cabell County*, p. 82.

402   Col. W.R. Brown to Capt. J.V. Young, dated "Head Quarters 13th Regt. V.V.I., April 2d, 1863," Young Civil War Papers.

403   Lieutenant Robert Brooks resigned from the U.S. service sometime after this scout was made. On June 25, 1863, Colonel Brown, commanding the 13th and his adjutant, John S. Cunningham, wrote to the Governor of West Virginia on behalf of Brooks to secure his back pay and recommend him for some appointment: "The undersigned have the honor to recommend to your notice the bearer of this, Robert Brook who enlisted in Capt. J.V. Young's Co G (attached to this Regt) as a private in December 1861 — he was promoted to 1st Lt in May 1862 and has served in the Army until recently when he resigned Owing to some informality in his muster he has not been paid for any part of his services from Dec 1861 to sometime this month. We respectfully recommend him to your notice for appointment as we consider him a brave moral and efficient officer, who has a general knowledge of West Virginia, particularly this part of the state, and has performed some difficult scouting whilst in the service — he has a family and the means he had to supply his family, have been exhausted whilst he was in the service." Colonel William R. Brown and Adjutant John S. Cunningham to Governor of West Virginia (dated "Head Quarters 13th Regt. Va. Vol. Infty, Hurricane Bridge, West Va., June 25, 1863"), Samuels-Pierpont Papers, WVSA.

404   Confederate cavalry commander Colonel John Clarkson is probably who is named here.

405   Colonel John H. Oley, 8th Virginia Infantry, U.S.A.

406   J.V. Young to Captain [Edgar B. Blundon?] (dated "Coals Mouth, April 5, 1863"), trans. p. 62.

407   Andrew Stiarwalt, Sr., Musician, Co. F 23rd Ohio Vol. Infantry Regiment, Diary 1861-64, Russell P. Hastings Papers, San Francisco, California, R.B. Hayes Papers Civil War Records (micro. Series 7/ roll No. 272), p. 6 of manu.

408   R.B. Hayes to his wife, Lucy (dated "Camp White, Sunday April 5, 1863"), *Diary and Letters*, p. 402.

409   Ibid., Hayes to Lucy, p. 402 and S.P.D. (dated "Camp Fayetteville, Virginia, April 15th, 1863"), "From the Second Virginia," *The Ironton Register*, April 23, 1863, p. 2.

410   S.P.D. (dated "Camp Fayetteville, Virginia, April 15th, 1863"), "From the Second Virginia," *Ironton Register*, April 23, 1863, p. 2.

411   Ibid., Hayes to Lucy (dated "Camp White, Sunday April 5, 1863"), *Diary and Letters*, p. 402.

412   Stiarwalt, Diary 1861-64, Russell P. Hastings Papers (micro. Series 7/ roll No. 272), p. 6 of manu.

413   Hayes to Lucy (dated "Camp White, April 10, 1863"), *Diary and Letters*, p. 404.

414   Robert C. Schenck, Maj. Gen., quoting Halleck (dated Baltimore, Md., March 31, 1863-12:30 a.m.), O.R. Vol. XXV Part I, as referenced by Poffenbarger, "Battle of Point Pleasant" in Poffenbarger Papers, Vol. 9, p. 47.

415   Anonymous typewritten document, "13th Regiment of Infantry," WVSA, p. 810.

416   Record of Events Field and Staff, Company Muster Rolls March and April 1863 (micro. #594 Roll 196); M508 Compiled Service Records of Volunteers (Union) West Virginia Station (micro. #M508 Roll 204), NARA; and Record of Events for the Thirteenth West Virginia Infantry, October 1862-June 1865 Field and Staff, Supp. to the O.R. Vol. LXXIV, p. 545

417   Ibid., Supp. O.R. Vol. LXXIV, p. 545.

418   Record of Events Field and Staff Special Muster Rolls Cos. B, C, D and F (dated "April 10, 1863" micro.# 594 Roll 196), NARA.

419   Hayes to Brown, by Avery (manu. dated "Hd Qtrs. 1st Brig. Dist. Kan., Camp White, Va., April 17, 1863"), RG94 Box 4839 Misc. unbound regimental papers 13th W.Va., NARA.

420   Col. Wm. R. Brown to Lieut. O.W. Griswold (manu. dated "13th Regt., Hurricane Bridge, Va., April 19, 1863"), RG94 13th Regt. Inftry. Vols. Order Book Co. H, [n.p.]

421   See however Supplement to the Official Records of the War of the Rebellion which states that Cos. B and G were stationed at Mud Bridge, W.Va. (Supp. O.R. Vol. LXXIV, p. 550.)

422   Field History 13th Regiment W.Va. Infantry (manu. dated "Head Quarters 13th Regt. W.Va. Vol. Inft., Barboursville, W.Va., Jan'y 13th, 1864"), Uncat. 13th W.Va. Inftry. Regt. Box, [n.p.]

423   Col. R.B. Hayes to Col. William R. Brown (manu. dated "Head Qtrs. 1st Brigade 3rd Division, Middle Dept., 8th Army Corps, Camp White, Va., April 25, 1863"), Registry #393 Part 2 Entry 1184 Press Copies of Letters Sent and Special Orders 1 of 2 Vol. 139, NARA; R.B. Hayes to Brown, by M.P. Avery (manu. dated "Hd. Qtrs. 1st Brig. 3d Division 8th A.C., Camp White, Va. April 25, 1863"), RG94 Box 4839 Misc. unbound regimental papers 13th W.Va., NARA.

424   Two sources show that Co. B was stationed at Mud River Bridge, W.Va. These were Co. B Rolls for April 30, 1863, and John Smoot, Military Pension File, NARA.

425   Col. R.B. Hayes to Lt. Col. James R. Hall (manu. dated "Head Qtrs. 1st Brigade 3rd Division, Dept 8th Army Corps, Camp White, Va., April 25, 1863"), Registry #393 Part 2 Entry 1184 #49 Letters Written to Lt. Col. James R. Hall H. 2 of 2 Vol. 139, NARA.

426   R.B. McCall Comdg. Detachment 5th Va. to Col. Hall, (manu. dated "Mud Bridge, April 28, 1863"), RG94 Box 4839 Misc. unbound regimental papers 13th W.Va., NARA.

427   William R. Brown, Col. Comdg, by John S. Cunningham, Lt. and Acting Adjutant to James R. Hall, Lt. Col. 13th Regiment, Special Order No. 10 (manu. dated "Head Quarters 13th Regt V.V.I. Hurricane Bridge April 30th 1863"), RG94 13th Regt. Inftry. Vols. Order Book, NARA, [n.p.]

428   Field History 13th Regiment W.Va. Infantry, [n.p.]

429   Col. R.B. Hayes to Capt. J.D. Carter (dated "Head Qtrs. 1st Brigade 3rd Division, Dept 8th Army Corps, Camp White, Va., April 25, 1863"), Registry #393 Part 2 Entry 1184 Letters written to Capt. J.D. Carter 1 of 2 Vol. 139, NARA.

430   West Virginia Clothing Books Co. C 13th Infantry, WVSA, [n.p.].

431   "The 13th Regiment West Va. Inf. Memoranda," WVSA, [n.p.]

432   Mark E. Robison to his father, John Robison (manu. dated "[Point Pleasant] April the 6 1863"), Mark E. Robinson, Military Pension File, NARA.

433   R.B. Hayes, Col., by M.P. Avery, Capt. and A.A.A.G. to Col. W.R. Brown (manu. dated "Hd. Qtrs. 1st Brig. Dist. Kanawha, Camp White, Va., April 6, 1863") RG94 Box 4839 Misc. unbound regimental papers 13th W.Va.

434   Burnside to Halleck (dated Headquarters Dept. of the Ohio, Cincinnati, Ohio, April 17, 1863), O.R. Series I Vol. XXIII Part II (Washington: Government Printing Office, 1889), p. 247.

435   Burnside to Halleck (dated Cincinnati, Ohio, April 20, 1863—2:30 p.m.), O.R. Vol. XXIII Pt. II, p. 259.

436   O.R. Vol. XXV Part II, pp. 652, 685, 710-712 and Festus P. Summers, "The Jones-Imboden Raid," *West Virginia History*, Vol. XLVII, 1988, p. 53.

437   Charles H. Ambler and Festus P. Summers, *West Virginia the mountain state* (Englewood, Cliffs, N. J.: Prentice Hall, 1958), p. 225.

438   Summers, "The Jones-Imboden Raid," *West Virginia History*, p. 62.

439   Ibid., Ambler and Summers, p. 53.

440   Burnside to Halleck (dated Cincinnati, Ohio, April 20, 1863—2:30 p.m.), O.R. Vol. XXIII Part II, p. 259.

441   Brown [to the State Adjutant General at Wheeling?] (manu. dated "Head Qtrs. 13th Hurricane Bridge, Va. April 15, 1863"), Uncat. 13th W.Va. Inftry. Regt. Box.

442   W.R. Brown, Col. Comdg, Spec. Order No. 6 (manu. dated "Head Quarters 13th Reg[iment] V[irgini]a Vol[unteer] Inf[antry], Hurricane Bridge, V[irgini]a April 20th, 1863"), RG94 13th Regt. Inftry. Vols. Order Book Co. H, [n.p.]

443   William R. Brown, Col. Comdg 13th Regt V.V.I., by William I. Mathews, Adj., Spec. Order No. 7 (manu. dated "Head Quarters 13th Regt V.V.I. Hurricane Bridge April 21st 1863"), RG94 Order Book Co. H, [n.p.]

444   John S. Cunningham Muster-In Roll as 1st Lieut. and Adjt.13th Regt., RG94 Box 4838 Misc. unbound regimental papers 13th W.Va.

445   W[illia]m R. Brown, Col[onel] Commanding, by William I. Mathews, Adj., Spec. Orders No. 1 (manu. dated "Head Quarters 13th Regt V.V.I. Hurricane Bridge April 15th 1863"), RG94 13th Regt. Inftry. Vols. Order Book Co. C, [n.p.]

446   William R. Brown, Col. Comdg, by William I. Mathews, Adj., Spec. Order No. 2 (manu. dated "Head Quarters 13th Regt V.V.I. Hurricane Bridge April 16th 1863"), RG94 Order Book Co. C, [n.p.]

447   Colonel W.R. Brown, by W.I. Mathews, Adj., Spec. Order No. 4 (manu. dated "Head Quarters 13th V[irgini]a V[olunteer] Inf[an]t[r]y, Hurricane Bridge April 15th, 1863"), Order Book Co. C, [n.p.]

448   W.R. Brown, Col. Comdg Reg., by W.I. Mathews Adj., Spec. Order No. 5 (manu. dated "Head Quarters 13th Regt Va Vol Inftry Hurricane Bridge April 18th 1863"). RG94 13th Regt. Inftry. Vols. Order Book, [n.p.] The prisoners referred to by Brown seem to have been all taken at the battle of Point Pleasant.

449   Col. R.B. Hayes to Capt James L. Botsford, Assist. A.G., Dist. Kanawha, Charleston, Va. (manu. dated "Head Qtrs. 1st Brigade District of Kanawha, Camp White, Va., April 14, 1863"), Registry #393 Part 2 Entry 1184 Press Copies of Letters Sent and Special Orders 1 of 2 Vol. 139, NARA.

450   R.B. Hayes, Col. Commdg. to Col. W.R. Brown, Comdg 13th Va. Inf., Hurricane Bridge, Va. (manu. dated "Head Quarters 1st Brig. Dist. Kan[awha], Camp White Va April 16, 1863"), RG94 Box 4839 Misc. unbound regimental papers 13th W. Va. Letters and Col. R.B. Hayes to Col Brown 13th Va. (manu. dated "Head Qtrs. 1st Brigade District of Kanawha, Camp White, Va., April 16, 1863"), Registry #393 Part 2 Entry 1184 Press Copies of Letters Sent and Special Orders 1 of 2 Vol. 139, NARA.

451   Capt. and A.A.A. General, H.P. Avery to Col. William R. Brown (manu. dated "Head Qtrs. 1st Brigade 3rd Division, Dept 8th Army Corps, Camp White, Va., May 4, 1863"), Registry #393 Part 2 Entry 1184 Press Copies of Letters Sent and Special Orders 1 of 2 Vol. 139, NARA.

452   Col. R.B. Hayes to Col. Brown 13th Va. (manu. dated "Head Qtrs. 1st Brigade District of Kanawha, Camp White, Va., April 23 1863"), Registry #393 Part 2 Entry 1184 Press Copies of Letters Sent and Special Orders 1 of 2 Vol. 139, NARA and RG94 Box 4839 Misc. unbound regimental papers 13th W.Va.

453   Col. R.B. Hayes to Col. Brown 13th Va. (manu. dated "Head Qtrs. 1st Brigade District of Kanawha, Camp White, Va., April 23 1863"), Registry #393 Part 2 Entry 1184 Press Copies of Letters Sent and Special Orders 1 of 2 Vol. 139, NARA.

454   J.D. Carter, Capt. Comdg. Post to J.S. Cunningham, Lieut. and A[sst]. Adjt. 13th V.V.I., Hurricane Bridge, Va. (manu. dated "Point Pleasant, Va., April 25, 1863"), RG94 Box 4839 Misc. unbound regimental papers 13th W.Va., NARA.

455   Ibid.

456   Wm. McKinley, Jr., 1st Lt. + A.A.Q.M. to S. Comstock R.Q.M. 13th Regt. Va. Vols. Pt. Pleasant (manu. dated "Hd. Qtrs. 1st Brig. 3rd Div. 8th A.C. Camp White, Va. April 23, 1863"), Registry #393 Part 2 Entry 1194 Register of letters received by Quartermasters, Letter 13, NARA, [n.p.]

457   Wm. McKinley, Jr., 1st Lt. + A.A.Q.M. to S. Comstock R.Q.M. 13th Regt. Va. Vols. et al (manu. dated "Hd. Qtrs. 1st Brig. 3rd Div. 8th A.C. Camp White, Va. April 25, 1863"), Registry #393 Part 2 Entry 1194 Register of letters received by Quartermasters, Letter 16, [n.p.]

458   Col. R.B. Hayes to Col. Brown at Hurricane Bridge; to Lt. Col. J.R. Hall at Mud Bridge; and to Col. A.A. Tomlinson at Barboursville (manu. dated "Head Qtrs. 1st Brigade 3rd Division, M.D. [Middle Dept.] 8th Army Corps, Camp White, Va., April 29, 1863"), Registry #393 Part 2 Entry 1184 Letters Written to Lt. Col. James R. Hall H. #601 of 2 of 2 Vol. 139 and Col. R.B. Hayes, by M.P. Avery to Brown, Order No. 19 (manu. dated "Head Quarters 1st Brigade 3d Division 8 A.C., Camp White, Va., April 29, 1863 Received April 30, 1863"), RG94 Box 4839 Misc. unbound regimental papers 13th W.Va.

459   William R. Brown, Col. Commanding, Spec. Orders No. 8 (manu. dated "Head Quarters 13th Regt Va Vol Infty Hurricane Bridge April 23rd 1863"), RG94 13th Regt. Inftry. Vols. Order Books, [n.p.]

460   William R. Brown, Col. Comdg, by John S. Cunningham, Lt. and Acting Adjutant, Spec. Order No. 9 (manu. dated "Head Quarters 13th Regt Va Vol Infty, Hurricane Bridge April 24th 1863"), RG94 13th Regt. Inftry. Vols. Order Books, [n.p.]

461   W.R. Brown, Col[onel] Com'd'g, by Lt. J.S. Cunningham Act[in]g Adj[utan]t, Special Order No. 8 (manu. dated "Head Quarters 13th Reg[imen]t V[irgini]a Inf[an]t[ry], Hurricane Bridge April 28th, 1863"), RG94 13th Regt. Inftry. Vols. Order Books, [n.p.] Articles of War (American Civil War Union Army) set down rules governing behavior and conduct for officers and soldiers and the consequences for breaking these rules. Article 49 provided that: "Any officer belonging to the service of the United States, who, by discharging of firearms, drawing swords, beating of drums, or by any other means whatsoever, shall occasion false alarms in camp, garrison, or quarters shall be punished, according to the nature of his offense, by the sentence of court martial." Sons of Union Veterans of the Civil War, https://suvcw.org., 2023.

462   RG94 Morning Reports Co. A 13th Regiment Virginia Vols. for April 1863, NARA, [n.p.]

463   RG94 Morning Reports Co. B 13th Regiment Virginia Vols. for April 1863, NARA, [n.p.] and RG94 Misc. Records Descriptive Roll Co. B 13th Regiment V.V.V.I., List of Non-Commissioned Officers and Register of Deaths, NARA, [n.p.]

464   Adjutant General's Office, War Dept., Wash. D.C., Patrick H. Caldwell, Military Pension File, NARA.

465   Ibid.

466   Milton Stewart, Captain, Co. B 13th Virginia Volunteers to Robert Caldwell (manu. dated "Coals Mouth Va April 20th/ 63"), Patrick H. Caldwell, Company B 13th W. Va. Regiment, Military Pension File, NARA. The $75.00 Federal bounty payed to the family after the soldier's death was used to buy his gravestone and to support his family. See letter from the Adj. Gen. Off. War Dept. Wash. D.C., Caldwell, Military Pension File.

467   Milton Stewart, Captain Co. B 13th Va. Vols. to Robert Caldwell (manu. dated "Coals Mouth, West Virginia, April 25, 1863"), Patrick H. Caldwell Pension File.

468   RG94 Morning Reports Co. B 13th Regiment Virginia Vols. for April 1863, NARA, [n.p.] and RG94 Misc. Records Descriptive Roll Co. B. 13th Regiment V.V.V.I. List of Non-Commissioned Officers and Register of Deaths, NARA, [n.p.]

469   Abraham Asberry, Private, Co. C 13th W. Va., Military History Cards on micro., WVU, Morgantown.

470   RG94 Morning Reports Co. C 13th Regiment Virginia Vols. for April 1863, NARA, [n.p.]

471   RG94 Morning Reports of Captain Simon Williams' Co. D 13th Regiment Virginia Vols. for April 1863, NARA [n.p.] These reports are unsigned and appear to have been a duplicate copy.

472   Lovina Burrows to her husband, David Burrows, Private, Co. F 13th Regiment Virginia Volunteers (letter received April 21, 1863), as quoted in Roush, *If Thou Wilt Remember,* p. 29.

473   David Burrows to his wife (dated April 22, 1863), Roush, *If Thou Wilt Remember,* pp. 29-30.

474   J.D. Carter, Captain 1[3] Va. V.I., Commanding Post, Special Order No. 39 (dated "Point Pleasant April 8, 1863") M508 Roll 260 Compiled Service Records of Volunteers 13th Regiment Infantry W.Va. Personal Papers arranged by organization, NARA.

475   RG94 Morning Reports of Co. F 13th Regiment Virginia Vols. for April 1863, NARA, [n.p.]

476   Brig. Genl Scammon to Col. W.R. Brown (manu. dated "Head Quarters Dist. Kananwha, Charleston Va., April 10, 1863"), RG94 Box 4839 Misc. unbound regimental papers 13th W.Va.

477   Col. R.B. Hayes to "Capt Young Co 13th Va at Coalsmouth Va." (manu. dated "Head Qtrs. 1st Brigade District of Kanawha, Camp White, Va., April 14, 1863"), Registry #393 Part 2 Entry 1184 Press Copies of Letters Sent and Special Orders 1 of 2 Vol. 139, NARA.

478   Capt. J.V. Young Com'dg. Detachment to Pauline "Marsh" (dated "Coals Mouth, April 19, 1863"), Young Civil War Papers (Vol. 31 Box 10), trans. p. 63.

479   Ibid.

480   For the foregoing quotations see John Hall to Governor F.H. Pierpont, at Wheeling (manu. dated "April 23, 1863"), Uncat. 13th W.Va. Inftry. Regt. Box.

481   George M. Kellog, Surgeon U.S.V., Medical Director District of Kanawha to F.[H.] Pierpont, Gov. of West Virginia (manu. dated "Medical Directors Office 3rd Division 8th Army Corps, Charleston, Va., April 28, 1863"), Samuels-Pierpont Papers, WVSA.

482   A.D. Williams was the son of Pryor Williams (born 1810; died 1846) and Anna Kern (born 1815; died 1895). The couple married July 7, 1834. After Pryor's death in 1846, Anna married a second time, to one Daniel Hall, a widower with two minor children. With Daniel Hall, Anna had four more children. One wonders if Daniel Hall was a relative of John Hall, Esq., of Point Pleasant, a coincidence which might explain how it came about that the Hoosier performed surgery upon John Hall's "cancer of the eye" and then was recruited and enrolled at Point Pleasant for the 13th West Virginia Volunteers.

483   1860 U.S. Census, Spice Valley Township, Lawrence County, Indiana taken August 7, 1860, p. 585.

484   Dr. Elkanah Williams to Abram Williams (dated "[Seminary Hospital Frederic] Nov. 12, [1862]"), trans. in "Williams Family History Elkanah Williams: Correspondence," Ancestry.com.

485   Jefferson Medical School had a long and prestigious history. It had been founded in 1825 by Dr. George McClellan upon the then revolutionary idea of clinical practice. Medical students would not just 'read medicine' but they would learn their profession by observing experienced doctors treating patients — an idea which continues today. There were few hospitals in 1825 but from its earliest days, Jefferson College maintained a teaching hospital. At first, in 1825, it was just a small infirmary at Tivoli Theater, Cannonsburg, Pennsylvania. Then, already by 1828, when the school moved into the Ely Building in Philadelphia, there was not only a hospital for clinical practice but a 700 seat surgical amphitheater, where students could watch master surgeons, such as Dr. Samuel Gross, perform operations. See the large (8 by 6.5 foot) painting by American Realist painter, Thomas Eakins of the 70-year-old Professor Gross, scalpel in hand, lecturing students in the surgical arena at Jefferson Medical College.

486   A.D. Williams, 1900 U.S. Census St. Louis, Missouri, Ward 21. The 1900 Census records that in 1900, Abram and Isabella had been married for 28 years.

487   Lizzie Ann was recorded as age 8 and Annie P. aged 5 in the 1880 U.S. Census St. Louis, Missouri.

488   A.D. Williams, M.D. St. Louis, Missouri (Formerly Lecturer on Otology in the Miami Medical College), *Diseases of the Ear Including The Necessary Anatomy Of The Organ. Illustrated With Numerous Woodcuts And One Lithograph* (Cincinnati: Robert Clarke & Co., 1873). This book is part of the public domain today. It is available in several rare book collections of medical libraries, has been scanned and is also available online. A superficial search of period publications also indicates that Williams published prolifically. In addition to his 310 page textbook *Diseases of the Ear* (1873), he wrote many articles for the *St. Louis Medical and Scientific Journal; British Guiana Medical Annotated; St. Louis Medical Fortnightly;* the *Kansas City Medical Record;* the *British Medical Journal;* and *St. Louis Annotated Ophthalmogy.*

489   In 1880, his address was given as 725 Chestnut (St. Louis City Directory for 1880) and in 1886 his address was published as 723 Chestnut, St. Louis, Missouri. (*Medical and Surgical Directory of the United States. Register,* R.L. Polk & Co., Publishers, 1886), p. 567.

490   "Woman's Medical College and Hospital," *Historical and Descriptive Review of St. Louis Her Enterprising Business Houses and Progressive Men* (St. Louis: Missouri Historical Society 63 Emilie Building John Lethem, [1894]), p. 222.

491   *Journal of the Indiana State Medical Association,* Vol. 5 No. 9, p. 410.

492   Willis Orville Nance, M.D. and Albert Henry Andrews, M.D., *Journal of Opthamology and Otalaryngology,* Vol. 6 (Chicago, 1912), p. 315 and "Dr. A.D. Williams Is Dead Physician Formerly Was Eye and Ear Specialist in St. Louis," uncited St. Louis, Missouri newspaper (dated August 19, 1912), courtesy of Archives Missouri History Museum. A.D. Williams was buried at Green Hill Cemetery, Bedford, Indiana.

493   Military History Cards on microfilm, Footnote.com/image/267950065; 267950067; 267950069; and 267950099.

494   Williams, Abram D. Asst. Surg., 13 Reg't W. Va. Inf., Military History Cards on microfilm, Footnote.com/image/267950101.

495   Williams, Footnote.com/image/267950107; 267950110; 267950112. Abrams' resignation (dated "Feb. 20, 1864") was forwarded to 1st Brigade HdQrs. with the notation by Col. Brown that "Asst Surgeon A.D. Williams feeling dissatisfied of the appointment of C D Dally to this Regt desires his resignation to be accepted." (Wm. R. Brown, dated "Hd Quarters 13 Regt VVI Barboursville Apr 6/64.") Resignation approved by Col. R.B. Hayes in a notation on the back, dated "Hd Qrs 1st Brig. 3 Division Camp White April 10, 1864."

496   Williams' leave of absence came in the form of Special Orders No. 124 (Brig. Gen. E.P. Scammon, by Jas. L. Botsford, with heading "Head Quarters 3d Division Dep't West Virginia" [no. date]). It read: "3rd Assistant Surgeon Abraham Williams 13th Regt Virginia Volunteer Infantry is hereby permitted to go to Cincinnati on business connected with the Civil Court. He will return in ten (10) days. By command of Brigadier General E.P. Scammon: Jas: L. Botsford Comdg. O. 13th V.V.I.Third Brig. Hd Quars." Williams, Footnote. com/image/267950129 and 267950134.

497   Williams, Footnote.com/image/267950144 and 267950148.

498   Williams, Footnote.com/image/267950152 and 267950159.

499   Telegraphic enclosure, Williams, Footnote.com/image/267950161.

500   L.L. Comstock, Surg., 7ᵗʰ W. Va.V.Cav., Act. Med. Director Kan. Val. Forces, Special Orders No. 1 (dated "Medical Directors Office Kanawha Valley Forces Charleston W. Va. July 17, 1862"), Footnote.com/image/267950123 and 2679501.

501   Williams, Abram D. Asst. Surg. 13 Reg't W. Va. Inf., Military History Cards on microfilm, Footnote.com/image/267950117 and 267950117.

502   These officers were Surgeon Charles D. Dalley and Assistant Surgeon Jacob Lallance. Dalley, of course was the unlicensed physician whose promotion over Dr. Williams prompted Williams first letter of resignation. Dr. Lallance was a licensed allopath, at least by 1881—West Virginia had at some point after the war and prior to 1881 begun to license their physicians. It should be noted in this context that Dr. Samuel G. Shaw of Point Pleasant and first Surgeon of the 13ᵗʰ regiment was also a licensed allopath. In the 1881 list, Shaw was shown as licensed. See George Worthington Adams, *Doctors in Blue. The Medical History of the Union Army in the Civil War* (Baton Rouge and London: Louisiana State Univ. Press), p. 1398, citing Directory of Deceased Physicians, Vol. I, p. 890.

503   Williams, Abram D. Asst. Surg. 13 Reg't W. Va. Inf., Military History Cards on microfilm, Footnote.com/image/267950164; 267950167; 267950171 and 297950173.

504   Williams, Footnote.com/image/267950137 and 267950140.

505   Colonel William R. Brown to Lieutenant O.W. Griswold, Commanding Co. H (manu. dated "Headquarters 13ᵗʰ Regt., Hurricane Bridge, Va., April 19, 1863"), RG94 Order Book Co. H, [n.p.]

506   RG94 Morning Reports Co. H 13ᵗʰ Regiment Virginia Vols. for April 1863, [n.p.]

507   RG94 Morning Reports Co. I 13ᵗʰ Regiment Virginia Vols. for April 1863.

508   Brigadier General B.S. Roberts at Clarksburg, Va. to Gov. Pierpont, [two telegrams] (dated "May 4, 1863"), *Calendar of the Frances Harrison Pierpont Letters and Papers in W.Va. Depositories* Prepared by the W.Va. Historical Records Survey Division of Professional and Service Projects, Works Projects Administration (Charleston, W.Va.: Historical Records Survey, October 1940), p. 241.

509   Col. R.B. Hayes to Col. W.R. Brown, Commanding 13ᵗʰ Va. at Hurricane Bridge and Lt. Col. James R. Hall, Commanding detachment of 13ᵗʰ at Mud Bridge and to Col. A.A. Tomlinson Commanding 5ᵗʰ Va. Vol. Inf. at Barboursville, Va. (manu. dated "Head Qtrs. 1ˢᵗ Brigade 3ʳᵈ Division, Dept 8ᵗʰ Army Corps, Camp White, Va., May 5, 1863"), H. #71, 72, 73 Registry #393 Part 2 Entry 1184 Press Copies of Letters Sent and Special Orders 1 of 2 Vol. 139, NARA and Col. R.B. Hayes, by M.P. Avery, Capt. + A A A G to Col. W.R. Brown, Comdg. Hurricane Bridge (manu. dated "Hd. Qrs. 1ˢᵗ Brigade 3 Div. 8 A.C. Camp White May 5, 1863"), RG94 Box 4839 Misc. unbound regimental papers 13ᵗʰ W.Va.

510   J.V. Young to Paulina (dated "Mud Bridge, Cabell Co., Va."), Young Civil War Papers, trans. p. 67.

511   John S. Cunningham to his wife (manu. dated "Head Quarters 13ᵗʰ Va Regt Hurricane Bridge May 5ᵗʰ, 1863"), Letters of John S. Cunningham, 13ᵗʰ W.Va., WVSA.

512   William R. Brown, Col Comdg, by John S. Cunningham, Lieutenant and Acting Adjutant to Capt. J.D. Carter, cmdg Co. E, 13ᵗʰ Regiment Virginia Infantry, Special Order No. 11 (manu. dated "Head Quarters 13ᵗʰ Regt Va Vol Infty, Hurricane Bridge West Va May 6ᵗʰ 1863"), RG94 Order Books 13ᵗʰ Regt. W.Va. Inftry., [n.p.]

513   William R. Brown, Col Comdg, by John S. Cunningham, Lt. and Acting Adjutant to Capt. James W. Johnson, Special Order No. 12 (manu. dated "Head Quarters 13ᵗʰ Regt V.V.I., Hurricane Bridge May 6ᵗʰ 1863"), RG94 Order Books, [n.p.]

514   [Untitled], reprinted from the Gallipolis *Journal* in the Point Pleasant *Weekly Register*, May 14, 1863, [p.1].

515   William R. Brown, Col Comdg, by John S. Cunningham, Lt and Acting Adjutant to Capt Van D. McDaniel, Special Order No. 13 (manu. dated "Head Quarters 13ᵗʰ Regt Va Vol Infty, Hurricane Bridge May 8ᵗʰ 1863"), RG94 Order Books, [n.p.]

516   Colonel W.R. Brown to Colonel R.B. Hayes, at Headquarters 1ˢᵗ Brigade, 3ʳᵈ Division, 8ᵗʰ Army Corps (manu. dated "H'dQ'rs. 13 Va. V.I. Hurricane Bridge W.Va. 8 May/63" Received at 1ˢᵗ Brigade Head Quarters on May 9, 1863), Letters Received Head Quarters 1ˢᵗ Brig. 3d Division 8ᵗʰ Army Corps continued from Book A, Letters Received Miscellaneous, Papers of R.B. Hayes (micro. Reel 272), Hayes Presidential Library, p. 347.

517   Lieutenant J.S. Cunningham, Adjutant 13ᵗʰ Virginia Volunteers to Headquarters 1ˢᵗ Brig. 3d Division 8ᵗʰ Army Corps (manu. dated "Head Quarters 13ᵗʰ Regiment Virginia Volunteer Infantry Hurricane Bridge, Virginia, May 8, 1863"), Letters Received Head Quarters 1ˢᵗ Brig. 3d Division 8ᵗʰ Army Corps. continued from Book A, Misc. Papers of R.B. Hayes (micro. Reel 272), HPL, p. 346. On May 4, Scammon ordered that "Commandants of Brigades and Detached Regts will see that Genl Order No 73 War Dept. March 24, 1863 is read at the head of every Co. in their respective commands." Brig. Genl. E.P. Scammon, by Botsford, Circular (manu. dated "Hd. Qtrs. 3d Div. 8ᵗʰ A.C., Charleston, W.Va., May 4, 1863"), RG94 Box 4838 Misc. unbound regimental papers 13ᵗʰ W.Va., Circulars.

518   Field History 13ᵗʰ Regiment W. Va. Infantry (manu. addressed to F. Pierpont, Adj. Genl. W. Va. and dated "Head Quarters 13ᵗʰ Regt. W. Va. Vol. Inft., Barboursville, W. Va., Jan'y 13ᵗʰ, 1864"), Uncat. 13ᵗʰ W.Va. Inftry. Regt. Box, WVSA, [n.p.] and Regimental Return 13ᵗʰ Va. Vol. Infantry for May 1863 (micro. #594 Roll 196), NARA.

519   Col. R.B. Hayes, by W.P. Avery, Capt. and A.A.A.G. to Col. Wm. R. Brown, 13ᵗʰ Va. at Hurricane Bridge, W. Va. and Lt. Col. James R. Hall commanding detachment 13ᵗʰ and 5ᵗʰ Virginia Volunteer Infantry at Mud Bridge (manu. dated "Head Qtrs. 1ˢᵗ Brigade 3ʳᵈ Division, Dept 8ᵗʰ Army Corps, Camp White, Va., May 9, 1863"), H. #82 Registry #393 Part 2 Entry 1184 Press Copies of Letters Sent and Special Orders 1 of 2 Vol. 139, NARA and RG94 Box 4838 Misc. unbound regimental papers 13ᵗʰ W.Va., Spec. Orders and Gen. Orders, NARA.

520   Annual Return of Casualties + c. 13ᵗʰ W.Va. Infy. Oct. 10/62 to Dec. 31/63 (manu. "Return completed Jan. 13, 1864"), RG94 Box 4838 Misc. unbound regimental papers 13ᵗʰ W.Va., NARA.

521   J.S. Cunningham, Lieutenant and Adjutant, 13ᵗʰ Regiment Virginia Volunteers to Head Quarters 1ˢᵗ Brigade, 3ʳᵈ Division, 8ᵗʰ Army Corps. (manu. dated "Coal Mouth Virginia May 10, 1863" received at Head Quarters May 10, 1863), Letters Received Head Quarters 1ˢᵗ Brig. 3d Division 8ᵗʰ Army Corps. continued from Book A, Miscellaneous, Papers of R.B. Hayes (micro. Reel 272), HPL, p. 346.

522   Regimental Return 13ᵗʰ Va. Vol. Infantry for May 1863 (micro. #594 Roll 196), NARA.

523   Annual Return of Casualties + c. 13ᵗʰ W. Va. Infy. Oct. 10/62 to Dec. 31/63 (manu. "Return completed Jan. 13, 1864"), RG94 Box 4838 Misc. unbound regimental papers 13ᵗʰ W.Va.

524   Field History 13ᵗʰ Regiment West Virginia Infantry, Uncat. 13ᵗʰ W.Va. Inftry. Regt. Box, [n.p.]

525   Burrows to Lovina (dated "May 15ᵗʰ, 1863"), Roush, *If Thou Wilt Remember,* pp. 31-31.

526   John S. Witcher, comdg. Co. G 3ʳᵈ Loyal Virginia Cavalry Regiment.

527   John V. Young, Capt., Co. G 13ᵗʰ Reg. to wife, Paulina (dated "Camp White, Kanawha, May 1863"), Young Civil War Papers (Vol. 31 Box 10), WVU, trans. p. 64.

528   Col. R.B. Hayes to Col. William R. Brown, Camp White; Tomlinson of 5ᵗʰ Va. V. I.; Capt. S.J. Simmones Battery Camp White; Capt. G.W. Gilmore 1ˢᵗ Va. Cav.; Capt. D. Delaney Co. A 1ˢᵗ Va. Cav; and Lt. John S. Witcher Co. 3ʳᵈ Va. Cav. (manu. dated "Head Qtrs. 1ˢᵗ Brigade 3ʳᵈ Division, Dept 8ᵗʰ Army Corps, Camp White, Va., May 12, 1863"), H. #88 Registry #393 Part 2 Entry 1184 Press Copies of Letters Sent and Special Orders Vol. 139, NARA.

529   Col. R.B. Hayes, by M.P. Avery, Capt. and A.A.A.G. to Col. W.R. Brown, Comdg. 13th Va.V.I., Special Orders No. 27 (manu. dated "Head Quarters 1st Brig., 3d Division 8th A.C., Camp White, Va., May 12, 1863"), RG94 Box 4838 Misc. unbound regimental papers 13th W.Va. Spec. Orders and Gen. Orders.

530   Col. R.B. Hayes to Colonel [W.R. Brown] (manu.dated "Head Quarters 1st Brig. 3rd Div. 8th Army Corps, Camp White, Va. May 12, 1863"), RG94 Box 4838 Spec. Orders and Gen. Orders. Another copy of this order exists which requests that "Fifteen men" report instead of "14." The order is also somewhat differently worded: "please direct a detail from your command for Guard duty tomorrow May 13, 1863 Two Corporals Fifteen men. Guard Mounting 8 o'clock A.M. They will report to these Hd. Qtrs." Colonel R.B. Hayes to Colonel William R. Brown (manu. dated "Head Qtrs. 1st Brigade 3rd Division, Dept 8th Army Corps, Camp White, Va., May 12, 1863"), H. #86 Registry #393 Part 2 Entry 1184 Press Copies of Letters Sent and Special Orders Vol. 139, NARA.

531   Colonel R.B. Hayes, by M.P. Avery, Capt. and A.A.A.G. to Col. Brown (manu. dated "Hd.Qtrs. 1st Brig., 3d Div. 8th A.C. Camp White, Va., May 13, 1863"), RG94 Box 4839 Misc. unbound regimental papers 13th W.Va. and Col. R.B. Hayes to Col. William R. Brown at Camp White (manu. dated "Head Qtrs. 1st Brigade 3rd Division, Dept 8th Army Corps, Camp White, Va., May 13, 1863"), H. #95 Registry #393 Part 2 Entry 1184 Press Copies of Letters Sent and Special Orders Vol. 139.

532   Col. R.B. Hayes to Colonel [W.R. Brown] (manu. dated "Head Quarters 1st Brig. 3rd Div. 8th Army Corps, Camp White, Va. May 13, 1863"), RG94 Box 4838 Misc. unbound regimental papers 13th W.Va., Spec. Orders and Gen. Orders, NARA.

533   Hayes, by M.P. Avery, Capt. and A.A.A.G. to Col. Brown (manu. dated "Hd.Qtrs.1st Brig., 3d Div. 8th A.C. Camp White, W. Va., May 13, 1863"), RG94 Box 4839 Misc. unbound regimental papers 13th W.Va.

534   Respectively, Hayes, by M.P. Avery, Capt. and A.A.A.G. to Col. Brown (manu.dated "Hd.Qtrs.1st Brig., 3d Div. 8th A.C. Camp White, W.Va., May 14, 1863") and Hayes, by M.P. Avery, Capt. and A.A.A.G. to Col. Brown (manu. dated "Hd.Qtrs.1st Brig., 3d Div. 8th A.C. Camp White, W.Va., May 14, 1863"), both in RG94 Box 4839 Misc. unbound regimental papers 13th W.Va., NARA.

535   R.B. Hayes to S[ardis] Birchard, dated "Camp White, May 17, 1863,", Williams, *Diary and Letters*, p. 410.

536   Hayes to Lucy Hayes, dated "Camp White, May 17, 1863," *Diary and Letters,* p. 409.

537   H[arrison] G[ray] O[tis], editor, "Personal Recollections of the War—No. III," *Santa Barbara Daily Press*, Saturday Evening, October 28, 1876, [n.p.], provided courtesy of Darl Stephenson.

538   E.P. Scammon, B.G. to Colonel [R.B. Hayes] (manu. dated "H. Qu. Dist. Kanawha, Tuesday 2 PM."), RG94 Misc. Dept. of W.Va. Papers, NARA.

539   R.B. Wilson, "Kanawha Division: Its Campaigns. [Part] I—cont.," *The National Tribune*, February 18, 1897, p. 2.

540   Milton W. Humphreys, Bryan's Battery, King's Artillery, C.S.A., *Military Operations 1861-1863 Fayetteville West Virginia* (Fayetteville, West Virginia: Charles A. Goddard,1926), pp. 24-25.

541   J.M. Merrill to his wife, Lydia, dated "Camp Piatt, May 21st, 1863," published as "From Kanawha," in *The Ironton Register*, May 28, 1863, p. 3.

542   Field History 13th Regiment West Virginia Infantry, Uncat. 13th W.Va. Inftry. Regt. Box, WVSA, [n.p.] and Regimental Return 13th Va. Vol. Infantry for May 1863 (micro. #594 Roll 196), NARA.

543   Col. R.B. Hayes, by M.P. Avery to Col. Brown (manu. dated "Hd. Qtrs. 1st Brig., 3d Div., 8 A.C., Camp White, May 17, 1863"), RG94 Box 4839 Misc. unbound regimental papers 13th W.Va. Another copy of this order to Col. Brown exists: "A steam boat will be at your Camp at 1 P.M. for you. [...]." See Col. R.B. Hayes to Col. William R. Brown (manu. dated "Head Qtrs. 1st Brigade 3rd Division, 8th Army Corps, Camp White, Va., May 17, 1863"), Registry #393 Part 2 Entry 1184 Press Copies of Letters Sent and Special Orders Vol. 139, NARA.

544   Stiarwalt, Sr., "Dairy entry for May 17, 1863," Diary 1861-64 (micro. Ser. 7/ Roll No. 272), p. 7 of manu.

545   Annual Return of Casualties + c. 13th W.Va. Infy. Oct. 10/62 to Dec. 31/63 (manu. "Return completed Jan. 13, 1864"), RG94 Box 4838 Misc. unbound regimental papers 13th W.Va., NARA.

546   Col. J.T. Toland Comdg., by E.W. Clark, Jr., Adj. (manu. dated "Hd. Qtrs. Gauley, W.Va., May 19, 1863"), RG94 Box 4839 Misc. unbound regimental papers 13th W.Va.

547   Letters of John S. Cunningham, 13th W.Va. Infantry Regiment, WVSA.

548   Col. Jno. T. Toland, Comdg., by E. W. Clark, Jr., Adjt. to Col. Brown (manu. dated "Hd. Qtrs. Gauley, W. Va., May 19, 1863"), RG94 Box 4839 Misc. unbound regimental papers 13th W.Va. The majority of people in Fayetteville and in the County of Fayette as a whole supported the Southern cause. "The enemy" referred to in the foregoing quote could have been used in the broadest sense, i.e., to include regular and irregular fighters.

549   Regimental Return 13th Va. Vol. Infantry for May 1863 (micro. #594 Roll 196), NARA.

550   An anonymous soldier correspondent writing to the *Ironton Register* also noted that "the 13th Virginia Inf. came up the night of the 19th to reinforce us [i.e., the 23rd, 12th and 91st O.V.I. Regiments]." See "Soldier of the 23d O.V.I. to the Ironton Register" (dated "May 25th, Camp Fayetteville"), *Ironton Register*, June 4, 1863, p. 2.

551   Field History 13th Regiment W.Va. Infantry, [n.p.]

552   Regimental Return 13th Va. Vol. Infantry for May 1863 (micro. #594 Roll 196), NARA. A bivouack was a simple camp. Its orderly arrangement approximated the order in which the command could would be drawn up in line of battle.

553   Ibid.

554   Humphreys, Bryan's Battery, King's Artillery, C.S.A., citing in his retelling of the engagement, the reports of Col. Carr B. "White and Hines," in *Military Operations 1861-1863 Fayetteville West Virginia*, pp. 24-25.

555   A soldier of the 12th Ohio Infantry Regiment, who with his unit had traveled with 91st Ohio and 13th (West) Virginia to Raleigh in pursuit of the McCausland's troops, wrote that at Raleigh, "[t]he 91st Ohio and the 13th loyal Va. were anxious to come up with the rebs and exchange a few Enfield civilities, but they could not accomplish their object." [Soldier correspondent belonging to the 12th Ohio Infantry] (dated "Fayette C.H., May 30, 1863") pub. as "Army Correspondence," *Gallipolis Journal*, June 11, 1863, p. 1.

556   Field History 13th Regiment W.Va. Infantry, [n.p.]

557   On May 20, according to one 23rd Ohio soldier: "It was 4 P.M. before we were aware that the enemy was retreating. As soon as we could cook 3 days rations, we started after them with a force composed of 8 companies of the 12th Ohio, 8 companies of 91st Ohio, 8 companies of the 13th Va, five companies 2d Va. Cav. and 6 pieces of artillery. We started at 1 o'clock P.M. of the 20th, and marched 8 miles that night, and next morning moved on and overtook them 4 miles beyond M'Coys. The 91st was in advance, they had quite a skirmish with them and finally routed them." "Soldier of the 23d O.V.I. to the *Ironton Register*" (dated "May 25th, Camp Fayetteville"), *Ironton Register*, June 4, 1863, p. 2.

558   Regimental Return 13th Va. Vol. Infantry for May 1863 (micro. #594 Roll 196), NARA.

559   In general, a forced march at this time was any march farther than fifteen miles a day.

560   John S. Cunningham to his wife (manu. dated "In Camp Near Fayetteville Va May 23rd 1863"), Letters of John S. Cunningham, WVSA.

561   Colonel C.B. White, Comm[an]ding 2nd Brigade, by J.W. Overturf, Lieutenant and A.A.D.C. to Colonel Brown, 13th Regt. Va. Vol. Inf'try. (dated "HeadQuarters 2nd Brig., 3rd Div. 8th Army Corps, Fayetteville, Va., May 24, 1863"), Field History of the 13th Regiment West Virginia Infantry Addendum B., Uncat. 13th W.Va. Inftry. Regt. Box, WVSA, trans. pp. 2-3.

562   Milton W. Humphreys, "Diary," as quoted in Humphreys, *Military Operations 1861-1863 Fayetteville*, pp. 23-24.

563   Col. C.B. White, Comdg. Brig., by J.W. Overturf Lt. + A.A.D.C., General Order No. 3 (manu. dated "Head Quarters 2d Brigade 3d Division 8th Army Corps, Fayetteville, Va., May 20, 1863"), RG94 Box 4838 Misc. unbound regimental papers 13th W.Va., Spec. Orders and Gen. Orders.

564   Col. Carr B. White stated in his report that continual skirmishing attended the pursuit of the Confederates towards Raleigh. See White's report of the expedition to Fayetteville, as referenced by Humphreys, *Military Operations 1861-1863 Fayetteville*, p. 25.

565   The 13th arrived at Fayetteville on May 23 and remained there two days to rest. (Field History of the 13th Regiment W.Va. Infantry, [n.p.])

566   Wm. R. Brown, Col. Comdg. 13th Va. Vol. Infty. to Col. C.B. White, Commanding 2nd Brigade, 3rd Division 8th A.C. (manu. dated "Head Quarters 13th Regt. Va. Vol. Infty, Camp near Fayetteville, May 24, 1863"), RG94 Box 4839 Misc. unbound regimental papers 13th W.Va. Letters A.G.O., NARA.

567   Diary of Milton W. Humphreys, as quoted in Humphreys, *Military Operations 1861-1863 Fayetteville*, p. 24.

568   White's report of the expedition to Fayetteville as referenced by Humphreys, *Military Operations 1861-1863 Fayetteville*, p. 25.

569   W.W. Harper to the Editor of the Point Pleasant *Weekly Register* (dated "Fayetteville, Fayette county, W.V. [Sunday] May 24th 1863"), "From the Thirteenth Va.," *Weekly Register*, May 28, 1863, Vol. II No. 11, [p. 2]. J.V. Young seems also to have referred to this one casualty when he wrote that aside from fatigue and "sore feet," the regiment passed through "without the loss of a man, but one got his finger shot off." See J.V. Young, Capt., Co. G 13th Regt.-Va. Vol. Inf. to wife, Paulina (dated "Fayetteville, May 24, 1863"), Young Civil War Papers, trans. p. 66.

570   RG94 Morning Reports of Co. F 13th Regiment Virginia Vols. for May 1863, [n.p.] Regimental records indicate that William Daniel had formerly been a resident of Raleigh and he may have straggled from the command to visit family or friends.

571   Captain W.I. Matthews, Company Order No. 2 (manu. dated "Head Quarters Company H, 13th Regt.Va. V.I., Hurricane Bridge, Va., June 1st, 1863"), RG94 Order Book Co. H 13th Regt. Inftry. Vols., [n.p.]

572   RG94 Misc. Records Descriptive Roll Co. H 13th Regiment V.V.V.I., List of Non-Commissioned Officers, NARA, [n.p.] On June 3, Col. Brown issued General Order No. 11 from Regimental Headquarters providing that: "Order No. [left blank] issued from Head Quarters Co. H reducing Thomas F. Hacker Sergea[n]t to the ranks and of promoting William Shannon (Private) to fill the vacancy caused thereby is hereby approved."W.R. Brown, Col. Com'd'g, by John S. Cunningham, Lt. + Actg. Adj., Gen. Order No. 11 (manu. dated "Head Quarters 13th Reg[imen]t V[irginia] Vol[unteer] Inf[an]t[r]ly, Hurricane Bridge, June 3rd, 1863"), RG94 Order Books, [n.p.]

573   Harper's statement that the 13th Regiment served as the rear guard of the brigade stands in contrast to Col. White's directions embodied in General Order No. 3 above, which explicitly placed the 12th Ohio in the position of rear guard.

574   W.W. Harper to the Editor of the Point Pleasant *Weekly Register* (dated "Fayetteville, Fayette county, W.V. [Sunday] May 24th, 1863"), "From the Thirteenth Va.," *Weekly Register*, May 28, 1863, Vol. II No. 11, [p. 2].

575   J.V. Young, Capt. Co. G 13th Regt.-Va. Vol. Inf. to his wife, Paulina (dated "Fayetteville, May 24, 1863"), Young Civil War Papers (Vol. 31 Box 10), WVU, trans. p. 66.

576   Col. C.B. White, comdg. Brig., by J.W. Overturf, Lt. + A.A.A.G. to Col. Brown (manu. dated "Hd. Qtrs. 2d Brig. 3d Division 8th A.C., Fayetteville, Va., May 24, 1863"), RG94 Box 4839 Misc. unbound regimental papers 13th W.Va.

577   The forgoing quotations are taken from Col. C.B. White, Comdg Brig., by J.W. Overturf, Lt. + A.A.D.C to Colonel Brown, 13 Regt. Va. Vol. (manu. dated "Head Quar 2d Brig 3d Div 8th Army Corps, Fayetteville, Va., May 24, 1863"), RG94 Box 4839 Misc. unbound regimental papers 13th W.Va.

578   To Col. William R. Brown 13th V.V.I.; Maj. J. McGrath 23rd O.V.I.; Tomlinson 5th V.V.I.; Battery under Simonds; Gilmore Delaney 1st V.V.I.; Witcher 3rd V.V. Cav.; and A.J. Austin Detachment Battery, Circular (manu. dated "Head Qtrs. 1st Brigade 3rd Division, Dept 8th, Camp White, Va., May 25, 1863"), Letter No. 114 Registry #393 Part 2 Entry 1184 Press Copies of Letters Sent and Special Orders Vol. 139, NARA.

579   Brig. Gen. E.P. Scammon, by J.M. Comly, Lt. Col. A[ssistant?] Insp. and Asst. Adjt. Genl., Special Order No. 34 (manu. dated "Head Quarters 3rd Division 8th A.C. Charleston, Va., May 25, 1864"), RG94 Box 4838 Misc. unbound regimental papers 13th W.Va., Spec. Orders and Gen. Orders.

580   R.B. Hayes, by M. Avery, Capt. + A.A.A.G. to Col. Brown (manu. dated "Charleston, May 25 [1863]"), RG94 Box 4839 Misc. unbound regimental papers 13th W.Va.

581   Col. C.B. White, comdg. Brigade, by J.W. Overturf, Lt. + A.A.D.C. to Col. Brown (manu.dated "Hd. Qtrs. 2 Brig. 3d Division 8th A.C., Fayetteville, Va., May 25, 1863"), RG94 Box 4839 Misc. unbound regimental papers 13th W.Va.

582   Col. C.B. White, comdg., by W.B. Nesbitt, A.A.A.G. to Col. Brown (manu. dated "Hd. Qtrs. 2 Brig. 3d Division 8th A.C., Fayette-ville, Va., May 25, 1863"), RG94 Box 4839 Misc. unbound regimental papers 13th W. Va.

583   R.B. Hayes to his uncle S. Birchard (dated "Camp White, May 25, 1863"), Williams, *Diary and Letters*, p. 411.

584   Dr. J.H. Rouse, Surgeon, 9th Regt. Va.Vol., *Horrible Massacre at Guyandotte, Va. and Journey to the Rebel Capital with a Description of Prison Life in a Tobacco Warehouse at Richmond, 1862*, p. 30.

585   R.B. Wilson, "Kanawha Division: Its Campaigns. [Part] I—cont.," *The National Tribune*, February 18, 1897, p. 2.

586   Annual Return of Casualties + c. 13th W. Va. Infy. Oct. 10/62 to Dec. 31/63" (manu. "Return completed Jan. 13, 1864"), RG94 Box 4838 Misc.unbound regimental papers 13th W.Va.

587   Brig. Gen. Julius White to Gen. Burnside (dated Catlettsburg, Ky., May 21, 1863), O.R. Vol. XXIII Part II, p. 252.

588   Stiarwalt, Sr., entry dated "May 28, 1863," Diary 1861-64, Russell P. Hastings (micro. Ser. 7/ Roll No. 272), p. 7 of manu.

589   Annual Return of Casualties + c. 13th W.Va. Infy. to Dec. 31/63 (manu. "Return completed Jan. 13, 1864"), RG94 Box 4838.

590   Hayes to Sardis Birchard (dated "Camp White, June 2, 1863"), Williams, *Diary and Letters*, p. 412.

591   Field History 13th Regiment W.Va. Inf., [n.p.]

592   Annual Return of Casualties + c. 13th W.Va. to Dec. 31/63, NARA.

593   Ibid.

594   W.R. Brown, Col. Commanding, by John S. Cunningham, Lt. + Actg. Adj., Gen. Order No. 9 (manu. dated "Head Quarters 13th Reg[imen]t V[irginia] V[olunteer] I[nfantry], Coalsmouth, V[irgini]a, May 30th, 1863"), RG94 Order Book Co. C 13th Regt. Inftry. Vols., NARA, [n.p.]

595   W.R. Brown, Col. Commanding, by John S. Cunningham, Lt. + Actg. Adj. Gen. Order No. 10 (manu. dated "Head Quarters 13th Reg[imen]t V[irginia] Vol[unteer] Inf[an]t[r]y, Coalsmouth, [Virginia], May 31st, 1863"), RG94 Order Book Co. C, [n.p.]

596   David Bailey/Baily, Private, Co. B 13th W.Va., Company Muster Roll, dated "Dec. 31, 1862 to April 30, 1863," Military History Cards on micro., WVU.

597   "MURDER NEAR POINT PLEASANT," Wheeling *Daily Intelligencer*, June 11 1863, p. 3.

598   Descendants of David Bailey believe he is buried in the Arbuckle-Craig Cemetery. See for the foregoing, Janice C. Veazy, "David Bailey," *History of Mason County, West Virginia 1987, 2nd ed.*, (Waynesville, N.C.: Walsworth Publishing Co., 1991), p. 10.

599   RG94 Morning Reports Co. A 13th Regiment of Virginia Volunteers Army for May 1863, [n.p.]; Record of Events Field and Staff Muster Roll Co. A 13th Va. Vol. Infantry for May and June 1863 (micro. #594 Roll 196); and Supp. O.R. Vol. LXXIV, p. 545.

600   RG94 "Remarks," Morning Reports of Co. A, [n.p.]

601   RG94 Morning Reports Co. A, [n.p.]

602   RG94 Morning Reports Co. B 13th Regiment Virginia Vols. for May 1863, NARA, [n.p.].

603   Colonel William R. Brown to Captain Van D. McDaniel, Spec. Order No. 13 (manu. dated "13th Regiment Head Quarters Hurricane Bridge, Virginia, May 8, 1863"), RG94 Order Book Co. C, [n.p.]

604   Meek would return from desertion on October 18, 1863. He was docked only one month's pay. See RG94 Misc. Records Descriptive Roll Co. C. 13th Regiment V.V.V.I., Register of Deserters, NARA, [n.p.]

605   RG94 13th Va. Hospital Cards Box 3849 Entry 534 Records of the A.G.O., NARA.

606   Oh. Reg. No. 301, Hosp. No. 46, p. 14, RG94 13th Va. Hospital Cards Box 3849 Entry 534.

607   RG94 "Remarks," Morning Reports Co. C 13th Regiment Va. Vols. for May 1863, [n.p.]

608   Ibid., "Remarks" Morning Reports Co. C for the Month of May 1863, [n.p.]

609   The page is folded near the bottom in the original obscuring the last five rows of totals for "absent sick."

610   RG94 Morning Reports Co. D 13th Regiment Va. Vols. and "Remarks" for May 1863, [n.p.]

611   RG94 Morning Reports Co. F 13th Regiment Va. Vols. for May 1863, NARA, [n.p.].

612   W.I. Mathews, Captain, Company H 13th Regiment, Co. H Order No. 1. (manu. dated "Head Quarters Co. H 13th Regt. Va. V. I., Camp near Charleston, May 11, 1863"), RG94 Order Book Co. H, [n.p.]

613   Annual Return of Casualties + c. 13th W. Va. Infy. to Dec. 31/63, RG94 Box 4838 Misc. unbound regimental papers 13th W.Va.

614   Ibid.

615   RG94 Morning Reports of Capt. W.I. Mathews Co. H 13th Regiment Va. Vols. for May 1863, [n.p.]

616   RG94 Morning Reports Co. I 13th Regiment Va. Vols. for May 1863, [n.p.]

617   Brig. Gen. E.P. Scammon to Colonel R.B. Hayes, Commanding Brigade (manu. dated "At Mr. Jeffries May 29, [1863]"), RG 94 Misc. Dept. of W.Va. Papers.

618   Gallipolis *Journal*, June 11, 1863, [p. 2].

619   Hayes to his uncle, S. Birchard, dated "Camp White, June 14, 1863," Williams, *Diary and Letters*, p. 413.

620   Proclamation of President Lincoln (dated June 15, 1863), as cited in "Proclamation of Governor Arthur I. Boreman," Point Pleasant *Weekly Register*, July 9, 1863, [p. 3].

621   Gov. F.H. Peirpoint, "By the Governor—Proclamation" dated "Wheeling, June 15, 1863," Point Pleasant *Weekly Register*, June 25, 1863, Vol. II No. 15, [p. 3].

622   "Governor Peirpoint's Valedictory," Point Pleasant *Weekly Register*, July 2, 1863, Vol. II No. 16, [p. 1].

623   Arthur I. Boreman, "Proclamation," Point Pleasant *Weekly Register*, July 9, 1863, [p. 3.]

624   Gen. Order No. 186 provided that "By direction of the President, that part of the Middle Department west of Hancock, including the adjacent counties of Ohio, will constitute the Dept. of W. Va." See Secretary of War E.M. Stanton, by E.D. Townsend, Assistant Adjutant General, War Department, A.G.O. (dated Washington D.C. June 24, 1863), O.R., Vol. XXIII Part II, pp. 454-55.

625   Reprint from the *Wheeling Intelligencer* in the Point Pleasant *Weekly Register*, July 2, 1863, [n.p.]

626   William R. Brown, Col. Comdg., Spec. Order No. 14 (manu. dated "Head Quarters 13th Regt Va Vol Infty Coalsmouth June 1st 1863"), RG94 Order Books 13th Regt. Inftry. Vols., [n.p.]

627   Col. R.B. Hayes to Col. William R. Brown (manu. dated "Head Qtrs. 1st Brigade 3rd Division, Dept 8th Army Corps, Camp White, Va., June 4, 1863"), H. #128 Registry #393 Part 2 Entry 1184 Press Copies of Letters Sent and Special Orders Vol. 139, NARA.

628   Col. R.B. Hayes, by M.P. Avery, Capt. and Act. Asst. Adjt. Genl. to Colonel W.R. Brown, Comdg. 13th V.V.I. (manu. dated "Head Qrs 1st Brig 3rd Divis 8th AC, Camp White West Va, June 5, 1863"), RG94 Box 4839 Misc. unbound regimental papers 13th W.Va.

629   Col. R.B. Hayes to Col. William R. Brown (manu. dated "Head Qtrs. 1st Brigade 3rd Division, Dept 8th Army Corps, Camp White, Va., June 5, 1863"), H. #129 Registry #393 Part 2 Entry 1184 Press Copies of Letters Sent and Special Orders Vol. 139; and Hayes, by M.P. Avery to Brown, Comdg. at Hurricane (manu. dated "Hd. Qtrs. 1st Brig. 3 Div. 8 A.C., Camp White, June 5, 1863") RG94 Box 4839 Misc. unbound regimental papers 13th W.Va., both sources held at the NARA.

630   Col. R.B. Hayes to Lt. Col. James R. Hall (manu. dated "Head Qtrs. 1st Brigade 3rd Division, Dept 8th Army Corps, Camp White, Va., June 5, 1863"), H. #130 Registry #393 Part 2 Entry 1184 Press Copies of Letters Sent and Special Orders Vol. 139.

631   Col. Brown to Hdqtrs. 1st Brig. 3d Div. 8th A. C[orps] (dated "13 Regt Va. V.I. Hurricane Bridge Va, June 6th 1863" Received at Hdqtrs. 1st Brig. 3d Div. 8th A. C[orps] on Monday June 8, 1863), Letters Received Head Quarters 1st Brig. 3d Division 8th Army Corps continued from Book A, Letters Received Miscellaneous, Papers of R.B. Hayes (micro. Reel 272), HPL, p. 352. Gen. Orders No. 113 pertains to the exchange of prisoners of war; flags of truce, etc.

632   E.P. Scammon, General Commanding 3rd Division to [Colonel R.B. Hayes commanding at] Head Quarters 1st Brig. 3d Div. 8th A.C., dated "June 7th , 1863," Letters Received Head Quarters 1st Brig. 3d Division 8th Army Corps continued from Book A in Misc. Papers of R.B. Hayes (micro. Reel 272), HPL, p. 352.

633   Colonel W.R. Brown to Hdqrs. 1st Brig. 3d Div. 8th A.C. (dated "Hd Qrs 13 Reg Va Vol. Inf. Hurricane Bridge Va June 8, 1863" Received at 1st Brigade Head Quarters June 11, 1863), Letters Received Head Quarters 1st Brig. 3d Division 8th Army Corps Book A, Letters Received Miscellaneous, Papers of R.B. Hayes (micro. Reel 272), p. 353.

634   J.V. Young to his wife and children, dated "Mud Bridge, June 11, 1863," Young Civil War Papers (Vol. 31 Box 10), trans. p. 68.

635   Ibid., Young to his wife and children (dated "Mud Bridge, June 15, 1863"), trans. p. 69.

636   Col. R.B. Hayes to Col. William R. Brown (manu. dated "Head Qtrs. 1st Brigade 3rd Division, Dept 8th Army Corps, Camp White, Va., June 17, 1863"), H. #145 + 146 Registry #393 Part 2 Entry 1184 Press Copies of Letters Sent and Special Orders Vol. 139, NARA and Col. R.B. Hayes, by M.P. Avery, Capt. and A.A.A.G. to Col. W.R. Brown, Comdg. 13th Va.V.I. (manu. dated "Head Qrs 1st Brig. 3rd Divis 8th A. C., Camp White W.Va., June 17, 1863"), noted at bottom of the page is: "Courier left Hd Qtrs (Brigade) at 12-1/2 O.C. P.M. June 17th 1863 Recd at Hd Qrts 13th Va Regt Stationed at Hurricane Bridge Va at 7—3/4 O.C. P.M. June 17, [18]63," RG94 Box 4839 Misc.

unbound regimental papers 13ᵗʰ W.Va., NARA. It had required 7-¼ hours from the time the army courier had left Charleston to reach Hurricane Bridge, a distance of somewhere between 22 and 26 miles depending on the route taken.

637   Lieut. Col. James R. Hall Commanding Detach. 13 and 5 Va. V.I. to Heaquarters 1ˢᵗ Brig, 3ʳᵈ Division 8ᵗʰ A.C., dated "June 18, 1863,", Letters Received Head Quarters 1ˢᵗ Brig. 3d Division 8ᵗʰ Army Corps continued from Book A, Letters Received Miscellaneous, Papers of R.B. Hayes (micro. Reel 272), HPL, p. 354.

638   Captain John T. Bowyer, of Putnam County, Virginia, had been appointed Commissioner for the State of West Virginia in March 1863, to take the absentee votes of soldiers stationed in camps across West Virginia in the all-important State elections of the first half of 1863. In August of 1863, we find him named in the local newspaper as "U.S. Commissioner for Putnam County." In 1864, he ran as candidate for representative to the West Virginia State Senate.

639   The man Joseph Tinsley, named by Young, was probably Sergeant Joseph Tinsley of Co. G 8ᵗʰ Virginia Cavalry, C.S.A. See Dickinson, *Tattered Uniforms and Bright Bayonets West Virginia's Confederate Soldiers,* p. 366.

640   J.V. Young to J.T. Bowyer (dated "Mud Bridge, June 26, 1863"), Young Civil War Papers, trans. pp. 70-71.

641   GROTIUS to Friend Harper (dated "Camp at Barboursville, Va., June 26ᵗʰ, 1863"), "Army Correspondence. For the Gallipolis Journal. From the 5ᵗʰ Va. Vol. Infantry," *Gallipolis Journal,* July 9, 1863, [p.1]. Sergeant Fuller in this letter is probably Stephen F. Fuller.

642   By command of Col. R.B. Hay[e]s to W.R. Brown, Comdg. 13ᵗʰ, Circular (dated "Hd. Qtrs. 1ˢᵗ Brig. 3d Div. 8ᵗʰ A.C., Camp White, W. Va., May 4, 1863"), RG94 Box 4838 Misc. unbound regimental papers 13ᵗʰ W.Va. Circulars.

643   Wm. McKinley Jr., 1ˢᵗ Lt. + A.A.Q.M. to S. Comstock R.Q.M. 13ᵗʰ Regt. Va. Vols. (manu. dated "Hd. Qtrs. 1ˢᵗ Brig. 3ʳᵈ Div. 8ᵗʰ A.C. Camp White, Va. June 18, 1863"), Registry No. 393 Part 2 Entry 1194 Register of letters received by Quartermasters, Letter 31, NARA, [n.p.]

644   Col. W.R. Brown Commanding 13ᵗʰ Va. V.I. to Headquarters 1ˢᵗ Brig., 3d Div. 8ᵗʰ A.C. (dated "Hurricane Bridge June 19ᵗʰ, 1863"), Letters Received Head Quarters 1ˢᵗ Brig., 3d Division 8ᵗʰ Army Corps continued from Book A, Letters Received Miscellaneous, Papers of R.B. Hayes (micro. Reel 272), p. 354.

645   W.R. Brown, Col. Commanding, by John S. Cunningham, Adj., General Order No. 13 (manu. dated "Head Quarters 13ᵗʰ Regiment V[irgini]a Vol[unteer] Inf[an]t[r]y], Hurricane Bridge, V[irgini]a, June 19ᵗʰ, 1863"), RG94 Order Book Co. C, [n.p.]

646   Col. Wm. R. Brown Commanding 13ᵗʰ Reg. Virginia to Hqrs. 1ˢᵗ Brig. 3ʳᵈ Div. 8ᵗʰ A.C. (dated "Hd Qrs. 13ᵗʰ Regt. Va.V.I. Hurricane Bridge June 20, 1863" Received at Headquarters 1ˢᵗ Brigade 3ʳᵈ Div., 8ᵗʰ Army Corps on June 22, 1863"), Letters Received Head Quarters 1ˢᵗ Brig. 3d Division 8ᵗʰ Army Corps, Letters Received Miscellaneous, Papers of R.B. Hayes (micro. Reel 272), trans. p. 11.

647   Col. Wm. R. Brown to Hdrs. 1ˢᵗ Brig. 3d Div. 8ᵗʰ A.C. (dated "H'd Qrs. 13ᵗʰ Regt. Va. V.I. Hurricane Bridge Va. June 20, 1863" Received June 20, 1863), Letters Received Head Quarters 1ˢᵗ Brig. 3d Division 8ᵗʰ Army Corps continued from Book A, Miscellaneous, Papers of R.B. Hayes (micro. Reel 272), p. 355.

648   William R. Brown, Col. Comdg by John S. Cunningham, Adj., Spec. Order No. 15 (manu. dated "Head Quarters 13ᵗʰ Regt Va Vol Infty, Hurricane Bridge June 20ᵗʰ 1863"), RG94 Order Books 13ᵗʰ Regt. Inftry. Vols., [n.p.]

649   Handley was a well-known secessionist. In November 1862, language in correspondence passed between John Bowyer and Capt. J.V. Young indicates that the extent of his partisan activities were suspected. Young wrote alleging that A.W. Handley was "a Troublesome bad rebel" with "2 or 3 sons in the Rebel army Provided for and Riged off by him for the purposes." Young recommended that Handley with "two or three others" be taken hostage to secure the safe return of one "Mr. Wood," a Union man, captured by the rebels near Scary, West Virginia, and forwarded to Richmond. See John Bowyer to Capt. John V. Young (dated "Putnam C H, Va., Nov. 4ᵗʰ, 1862"), Young Civil War Papers (Vol. 31 Box 10), WVU.

650   Col. R.B. Hayes to Col. William R. Brown (manu. dated "Head Qtrs. 1st Brigade 3rd Division, Dept 8th Army Corps, Camp White, Va., June 19, 1863"), H. #150 Registry #393 Part 2 Entry 1184 Press Copies of Letters Sent and Special Orders Vol. 139 and M.P. Avery, Capt. and A.A.A.G to Col. W.R. Brown, Comdg. 13th Regt Va.V.I., Hurricane Bridge, West Va. RG94 Box 4839 Misc. unbound regimental papers 13th W.Va., both letters, NARA.

651   "Report of the Search for Rebel Mail made June 22, 1863," to Capt. M.P. Avery, A.A.A.G. (frag. manu. dated "Head Quarters 13th Regt Va Vol Infty, Hurricane Bridge, Va., June 24, [18]63"), RG94 Box 4839, NARA.

652   Col. R.B. Hayes, by M.P. Avery to Col. W.R. Brown (manu. dated "Hd.Qtrs.1st Brig., 3d Div. 8 A.C., Camp White, June 24, 1863"), RG94 Box 4839 and Col. R.B. Hayes to Col. William R. Brown (manu. dated "Head Qtrs. 1st Brigade 3rd Division, Dept 8th Army Corps, Camp White, Va., June 25, 1863"), H. #155 Registry #393 Part 2 Entry 1184 Press Copies of Letters Sent and Special Orders Vol. 139, NARA.

653   W.R. Brown, Col. Com'd'g., by John S. Cunningham, Adj., Gen. Order No. 14 (manu. dated "Head Quarters 13th Regt V V Inft Hurricane Bridge June 21st 1863"), RG94 Order Book Co. C, [n.p.]

654   Col. W.R. Brown to Col. R.B. Hayes, Commanding at Hdqtrs. 1st Brigade, 3d Division, 8th Army Corps (dated "Hd Qrs. 13th Regt. Va. V.I. Hurricane Bridge Va. June 22/63"), Letters Received Head Quarters 1st Brig. 3d Division 8th Army Corps, Letters Received Miscellaneous, Papers of R.B. Hayes (micro. Reel 272), trans. p. 11.

655   Col. Hayes to Col. Brown, cmdg. 13th at Hurricane Bridge (manu. dated "Hd. Qrs. 1st Brig. 3rd Div. 8th A.C., Camp White, Va., June 25, 1863"), RG94 Box 4839 Misc. unbound regimental papers 13th W.Va. and Col. R.B. Hayes to Col. William R. Brown (manu. dated "Head Qtrs. 1st Brigade 3rd Division, Dept 8th Army Corps, Camp White, Va., June 23, 1863"), H. #152 Registry #393 Part 2 Entry 1184 Press Copies of Letters Sent and Special Orders Vol. 139.

656   RG94 Morning Reports Co. A 13th Regiment Va. Vols. for June 1863, [n.p.]

657   RG94 Morning Reports Co. B 13th Regiment Va. Vols. for June 1863, NARA, [n.p.]

658   Christopher C. Barnett to his father and mother [James and Rebecca A. Barnett] (manu. dated "Mud Bridge, June 8th, 1863"), C.C. Barnett, Military Pension File, NARA.

659   Christopher C. Barnett to his father [James Barnett] (manu. dated "Mud Bridge, Cabell County, June 23rd, 1863"), C.C. Barnett, Pension File.

660   RG94 Carded Medical Records Vols. 1846-1865 Entry 534 Box 3849 Records of the A.G.O.

661   RG94 Morning Reports Company C 13th Regiment Va. Vols. for June 1863 signed by 1st Sergt. John P. Wood and Capt. Van D. McDaniel, [n.p.]

662   RG94 Morning Reports Co. D 13th Regiment Va. Vols. for June 1863 signed by 1st Sergt. John Jones, [n.p.]

663   Mark E. Robison to John Robison (manu. dated "Camp Mud Bridge, June 20, 1863"), Mark E. Robinson, Military Pension File, NARA.

664   Measures were taken by the army at this time to issue provisions to the families of soldiers, who were in destitute circumstances. See letter from Lieut. William [Feazel?], 13th Va. V.I. (dated "Camp at Coals Mouth Va June 2, 1863" received June 2, 1863 at Head Quarters 1st Brig. 3d Division 8th Army Corps), requesting information regarding the order as to "distributing provisions to Soldiers families who are in destitute circumstances." Letter 128, Letters Received Head Quarters 1st Brig. 3d Division 8th Army Corps continued from Book A, Miscellaneous, Papers of R.B. Hayes (micro. Reel 272), p. 351 and Miscellaneous Endorsements Book B — pg. 2- No. 7-1863, Papers of R.B. Hayes, both at HPL.

665   RG94 Morning Reports Company F 13ᵗʰ Regiment of Va. Vols. and "Remarks" for June 1863, signed by 1ˢᵗ Sergt. A.C. Mason and Capt. A.F. McCown, [n.p.]

666   Colonel W.R. Brown to 1ˢᵗ Brigade Head Quarters, 3ʳᵈ Division 8ᵗʰ Army Corps (dated "Head Quarters 13 Regiment Virginia Volunteer Infantry, Hurricane Bridge, Virginia June 3, 1863"), Letters Received Head Quarters 1ˢᵗ Brig. 3d Division 8ᵗʰ Army Corps continued from Book A, Letters Received Miscellaneous, Papers of R.B. Hayes (micro. Reel 272), p. 351 and Endorsements (47-8-1863 for orders discharges + L-13-1863, Papers of R.B. Hayes.

667   RG94 Morning Reports Co. H 13ᵗʰ Regiment Va. Vols. for June 1863, [n.p.]

668   Albert C. Jamison, "Muster-In Roll of Detachment of Recruits for the 13ᵗʰ Regiment of Va Infty Volunteers commanded by Colonel William R. Brown called into the service of the United States by the President from the 11ᵗʰ day of December 1862, (date of this muster,) for the term of three years unless sooner discharged." Uncat. 13ᵗʰ W.Va. Inftry. Regt. Box, WVSA and RG94 Records of the A.G.O. Carded Medical Records Vols. 1846-1865 Entry 534 Box 3849, NARA.

669   James H. Brown (sealed), Court Order (manu. dated "Circuit Court of Kanawha County, Charleston, December 31ˢᵗ, 1862, in the 87ᵗʰ year of the [Virginia] Commonwealth") copied by John S. Cunningham, Adjutant, *In Case of Albert C. Jamison on Writ of Habeas Corpus*, RG94 Order Book Co. C 13ᵗʰ Regt. Inftry. Vols., NARA, [n.p.]

670   *In Case of Albert C. Jamison on Writ of Habeas Corpus.*

671   Transcript from the court records of the Kanawha County Circuit Court, *In Case of Albert C. Jamison on Writ of Habeas Corpus*, copied from the orig. by John S. Cunningham. RG94 Order Book Co. C 13ᵗʰ Regt. Inftry. Vols., NARA, [n.p.]

672   By order W.R. Brown, Col. Commanding, by John S. Cunningham, Lt + Actg Adjutant, Gen. Order No. 12 (manu. dated "Head Quarters 13ᵗʰ Regt V V I, Hurricane Bridge, [Virginia], June 6ᵗʰ, 1863"), RG94 Order Book Co. C, [n.p.]

673   By order of James R. Hall, Lt. Col. Commanding, by John S. Cunningham, Adj., *In Case of Albert C. Jamison on Writ of Habeas Corpus*, RG94 Order Book Co. C, [n.p.]

674   Levett Perdew to his father and mother (manu. dated "Hurricane Bridge, Putnam County, June 10ᵗʰ 1863"), Levett Perdew, Military Pension File, NARA.

675   J.T.B. [J.T. Bowyer?], letter of authority (manu. dated "June 27, 1863"), Uncat. 13ᵗʰ W.Va. Inf. Regt. Box, WVSA.

676   RG94 Morning Reports Co. I 13ᵗʰ Regiment Va. Vols. for June 1863, [n.p.]

677   Record of Events Field and Staff Muster Roll Co. I 13ᵗʰ Va. Vol. Infantry Nov. and Dec. 1863 (micro. #594 Roll 196), NARA.

678   William E. Feazel to F.H. Pierpoint, Adjt. Genl W.Va. (dated "Charleston W. Va. Sept. 9ᵗʰ, 1864"), Samuels-Pierpont Papers, WVSA.

679   Other amusements for boys and young men at this time were "fighting, pitching quoits and footracing." See John L. Mason. "A Reminiscence. West Columbia of a Half Century Ago Contrasted With the West Columbia of Today," Second Letter to the *State Gazette*, Point Pleasant, W.Va., June 24, 1909, in History of Mason County, W.Va. Accession #203668, Cabell Co. Pub. Lib., Huntington, W.Va. The game of pitching quoits involved tossing iron rings at an iron stake, very like our game of pitching horseshoes today.

680   W[illiam] H. H[arper] to the Editor of the Point Pleasant *Weekly Register* (dated "Hurricane Bridge, West Va., [Tuesday] June 23d, 1863," "From the Thirteenth Va.," *Weekly Register*, July 2, 1863, Vol. II No. 16, [p. 2].

681   Colonel William R. Brown to 1ˢᵗ Brigade Headquarters 3ʳᵈ Div. 8ᵗʰ A.C. (dated "Head Quarters 13ᵗʰ Va.V.I., Hurricane Bridge, Va., June 23, 1863"), Letters Received Head Quarters 1ˢᵗ Brig. 3d Division 8ᵗʰ Army Corps, Letters Received Miscellaneous, Papers of R.B. Hayes (micro. Reel 272), trans. p. 11.

682   Col. R.B. Hayes to Col. William R. Brown (manu. dated "Head Qtrs. 1ˢᵗ Brigade 3ʳᵈ Division, Dept 8ᵗʰ Army Corps, Camp White, Va., June 25, 1863"), H. #154 Registry #393 Part 2 Entry 1184 Press Copies of Letters Sent and Special Orders Vol. 139 and RG94 Box 4839 Misc. unbound regimental papers 13ᵗʰ W.Va.

683   Brig. Gen. E.P. Scammon to Col. Hayes, *Confidential* (manu. dated "Head Quarters 3ʳᵈ Division 8ᵗʰ Army Corps, Charleston June 23, 1863") RG94 Misc. 13ᵗʰ W.Va. Papers, NARA.

684   Wm. McKinley, Jr., 1ˢᵗ Lt. + A.A.Q.M. to S. Comstock R.Q.M. 13ᵗʰ Regt. Va. Vols. Hurricane Bridge (manu. dated "Hd. Qtrs. 1ˢᵗ Brig. 3ʳᵈ Div. 8ᵗʰ A.C. Camp White, Va. June 23, 1863"), Registry #393 Part 2 Entry 1194 Register of letters received by Quartermasters, Letter 33, [n.p.].

685   Col. Wm. R. Brown, Comdg. 13 Va V.I. to 1ˢᵗ Brigade Hdqtrs. 8ᵗʰ A.C. (dated "H'd Qrs.13 Va V I Hurricane Bridge Va June 25, 1863" Received at 1ˢᵗ Brigade Headquarters June 28, 1863), Letters Received Head Quarters 1ˢᵗ Brig. 3d Division 8ᵗʰ Army Corps, Letters Received Miscellaneous, Papers of R.B. Hayes (micro. Reel 272), HPL, trans. p. 13.

686   Col. R.B. Hayes to the Commanding Officer at Mud Bridge (manu. dated "Head Qtrs. 1ˢᵗ Brigade 3ʳᵈ Division, Dept 8ᵗʰ Army Corps, Camp White, Va., June 28, 1863"), H. #159 Registry #393 Part 2 Entry 1184 Press Copies of Letters Sent and Special Orders Vol. 139; Col. R.B. Hayes to Capt. Milton Stewart (manu. dated "Head Quarters 1ˢᵗ Brigade, 3ʳᵈ Division, 8ᵗʰ Army Corps, Camp White, June 28ᵗʰ, 1863"), RG94 Order Book Co. B 13ᵗʰ Regt. Inftry. Vols., [n.p.]; and Captain W.T. McQuigg, Commanding post at Mud Bridge, Virginia, to Capt. Milton Stewart (manu. dated "June 29, 1863"), Order Book Co. B, [p. 1].

687   Col. R.B. Hayes, by M.P. Avery A.A.A.G. to Col. W.R. Brown, Comdg 13ᵗʰ Va Vol Hurricane Bridge" (manu. dated "Head Quarters 1ˢᵗ Brigade 3rd Div[ision], 8 A.C., Camp White, June 28, 1863") RG 94 Box 4839 Misc. unbound regimental papers 13ᵗʰ W.Va. and Col. R.B. Hayes to Col. William R. Brown (manu. dated "Head Qtrs. 1ˢᵗ Brigade 3ʳᵈ Division, Dept 8ᵗʰ Army Corps, Camp White, Va., June 28, 1863"), H. #161 Registry #393 Part 2 Entry 1184 Press Copies of Letters Sent and Spec. Orders Vol. 139, both documents at the NARA.

688   Col. R.B. Hayes to Lt. Col. Tomlinson (manu. dated "Head Qtrs. 1ˢᵗ Brigade 3ʳᵈ Division, Dept 8ᵗʰ Army Corps, Camp White, Va., June 28, 1863"), H.#160 Registry # 393.

689   Field History 13ᵗʰ Regiment W.Va. Infantry, [n.p.]

690   Regimental Return 13ᵗʰ Va. Vol. Infantry for August 1863, dated "Barboursville, W.Va. Sept. 9, 1863" (micro. #594 Roll 196), NARA.

691   West Virginia Clothing Books Co. C 13ᵗʰ Inftry., WVSA, [n.p.].

692   Hazard Farley, application for military pension No. 895211. No Certificate, NARA.

693   Stiarwalt, Diary 1861-64, manu. p.7.

694   A.I. Boreman, Governor of West Virginia [inaugural speech], "Governor's Message," Point Pleasant *Weekly Register*, July 9, 1863, Vol. II No. 17, [pp. 1-2].

695   H. Samuels to Maj. Gen. Crook (dated Guyandotte, W.Va., Dec. 3, 1864), O.R. Vol. XLIII (Ser. I) Part II, p. 737.

696   Brig. Gen'l Scammon to [Col. R.B. Hayes], at Hdqtrs 1ˢᵗ Brig. 3ʳᵈ Div. 8ᵗʰ A.C. (dated "H'd Qrs 3 Divis. 8ᵗʰ A.C. Charleston Va July 1ˢᵗ 1863"), Letters Received Miscellaneous, "Papers of R.B. Hayes" (micro. Reel 272), HPL, trans. p.13.

697   Col. R.B. Hayes to Col. William R. Brown (manu. dated "Head Qtrs. 1ˢᵗ Brigade 3ʳᵈ Division, Dept 8ᵗʰ Army Corps, Camp White, Va., July 1, 1863"), H #165 Registry #393 Part 2 Entry 1184 Press Copies of Letters Sent and Special Orders Vol. 139, NARA.

698   Col. R.B. Hayes, by R. Hastings Capt. and A[cting] A[ssistant] A[djutant] G[eneral], Orders (manu. dated "Head 1st Brigade Army of Kanawha, Camp Crook West Va., July 3, 1864"), RG94 Box 4838 Misc. unbound regimental papers 13th W.Va., Spec. Orders and Gen. Orders, NARA.

699   John S. Cunningham, "Muster-In Roll" (dated "June 18, 1864"), RG94 Box 4838 Misc. unbound regimental papers 13th W.Va. Inf. Muster-In Jud'l 1862-1865, NARA.

700   J.V Young to his daughter, Emma (dated "Cotton Hill, July 10, 1863"), Young Civil War Papers (Vol. 31 Box 10), trans. pp. 73-74.

701   Brig. Gen'l Scammon to Hdqtrs 1st Brig. 3rd Div. 8th A. C. (dated "H'd Qrs 3 Divis 8th A C Charleston W. Va. July 8th/63"), Letters Received Miscellaneous, Papers of R.B. Hayes (micro. Reel 272), trans. p.15.

702   Williams, *Diary and Letters*, p. 417 and "History of 23rd Ohio Regiment," Comly Papers (micro. Reel 1), OHC, pp. 389-90.

703   Fourth Regiment Infantry Roster of the Field, Staff and Company Officers from the date of original organization to the date of consolidation with the First West Virginia Infantry Dec. 10, 1864, *Annual Report of the Adjutant General of the State of West Virginia for the Year Ending December 31, 1865* (Wheeling: John Frew, Public Printer, 1866), p. 32.

704   After the war, Columbus returned to Mason County to farm and serve as county sheriff from 1867 to 1871. See Comstock, *West Virginia Heritage Encyclopedia. Supp. Hardesty's*, Vol. 5 "Mason County" (Richwood, W. Va.: Jim Comstock), pp. 71-72.

705   C. Shrewsbury to his wife (dated "Coals Mouth Sunday July 5, 1863"), A & M 3216 Lt. Columbus Shrewsbury Co A 4th WVI, Roy Bird Cook Collection, WVU.

706   W.H.H[arper] to the Editor of the Point Pleasant *Weekly Register* (dated "Charleston, W. Va. July 8, 1863"), "From the Thirteenth," *Weekly Register*, July 16, 1863. Vol. II No. 18, [n.p.]

707   Henrietta Fitzhugh Barr, *Civil War Diary 1862-1863*, p. 26.

708   D.S., "Letter to the Cincinnati *Gazette*" (dated "July 4, [1863]"), "From the Kanawha Valley," reprinted in the Point Pleasant *Weekly Register*, July 16, 1863, Vol. II No. 18, [p. 1].

709   Ibid.

710   Ibid.

711   Paraphrase of D.S., "Letter to the Cincinnati *Gazette*" of "July 4, [1863]."

712   Hayes to his wife, dated "July 6, 1863," *Diary and Letters*, p. 416.

713   Annual Return of Casualties + c. 13th W. Va. Infy. to Dec. 31/63 (manu. "Return completed Jan. 13, 1864"), RG94 Box 4838 Misc. unbound regimental papers 13th W.Va., NARA.

714   Brig. Gen. Scammon to Hdqrs. 1st Brig. 3rd Div. 8th A.C. (dated "H'd Qrs. 3rd Division 8th A. C., Fayetteville Va, July 11 '63"), Letters Received Miscellaneous, Papers of R.B. Hayes (micro. Reel 272), trans. p. 16.

715   R.B. Hayes, diary entry (dated "Camp Joe Webb, Near Fayetteville, West Virginia, Sunday, July 12, 1863"), *Diary and Letters*, p. 418.

716   Annual Return of Casualties + c. 13th W. Va. Infy. to Dec. 31/63 (manu. "Return completed Jan. 13, 1864"), RG 94 Box 4838 Misc. unbound regimental papers 13th W.Va.

717   Brig. Gen.Scammon to Hdqtrs. 1ˢᵗ Brig. 3 Div. 8ᵗʰ A.C. (dated "Head Qrs. 3 Divis. 8ᵗʰ AC. Charleston W. Va. July 13 '63.") Orders for 1ˢᵗ Brigade to march on July 14, 1863, were received at 1ˢᵗ Brig. from division headquarters on July 13. Letters Received Miscellaneous, Papers of R.B. Hayes (micro. Reel 272), HPL, trans. p.16.

718   Letter written from Columbus, Ohio to the *New York Times*, quoted and corrected by H.G.O. [Harrison Gray Otis], "Personal Recollections of the War—No. III," *Santa Barbara Daily Press*, October 28, 1876, courtesy of Darl Stephenson.

719   Regimental Return 13ᵗʰ Va. Vol. Infantry for August 1863, dated "Barboursville, W.Va. Sept. 9, 1863" (micro. #594 Roll 196), NARA.

720   Comly, Diaries, entry dated "Monday, July 13, 1863" through "Tuesday, July 14, 1863," (micro. Roll 1), OHC, [n.p.]

721   RG94 Morning Reports of McCown's Co. F 13ᵗʰ Regiment Virginia Vols., NARA, [n.p.] The 23ʳᵈ Ohio Infantry was also among those who advanced into the town of Raleigh according to "History of 23ʳᵈ Ohio Regiment," Comly Papers (micro. Reel 1), OHC, p. 388.

722   Ibid., "History of 23ʳᵈ Ohio Regiment," p. 388.

723   Hayes to his wife, Lucy (dated "Fayetteville, July 16 P.M., 1863"), *Diary and Letters*, pp. 418-19.

724   Comly, Diaries, entry dated "Monday, July 13, 1863 through Tuesday, July 14, 1863"), (micro. Roll 1), [n.p.]

725   Comly, Diaries, entry dated "Wednesday July 15, 1863 through Thursday, July 16, 1863," [n.p.]

726   Regimental Return 13ᵗʰ Va. Vol. Infantry for August 1863 (micro. #594 Roll 196), NARA.

727   Annual Return of Casualties + c. 13ᵗʰ W. Va. Infy. to Dec. 31/63, RG94 Box 4838 Misc. unbound regimental papers 13ᵗʰ W.Va.

728   Twenty-third, "From West Virginia. The Expedition to Wytheville, Virginia.—How it Happened and What was Accomplished—Death of Col. Toland—Full Particulars of the Affair," *The Ironton Register*, August 6, 1863, p. 1.

729   Cunningham to his wife (manu. dated "Head Quarters 13ᵗʰ Va Regt Vol Infty, Camp Laurel Creek South side Cotton Mounain July 14, [1863]"), Letters of John S. Cunningham, WVSA. Laurel Creek is a tributary of the New River. It flows down out of the mountains in northern Fayette County, W. Virginia.

730   Scammon permitted his cavalry to push on to make a strike at Wytheville, Virginia. This expedition was continued until about July 25. It was concluded to Union advantage.

731   E.P. Scammon, Brigadier General, Commanding Division, by Capt. T. Melvin, Asst. Adjt. Gen., Dept. of West Virginia, Report of Brig. Gen. E. Parker Scammon, U.S. Army, Commanding Third Division, Eighth Army Corps (dated Hd Qrs. 3d Div., 8ᵗʰ A.C., Charleston, W.Va., July 23, 1863), O.R. XXIII (Ser. I), Part I, p. 677.

732   Regimental Return 13ᵗʰ Va. Vol. Infntry. for August 1863, dated "Barboursville, W.Va. Sept. 9, 1863" (micro. #594 Roll 196), NARA.

733   "History of 23ʳᵈ Ohio Regiment," Comly Papers (micro. Reel 1), OHC, pp. 390-92.

734   RG94 Morning Reports Co. F 13ᵗʰ Regiment Va. Vols., [n.p.]

735   Regimental Return 13ᵗʰ Va. Vol. Infantry for August 1863 (micro. #594 Roll 196), NARA and Field History 13ᵗʰ Regiment W. Va. Inftry., WVSA, [n.p.]

736   Written by an unnamed gentleman of Cincinatti, Ohio. Reprinted from the Cincinnati *Times* of 1875 as "A War Reminiscence. General Hayes and the Morgan Raid—Another Account," *Santa Barbara Daily Press*, October 30, 1876, courtesy of Darl Stephenson.

737   This was a period of great optimism for the South. Gen. Robert E. Lee had decided to take the war into the Northern States and invaded Maryland and Pennsylvania. Brig. Gen. John H. Morgan launched a similar strike of his own raiding in Indiana and Ohio. About July 1, Morgan, at the head of a large cavalry force (estimates vary from something less than 2500 to about 3,000 men of the Kentucky and Tennessee cavalry and six pieces of artillery) crossed the Cumberland River in the vicinity of Burkesville, Kentucky. This force commenced moving northwards in the direction of Columbia. Morgan moved rapidly destroying railroads and telegraph lines. Federal forces were engaged in the larger theaters of war and consequently, there were few regular troops that could be spared to intercept Morgan. This left the towns in Morgan's path pretty much on their own to defend themselves as best they could. While available forces were disposed so that they might check Morgan's advance, Morgan's destruction of railroads and telegraph lines as he moved very much hampered Federal means of determining his movements and he moved so quickly securing fresh horses along the way as needed, that he could not be blocked nor intercepted. As a result, he passed unchallenged from Kentucky through Indiana and Ohio to West Virginia before he was stopped. From Burkesville, Kentucky, Morgan passed through Columbia and Lebanon and reached Green River on July 4. From Green River, he moved towards Louisville but then turned left before reaching Louisville and struck the Ohio River, where he captured some steamers and slipped into Indiana before he could be overtaken by Generals Hobson, Judah, Carter and Colonel Wolford, who were now in pursuit of the raiders, with parts of their divisions. Federal cavalry pursued Morgan into Indiana and Ohio. So pressed were they, however, to keep pace with Morgan, that they had scarcely time to rest and feed their animals. Many of these pursuing troops had to fall out of the chase and halt in Cincinnati "on account of the breaking down of their horses." (Burnside, Report, O.R., p. 14.) Meanwhile, Morgan secured a constant fresh supply of horses for his command by seizing all horses on his path. As one correspondent wryly remarked: it was only as Hobson and Judah also began to seize horses that they began to gain on Morgan's 'rough riders.' No serious resistance was offered the raiders as they passed through Indiana and Ohio and in these States, they destroyed and looted much public and private property and killed a number of inhabitants. Morgan had until this point in time, succeeded in escaping at every juncture before pursuing forces could be gotten into position.

738   Comly, Diaries, entry dated "Thursday, July 16, 1863" (micro. Roll 1), [n.p.]

739   Written by an unnamed Cincinatti gentleman reprinted from the Cincinnati *Times* (1875), as "A War Reminiscence. General Hayes and the Morgan Raid—Another Account," *Santa Barbara Daily Press*, Monday Evening, October 30, 1876, courtesy of Darl Stephenson.

740   Letter written from Columbus, Ohio to the New York *Times*, quoted and corrected by H.G. O[tis], "Personal Recollections of the War—No. III," *Santa Barbara Daily Press*, October 28, 1876, courtesy of Darl Stephenson.

741   J.Q. Howard, *The Life Public Services and Select Speeches of Rutherford B. Hayes* (Cincinnati: Robert Clark & Co, 1876), p. 36. As indicated, few forces could be spared to go after Morgan. In anticipation of Morgan's crossing of the Ohio River to make good his escape from the North, Brig. Gen. H.M. Judah requested Capt. Leroy Fitch, comdg. a fleet of 'tinclads' on the Ohio River, to move with his gunboats post haste to Gallipolis and Pomeroy on Wednesday, July 15, to head off Morgan. Judah also sent boats to Col. Carr B. White with directions that cavalry and some infantry be shipped under convoy of the gunboats to Gallipolis or Pomeroy, as might be directed. (O.R. Ser. 1 Vol. 25, p. 252.) Whether the boats ordered up river by Gen. Judah to meet Kanawha forces in Fayette Co. were the ones which were in fact sent to pick up Hayes' troops at Loup Creek is not known. The two ships which Hayes' 23rd Ohio and 13th W. Va. regiments did in fact board at Loup Creek were the government steamers, the *Victor No. 2* and the *B.C. Levi*.

742   Hayes, diary entry (dated "July 22, 1863"), *Diary and Letters*, p. 420. From early days there had been friction between Scammon, then Colonel of the 23rd Ohio, and Hayes, his immediate subordinate. When Scammon was captured in early February 1864, Hayes was clearly glad to have him gone.

743   Howard, *Life Public Services and Select Speeches of Rutherford B. Hayes*, p. 36.

744   H.G.O., "Personal Recollections of the War—No. III," *Santa Barbara Daily Press*, October 28, 1876, courtesy of Darl Stephenson.

745   E.P. Scammon, Brigadier General, Commanding Division, by Capt. T. Melvin, Asst. Adjt. Gen., Dept. of West Virginia, Report of Brig. Gen. E. Parker Scammon, U.S. Army, Commanding Third Division, Eighth Army Corps (dated "Hd Qrs. 3d Div., 8th A.C., Charleston, W.Va., July 23, 1863"), O.R. XXIII (Ser. I) Part I, p. 677.

746   O[tis], "Personal Recollections No. III," *Santa Barbara Daily Press* October 28, 1876, and "A War Reminiscence. General Hayes and the Morgan Raid," *Santa Barbara Daily Press*, October 30, 1876.

747   Point Pleasant *Weekly Register*, July 16, 1863, Vol. II No. 18, [n.p.]

748   Howard, *Life Public Services and Select Speeches of Rutherford B. Hayes*, p. 36.

749   Letter written from Columbus, Ohio to the New York *Times*, bracketed words such as "Gauley Bridge" represent a correction made by Otis from "Fayetteville" a clear mistake in the Ohio account. H.G. O[tis], ed., "Personal Recollections of the War—No. III," *Santa Barbara Daily Press*, October 28, 1876, [n.p.], courtesy of D. Stephenson.

750   "History of 23rd Ohio Regiment," Comly Papers (micro. Reel 1), pp. 390-92.

751   RG94 Morning Reports Capt. A.F McCown's Co. F 13th Regiment Va. Vols. July 1863, [n.p.]

752   Regimental Return 13th Va. Vol. Infantry for August 1863 (micro. #594 Roll 196), NARA.

753   Comly, Diaries, entry dated "Friday, July 17, 1863 through Thursday, July 21, 1863" (micro. Roll 1), [n.p.].

754   Annual Return of Casualties + c. 13th W. Va. Infy. to Dec. 31/63, RG94 Box 4838 Misc. unbound regimental papers 13th W.Va.

755   Comly, cmdg. 23rd Ohio Infantry Regiment, noted in his diary that Gallipolis was reached at 3 a.m. on the morning of July 18. Comly, Diaries, entry dated "Friday, July 17, 1863 through Thursday, July 21, 1863," (micro. Roll 1), [n.p.]

756   Annual Return of Casualties + c. 13th W.Va. Infy., RG94 Box 4838.

757   Captain Sidney P. Cunningham of Morgan's Cavalry Division wrote in his report of the raid that in southeastern Ohio they had begun to feel the resistance of the population. Trees were felled by the locals to blockade roads and the woods were alive with militia, who ambushed and bushwhacked the raiders. Roads were "bad" which required the command to make detours which brought them close to Chillicothe and Hillsboro on the north and Gallipolis on the south. (Cunningham, Report, Supp. O.R. Part I, Reports Vol. 4 Ser. No. 4, pp. 222-25.)

758   Captain A.A. Hunter, commanding Post at Gallipolis to Gen. Burnside (dated Gallipolis, July 18, 1863), O.R. Vol. XXIII Part II, p. 770.

759   H.G. O[tis], "Personal Recollections of the War—No. III," *Santa Barbara Daily Press*, October 28, 1876, [n.p.].

760   Hayes, diary entry (dated "July 22, 1863"), *Diary and Letters*, p. 420.

761   E.P. Scammon, Brig. Gen., Commanding Division, by Capt. T. Melvin, Asst. Adjt. Gen., Dept. of West Virginia, Report of Brig. Gen. E. Parker Scammon, U.S. Army, Commanding Third Division, Eighth Army Corps (dated Hd Qrs. 3d Div., 8th A.C., Charleston, W.Va., July 23, 1863), O.R. Vol. XXIII Part I, p. 677.

762   Written by a Cincinatti gentleman and reprinted from the Cincinnati *Times* (1875) as "A War Reminiscence. General Hayes and the Morgan Raid—Another Account," *Santa Barbara Daily Press*, October 30, 1876, [n.p.]

763   "History of 23rd Ohio Regiment," Comly Papers (micro. Reel 1), OHC, pp. 392-93.

764   Scammon would have been and definitely was, on the 18th, authorized to "use any transportation" that could carry his troops up the river as quickly as possible — on water or over land. (Burnside to Scammon, dated [probably Cincinnati] July 18, 1863, O.R. XXIII Part II, p. 773.) Lieut. Col. James M. Comly stated in his report that the 23rd Ohio left Gallipolis for Pomeroy in the steamer *B.C.*

Levi. See Report of Lieut. Col. James M. Comly Twenty-third Ohio Infantry, First Brigade (dated Hdqrs. Twenty-third Regiment Ohio Volunteers, Camp White, W. Va.., July 22, 1863), O.R. XXIII Ser. I Part I, pp. 678-79. The 13th West Virginia may have again boarded the *Victress, Victor No. 2* and *General Meigs* for the trip to Pomeroy.

765   Regimental Return 13th Va. Vol. Infantry for August 1863, dated "Barboursville, W.Va. Sept. 9, 1863" (micro. #594 Roll 196), NARA.

766   Annual Return of Casualties + c. 13th W.Va. Infy. to Dec. 31/63, RG94 Box 4838 Misc. unbound regimental papers 13th W.Va.

767   Quoted and published by Harrison Gray Otis, editor, several years previous in the *Grand Army Journal* (Washington) and reprinted in Otis' series "Personal Recollections of the War—No. III," *Santa Barbara Daily Press*, October 28, 1876.

768   General Basil W. Duke, "Sketch of General John H. Morgan," *Confederate Veteran Magazine*, Vol. XIX (1911), p. 569.

769   A machinist was "[a] constructor of machines and engines or one well versed in the principles of machines." Noah Webster, *An American Dictionary of the English Language,* revised and enlarged by Chauncey A. Goodrich, Professor in Yale College (Spring-field, Massachusetts: George and Charles Merriam, 1858), p. 684.

770   See the 1860 Census for Meigs County, Ohio, Salisbury Township, town of Pomeroy, page 82 which lists William R. Brown, age 34, as a machinist by profession, with $700 in real estate and $100 personal estate. Brown's wife, Violetta, age 24, son, Allen S., eight years old; daughter Sally, four years old; and daughter Malitta, one year old are all listed in the 1860 Census.

William R. Brown was a native of Pennsylvania. His obituary contains quite a bit of information about the man. He was, according to this source, born in 1823. He died on the evening of Tuesday, March 24, 1891, "at his residence just east of [Independence, Kansas], from a severe attack of pneumonia." His age at death was given as "sixty five years and nine months." The article makes plain that he was in every sense of the word a self-made man. He began to earn his living when still a boy. He was "possessed all those sterling virtues which characterize a man, who from his boyhood wins his way to fame and fortune by hard licks. He was quick to recognize his duty to society and humanity and rigorously performed it. At the breaking out of the war he occupied a lucrative position as foreman in large machine shops at Pomeroy O. As his state's quota was full, he took a company of men from the shops and went over into West Virginia and enlisted. He was made captain of the company, [and assigned to the 4th V.V.I.] but later, on account of his distinguished services, was promoted to colonel of the 13th West Va infantry, which was a regiment of the brigade of Gen. R.B. Hayes." Brown moved to Kansas in 1874. Here he served in several public capacities: as county commissioner, probate judge and at the time of his death, he was serving on the board of education. [Obituary of General William R. Brown], *Star and Kansan*, Friday, March 27, 1891, [n.p.] and "Mustered Out. Brigadier- General, W.R. Brown, Entered into Rest," *South Kansan Tribune*, Wednesday, March 25, 1891, [n.p.], courtesy of Roger D. Hunt.

771   "Latest Telegraphic News. Special Dispatches exclusive to Cincinnati Daily Commercial. Telegraphic Correspondence Daily Commercial. From Columbus. The Pursuit of Morgan—700 Prisoners, 700 Horses and Three Pieces Artillery Captured at Buffing-ton—The Attempt to Cross at Bealeville and How it was Frustrated—Nearly all of Morgan's Command Gobbled Up. Columbus, O., July 20," *Cincinnati Daily Commercial*, July 21, 1863, p. 3. and L.W.C., "The Morgan Raid. Our Special Correspondence" (dated "Marietta, July 24, 1863"), *Cincinnati Daily Commercial*, July 27, 1863, p. 1.

772   *Meigs County, Ohio from Hardesty's Historical And Geographical Encyclopedia 1883.* Reprint (Defiance Ohio: The Hubbard Com-pany, 1982), p. 276.

773   L.W.C., "The Morgan Raid. Our Special Correspondence" (dated "Marietta, July 24, 1863"), *Cincinnati Daily Commercial*, July 27, 1863, p. 1.

774   Ibid.

775   Ibid.

776   *Meigs County, Ohio from Hardesty's 1883.* Reprint, p. 276.

777   Paraphrase, ibid., p. 276.

778   All preceding quotations in this paragraph are taken from "History of 23rd Ohio Regiment," Comly Papers (micro. Reel 1), OHC, pp. 392-93. This writing was apparently a fleshing out of Hayes' terse diary entry for July 22, 1863 ("formed lines of battle. Morgan ditto.")

779   H.G. O[tis], "Personal Recollections of the War—No. III," *Santa Barbara Daily Press*, October 28, 1876, [n.p.]

780   Col. A.B. Jones, assistant inspector general on General Scammon's staff accompanied the troops up the Ohio River and "rendered important service" assisting Scammon "in the movement of troops on the field." See E.P. Scammon, Brigadier General, Commanding Division, by Capt. T. Melvin, Asst. Adjt. Gen., Dept. of West Virginia, Report of Brig. Gen. E. Parker Scammon, U.S. Army, commanding Third Division, Eighth Army Corps (dated Hd Qrs. 3d Div., 8th A.C., Charleston, W.Va., July 23, 1863), O.R. XXIII Part I, p. 677.

781   Col. R.B. Hayes to Capt. James L. Bottsford, A.A.G. 3rd Division 8th A.C., Report (manu. dated "Head Qtrs. 1st Brigade 3rd Division, Dept 8th Army Corps, Camp White, Va., July 23, 1863"), H.178 Registry #393 Part 2 Entry 1184 Press Copies of Letters Sent and Special Orders Vol. 139, NARA.

782   Regimental Return 13th Va. Vol. Infantry for August 1863, dated "Barboursville, W.Va. Sept. 9, 1863," (micro. #594 Roll 196), NARA.

783   One summary of the 13th Regiment's field service noted that on July 18 about "two miles" from Pomeroy, the 13th "had a severe skirmish with the enemy." (The 13th Regiment West Va. Inf. Memoranda, Field History, WVSA), [n.p.]

784   RG94 Morning Reports of Co. A 13th Regiment Va. Vols. for July 1863, [n.p.]

785   RG94 Morning Reports of Co. F 13th Regiment Va. Vols. for July 1863, [n.p.]

786   J.V. Young to Paulina (dated "Point Pleasant, July 24, 1863, 3 o'clock 20 min. P.M."), Young Civil War Papers, trans. p. 77.

787   RG94 Morning Reports of Co. D 13th Regiment Va. Vols. for July 1863, [n.p.]

788   History is not clear whether Hayes' troops skirmished with Morgan's advance upon the town of Pomeroy or with Morgan's rear as they made their escape up river to attempt crossing into West Virginia higher up. Official reports suggest that Hayes' men waited for Morgan to advance upon Pomeroy, which they did at about midday. Hardesty's history of Meigs County, however, states that Federal troops skirmished with Morgan's rear at Pomeroy. Andrew Stiarwalt, musician, 23rd Ohio, also states this in his war journal. Stiarwalt, an eye witness, wrote that at Pomeroy "our force move out on the hills in Ohio or near Pomroy and had some scurmishing with morgan's Reer but he was making all possible hast to escape us […]." (Andrew Stiarwalt, Sr., Diary 1861-64, R.B. Hayes Papers Civil War Records, micro. Ser. 7/ Roll No. 272, pp. 8-9 of manu.) Perhaps, Hayes' regiments skirmished with both advance and rear. That is, hypothetically, with Morgan's advance as it moved to cross the Ohio River at Pomeroy and then skirmishing with the rear guard when the raiders veered away from Pomeroy to try attempt to cross again farther up river.

789   Regimental Return 13th Va. Vol. Infantry for August 1863, dated "Barboursville, W.Va. Sept. 9, 1863") micro. #594 Roll 196, NARA.

790   Report of Lieut. Col. James M. Comly, Twenty-third Ohio Infantry, First Brigade (dated Hdqrs. Twenty-third Regiment Ohio Volunteers, Camp White, W.Va.., July 22, 1863), O.R. XXIII (Ser. I) Part I, pp. 678-79. Brig. Gen. Scammon used the same language as Comly, writing that after their arrival at Pomeroy his troops had "a slight skirmish" with Morgan's men after which the Confederates "hastily retreated up the river." See Report of Brig. Gen. E. Parker Scammon, U.S. Army, Commanding Third Division, Eighth Army Corps (dated Hd Qrs. 3d Div., 8th A.C., Charleston, W.Va., July 23, 1863), O.R. XXIII Part I, p. 677.

791   Report of Brig. Gen. E. Parker Scammon, p. 677.

792   Regimental Return 13th Va. Vol. Infantry for August 1863 (micro. #594 Roll 196), NARA.

793   McCreary probably refers to militia forces here.

794   Diary of Major James Bennett McCreary, 11th Ky. Cav., on Morgan's Raid in Kentucky June 16-July 1863, Reports, Supp. O.R. Part I, Ser. No. 4 Vol. 4, pp. 215-17.

795   In a post-war interview, Duke laid greater emphasis on the danger they had passed through behind Pomeroy writing: "We had to run a terrible gauntlet for nearly five miles, through a ravine, on the gallop." See [Interview with Basil Duke], Martin R. Andrews, ed. and comp., *History of Marietta's Washington County, Ohio & Representative Citizens* (Chicago, Illinois: Biographical Publishing Co., 1902, p. 596 and Basil W. Duke, *A History of Morgan's Cavalry,* Cecil Fletcher Holland, ed. (Bloomington: Indiana University Press, 1960), pp. 445-46.

796   Colonel J. Warren Grigsby commanded 6th Kentucky Cavalry. Major Thomas Webber commanded 2nd Kentucky Cavalry.

797   Duke, *History of Morgan's Cavalry,* pp. 445-46.

798   Allan Keller, *Morgan's Raid* (Indpls. & New York: The Bobbs-Merrill Co., 1961), p. 188.

799   Anna Starr to her husband, W.C. Starr (dated "Pomeroy, July 27, 1863"), SC1400 Civil War Files of William C. Starr, Manuscript Coll. Indiana Historical Society, Indianapolis.

800   E.H., correspondent to the *Cincinnati Commercial* writing from the *Imperial*, the dispatch boat of the Federal gunboat *Moose*, "The Naval Engagement at Buffington Island. Termination of the Morgan Raid. Sixteen Hundred Rebels, Seven Pieces of Artillery and Immense Numbers of Horses and Quantities of Supplies Taken—Basil Duke and Dick Morgan Among the Prisoners—The Pursuit. Correspondence of the Cincinnati Commercial" (dated "Cincinnati, July 21, 1863"), *Cincinnati Daily Commercial Gazette*, Wednesday, July 22, 1863, p. 1.

801   Ibid.

802   Rear Admiral David D. Porter (supreme commander of all U.S. naval forces on the Mississippi River) reorganized his command in May 1863 into six divisions. The young Hoosier born Lieutenant Leroy Fitch, a regular naval line officer, was in command of the Sixth Division. This division operated in the area embraced by the Cumberland River and the Ohio River up from Smithville, Kentucky. Fitch commanded a fleet of 'tinclad' gunboats. These were lighter than ironclad ships and thus able to penetrate higher up river, where the heavier ironclads could not penetrate. His boats were the *Brilliant*; the *St. Clair*; *Fairplay*; *Silver Lake*; the *Moose*, Fitch's flagship at the battle of Buffington Island; the *Springfield*; *Reindeer*; *Victory*; and the *Naumkeag*. See Mark F. Jenkins, "Operations of the Missippi Squadron during Morgan's Raid," http://www.wideopenwest.com/~jenkins/ironclads/buffint.htm (Copyright 1999), p. 2.

803   W.P. Anderson, Aide to Gen. Burnside (telegraph dated Pomeroy, July 17, 1863) O.R. XXIII Part II, p. 764.

804   Anna Starr to her husband, William C. Starr (manu. dated "Mason City, July 20, 1863"), SC1400 Civil War Files of William C. Starr, IHS.

805   Ibid.

806   All quotations in the foregoing paragraph from R.B. Hayes to his wife, Lucy (dated "Camp White, July 22, 1863"), *Diary and Letters*, p. 421.

807   Scammon, Report, O.R. XXIII Part I, p. 677.

808   Lieut. Col. James M. Comly, Report, O.R. XXIII Part I, pp. 678-79.

809   Record of Events Field and Staff Company Muster Roll July and Aug. 1863 for Co. F (micro. #594 Roll 196), NARA.

810   *Diary of Henrietta Barr* quoted in Dean W. Moore, *Washington's Woods. A History of Ravenswood and Jackson County, West Va.* (Parsons, W.Va.: McClain Printing Co., 1971), p.129.

811   RG94 Morning Reports of Co. F 13th Regiment Va. Vols. for July 1863, [n.p.]

812   Hayes was a day ahead of himself both in his personal journal and in his official report of events, which may not have been included in the U.S. Printing Office's publication of the *War of the Rebellion* documents because of the confusion this small mistake created in the overall reading of what is othewise a valuable report. Specific times of movements and events during the chase after Morgan tend to be somewhat off in official accounts that come down to us today. This seems due to the fact that Scammon's troops had just come off the expedition to Fayetteville; were exhausted probably sleeping at odd times; and were getting their days and nights mixed up.

813   Duke, *A History of Morgan's Cavalry,* p. 450. This is the high estimate. Duke also wrote that by the time the raiders reached Buffington their strength was something "less than 1,800." See General Basil W. Duke, "Sketch of General John H. Morgan," *Confederate Veteran Magazine*, Vol. XIX (1911), p. 569.

814   With Fitch above Pomeroy were only the *Moose*, Fitch's flagship; the *Springfield*; *Reindeer*; *Victory* and the *Imperial*, Fitch's dispatch boat. See Jenkins, "Operations of the Mississippi Squadron during Morgan"s Raid," p. 2.

815   [Editorial], *Cincinnati Commercial Gazette*, July 20, 1863, p. 2.

816   Martin R. Andrews, ed. and comp., [Interview with Colonel Basil Duke], *History of Marietta's Washington County, Ohio and Representative Citizens* (Chicago, Illinois: Biographical Publishing Co., 1902), p. 596.

817   Basil W. Duke, *History of Morgan's Cavalry,* p. 446.

818   Ibid.

819   Ibid., p. 447.

820   The Kentucky infantry to which Duke refers were probably Scammon's regiments.

821   Duke, *History of Morgan's Cavalry,* p. 447.

822   Ibid.

823   Duke, "Sketch of General John H. Morgan," *Confederate Veteran Magazine*, Vol. XIX (1911), p. 569. Myron J. Smith, Jr. stated in his article about the U.S. Navy's part of the battle at Buffington that about ten days before Morgan reached Portland, the water level at the islands of Buffington and Blennerhassett was reported by visitors, who had traveled from Pittsburg as being about two feet deep, ideal for fording. Four days later (about July 13), however, in consequence of the heavy rains in the West Virginia mountains, the water level at Pittsburg rose rapidly and "the surge continued down river. This rise [was] estimated at five and one half feet in some places." Smith, citing the Cincinatti *Daily Commercial*, July 9-13, 1863, in "Gunboats at Buffington. The U.S. Navy and Morgan's Raid," *West Virginia History*, Vol. 44 No. 2, p.105. Contemporary newspaper journalists (Cincinnatti *Daily Commercial*) and later writers also observed that the unusual rise in the Ohio River was the highest in twenty years and "the first of its kind twenty years," (Church M. Matthews, "Cadet United States Military Academy," *Confederate Veteran Magazine*, Vol. XXXVI 1928, p. 178.) The high water level made crossing the river a dangerous matter even in daylight and not a thing to be undertaken in the dark.

824   Duke, *History of Morgan's Cavalry,* p. 447.

825   Ibid., p. 448.

826   Andrews, [Interview with Col. Basil Duke], *History of Marietta's Washington County, Ohio,* p. 596.

827   Duke, *History of Morgan's Cavalry,* p. 449.

828   E.H., "The Naval Engagement at Buffington Island Correspondence of the Cincinnati Commercial" (dated "Cincinnati, July 21, 1863"), *Cincinnati Daily Commercial Gazette,* July 22, 1863, p. 1.

829   "Johnie Chandler of the 7th Ohio Cavalry, Hobson's command, "Account which appeared in an Ohio newspaper in September 1922," reprinted with commentary as "How Far Did Morgan Get?" *Confederate Veteran Magazine,* Vol. XXXI (1923), p. 171 and Duke "Sketch of General John H. Morgan," *Confederate Veteran Magazine,* Vol. XIX (1911), p. 569.

830   Ibid., Duke, "Sketch of General John H. Morgan," p. 569.

831   Smith, "Gunboats at Buffington. The U.S. Navy and Morgan's Raid" *West Virginia History,* Volume 44 No. 2, p. 106.

832   LeRoy Fitch, Lieutenant Commander to Acting Rear Admiral David D. Porter, Commanding Mississippi Squadron, O.R. (Ser. 1) Vol. XXV, p. 255. Myron J. Smith, Jr. ("Guns at Buffington," *West Virginia History*) counted "[f]our major fords and a number of minor fords" thus patrolled.

833   Smith, "Gunboats at Buffington," *West Virginia History,* fn. 16, p. 106, quoting Burnside to the Adj. Gen. U.S. Army (dated November 13, 1863), O.R. Vol. XXV, p. 259; Burnside to Fitch, dated "July 16, 1863," O.R. Vol. XXIII, p.781; and Fitch to Porter, dated November 5, 1863, O.R. Vol. XXV, p. 318.

834   Jenkins in his internet article about the 6th Mississippi Squadron indicated the position of the tinclads comprising the blockade: the "*Reindeer* was guarding Goose Island crossing. *Naumkeag* was guarding a ford at Eight-Mile Island, and *Victory* and *Springfield* were guarding Pomeroy and Wolf's Shoals." (Jenkins, "Operations of the Mississippi Squadron during Morgan's Raid," pp. 4-5, Jenkins citing to O.R. Vol. XXV, pp. 241-45.)

835   Smith, "Gunboats at Buffington," p. 106.

836   Capt. Sidney P. Cunningham, Morgan's Cavalry Division, Report, Supp. O.R. Part I, Vol. 4 No. 4, pp. 222-25.

837   McCreary, Major, 11th Ky. Cav. Morgan's Div., diary entry (dated "July 18, [1863]"), Diary of Major James Bennett McCreary, 11th Kentucky Cavalry, on Morgan's Raid in Kentucky June 16-July 1863, Supp. O.R., Part I Reports, Vol. 4 Ser. No. 4, pp. 215-217.

838   E.H., "Naval Engagement at Buffington Island. [...] The Pursuit. Correspondence of the Cincinnati Commercial" (dated "Cincinnati, July 21, 1863"), *Cincinnati Daily Commercial Gazette,* July 22, 1863, p. 1. Morgan's artillery consisted of not more than five pieces. No less than three of these (all brass six pounders) must have been on the heights along the river bank as they were found abandoned there after the fight.

839   Cunningham stated in his report that at 4 a.m., two companies were "thrown" across the river and "instantly opened upon" by Federal forces. (Cunningham, pp. 222-25.)

840   Duke, *History of Morgan's Cavalry,* p. 448.

841   Cunningham, Report, Supp. O.R. Part I ,Vol. 4 Ser. No. 4, pp. 222-25.

842   Brig. Gen. Henry M. Judah to Maj. Gen. Ambrose E. Burnside (dated Headquarters United States Forces, Buffington Bar, July 19, 1863-10 a.m.), O.R. XXIII Part I, p. 776.

843   Brig. Gen. Henry M. Judah, U.S. Army, Commanding Third Division, Twenty-Third Army Corps, Report, O.R. XXIII Part I, p. 656. Judah reported that he and his command "reached the last descent to the river bottom at Buffington Bar at 5:30 a.m. on Sunday July 19th [...] in a very dense fog." (Report of Brig. Gen. Henry M. Judah, U.S. Army, Commanding Third Division, Twenty-Third Army Corps, O.R. XXIII Part I, p. 656.)

844   West Virginia historian, Boyd B. Stutler wrote that visibility was limited to "less than 50 yards." Stutler, *West Va in the Civil War* (Charleston, W.Va.: Education Foundation, 1963), pp. 232-33.

845   Judah to Burnside (dated Buffington, July 19[?], 1863, 1.30 a.m.), O.R. XXIII Part I, pp. 775-76.

846   E.H., writing from the *Imperial*, the dispatch boat of the U.S. gunboat *Moose*. E.H.'s remarks concerning the engagement between forces on land, while reliable are "of a general character" inasmuch as "their operations were not witnessed to any extent from the boats, on account of the high banks, over which the shots from the Moose had to be thrown." "The Naval Engagement at Buffington Island" (dated "Cincinnati, July 21, 1863"), *Cincinnati Daily Commercial Gazette*, July 22, 1863, p. 1.

847   Ibid.

848   Duke, *History of Morgan's Cavalry*, p. 450.

849   The *Moose* carried six twelve pound guns. Two were at each side of the ship and two were at the bow. E.H. "The Naval Engagement at Buffington Island," dated "Cincinnati, July 21, 1863," *Cincinnati Gazette*, July 22, 1863, p. 1.

850   Smith, "Gunboats at Buffington," *West Virginia History*, Vol. 44 No. 2, p. 107.

851   McCreary, Diary of Major James Bennett McCreary, 11th Kentucky Cavalry, on Morgan's Raid in Kentucky June 16-July 1863 (entry for "July 18, [1863]"), Supp. O.R. Part I, Vol. 4 No. 4, pp. 215-17.

852   Ibid.

853   Sidney Cunningham (2nd Brigade, Morgan's Div.), Report, pp. 222-25.

854   Ibid., McCreary, pp. 215-217. We cannot know whether McCreary's reference to the landing of the infantrymen refers to Gen. Judah's infantrymen or to the Kanawha troops or both.

855   Ibid., Cunningham, pp.222-25.

856   Ibid., McCreary, pp. 215-17.

857   Duke, *History of Morgan's Cavalry*, p. 452.

858   Ibid.

859   Johnie Chandler, 7th Ohio Cavalry, account which appeared in an Ohio newspaper in September 1922, reprinted as "How Far Did Morgan Get?," *Confederate Veteran Magazine*, Vol. XXXI (1923), p. 171.

860   Ibid.

861   "Latest Telegraphic News. Special Despatches Exclusive to Cincinnati Daily Commercial. Telegraphic Correspondence Daily Commercial. From Columbus. Morgan's Attempt to Cross the Ohio at Buffington—He is Hemmed in by Gens. Hobson and Judah and the Gunboats—Result of the Engagement—1,000 Prisoners Taken—A Dash Through our Lines—Morgan Reported Crossing at Beallville—Forces Moving to Intercept Him. Columbus, O., July 19," *Cincinnati Commercial Gazette*, July 20, 1863, p. 3.

862   Colonel Basil Duke indicated that there were two routes of escape: "roads" at the upper end of the valley. One route was certainly the river road, a narrow precarious thing passing along the precipice presented by the rocky ridge at the river adjacent to which was a ravine leading down to the Ohio River. See Duke, *History of Morgans Cavalry*, p. 452.

863   Ibid.

864   Smith, "Gunboats at Buffington," *West Virginia History*, pp. 107-08.

865   Cunningham (2nd Brigade, Morgan's Division), Report, Supp. O.R. Part I Reports, Vol. 4 Ser. No. 4, pp. 222-25.

866   LeRoy Fitch, Lieutenant Commander to Acting Rear Admiral David D. Porter, Commanding Mississippi Squadron, O.R. Vol. XXV, p. 255.

867   Ibid., pp. 254-55.

868   Ibid., p. 255.

869   In addition to various contemporary Northern writers and combatants whose writings suggest that the battle was over in something less than an hour, see Brig. Gen. H.M. Judah, who reported to Maj. Gen. Ambrose E. Burnside from Buffington Island that "in less than half an hour" after hostilities had commenced at Buffington, Morgan was driven and fleeing. (Report, dated Buffington, July 19[?], 1:30 a.m., O.R. XXIII Part I, pp. 775-76.)

870   Burnside, O.R., p. 14.

871   In *History of Morgan's Cavalry,* Duke stated that Morgan's Division arrived at Buffington Island with under nineteen hundred (p. 450). In an earlier publication, Duke indicated that by the time the raiders reached Buffington, their strength was something "less than 1,800." Duke, "Sketch of General John H. Morgan," *Confederate Veteran Magazine*, Vol. XIX (1911), p. 569.

872   Duke, *History of Morgan's Cavalry*, p. 453.

873   *Meigs County, Ohio from Hardesty's 1883,* reprint, p. 276.

874   Regimental Return 13th Va. Vol. Infantry for August 1863, dated "Barboursville, W.Va. Sept. 9, 1863" (micro. #594 Roll 196), NARA and "History of 23rd Ohio Regiment," Comly Papers (micro. Reel 1), OHC, pp. 392-93.

875   Annual Return of Casualties + c. 13th W.Va. Infy. to Dec. 31/63 (manu. "Return completed Jan. 13, 1864"), RG94 Box 4838 Misc. unbound regimental papers 13th W.Va., NARA.

876   Field History 13th Regiment W.Va. Inftry., [n.p.]

877   A summary tribute to the 13th West Virginia published in at least a couple of W.Va. papers in the days after the regiment was mustered out at Wheeling, June 22, 1865, included in their list of engagements Buffington Island with the note that the regiment skirmished there with Morgan's forces during the battle on July 19. ("The Thirteenth West Va. Infantry,"Wheeling *Daily Intelligencer,* June 24, 1865, p. 3 and *The Journal*, Charleston, W. Va., July 5, 1865, p. [2].)

878   RG94 Morning Reports of Captain A.F. McCown's Co. F 13th Regiment Va. Vols. for July 1863, [n.p.] "13th Regiment Memoranda" carries that the 13th was "actively engaged" in the fight at Buffington. "The 13th Regiment West Va. Inf. Memoranda," Field History, WVSA, [n.p.] Allan Keller stated in his study that the 23rd Ohio was used to defend on Ohio soil and 13th W. Va. was chiefly employed on the West Virginia river bank to keep Morgan's men from escaping south over the Ohio into West Virginia and that when Morgan stopped trying to cross at the bend of the river near Buffington Island, Scammon's force was ferried from the West Virginia bank to the Ohio side of the river. See Allan Keller, *Morgan's Raid* (Indpls. & New York: The Bobbs-Merrill Co., 1961), p. 155.

879   J.V. Young to wife, Paulina and children (dated "Point Pleasant, Mason Co. July 23, 1863"), Young Civil War Papers, trans. p. 76.

880   Record of Events Field and Staff Muster Roll Co. B 13th Va. Vol. Infantry July and Aug. 1863 (micro. #594 Roll 196), NARA.

881   Stiarwalt, Diary 1861-64, Russell P. Hastings Papers (micro. Ser. 7/ Roll No. 272), pp. 8-9 of manu.

882   Hayes, diary (dated "July 22, 1863"), *Diary and Letters*, p. 420. Colonel Hayes' wrote in his report to division headquarters that Morgan "attempted to cross but was defeated by the cavalry under General Judah, and the Gun Boats under Capt. Fitch." Report of Col. R.B. Hayes to Capt. James S. Bottsford, A.A.G., 3rd Division 8th AC" (manu. dated "Head Qtrs. 1st Brigade 3rd Division, Dept 8th Army Corps, Camp White, Va., July 23, 1863"), H. #178 Registry #393 Part 2 Entry 1184 Press Copies of Letters Sent and Spec. Orders Vol. 139, NARA.

883   Respectively, Comly, Diaries, "Friday, July 17, 1863 through Thursday, July 21, 1863," (micro. Roll 1), OHC, [n.p.] and Report of Lieut. Col. James M. Comly. Twenty-third Ohio Infantry, First Brigade (dated Hdqrs. Twenty-third Regiment Ohio Volunteers, Camp White, W.Va. July 22, 1863), O.R. XXIII Part I, pp. 678-79.

884   E.H., "The Naval Engagement at Buffington Island" (correspondence dated "Cincinnati, July 21, 1863"), *Cincinnati Daily Commercial Gazette*, July 22, 1863, p. 1.

885   LeRoy Fitch, Lieutenant Commander to Acting Rear Admiral David D. Porter, Commanding Mississippi Squadron, O.R. Vol. XXV, pp. 254-55.

886   "Cincinnati gentleman," reprinted from the Cincinnati *Times* of 1875, as "A War Reminiscence General Hayes and the Morgan Raid—Another Account," *Santa Barbara Daily Press*, Monday Evening, October 30, 1876.

887   Basil Duke's account as given to Whitelaw Reid for his *Ohio in the War*; published in abridged form in James Q. Howard, "A Glance at Ohio History and Historical Men" in Henry Howe's *Historical Collections of Ohio*, Vol. I (Cols.: Henry Howe and Son, 1890), p. 456.

888   E.H., "Naval Engagement at Buffington Island" (dated "Cincinnati, July 21, 1863"), *Cincinnati Daily Commercial Gazette*, July 22, 1863, p. 1.

889   Col. Hayes wrote in his official report to division headquarters that "[t]he infantry under my command were present in time to witness the defeat of the enemy, but not in time to render important assistance." See Report of Col. R.B. Hayes to Capt. James S. Bottsford, A.A.G. 3rd Division 8th A.C. (manu. dated "Head Qtrs. 1st Brigade 3rd Division, Dept 8th Army Corps, Camp White, Va., July 23, 1863"), H. 178 Registry #393 Part 2 Entry 1184 Press Copies of Letters Sent and Special Orders Vol. 139, NARA.

890   Regimental Return 13th Va. Vol. Infantry for August 1863, dated "Barboursville, W.Va. Sept. 9, 1863," (micro. #594 Roll 196), NARA and Field History 13th Regiment W.Va. Infantry, [n.p.]

891   It is apparent from the historical record that after the break up of Morgan's force at Buffington, the desperate raiders made a number of attempts at crossing the river at the shoals above that island and below Hockingport. Estimated mileages within the area of operations are provided here indicating a general vicinity for Hayes' troops part in the mopping up phase of the 'great raid.' Place names which are mentioned in this connection are Swan Bar; Belleville (also called Bealeville), West Virginia; and Craig's Bar, presumably on the Ohio side opposite Belleville; Indian Run; Lee's Creek; Hocking River; and Hockingport, Ohio. Some of the place names are no longer in use today but using a DeLorm map of West Virginia one can gain a sense of this stretch of river. One finds that the distance from Parkersburg to Hockingport is about 15 miles following the river (not as the crow flies) and Parkersburg to Belleville is about 19 miles (also following the meanderings of the river). Hockingport is distant up river from Belleville, as the crow flies, about 4.2 miles and distant from Buffington Island about eighteen miles. Lee's Creek is distant from Belleville just over two miles and from Hockingport about three miles.

892   Annual Return of Casualties + c. 13th W.Va. Infy., RG 94 Box 4838 Misc. unbound regimental papers 13th W.Va.

893   "History of 23rd Ohio Regiment," Comly Papers, pp. 392-93.

894   Brig. Gen. H.M. Judah reported to Burnside after the battle: "The enemy has scattered in the woods. Hobson has a brigade of his force after them. We both move upon him immediately. Scammon is here with two regiments of infantry. At my suggestion, he sends them upon boats, under convoy of a gunboat, to Blennerhassett's Island." Brig. Gen. H.M. Judah to Major General Ambrose E. Burnside, dated "Buffington, [July 19], 1863, 1.30 [p].m.," O.R. XXIII Part 1, pp. 775-76.

895   Colonel William Wallace to General Burnside (dated "Foot of Blennerhassett's Island, July 19th, 1863"), O.R. XXIII Part I, p. 783.

896   Report of Col. R.B. Hayes to Capt. James S. Bottsford, A.A.G. 3rd Division 8th AC (manu. dated "Head Qtrs. 1st Brigade 3rd Division, Dept 8th Army Corps, Camp White, Va., July 23, 1863"), H.178 Registry #393 Part 2 Entry 1184 Press Copies of Letters Sent and Special Orders Vol. 139, NARA.

897   R.B. Hayes, diary entry dated "July 22, 1863," *Diary and Letters*, p. 420.

898   Respectively, James Comly, Diaries, dated "Friday, July 17, 1863 through Thursday, July 21, 1863" (micro. Roll 1), [n.p.] and RG94 Morning Reports of Co. A 13th Regiment Va. Vols. for July 1863, [n.p.]

899   H. "The Battle of Buffington Island By One of the Pursuers Correspondence to Cincinnati *Commercial*," *Cincinnati Daily Commercial*, Aug, 4, 1863, p. 1.

900   J.V. Young to wife, Paulina (dated "Point Pleasant, July 24, 1863, 3 o'clock 20 min. P.M."). Young Civil War Papers (Vol. 31 Box 10), WVU, trans. p. 77.

901   "History of 23rd Ohio Regiment," Comly Papers (micro. Reel 1), OHC, pp. 392-93.

902   H.G O[tis], "Personal Recollections of the War—No. III," *Santa Barbara Daily Press*, Saturday Evening, October 28, 1876.

903   Other eyewitness sources stated that the location of the second attempted crossing in miles distant from Buffington was fourteen to fifteen and up to twenty miles.

904   A significant body of Confederates under Johnson and Morgan attempted to cross at Belleville. Three hundred and thirty men reportedly made it over to the W. Virginia shore.

905   It seems that Lee's Creek, West Virginia, also came to be guarded by Scammon's men because of this attack on the *Moose*. Myron Smith, Jr. described this last "shoot" between raiders and Federal warships. He stated, however, that instead of occurring on the 19th, as E.H. seems to suggest, it occurred on Monday, the 20th. Not all Confederates, wrote Smith, who had escaped into West Virginia had immediately left the area. Some of Johnson's men, who had escaped at the head of Belleville Island planned a last salute, a parting shot to the ships patrolling the river. They lay in ambush for the gunboats "[n]ear the Wells farm, on the upper side of Lee Creek two miles above Belleville and when the steamer approached within a range of less than twenty yards," the Confederates fired two volleys of musketry into them. The *Moose*, wrote Smith, replied with fire from her starboard battery and killed nine of the bushwhackers. Then the *Moose* and *Allegheny Belle* began a general shelling of the area, concentrating fire especially on roads leading inland from the river. This may have been the point at which the 13th was subjected to fire from the *Allegheny Belle* as described by Roush in *If Thou Wilt Remember*, his book about about his grandfather Private David Burrows, Co. E 13th West Virginia. After this episode, wrote Smith, "the Rebels departed continuing over the hills leading to the Little Kanawha Valley and the two gunboats, finding the water upstream too shallow to proceed, rounded about and returned to Buffington Island." See Smith, "Gunboats at Buffington," *West Virginia History*, Vol. 44 No. 2, p. 109, fn. 21, citing as authorities for his statement, the Richmond *Enquirer*, July 31, 1863; the Cincinnatti *Daily Gazette* quoted in the New York *Herald*, July 23, 1863, as "The Battle of Buffington Island;" and Charles R. Rector, "Morgan 'Goes-a-Raiding' and Views West Virginia," *West Virginia Review*, 6 May 1929, p. 311.

906   This attack on the *Moose* from the Virginia side off Belleville Island was identified as the 'parting-shot' fired by some of Col. Johnson's men.

907   E.H., "The Naval Engagement at Buffington Island Correspondence of the Cincinnati Commercial" (dated "Cincinnati, July 21, 1863"), *Cincinnati Daily Commercial Gazette*, July 22, 1863, p. 1.

908   "Latest Telegraphic News. Special Dispatches exclusive to Cincinnati Daily Commercial. Telegraphic Correspondence Daily Commercial. From Columbus. The Pursuit of Morgan—700 Prisoners, 700 Horses and Three Pieces Artillery Captured at Buffington—The Attempt to Cross at Bealeville and How it was Frustrated—Nearly all of Morgan's Command Gobbled Up" (dated "Columbus, O., July 20"), *Cincinnati Daily Commercial*, July 21, 1863, p. 3.

909   Ibid.

910   James Montgomery, a member of Morgan's Division, who escaped with him from Buffington, "Gen. John Morgan Would Not Leave His Men," *Confederate Veteran Magazine*, Vol. XIX (1911), p. 384.

911   N.B. Stanfield, "Harships of a Morgan Raider," *Confederate Veteran Magazine*, Vol. VII (1899), pp. 60-61.

912   At Salineville, Morgan surrendered his command. Burnside continued his report to state that prisoners captured near Buffington Island together with those taken at Salineville totaled about 3,000. All were conveyed to Cincinnati. Maj. Gen. Ambrose E. Burnside, U.S. Army, Cmdg. Department of the Ohio, Report of operations March 25—August 10, 1863, O.R., pp. 12-14.

913   Cunningham, Report, Supp. O.R. Part I, Reports Vol. 4 Ser. No. 4, pp. 222-25.

914   Ibid.

915   Indian Run is a small tributary of the Ohio River on the Ohio side. It enters the Ohio River about one and a half to two miles south of where Hocking River enters.

916   Keller, *Morgan's Raid*, p. 188.

917   Special Correspondence of the Cincinatti *Gazette* (dated "National Fleet on Ohio River, Below Buffington Island, July 20, [1863]"), reprinted as "End of the Morgan Raid," *Baltimore American*, July 25, 1863, p. 1.

918   Ibid.

919   *Jackson Countians in America's Wars, 1775-1918* (Jackson County Historical Society of W.Va. Ripley, W.Va., 1978), p. 6.

920   Leroy Fitch, Lieutenant Commander, U.S. Navy to Rear Admiral D.D. Porter, Commanding Mississippi Squadron, Vicksburg, (dated "U.S.S. Moose Above Buffington Island Ohio River, July 19, 1863"), O.R. Vol. XXV, pp. 254-55.

921   Fitch to Porter (dated "U.S.S. Moose Above Buffington Island Ohio River, July 19, 1863"), *O.R.* Vol. XXV, p. 254.

922   E.H., "Naval Engagement at Buffington Island," dated "Cincinnati, July 21, 1863," Cincinnati *Daily Commercial Gazette*, July 22, 1863, p. 1.

923   "History of 23rd Ohio Regiment," pp. 392-93.

924   Scammon stated that Hayes' 23rd Ohio and 13th Virginia Volunteer Infantry were deployed on the Virginia shore for a distance of 5 miles and that "[t]he Second Brigade, under Colonel White, was also deployed in guarding the passage of the river, and did it effectually." Report of Brig. Gen. E.P. Scammon, U.S. Army, Commanding Third Division, Eighth Army Corps (dated HDQRS. 3d Div., 8th A.C., Charleston, W.Va., July 23, 1863), O.R. XXIII Part I, p. 677.

925   Roush, *If Thou Wilt Remember*, pp. 34-36.

926   Record of Events Field and Staff, Company A Muster Rolls July and August 1863 (micro. #594 Roll 196), NARA.

927   RG94 Morning Reports of Capt. W.I. Mathews Co. H 13th Regiment Va. Vols. for July 1863, NARA, [n.p.]

928   Confederates, who had surrendered at Buffington, were on July 20 started down the river to Cincinnati. ("Reminiscences of Morgan"s Men," The Southern Bivouac, Louisville, Ky., July 1883 Vol. 1 No. 11, p. 410.)

929   RG94 Morning Reports of Capt. Simon Williams' Co. D 13th Regiment Va. Vols. for July 1863, [n.p.]

930   Brig. Genl. Scammon, by Ja[me]s L. Botsford, Asst. Adjt. Genl[eral] to Col. R.B. Hayes Comdg. Brigade (manu. dated "Head Quarters, 3rd Division, 8th Army Corps, Hockingsville O. July 19, 1863"), RG94 Misc. Dept. of W.Va. Papers, NARA.

931   23rd Ohio Volunteer Infantry sources suggest that the Kanawha troops were following up reports of sightings of the escaped raiders at various locations. See for example, Report of Lieut. Col. James M. Comly, dated July 22, 1863, O.R. XXIII Part 1, pp. 678-79; Comly's diary entry for "Friday, July 17, 1863 through Thursday, July 21, 1863;" and Stiarwalt's Diary 1861-64, pp. 8-9. Comly's writings describe a scout undertaken in the afternoon of the 19th by a portion of the 23rd Ohio Regiment (termed by Comly the "left wing" of scouting operations) under command of Captain Zimmerman. This group was sent over the Ohio River to capture a party of the raiders, which had crossed over into West Virginia before Scammon's forces had arrived at Hockingport. "This expedition returned late in the evening, unsuccessful, the party having been notified of Captain Zimmerman's approach, by a disloyal citizen." After the "left wing" had started, another detachment of the 23rd was ordered out to scout that afternoon. Scammon ordered James Comly "to attempt the capture of a party of rebels supposed to be lurking in the hills below Hocking." Comly sent his men to scout through the woods in every direction. A citizen was brought in, who had been captured that morning by the enemy and stripped of everything but "shirt, pants, and an old hat." He agreed to guide Comly through the woods to where the party of raiders had halted to feed their horses. After a "circuitous" and "very laborious" march of about four or five miles through the bushes and "ravines filled with tangled underbrush," the 23rd reached the enemy camp in the evening. Comly deployed his men so as to entirely cover the rear of the camp and sent in three companies of skirmishers. These latter approached to within "30 yards" of the camp before being discovered. The skirmishers opened fire on the Confederates, who "surrendered immediately, without firing a shot." Six officers and forty-three men (non-commissioned officers and privates) were captured. The prisoners according to Stiarwalt, were brought "down to the Steam Boats [in] the evening then we kept a good lookout up and down the Ohio for two or three days." Stiarwalt continued: "on 20th to the 22 we patrolled the Ohio River to the mouth of Kanawha River on the 22d July/63 we arrived at Camp White as did the 12th OVI + 13th VAI."

932   The 13th Regiment West Va. Inf. Memoranda, Field History, [n.p.]

933   See "The 13th West Va. Infantry," *The Journal* (Charleston, W. Va.), July 5, 1865, [p. 2] or "The Thirteenth West Va. Infantry," Wheeling *Daily Intelligencer*, June 24, 1865, p. 3.

934   Boyd B. Stutler, *West Va in the Civil War*, (Charleston, W.Va.: Education Foundation, 1963), pp. 233-34.

935   Young to Paulina (dated "Point Pleasant, July 24, 1863, 3 o'clock 20 min. P.M."), Young Civil War Papers (Vol. 31 Box 10), trans. p. 77.

936   Report of Col. R.B. Hayes to Capt. James S. Bottsford, A.A.G., 3rd Division 8th A.C., (manu. dated "Head Qtrs. 1st Brigade 3rd Division, Dept 8th Army Corps, Camp White, Va., July 23, 1863"), H.178 Registry #393 Part 2 Entry 1184 Press Copies of Letters Sent and Spec. Orders Vol. 139, NARA.

937   Ibid.

938   "Morgan Still at Large," *Cincinnati Commercial Gazette*, July 21, 1863, p.2.

939   Hobson to Col. Lewis Richmond (dated Buffington Bar, July 20, 1863), O.R. XXIII Part I, p. 780.

940   Duke, *A History of Morgan's Cavalry*, p. 463.

941   Ibid.

942   Ibid.

943   "Morgan Still at Large," *Cincinnati Commercial Gazette*, July 21, 1863, p. 2.

944   Ibid.

945   O[tis], "Personal Recollections of the War—No. III," *Santa Barbara Daily Press*, October 28, 1876.

946   Field History 13th Regiment W.Va. Infantry, [n.p.]

947   Annual Return of Casualties + c. 13th W.Va. Infy. to Dec. 31/63, RG94 Box 4838 Misc. unbound regimental papers 13th W.Va.

948   RG94 Morning Reports of Co. H 13th Regt. Va. Vols. for July 1863, [n.p.]

949   RG94 Morning Reports of Co. A 13th Regt. Va. Vols. for July 1863, [n.p.]

950   RG94 Morning Reports Co. F 13th Regt. Va. Vols. for July 1863, NARA, [n.p.]

951   Annual Return of Casualties + c. 13th W.Va. Infy. to Dec. 31/63, NARA.

952   Judah to Burnside (dated Pomeroy, July 22, 1863), O.R. XXIII Part I, p. 788. R.B. Wilson wrote a history of the Kanawha Division for the veterans' newspaper, the *National Tribune*. He stated that Hayes' men arrived "at Buffington Island in time to be in, at the surrender of the larger part of Morgan's force, which as prisoners it escorted down the Ohio on boats to Gallipolis." (Wilson, "Kanawha Division: Its Campaigns. [Part] I—cont.," *The National Tribune*, February 18, 1897, p. 2.)

953   Annual Return of Casualties + c. 13th W.Va. Infy. and Regimental Return 13th Va. Vol. Infantry for August 1863 (dated "Barboursville, W.Va. Sept. 9, 1863"), micro. #594 Roll 196, NARA.

954   Stiarwalt, Diary 1861-64, pp. 8-9 of manu.

955   Morgan's officers were sent to the city prison at Cincinnati where they remained three days before being sent to military prison at Johnson's Island. Privates were immediately sent to camps Morton, in Indianapolis, Indiana and Douglass, in Chicago, Illinois.

956   RG94 Morning Reports of Capt. James W. Johnson's Co. A for July 1863, [n.p.]

957   "History of 23rd Ohio Regiment," Comly Papers (micro. Reel 1), p. 394.

958   Report of Col. R.B. Hayes to Capt. James S. Bottsford A.A.G., 3rd Division 8th A.C. (dated "Head Qtrs. 1st Brigade 3rd Division, Dept 8th Army Corps, Camp White, Va., July 23, 1863"), H178 Registry #393 Part 2 Entry 1184 Press Copies of Letters Sent and Spec. Orders Vol. 139, NARA and E.P. Scammon, Brig. Gen. Commanding Division, by Capt. T. Melvin, Asst. Adjt. Gen., Dept. of West Virginia, "Report" (dated Hd Qrs. 3d Div., 8th A.C., Charleston, W.Va., July 23, 1863), O.R. XXIII Part I, p. 677.

959   "History of 23rd Ohio Regiment," p. 394.

960   George Wilding, Staff member of the Mason City (or New Haven) Home Guard Company, as quoted in Mildred Chapman Gibbs, *Mason City, W.Va. The History of the Town and It's People*, (Middleport, Ohio: Quality Print Shop, 1978), p. 36.

961   Roush, *If Thou Wilt Remember*, pp. 34-36. The towns of Mason and Hartford City were the location of several salt works. These provided steady employment to a large number of men many of whom joined the 13th Loyal Virginia Regiment.

962   Field History, [n.p.]

963   RG94 Reports of Co. F for the month of July 1863, NARA, [n.p.] "12 o'clock m." likely meaning twelve o'clock midday or noon.

964   The designation "Ten Mile" probably referred to the country around Ten Mile Creek, a tributary of the Kanawha River, located about ten to twelve miles up the Kanawha River from Point Pleasant.

965   RG94 Morning Reports of Co. F, [n.p.] Some of Morgan's men, had sought to escape by back-tracking westward along the Ohio. Beside the captures made by Sergt. Allen C. Mason (future Captain of Co. G 13th W. Va. Regiment), Gen. Scammon reported that Brig. Gen. White's command and Col. John A. Turley, 91st Ohio Volunteer Infantry had captured "two lieutenants and thirty privates, with horses and equipments." These latter captures were made attempting to cross into West Virginia at Guyandotte. See Report of Brig. Gen. E.P. Scammon, U.S. Army, Cmdg. Third Division, Eighth Army Corps dated (Hd Qrs. 3d Div., 8th A.C., Charleston, W. Va., July 23, 1863), O.R. XXIII Part I, p. 677.

966   Inasmuch as Co. D formed a part of the military escort detailed to guard prisoners on their way to Cincinnati, it is perhaps not amiss to include some reference to the scene at Cincinnati as prisoners and guards disembarked and proceeded to move through the town to the city prison or to the train station for conveyance farther north. Confederate prisoners taken at Buffington Island arrived at Cincinnati about 11 a.m. on July 23. Their arrival was chronicled by a number of eyewitnesses in papers of the day. The officers, including Dick Morgan, John's brother and Basil Duke, were disembarked from the steamer *Starlight* and taken by ferry-boats to the foot of Main Street. A carriage was provided, at least for a time, for Morgan and Duke to travel in as Morgan was wounded and Duke was lame. The balance of the officers were formed into two ranks behind Morgan and Duke and "the column, strongly guarded, moved through the city to the city prison, on Ninth-street. The boats containing the privates then proceeded down the river to the foot of Fifth-street, where the prisoners were marched to a special train on the Indianapolis and Cincinnati railroad, and sent to Indianapolis," to Camp Morton, on the site of what is today the Indiana State Fairgrounds on 38th Street. A large crowd had assembled at the levee at Cincinnati and train depot to see the cavalrymen and as the prisoners were moved the crowd of spectators grew until an immense number of men, women and children, thronged the column, in their eagerness to catch sight of the prisoners. Many sympathizers—friends and family of the prisoners were present. Some of these attempted to speak to or give the prisoners things such as money, refreshments, etc., but the guards soon put a stop to such efforts and none were permitted to communicate with the prisoners. Another account highlighted the volatile nature of the crowd relating that as the prisoners were marched down Ninth Street to Ninth Street Prison not less than five thousand surrounded the cavalrymen. Some of the on-lookers flourished pistols and cried, 'Hang the cut-throats,' 'bully for the horse-thieves,' but the guard quickly drove these away. (Account of the arrival at Cincinnati of the prisoners taken in the fight at Buffington Island, from one of the newspapers of the day and quoted by Henry Howe in *The Times of the Rebellion in the West: A Collection of Miscellanies, [etc.]*, Ohio, 1867, pp. 16 and 19.

967   RG94 Morning Reports of Co. D for July 1863, NARA, [n.p.]

968   Total numbers for Confederate killed, wounded and captured were variously reported. Confederate numbers underestimated and Union numbers over-estimated captures as was typical throughout the war when a battle resulted in Union advantage. A selected few who provided first hand estimates were the following. Capt. Sidney Cunningham (Morgan's Division) reported officially that "three companies of dismounted men and perhaps two hundred sick and wounded" were left behind and captured. Duke wrote that Morgan's brother, Richard and Colonels William W. Ward, Duke himself and Col. John Huffman with about seven hundred prisoners were taken at the Buffington fight. (Duke, *History of Morgan's Cavalry*, p. 453.) Reports generally agree that most captures were made at Buffington Island, and that after the successful crossing of Morgan's men at Belleville was interrupted by Lieut. Fitch more men scattered and straggled and "portions of the command were captured at different places." (N.B. Stanfield, "Hardships of a Morgan Raider," *Confederate Veteran Magazine*, Vol. VII 1899, pp. 60-61.) Morgan reportedly surrendered with between three hundred and fifty to four hundred men at Salineville, Ohio, on July 26. Dispatches to Maj. Gen. Ambrose E. Burnside, cmdg. Dept. of Ohio, reported that at Buffington bar upwards of 120 to 200 men had been killed, wounded and drowned and 1,000 were reported taken prisoner, among the latter 48 commissioned officers. ("Official Dispatches The Morgan Raid—Morgan Defeated and Most of His Force Captured," *Cincinnati Commercial Gazette*, July 20, 1863, p. 2.) Also "Morgan's Ohio Raid," *Battles and Leaders of the Civil War*, Vol. III, p. 635; Maj. Gen. Geo. L. Hartsuff to Lt. Col. Drake (dated Cincinnati, July 19, 1863), O.R. XXIII Part I, p. 775; and "Latest Telegraphic News Special Despatches Exclusive to Cincinnati Daily Commercial," *Cincinnati Commercial Gazette*, July 20, 1863, p. 3.

969   E.H., Correspondent to the *Cincinnati Commercial*, aboard the *Imperial*, dispatch boat of the Federal gunboat *Moose*, "The Naval Engagement at Buffington Island Termination of the Morgan Raid Sixteen Hundred Rebels, Seven Pieces of Artillery and Immense Numbers of Horses and Quantities of Supplies Taken—Basil Duke and Dick Morgan Among the Prisoners—The Pursuit. Correspondence of the Cincinnati Commercial" (dated "Cincinnati, July 21, 1863"), *Cincinnati Daily Commercial Gazette*, Wednesday, July 22, 1863, p. 1.

970   Record of Events Field and Staff, Company Muster Roll July and Aug. 1863 for Co. F (micro. #594 Roll 196) and Regimental Return 13th Va. Vol. Infantry for August 1863, dated "Barboursville, W.Va. Sept. 9, 1863" (micro. #594 Roll 196).

971   Respectively, Hayes to Bottsford (manu. dated "Head Qtrs. 1ˢᵗ Brigade 3ʳᵈ Division, Dept 8ᵗʰ Army Corps, Camp White, Va., July 23, 1863"), H 178 Registry #393 Part 2 Entry 1184 Press Copies of Letters Sent and Spec. Orders Vol. 139 and Scammon's Report (dated HdQrs. 3d Div., 8ᵗʰ A.C., Charleston, W. Va., July 23, 1863), O.R. XXIII Part I, p. 677.

972   J.V. Young to Paulina and children (dated "Point Pleasant, Mason Co. July 23, 1863") Young Civil War Papers (Vol. 31 Box 10), trans. p. 76.

973   Report of Lieut. Col. James M. Comly (dated Hdqrs. Twenty-third Regiment Ohio Volunteers, Camp White, W.Va. July 22, 1863), O.R. XXIII Part 1, pp. 678-79.

974   O.B. Norvell, originally from Lynchburg, Virginia, wrote an interesting account of his part in the planning and successful escape from prison (Camp Douglas, Chicago, Illinois) undertaken by himself and six former members of Morgan's command. The escape party consisted of O.B. Norvell, A.W. Cockrell, Harmon H. Bartow, Clayton Anderson, John H. Waller, E.M. Henderson and Winder Monroe. See Otway Bradfute Norvell, *Confederate Veteran Magazine*, Vol. 11, pp. 168-71. See also a group photograph of Confederate prisoners of war taken at Camp Douglass, Illinois, sent in by "Mrs. L.R. Swan of New Orleans." The men posing are identified in the caption. Among them is Otway Bradfute Norvell. "Some Men Captured in Ohio," *Confederate Veteran Magazine*, Vol. 10 1902, p. 494.

975   Colonel D.W. Chenault had been in command of the 11ᵗʰ Kentucky Cavalry.

976   Charles Haddox was singled out for mention by Basil Duke in his popular book about Morgan's cavalry. Describing the day to day experience of the men while on the great raid, Duke referred to Haddox's description of the daily capture of local militiamen, who hung about the Confederate column waiting to give fight, or for an opportunity to wound or kill one of the raiders. Usually, began Duke, these militia men were captured by the advance videttes and then released after their guns were destroyed. Duke related that "[o]n one occasion, a very gallant fellow of the Second Kentucky, Charlie Haddox, came upon five of them, who had made some of the command prisoners. He captured them, in turn, and brought them in." (Duke, *History of Morgan's Cavalry*, p. 445.) See also "Some Men Captured in Ohio," *Confederate Veteran Magazine*, Vol. 10 1902, p. 494, which publishes a photograph taken at Camp Douglass, Illinois. Photograph submitted for publication by Mrs. L.R. Swan of New Orleans. This photograph shows Charles Haddox; Otway Bradfute Norvell (O.B. Bradfute Norvell, prisoner number 18); and S.G. Adams of Bardstown, Kentucky (George Adams, prisoner number 46?)

977   [John S. Cunningham, Lieutenant and Adjutant 13ᵗʰ West Virginia Infantry Regiment], "List of Prisoners of War Captured by the 23ʳᵈ Ohio + 13ᵗʰ Va at Buffington + near the mouth of Big Hocking on Sunday July 19ᵗʰ 1863," WVSA.

978   In February 1863, Morgan's command was regularly brigaded. The Second, Fifth, and Ninth Kentucky Cavalry and Ninth Tennessee Cavalry were brigaded together to form the First Brigade and the Third, Eighth, Eleventh and Tenth Kentucky Cavalry Regiments composed the Second Brigade. (Duke, *History of Morgan's Cavalry*, p. 359.)

979   Col. Grigsby and Capt. Byrnes had crossed the river at Buffington Island and succeeded in escaping. (Duke, p. 453.)

980   Elijah F. Newell, in his early forties, was eventually discharged on account of disability on December 7, 1864, at U.S.A. General Hospital, Gallipolis, Ohio. RG94 Misc. Records Descriptive Roll Co. A. 13ᵗʰ Regiment V.V.V.I., Register of Men Discharged, [p. 4] and Misc. Records Descriptive Roll Co. A First List, [pp. 14–15] and Second List, [pp. 37-38].

981   Letter received from Capt. Jas. W. Johnson, 13ᵗʰ Va.V.I. Comdg. at Coals Mouth, notifying that he had "received a leave of absence from the Sec. of War &c." (dated "July 1, 1863 Hdqrs. 1ˢᵗ Brig. 3ʳᵈ Div. 8ᵗʰ A.C."), Letters Received Head Quarters 1ˢᵗ Brig. 3d Division 8ᵗʰ Army Corps, Letters Received Miscellaneous, Papers of R.B. Hayes (micro. Reel 272), HPL, p. 14.

982   RG94 Morning Reports of Capt. James W. Johnson's Co. A 13ᵗʰ Regiment Va. Vols. for July 1863, [n.p.]

983   James [W.] Johnson, 13ᵗʰ Regiment Infantry, Roster of the Regiment, WVSA, p. [4].

984   James R. Hall, Lt. Col. Comdg. 13th V.V.I., by John S. Cunningham, Adj., Special Order No. 18 (manu. dated "Head Quarters 13th Regt Va Vol Infty Point Pleasant West Va July 27th 1863"), RG94 Order Book Co. C 13th Regt. Inftry. Vols., [n.p.]

985   RG94 Morning Reports of Capt. Milton Stewart's Co. B 13th Regiment Va. Vols. for July 1863, [n.p.]

986   James R. Hall, Lieutenant Colonel, 13th Regiment West Virginia Infantry Volunteers to Mrs. [Rebecca A.] Barnett (manu. dated "Barboursville, West Virginia, July 20th, 1863"), Mother's Application, C.C. Barnett, Military Pension File, NARA.

987   Brig. Gen'l. Scammon to Hdqts 1st Brig 3rd Div. 8 A.C. (dated "H'd Qrs 3 Division 8th A.C. Charleston W.Va. July 2, 1863"), Letters Received Head Quarters 1st Brig. 3d Division 8th Army Corps, Letters Received Miscellaneous, Papers of R.B. Hayes (micro. Reel 272), trans. p. 14.

988   Col. R.B. Hayes to Capt. Milton Stewart, Order No. 48 (manu. dated "Head Quarters 1st Brigade 3rd Div. 8th Army Corps, Camp White, Virginia July 2, 1863"), RG94 Order Book Co. B 13th Regt. Inftry. Vols., [p.1].

989   "Married," Point Pleasant *Weekly Register*, Vol. II No. 20, August 6, 1863, [p. 3] and RG94 Misc. Records Descriptive Roll Co. B, List of Non-Commissioned Officers, [n.p.]

990   "Va. Reg. No. 28; Hos No 557 p. 95," as cited in RG94 13th Va. Hospital Cards Box 3849.

991   RG94 Morning Reports of Capt. Van D. McDaniels' Co. C 13th Regiment Va. Vols. for July 1863, [n.p.]

992   Brig. Gen. John Echols to Gen. S. Cooper (dated "Hd. Qtrs Army of Western Va. Warner's, 5 miles east of Falls of Kanawha, Oct. 29, 1862"), O.R. XIX Part II, p. 688.

993   RG94 Morning Reports of Capt. Simon Williams' Co. D 13th Regiment Va. Vols. for July 1863, [n.p.]

994   RG94 13th Va. Hospital Cards.

995   RG94 Morning Reports of Capt. Simon Williams' Co. D 13th Regiment Va. Vols. for July 1863, [n.p.]

996   RG94 Carded Medical Records Vols. 1846-1865 Entry 534 Box 3849, NARA

997   Brig. Genl. Scammon to Hdqtrs 1st Brig. 3rd Div. 8th A.C. (dated "Head Qrs. 3 Division 8th A.C. Charleston W. Va. July 7, 1863") Letters Received Head Quarters 1st Brig. 3d Division 8th Army Corps, Miscellaneous, Papers of R.B. Hayes (micro. Reel 272), trans. p. 14.

998   William R. Brown, Col. Comdg, by John S. Cunningham, Adj. to Albert F. McCown Co. F, Spec. Order No. 17 (manu. dated "Head Quarters 13th Regt Va Vol Infty, Camp near Charleston W Va July [month and day omitted] [1863]") RG94 Order Book Co. C 13th Regt. Inftry. Vols., [n.p.]

999   Brown's order originally specified that Co. F would be taken to Loup Creek "by transport Sumpter." Company records, however, state that it was the *General Meigs* and not the *Sumpter* which transferred them to their new post.

1000   RG94 Morning Reports of Capt. A.F. McCown's Co. A 13th Regiment Va. Vols. for July 1863, [n.p.]

1001   Record of Events Field and Staff, Muster Roll Co. F 13th Va. Vol. Infantry (micro. #594 Roll 196), NARA.

1002   Two musicians were on duty with Co. F at this time. These were Hazard Farley and James King. Both men were listed as drummers on descriptive rolls.

1003   July 7, 1863, Col. Brown issued Special Order No. 16 directing that "Commanding Officer Co F will detail Private David Forbes for Extra duty in A.Q.M. Dept. He will report immediately to Brigade Hd Qrts for instructions." William R. Brown, Col. Comdg., by John S. Cunningham, Adj., Special Order No. 16 (manu. dated "Head Quarters 13th Regt Va Vol. Infty, Camp near Charleston July 7th 1863"), RG94 Order Books 13th Regt. Infry. Vols., [n.p.]

1004   RG94 Morning Reports Co. A 13th Regiment Va. Vols. for July 1863, [n.p.] and Record of Events Field and Staff, Muster Roll Co. F 13th Va. Vol. Infantry (micro. #594 Roll 196), NARA.

1005   Handleys may have been one of the strongly partisan pro-South families in the valley. A Monroe Handley from Putnam County served in the 36th Virginia Inf.; one Charles W. Handley served in the 14th Virginia Cavalry; and Bennah F. Handley from Kanawha Co. served in the 22nd Va. Inf.

1006   Capt. J.V. Young to his wife, Paulina (dated "Camp White, July 6, 1863"), Young Civil War Papers, trans. p. 72.

1007   Young to his daughter, Emma (dated "Cotton Hill, July 10, 1863"), Young Civil War Papers, trans. pp. 73-74.

1008   Young to his wife (dated "Cotton Hill, July 10, 1863"), Young Civil War Papers (Vol. 31 Box 10), WVU, trans. p. 75.

1009   Young to Emma, dated "Cotton Hill, July 10, 1863," trans. pp. 73-74.

1010   RG94 Carded Medical Records Vols.

1011   Young to Paulina (dated "Point Pleasant, July 24, 1863, 3 o'clock 20 min. P.M."), Young Civil War Papers, trans. p. 77.

1012   Young to Paulina (dated "Point Pleasant, July 28, 1863"), trans. p. 78.

1013   RG94 Morning Reports of Capt. W.I. Mathews' Co. H 13th Regiment Va. Vols. for July 1863, NARA, [n.p.]

1014   Muster-in roll of Lieut. W.E. Feazel's Co. in the 13th Regt. of Va. Inf. Vols., commanded by Col. W.R. Brown, "called into service of the U.S. by the President from the dates set opposite their respective names for the term of 3 yrs., unless sooner discharged. Cyrus S. Roberts, Lieut. 100th N.Y. Vols. Assistant Commissary of Musters, 3rd Div., 8th A.C., Mustering Officer." Supp. O.R., Vol. 74, p. 571.

1015   Record of Events Field and Staff, Co. I Muster Roll July 1863 (micro. #594 Roll 196), NARA.

1016   Col. R.B. Hayes to the "Commanding Officer Co. I 13th Regt. Va.V.I." (manu. dated "Head Qtrs. 1st Brigade 3rd Division, Dept 8th Army Corps, Camp White, Va., July 26, 1863"), Despatch 186 Registry #393 Part 2 Entry 1184 Press Copies of Letters Sent and Special Orders 139, NARA.

1017   RG94 Morning Reports of Lieut. William Feazel's Co. I 13th Regiment Va. Vols. for July 1863, [n.p.]

1018   A.A. Tomlinson, Colonel, Comdg. Post to Lt. Feazel (manu. dated "Charleston Va July 16 [1863]"), RG94 Order Books 13th Regt. Inftry. Vols., [n.p.]

1019   The 13th Regiment West Va. Inf. Memoranda, Field History, WVSA, [n.p.]

1020   Young to Paulina (dated "Point Pleasant, July 24, 1863, 3 o'clock 20 min. P.M."), trans. p. 77.

1021   Lt. Col. J.R. Hall, Comdg 13 Va. V.I. to Hdqrs. 1st Brig. 3 Div. 8th A.C., Letters Received Head Quarters 1st Brig. 3d Division 8th Army Corps, Letters Received Miscellaneous (micro Reel 272), HPL, trans. p. 17.

1022   James R. Hall, Lt. Col. Commanding, by John S. Cunningham, Adj., Order No. 15 (manu. dated "Head Quarters 13th Regt Va Vol Infantry Point Pleasant West Va July 27th 1863"), RG94 Order Book Co. C 13th Regt. Infry. Vols., [n.p.]

1023   James R. Hall, Lt. Col. Comdg Post, by John S. Cunningham, Adj., Spec. Order No. 19 (manu. dated "Head Quarters 13th Regt Va Vol Infty, Point Pleasant West Va July 28th [1863]"), RG94 Order Books 13th Regt. Inftry., [n.p.]

1024   O[tis], Editor, "Personal Recollections of the War—No. IV. Fighting Bushwhackers," *Santa Barbara Daily Press*, October 30, 1876.

1025   R.B. Wilson, "Kanawha Division: Its Campaigns. [Part] I—cont.," *The National Tribune*, February 18, 1897, p. 2.

1026 O[tis], "Personal Recollections of the War—No. IV. Fighting Bushwhackers," *Santa Barbara Daily Press*, October 30, 1876.

1027 R.B. Hayes to his uncle, S[ardis] Birchard (dated "Camp White Aug. 25, 1863"), *Diary and Letters*, p. 430.

1028 "Summary of the Week. Virginia," Frank *Leslie's Illustrated Newspaper*, Aug. 8, 1863, p. 310.

1029 Delta, "The Smith Raid Correspondence Cincinnati Commercial" (dated "Camp Louis, September 11, 1863"), *Cincinnati Daily Commercial*, Sept. 17, 1863, p. 3.

1030 Annual Return of Casualties + c. 13th W. Va. Infy., RG94 Box 4838 Misc. unbound regimental papers 13th W.Va.

1031 Brig. Gen'l Scammon to Hdqrs. 1st Brig. 3 Div. 8th A.C. (dated "Head Qt. 3 Divis. 8th A. C. Charleston W. Va. Aug. 4, 1863"), Letters Received Head Quarters 1st Brig. 3d Division 8th Army Corps, Letters Received Miscellaneous, Papers of R.B. Hayes (Reel 272), trans. p. 17.

1032 Col. R.B. Hayes to Col. William R. Brown (manu. dated "Head Qtrs. 1st Brigade 3rd Division, Dept 8th Army Corps, Camp White, Va., Aug. 4, 1863"), Despatch 194 Registry #393 Part 2 Entry 1184 Press Copies of Letters Sent and Special Orders Vol. 139 and RG94 Box 4839 Misc. unbound regimental papers 13th W.Va., NARA.

1033 Field History 13th Regt. W.Va. Inftry., [n.p.]

1034 Col. R.B. Hayes to Col. William R. Brown Commanding 13th V.V.I. and Capt. Comly Commanding 23rd O.V.I., by M.P. Avery, Capt. and A.A.A.G. (manu dated "Hd.Qtrs.1st Brig., 3d Div. 8th A.C., Camp White W. Va., Aug. 5, 1863"), RG94 Box 4839 Misc. unbound regimental papers 13th W.Va. Letters A.G.O. and Registry #393 Part 2 Entry 1184 Press Copies of Letters Sent and Special Orders Vol. 139.

1035 Col. R.B. Hayes, by M.P. Avery, Capt. and A.A.A.G., Spec. Orders No. 56 (manu. dated "Head Quarters 1st Brig., 3d Division 8th A.C., Camp White, W.Va., Aug. 5, 1863"), Return completed Jan. 13, 1864 in RG94 Box 4838 Misc. unbound regimental papers 13th W.Va.

1036 James R. Hall, Lt. Col. Commanding, by John S. Cunningham, Adj., Gen. Order No. 16 (manu. dated "Head Quarters 13th Regt Va Vol Infty Point Pleasant West Va July 29th 1863"), RG94 Order Book Co. C 13th Regt. Inftry. Vols., [n.p.]

1037 William R. Brown, Col. Comdg, by John S. Cunningham, Adj., Spec. Order No. 20 (manu. dated "Head Quarters 13th Regt Va Vol Infty Point Pleasant West Va July 30th 1863"), RG94 Order Books Co. C 13th Regt. Inftry. Vols., [n.p.]

1038 Regimental Return 13th Va. Vol. Inftry. for July 1863, dated "Barboursville, W.Va. Sept. 9, 1863 (micro. #594 Roll 196), NARA.

1039 Record of Events Field and Staff, Company Muster Rolls for July and August for Cos. A, B, C, F and H. 13th Va. Vol. Inftry. (micro. #594 Roll 196), NARA.

1040 Annual Return of Casualties + c. 13th W.Va. Infy. to Dec. 31/63 (manu. "Return completed Jan. 13, 1864"), RG 94 Box 4838 Misc. unbound regimental papers 13th W.Va., NARA.

1041 Co. A 13th Va. Vol. Inftry. Muster Roll July and August 1863 (micro. #594 Roll 196), NARA.

1042 W[illiam] W. H[arper] to the Editor of the Point Pleasant *Weekly Register* (dated "Camp Defiance, Kanawha, August 8th, 1863"), "From the Thirteenth Va.," *Weekly Register*, Aug. 13, 1863, Vol. II No. 21, [n.p.]

1043 Young to Paulina Marshall Franklin Young, dated "Camp Defiance, August 6, 1863," Young Civil War Papers (Vol. 31 Box 10), trans. p. 79.

1044 David Burrows to his wife (dated Aug. 17, 1863), Roush, *If Thou Wilt Remember*, p. 37.

1045   R.B. Hayes to his uncle, S[ardis] Birchard (dated "Camp White, Charleston, West Virginia, Aug. 6, 1863"), *Diary and Letters*, p. 427.

1046   Brig. Gen'l Scammon to Hdqtrs. 1st Brig. 3rd Div. 8th A.C. Letters Received Head Quarters 1st Brig. 3d Division 8th Army Corps, Letters Received Miscellaneous, Papers of R.B. Hayes (micro. Reel 272), HPL, p. 19. Undated but probably written on August 6, 1863. See the letter from Hayes to Col. Brown, dated "Camp White August 6" regarding a "considerable" force of enemy cavalry, reported at the "marshes of Coal." Letter #200 Registry #393 Part 2 Entry 1184 Press Copies of Letters Sent, NARA.

1047   Col. R.B. Hayes to Lt. John S. Witcher, Comdg. Co. 3rd Va. Cavalry (manu. dated "Head Qtrs. 1st Brigade 3rd Division, Dept 8th Army Corps, Camp White, Va., Aug. 6, 1863"), Letter #198 Registry #393 Part 2 Entry 1184 Press Copies of Letters Sent and Special Orders Vol. 139.

1048   Col. R.B. Hayes to Col. W.R. Brown, by Avery (manu. dated "Hd. Qtrs. 1st Brig. 3 Div. 8th A.C., Camp White W.V, Aug. 6, 1863"), RG94 Box 4839 Misc. unbound regimental papers 13th W.Va. and Col. R.B. Hayes to Col. Wm.R. Brown (manu. dated "Head Qtrs. 1st Brigade 3rd Division, Dept 8th Army Corps, Camp White, Va., Aug. 6, 1863"), Letter #200 Registry #393 Part 2 Entry 1184 Press Copies of Letters Sent and Special Orders Vol. 139.

1049   Col. Wm. R. Brown to Hd Qtrs. 1st Brig. 3rd Div. 8th A.C., dated respectively "Head Qtrs 13th Regt. Va. V.I. Coals Mouth Aug. 9/63;" Head Qts 13th Regt Va.V.I. Camp near Coals Mouth Aug. 10/63" and "13th Regt. Va. V.I. Coals Mouth W.Va. Aug. 17, 1863." Both in Letters Received Head Quarters 1st Brig. 3d Division 8th Army Corps, Letters Received Miscellaneous, Papers of R.B. Hayes (micro. Reel 272), trans. pp. 19-20.

1050   William R. Brown, Col. Comdg, by John S. Cunningham, Adj., Special Order Nos. 23, 25, 26, 27, 28 and 29 (manuscripts dated "Head Quarters 13th Regt Va Vol Infty Camp Defiance W Va August 9th 1863"), Order Books 13th Regt. Inftry. Vols., [n.p.]

1051   William R. Brown, Col. Comdg, by John S. Cunningham, Adj., Special Order No. 32 (manu. dated "Head Quarters 13th Regt V.V.I. Camp Defiance August 10th 1863") Order Books, [n.p.]

1052   Brig. Gen. E.P. Scammon, Special Order No. 73 (manu. dated "Hd. Qtrs. 3rd Div. 8th A.C. Charleston, W.Va., Aug. 10, 1863"), provenance lost.

1053   Perhaps Rucker refers to Private John M. Chapman, of Cabell County, Co. C 36th Battalion Virginia Cavalry. See Dickinson, *Tattered Uniforms and Bright Bayonets*, p. 119.

1054   Isaac M. Rucker, Captain, comdg. Co. A 181st Regiment W. Va. Volunteer Militia to Col. W.R. Brown (manu. dated "Winfield West Va. August 10, 1863"), RG94 Box 4839 Misc. unbound regimental papers 13th W.Va.

1055   Capt. Rucker may have had a checkered history. Charles B. Waggoner, John S. Miller, B.J. Redmond and others wrote to Gov. A.I. Boreman in a letter dated (probably Point Pleasant) **"Aug. 20, 1863"** petitioning that a company numbering some 40 to 50 men, who have organized themselves as "Home Guards" and who are commanded by Isaac M. Rucker be placed under U.S. forces or be disbanded and disarmed. These men, continued petitioners, reside at different points between the Ohio and Kanawha Rivers in the counties of Mason and Putnam and "have been acting for some months past as one organized company of 'Home Guards.' The company is of little or no benefit in repelling the enemy," and petitioners believed that a number of them retreat to the Ohio side of the river, where many of their families moved. "They are in the habit of descending upon the farms along the Ohio and Kanawha, impressing horses of peaceable, quiet men, the undersigned believe, in order to carry off bacon, flour etc. They have in fact done these things." The petition was signed by Charles B. Waggoner, John S. Miller, B.J. Redmond, G.W. Setszer and J.P.R.B. Smith. See *Calendar of the Frances Harrison Pierpont Letters and Papers in W. Va. Depositories* Prepared by the W.Va. Historical Records Survey Division of Professional and Service Projects, Works Projects Administration (Charleston, W.Va.: Historical Records Survey, October 1940), p. 9, entry 39. But see George C. Bowyer's correspondence to Gov. A.I. Boreman dated "Charleston, W.Va., Oct. 17, 1863," in which Bowyer wrote in reply to Boreman's letter concerning the disbanding of Capt. Rucker's Co. Bowyer says that he saw Col. Smith at Point Pleasant and Smith had not heard of the company doing anything "out of the way." John Hall and C.C. Miller, citizens of

Mason county, had likewise heard no complaints "of late" against the company. These men did not think the company should be disbanded and Judge Polsey was of the same opinion. In fact, they opposed disbanding it because it would leave that section of the country without protection against "confederate thieves." Bowyer was of the opinion, that if Southern sympathizers would treat Rucker's men more kindly there would be "but little cause for complaint." (Ibid., p. 15.)

1056   John Bowyer, U.S. Commissioner for Putnam County to Col. Brown, comdg. 13th Regt. near Coalsmouth, West Va. (manu. dated "Winfield Putnam C.H. Aug. 23, 1863"), RG94 Box 4839 Misc. unbound regimental papers 13th W.Va.

1057   Writer unknown, RG94 Box 4839 Misc. unbound regimental papers 13th W.Va.

1058   R.B. Hayes to Col. W.R. Brown (manu. dated "Hd Qtrs. 1 Brig. 3 Div. 8 A.C., Camp White, W.Va., Aug. 13, 1863"), RG94 Box 4839 Misc. unbound regimental papers 13th W.Va.

1059   R.B. Hayes to Lucy (dated "Camp White, Aug. 15 Saturday afternoon, 1863"), *Diary and Letters*, p. 428. Hayes probably refers to the 23rd Ohio regimental band.

1060   Col. Wm. R. Brown Comdg. 13 Va.V.I. to Hdqrs. 1st Brig. 3rd Div. 8th A.C. (dated "Head Qrs. 13th Va. V. I. Camp Coals Mouth W. Va. Aug. 18, '63"), Letters Received Miscellaneous, Papers of R.B. Hayes (micro. Reel 272), trans. p. 22.

1061   William R. Brown, Col. Comdg, by John S. Cunningham, Adj., Spec. Order No. 33 (manu. dated "Head Quarters 13th Regt V.V.I. Camp Defiance August 20th 1863"), RG94 Order Books 13th Regt. Infry. Vols., [n.p.]

1062   Col. Wm. R. Brown, Col. Comd'g, by John S. Cunningham, Adj., Gen. Order No. 22 (manu. dated "Head Quarters, 13th Regt Va Vol Infty Camp Defiance near Coals Mouth, [Va.,] August 25, 1863"), RG94 Order Books, [n.p.]

1063   William R. Brown, Col. Comdg, by John S. Cunningham, Adj. (manu. dated "Head Quarters 13th Regt V.V.I. Camp Defiance West Va August 27th 1863"), RG94 Order Books 13th Regt. Inftry. Vols., [n.p.]

1064   William R. Brown, Col. Comdg, by John S. Cunningham, Adj., (manu. dated "Head Quarters 13th Regt V. V. I. Camp near Coalsmouth W. Va. Aug. 27th /63") and Spec. Order No. 37 (manu. dated "Head Quarters 13th Regt V. V. I. Camp near Coalsmouth, W.Va. Aug. 27th /63"), RG94 Order Books, [n.p.]

1065   Field History 13th Regt. W.Va. Inftry., [n.p.]

1066   West Virginia Clothing Books Co. C, [n.p.]

1067   Transfer effected pursuant to Order No. 335 (dated "July 20, 1863"), War Department, Washington.

1068   W.R. Brown, Col. Comdg, by John S. Cunningham, Adj., Gen. Order No. 21 (manu. dated "Head Quarters 13th Regt Va Vol Infty Camp Defiance near Coalsmouth August 24th 1863"), RG94 Order Books 13th Regt. Inftry. Vols., [n.p.]

1069   August 9, Col. Brown issued Special Order No. 23 providing that: "Sergeant Grinstead and four men of Co A are hereby detailed to proceed to Charleston W Va for the purpose of making an effort to raise some recruits for this Regt.—They will report at these Hd Quarters on Wednesday evening next—" See William R. Brown, Col. Comdg, by John S. Cunningham, Adj., Spec. Order No. 23 (manu. dated "Head Quarters 13th Regt Va Vol Infty Camp Defiance W Va August 9th 1863"), RG94 Order Books, [n.p.]

1070   RG94 Morning Reports of Capt. James W. Johnson's Co. A 13th Regiment Va. Vols. for August 1863, [n.p.]

1071   RG94 Morning Reports of Capt. Greenbury Slack's Co. A 13th Regiment Va. Vols. for September 1863, [n.p.]

1072   Letter from the "War Dept. Surgeon General's Office, Record and Pension Division Washington D.C." (manu. dated "March 14, 1887"), regarding Robert H. Snodgrass, Co. A 13th West Va. Vol. Inftry. This letter was with Eliza A. Snodgrass, Mother's Application for military pension No. 332378, which was for some reason filed together with Carey Toney, Father's Application for pension, under soldier, Lawson Toney, Co. C 7th West Va. Cavalry, Military Pension File, NARA. It is not altogether clear why this application was filed as it was but I believe that mother Eliza had remarried and to Carey Toney.

1073   RG94 Morning Reports of Capt. James W. Johnson's Co. A 13th Regiment Va. Vols. for August 1863, NARA, [n.p.]

1074   RG94 Morning Reports of Capt. Milton Stewart's Co. B 13th Regiment Va. Vols. for August 1863, NARA, [n.p.]

1075   W.R. Brown, Col. Com'd'g, by John S. Cunningham, Adj., Spec. Order No. 17 (manu. dated "Head Quarters 13th Regt Va. Vol. I Point Pleasant [West Virginia] August 1st 1863"), RG94 Order Book Co. C, [n.p.]

1076   Using their data, the total number of enlisted men should have been eighty-four.

1077   William R. Brown, Col. Comdg, by John S. Cunningham, Adj., Spec. Order No. 28 (manu. dated "Head Quarters 13th Regt Va Vol Infty Camp Defiance August 9th 1863"), Order Book Co. C, [n.p.]

1078   David Eads, Co. C, 13th Regiment Virginia Infantry Volunteers, "Oh Reg No. 30 Hos. No. 50 p. 14 and Va. Reg. No. 28; Hos No 557 p. 95," as cited in RG94 13th Va. Hospital Cards Box 3849 Entry 534.

1079   RG94 Morning Reports of Capt. Van D. McDaniel's Co. C 13th Regiment Va. Vols. for August 1863, [n.p.]

1080   August 10, Col. Brown issued Gen. Order No. 18 authorizing the reduction of Sergeant John W. Graham (as put forward in Spec. Order No. 2) and promotion of Private Henry C. Williamson, the latter to be "respected and obeyed accordingly." See W.R. Brown, Col. Com'd'g, by John S. Cunningham, Adj., Gen. Order No. 18 (manu. dated "Head Quarters 13th Regt V.V. Infty Camp Defiance Aug. 10th 1863"), RG94 Order Book Co. C, [n.p.]

1081   RG94 Morning Reports of Capt. Simon Williams' Co. D 13th Regiment Va. Vols. for August 1863, [n.p.]

1082   William R. Brown, Col. Comdg, by John S. Cunningham, Adj., Spec. Order No. 21 (manu. dated "Head Quarters 13th Regt Va Vol Infty, Camp Defiance West Va August 7th 1863"), RG94 Order Books, [n.p.]

1083   It is not clear who Darnold Kimes is. There were at this time, two men with surname Kimes in Co. D: John C. and Samuel R. Perhaps "Darnold" should be read "Samuel."

1084   William R. Brown, Col. Comdg, by John S. Cunningham, Adj., Spec. Order No. 26 (manu. dated "Head Quarters 13th Regt Va Vol Infty Camp Defiance August 9th 1863"), RG94 Order Books, [n.p.]

1085   Lt. Col. James R. Hall, Col. Comdg, by John S. Cunningham, Adj., Spec. Order No. 33 (manu. dated "Head Quarters 13th Regt V.V.I. Camp Defiance August 13th 1865), RG94 Order Books, [n.p.]

1086   RG94 Morning Reports Co. D 13th Regiment Va. Vols. for August 1863, [n.p.]

1087   William R. Brown, Col. Comdg, by John S. Cunningham, Adj., Spec. Order No. 22 (manu. dated "Head Quarters 13th Regt V.V.I. Camp Defiance West Va August 7th 1863"), RG94 Order Books, [n.p.]

1088   W.R. Brown, Col. Com'd'g., by John S. Cunningham, Adj., Gen. Order No. 19 (manu. dated "Head Quarters 13th Regt Va Vol Infantry Camp Defiance near Coalsmouth West Va Aug 21/1863"), RG94 Order Books, [n.p.]

1089   Private Young was promoted pursuant to Gen. Order No. 20 issued by Col. Brown from regimental headquarters at Camp Defiance near Coalsmouth on August 22. The order read as follows: "General Order No 1 issued from Head Quarters Co F to fill vacancy occasioned by the death of Corporal Russel B Shrewsbury and promoting private John M Young to Corporal vice Russell B Shrewsbury deceased is hereby approved. — He will be obeyed and respected accordingly." William R. Brown, Col. Comdg, by John S. Cunningham, Adj., Gen. Order No. 20 (manu. dated "Head Quarters 13th Regt Virginia V.I. Camp Defiance near Coalsmouth West Va Aug 22/63"), RG94 Order Books, [n.p.].

1090   William R. Brown, Col. Comdg by John S. Cunningham, Adj., Spec. Order Nos. 24, 25, 27 and 29, (manuscripts all dated "Head Quarters 13th Regt Va Vol Infty Camp Defiance W Va Aug 9th 1863"), RG94 Order Books, [n.p.]

1091   Wm. R. Brown, Col. Cmdg, by John S. Cunningham, Adj., Spec. Order No. 34 (manu. dated "Hd Qtrs 13th Regt Camp Defiance Aug. 21, 1863"), Order Books, [n.p.]

1092   Brig. Genl. E.P. Scammon, by Rigdon Williams, Capt. and Provost Marshal (manu. dated "Head Quarters, 3rd Division, 8th Army Corps, Provost Marshal's Office, Charleston, West Va., Aug. 27, 1863"), RG94 Box 4839 Misc. unbound regimental papers 13th W.Va.

1093   J.V. Young to wife, Paulina (dated "Headquarters Company G Camp Defiance Aug. 22, 1863"), Young Civil War Papers (Vol. 31 Box 10), trans. p. 81.

1094   John V. Young, Captain, Co. G, to Pierpont, late Governor now at this date Adj. Gen., W.Va., (manu. dated "Head Quarters Co G, 13th Regiment V.V.I., Camp Defiance, August 12, 1863") Uncat. 13th W.Va. Inftry. Regt. Box, WVSA, [n.p.]

1095   13th W.Va. Infantry Muster In Detachments of Recruits 1862-63 in RG94 Box 4839 Misc. unbound regimental papers 13th W.Va. and Annual Return of Casualties + c. 13th W.Va. Infy. to Dec. 31/63, RG94 Box 4838 Misc. unbound regimental papers 13th W.Va.

1096   William R. Brown, Col. Commanding the 13th V.V.I. to Governor [A.I.] Boreman, Gov. of West Virginia (manu. dated "Head Quarters 13th Regt. V.V.I., Camp Defiance Aug. 22, 1863"), Uncat. 13th W.Va. Inftry. Regt. Box, WVSA.

1097   RG94 Morning Reports of Capt. W.I. Mathews' Co. H 13th Regiment Va. Vols. for August 1863, [n.p.]

1098   RG 94 Misc. Records Descriptive Books 13th Regt. W.Va. Inf. Co. H, [n.p.]

1099   On August 25, Col. Brown issued Spec. Order No. 36 providing that: "William Dunlap, Private Co. 'C' and Samuel Waran Private Co. 'H' are hereby detailed permanently as Teamsters and They will report to Lt. S. Comstock R[egimental] Q[uarter] M[aster] without delay." Col. Wm. R. Brown, Col. Comdg, by John S. Cunningham, Adj., Spec. Order No. 36 (manu. dated "Head Quarters 13th Regt. V.V.I., Camp Defiance West Va August 25, 1863"), RG94 Order Book Co. C 13th Regt. Inftry. Vols., [n.p.]

1100   RG94 Morning Reports Co. H 13th Regiment Va. Vols. and "Remarks" for the month of August 1863, [n.p.]

1101   Among those absent without leave may have been Irwin Lowe. On August 9, Col. Brown issued Special Order No. 24 regarding the absent Lowe and directing that Sergeant Darnold (Darnel) of Company F and Private James H Goal (Goad) of Company I were to "proceed to Letart Falls Mason Co and arrest Irwin Lowe a deserter from Said Co [Company I] and will report to these Hd Quarters without unnecessary delay." (Special Order No. 24, by order of William R. Brown, Col. Comdg, by John S. Cunningham, Adj., manu. dated "Head Quarters 13th Regt Va Vol Infty Camp Defiance W Va Aug 9th 1863," RG94 Orders Books.) Irwin Lowe (full name probably John Irwin Lowe) was eventually cleared of charges of desertion for the period of July 26, 1863, to September 12, 1863. The charges against him were dropped and removed from his record. He was absent again without leave from Aug. 10, 1864, to Nov. 17, 1864, but was apparently cleared of these charges also. Certificate signed by Capt. and Asst. Surgeon U.S. Army [name illeg.], letter from War Dept. Washington City, to Adjt. Gen. State of W.Va., dated "Jan. 25, 1890," Uncat. 13th Box, WVSA.

1102   Wm. R. Brown, Col. Cmdg, by John S. Cunningham, Adj., Spec. Order No. 30 (manu. dated "Hd Qtrs 13th Regt V.V.I. Camp Defiance Aug. 9th, 1863"), RG94 Order Books, [n.p.]

1103   Wm. R. Brown, Col. Cmdg, by John S. Cunningham, Adj., Spec. Order No. 31 (manu. dated "Hd Qtrs 13th Regt V.V.I. Camp Defiance August 10th, 1863"), RG94 Order Books, [n.p.]

1104   Wm. R. Brown, Col. Cmdg, by John S. Cunningham, Adj., Spec. Order No. 34 (manu. dated "Hd Qtrs 13th Regt Camp Defiance Aug. 21, 1863"), RG94 Order Books, [n.p.]

1105   RG94 Morning Reports of Lieut. William Feazel's Co. I 13th Regiment Va. Vols. for August 1863, NARA, [n.p.]

1106   West Virginia Clothing Books, WVSA, [n.p.].

1107   David Burrows to his wife, Lovina (dated Aug. 28, 1863), Roush, *If Thou Wilt Remember,* p. 37.

1108   The organization of the 11th Virginia Volunteer Infantry was begun in December 1861 but was not completed until December 1862.

1109   Col. W.R. Brown to Gov. Pierpont (manu. dated "Head Quarters 13th Regt. V.V.I., Camp near Coalsmouth West Va., August 28, 1863"), Uncat. 13th W.Va. Inftry. Regt. Box, WVSA, [n.p.] and Copy of letter sent by Col. William R. Brown, Comdg 13th Regt Va Vol Inf to F.H. Pierpont, Adjt. Genl. W.Va. (manu. dated "Head Quarters 13th Regt Va Vol Infy Camp Near Coalsmouth August 28, 1863"), RG94 Box 4839 Misc. unbound regimental papers 13th W.Va. Letters A.G.O., NARA.

1110   Uncat. 13th W.Va. Inftry. Regt. Box, WVSA.

1111   Major General William S. Rosecrans had recently maneuvered General Braxton Bragg out of the strategically important city Chattanooga, Tennessee. Rosecrans had taken possession of the city and occupied it. By September 19, however, Bragg engaged Rosecrans at Chickamauga and defeated Federal forces. Rosecrans was replaced as commander by Gen. George H. Thomas, the "Rock of Chickamauga."

1112   Brig. Gen'l. Scammon to Hdqtrs. 1st Brig. 3rd Div. 8th A.C. (dated "Head Qrs. 3 Divis. 8th A.C. Charleston W.Va. Sept. 3 1863"), "Order for the 13th Va.V.I. and Lieut. Witchers Co - 3 Va Cav. to Barboursville," Letters Received Head Quarters 1st Brig. 3d Division 8th Army Corps, Letters Received Miscellaneous, Papers of R.B. Hayes (micro. Reel 272), trans. p. 24.

1113   Hayes to Brown, by Avery (manu. dated "Hd. Qt. 1st Brig. 3d Div. 8 AC, Camp White, W.Va., Sept. 3, 1863"), RG94 Box 4839 Misc. unbound regimental papers 13th W.Va. and Col. R.B. Hayes to Col. Wm. R. Brown (manu. dated "Head Qtrs. 1st Brigade 3rd Division, Dept 8th Army Corps, Camp White, Va., Sept. 5, 1863"), Letter #256 Registry #393 Part 2 Entry 1184 Press Copies of Letters Sent and Special Orders Vol. 139.

1114   Field History 13th Regiment W.Va. Infantry, [n.p.]

1115   Col. Wm. R. Brown to Hdqtrs. 1st Brig. 3d Div. 8th A.C. (dated "Head Qrs. 13th Regt. Va. V.I. Coals Mouth Va. Sept. 4th 1863"), Letters Received Head Quarters 1st Brig. 3d Division 8th Army Corps, Letters Received Miscellaneous, Papers of R.B. Hayes (micro Reel 272), trans. p. 25.

1116   Annual Return of Casualties + c. 13th W.Va. Infy. to Dec. 31/63, RG94 Box 4838 Misc. unbound regimental papers 13th W.Va.

1117   Ibid.

1118   Regimental Return 13th Va. Vol. Infantry for Sept. and Oct. 1863 (micro. #594 Roll 196) Record of Events Field and Staff, 13th W.Va. Infantry, NARA and Field and Staff Muster Roll Record of Events in Compiled Service Records of Volunteers (Union) West Virginia Station: Winchester March and April 1865 (micro. #M508 Roll 204), NARA.

1119   Companies A, B, C, D, E, F and H were reported stationed at Barboursville for the months of September and October. Co. I was reported stationed at Guyandotte. Company Muster Rolls 13th Va. Vol. Inftry. Sept. and Oct. 1863 (micro. #594 Roll 196) and Co. I was stationed at Guyandotte Sept. 1863-Feb. 1864. (Supp. O.R. Vol. 74, p. 571.)

1120   William R. Brown, Col. Comdg by John S. Cunningham, Adj. to Lt. William E. Feazel Co. I, Special Order No. 38 (manu. dated "Head Quarters 13th Regt Va Vol Infty Barboursville West Va Sept 9th /63"), RG94 Order Books, [n.p.]

1121   Field History of the 13th Regiment, trans. p. 4.

1122   RG94 Box 4839 Misc. unbound regimental papers 13th W.Va.

1123   Col. Wm. R. Brown to Hdqtrs. 1st Brig. 3d Div. 8th A.C. (dated "H'd Qrs. 13 Regt. Va. V.I. Barboursville W. Va. Sept. 7th, 1863"), Letters Received Head Quarters 1st Brig. 3d Division 8th Army Corps, Letters Received Miscellaneous, Papers of R.B. Hayes (micro. Reel 272), trans. p. 27.

1124   R.B. Hayes, Col. Comdg. to Col. Wm. R. Brown, 13th Va Vols. (manu. dated "Hd Qrs 1st Brig. 3d Div. 8th A C, Camp White, Sept. 8, 1863"), RG94 Box 4839 Misc. unbound regimental papers 13th W.Va.

1125   R.B. Hayes, Col. Comd'g to "Col. Wm. R. Brown Comdg 13 Va.V.I., Barboursville, Va." (manu. dated "Head Qure 1 Brig. 3d Divis 8 A.C., Camp White W. Va., Sept. 8, 1863"), RG94 Box 4839.

1126   William R. Brown, Col. Comdg, by John S. Cunningham, Adj., Spec. Order No. 40 (manu. dated "Head Quarters 13th Regt V.V.I. Barboursville W. Va. Sept. 14th /63"), RG94 Order Books, [n.p.]

1127   William R. Brown, Col. Commanding, by John S. Cunningham, Adj., Gen. Order No. 23 (manu. dated "Head Quarters 13th Regt Va Vol Infantry Barboursville West Virginia September 10th 1863"), RG94 Order Book Co. C 13th Regt. Inftry. Vols., [n.p.]

1128   William R. Brown, Col. Comdg by John S. Cunningham, Adj., Spec. Order No. 39 (manu. dated "Head Quarters 13th Regt V.V.I. Barboursville West Va. Sept 13th /63"), RG94 13th Va. Order Books, [n.p.]

1129   Wm. R. Brown, Col. Comdg by John S. Cunningham, Adj., Spec. Order No. 41 (manu. dated "Head Quarters 13th Regt V.V.I. Barboursville West Va. Sept. 14th /63"), RG94 Order Books, [n.p.]

1130   Wm. McKinley, Jr., 1st Lt. + A.A.Q.M. to S. Comstock, R.Q.M., 13th Regt. Va. Vols. at Barboursville (manu. dated "Hd. Qtrs. 1st Brig. 3rd Div. 8th A.C. Camp White, Va. Sept. 11, 1863"), Letter 53 Registry No. 393 Part 2 Entry 1194 Register of letters received by Quartermasters, NARA, [n.p.]

1131   Thomas M. Vincent, Assistant Adjutant General to Brigadier General B.F. Kelley, Commanding Department of West Va., Clarksburg, Virginia, Official Copy for C.O. 13th Va. Vols. through Genl Scammon, by T. Melvin (manu. dated "War Department, Adjutant Generals Office, Washington D.C. September 11, 1863"). Copy sent by Col. William R. Brown, Comdg 13th Regt Va Vol Inf to F.H. Peirpoint, Adjt Genl of West Va. (manu. dated "Head Quarters 13th Regt V.V.I. Barboursville, West Virginia Sept 15, [18]63"), both in RG94 Box 4839 Misc. unbound regimental papers 13th W.Va.

1132   Wm. R. Brown, Col. Comdg 13th Regt Va. Vol. Infty to Hon. H.J. Samuels, Judge of Circuit Court of Cabell Co. at Guyandotte (manu. dated "Head Quarters 13th Regt V.V.I., Barboursville, West Virginia, Sept. 10, 1863"), RG94 Box 4839 Misc. unbound regimental papers 13th W.Va.

1133   W.R. Brown, Col. Comd'g to Rigdon Williams, Capt. and Provost Marshall (manu. dated "Head Quarters 13th Regt V.V.I., Barboursville, Cabell Co. Sept. 14, 1863"), RG94 Box 4839 Misc. unbound regimental papers 13th W.Va.

1134   Col. Hayes to Col. Brown (manu. dated "Hd. Qrs. 1st Brig. 3rd Div. 8th A.C., Camp White, Va., Sept. 17, 1863"), RG94 Box 4839 Misc. unbound regimental papers 13th W.Va.

1135   The recruits at Roane County would become part of Co. K 13th W. Virginia Infantry Regiment.

1136   William R. Brown, Col. Comdg, by John S. Cunningham, Adj., Special Order No. 42 (manu. dated "Head Quarters 13th Regt V.V.I. Barboursville West Va Sept 21st /63"), RG94 Box 4839.

1137   "A Convalescent" to the Editor (dated "The Meeting U.S. Gen[eral] Hospital, Sept. 21"), Point Pleasant *Weekly Register*, September 24, 1863, [n.p.]

1138   W[illia]m R. Brown, Col. 13 Regt Va Vol I to G.M. Bascom, Capt. and A.A. Gen. (manu. dated "Head Quarters 13th Regt. V. V. Infty, Barboursville, Sept. 23, 1863"), RG94 Box 4839 Misc. unbound regimental papers 13th W.Va. Brown's letter was endorsed on September 29, 1863, and forwarded to Gen. Scammon by way of Brig. Gen. J.D. Cox, by G.M. Bascom, Major + A.A.G.

1139   M.R. Avery, Capt. and A.A.A.G. to Col. W.R. Brown, Comdg. 13th (manu. dated "Hd.Qtrs.1st Brig., 3d Div. 8th A.C. Camp White, Va., Sept. 24, 1863"), RG94 Box 4839 Misc. unbound regimental papers 13th W.Va.

1140   Col. Wm. R. Brown to Hdqrs. 1st Brig 3rd Div. 8th A.C. (dated "Camp at Coals Mouth Va. Sept. 27, [1863]"), Letters Received Head Quarters 1st Brig. 3d Division 8th Army Corps, Letters Received Miscellaneous, Papers of R.B. Hayes (micro. Reel 272), trans.p. 24.

1141   William R. Brown, Col. Comdg, by John S. Cunningham, Adj., Spec. Order No. 51 (manu. dated "Head Quarters 13th Regt V.V.I. Barboursville West Va. Sept 28th /63"), RG94 Order Books 13th Regiment Infantry Vols., [n.p.]

1142   Col. R.B. Hayes to Col. Wm. R. Brown (manu. dated "Head Qtrs. 1st Brigade 3rd Division, Dept 8th Army Corps, Camp White, Va., Sept. 29, 1863"), Registry #393 Part 2 Entry 1184 Press Copies of Letters Sent and Special Orders Vol. 139, NARA.

1143   West Virginia Clothing Books, [n.p.]

1144   Col[onel] W.R. Brown to Hd Qtrs. 1st Brig. 3rd Div. 8th A.C. Three identical letters were sent from Colonel Brown at 13th Head Quarters. The first two letters (at least according to the ledger recording the receipt of letters at 1st Brigade Head Quarters) had no date only "Head Qrs. 13th Va. V. I. Barboursville Va." Only the third of these (designated letter 305) was dated in terms of month, day and year: "Head Qrs. 13th Va. V.I. Barboursville Va., Sept. 30, 1863," Letters Received Head Quarters 1st Brig. 3d Division 8th Army Corps, Letters Received Miscellaneous, Papers of R.B. Hayes (micro. Reel 272), trans.p. 31.

1145   RG94 Morning Reports of Capt. Greenbury Slack's Co. A 13th Regiment Va. Vols. for September 1863, [n.p.]

1146   W.R. Brown, Col. Com'dg, by John S. Cunningham, Adj., Spec. Order No. 49 (manu. dated "Head Quarters 13th Regt V.V.I. Barboursville West Va. Sept 24th /63"), RG94 Order Books 13th Regiment Infantry Vols., [n.p.]

1147   RG94 Morning Reports Co. A 13th Regiment Va. Vols. for September 1863, [n.p.]

1148   R.H. Snodgrass, Military Pension File.

1149   General Affidavit of William A. Harliss and George W. Turley (dated "Kanawha Co. Charleston, May 3, 1883"), Robert H. Snodgrass, Military Pension File (Mother: Eliza A. Snodgrass, Mother's Application in Carey Toney, father of Lawson Toney, Co. C 7th West Va. Cavalry, Military Pension File), NARA.

1150   W.R. Brown, Col. Com'dg, by John S. Cunningham, Adj., Spec. Order No. 48 (manu. dated "Head Quarters 13th Regt V.V.I. Barboursville Sept 24th /63"), RG94 Order Books 13th Regiment Inftry. Vols., [n.p.]

1151   RG94 Morning Reports of Capt. Milton Stewart's Co. B 13th Regiment Va. Vols. for September 1863, [n.p.]

1152   Special Order No. 45 authorizing recruiting detachment was the following: "Capt Stewart Co B is hereby ordered to take charge of eight men and will proceed with them to Thirteen mile Creek in Mason Co. or to any other point in said Co. that may seem to him to be proper for the purpose of finding recruits for the 13th Regt V.V.I. Capt. Stewart will report to these Head Quarters weekly giving a statement of his success in the enterprise." W.R. Brown, Col. Com'dg, by J.S. Cunningham, Adj., Spec. Order No. 45 (manu. dated "Head Quarters 13th Regt V.V.I. Barboursville West Va. Sept 24th 1863"), RG94 Order Books 13th W.Va. Regt., [n.p.]

1153   RG94 Morning Reports Co. B 13th Regiment Va. Vols. for September 1863, [n.p.]

1154   [Memorial tribute to Captain Van D. McDaniel], dated "Headquarters 13th [Virginia] Vols., Barboursville, W.V., Sept. 13th," "Tribute of Respect," Point Pleasant *Weekly Register*, Sept. 24, 1863, Vol. II No. 27, [p. 2].

1155   Van D. McDaniel had served as Surveyor for Mason County Road District No. 34 before his entry into military service. See Mason County Court Records. November Term 1862, pp. 326, Mason County Court House, Point Pleasant, West Virginia.

1156   [George W. Tippett or E.W. Fitzgerald, ed.], "Capt. V.B. McDaniel," Point Pleasant *Weekly Register*, October 1, 1863, [p. 2].

1157   RG94 Morning Reports of Capt. Van D. McDaniels' Co. C 13th Regiment Va. Vols. for September 1863, [n.p.]

1158 Entries for September 15 and 18 were written in very poor script and also seem to be out of sequence. I include them here in footnote only because of their location in the original manuscript. The men reported enlisted — John Mackentire and J. Canes Harses — were apparently not mustered in. The entries in question are the following: "Sept. 15. John Mackentire joined went in roster Monday this 19ᵗʰ day of May 18[6]4" and "Sept. 18 J. Canes Harses went in Roster Wed. the 25 of June this Boy were taken out July the 1 five days." RG94 Morning Reports of Capt. Simon Williams' Co. D 13ᵗʰ Regiment Va. Vols. for September 1863, [n.p.]

1159 C.D. Dally, Asst. Surg., 5ᵗʰ V.V. Infty. in charge to Col. Brown, 13ᵗʰ V.V.I. (manu. dated "Sept. 20, 1863"), RG94 Box 4839 Misc. unbound regimental papers 13ᵗʰ W.Va.

1160 Lewis C. Barnes, Military Pension File (Jane Barnes, Mother's Application for Pension #296929 filed Sept. 25, 1882; Certificate # 205598), NARA; L.C. Barnes, 13ᵗʰ W.Va. Military History Cards on micro., WVU; and *Medal List. Unclaimed. G.A.R.* 13ᵗʰ Infantry D., WVSA, [n.p.]

1161 Wm R. Brown, Col. Comdg, by John S. Cunningham, Adj., Special Order No. 52 (manu. dated "Head Quarters 13ᵗʰ Regt Va Vol Inftry Barboursville West Va. Sept 28ᵗʰ /63"), RG94 Order Books, [n.p.].

1162 RG94 Morning Reports of Capt. Simon Williams' Co. D 13ᵗʰ Regiment Va. Vols. for September 1863, [n.p.]

1163 Brig. Gen'l Scammon to 1ˢᵗ Brigade Head Quarters (dated "Hd Qrs. 3 Divis 8ᵗʰ A.C. Charleston W. Va. Sept. 19ᵗʰ 1863"), Letters Received Head Quarters 1ˢᵗ Brig. 3d Division 8ᵗʰ Army Corps, Letters Received Miscellaneous, Papers of R.B. Hayes (micro. Reel 272), trans. p. 28.

1164 Col. Wm. R. Brown Comdg 13ᵗʰ Regt. Va. V.I. to Hd qtrs. 1ˢᵗ Brig. 3d Div. 8ᵗʰ A.C. (dated "H'd Qts. 13 Regt. Va Vol Inf. Barboursville W. Va. Sept. 26 1863"), Letters Received Head Quarters 1ˢᵗ Brig. 3d Division 8ᵗʰ Army Corps, Letters Received Miscellaneous, Papers of R.B. Hayes (micro. Reel 272), trans.p. 30.

1165 West Virginia Clothing Books, WVSA, [n.p.]

1166 Mark E. Robinson to "The Cot[t]age Home" (manu. dated "Camp Barbers Ville Cabbille Co West Va September the 25 1863"), Mark E. Robinson, Military Pension File, NARA.

1167 C. Shrewsbury to his wife (manu. "Barboursville, Cabell Co., W. Va, [no date]"), A & M 3216 — Lt. Columbus Shrewsbury Co A, 4ᵗʰ WVI, Roy B. Cook Collection, WVU, Morgantown.

1168 RG94 Misc. Records in Descriptive Books 13ᵗʰ Regt. W. Va. Inf. Co. E, NARA, [n.p.]

1169 RG94 Morning Reports of Capt. A.F. McCown's Co. F 13ᵗʰ Regiment Va. Vols. Army of the U.S. and "Remarks" for September 1863, [n.p.] Col. Brown's orders (Special Order No. 47) detailing the recruiting detachment from Co. F was as follows: "Lt. Russel Co F will take charge of eight men and will proceed with them on a recruiting expedition to Mason Co West Va Lt. Russel will report to these Head Quarters weekly the success attending his efforts." See W.R. Brown, Col. Com'dg, by J.S. Cunningham, Adj. (manu. dated "Head Quarters 13ᵗʰ Regt V.V.I. Barboursville Sept 24ᵗʰ /63"), RG94 Order Books 13ᵗʰ Regiment Infantry Vols., [n.p.]

1170 RG94 Morning Reports Co. F 13ᵗʰ Regiment of Va. Vols. for September 1863, [n.p.]

1171 David Burrows to his wife, Lovina, dated "Sept. 7, 1863," Roush, *If Thou Wilt Remember,* pp. 38-39.

1172 William R. Brown, Col. Commanding, 13ᵗʰ Regt. V.V.I. to F.H. Peirpont, Adj. Gen. of W.Va. (dated "Head Quarters 13ᵗʰ Regt. Va. Vol. Infty., Barboursville, West Va., Sept. 15ᵗʰ, 1863"), Samuels-Pierpont Papers, WVSA.

1173 J.V. Young to wife, Paulina and children (dated "Barboursville, Cabell Co. W. Va. Sept. 10, 1863"), Young Civil War Papers (Vol. 31 Box 10), WVU, trans. p. 82.

1174 Hamiline is probably present day Hamlin, located on Mud River, Lincoln County.

1175   William R. Brown, Col. Com'dg, by John S. Cunningham, Adj., Special Order No. 43 (manu. dated "Head Quarters 13th Regt V.V.I. Barboursville West Va. Sept 21 /63"), RG94 Order Books, [n.p.]

1176   RG94 Morning Reports of Capt. W.I. Mathews' Co. H 13th Regiment Va. Vols. for September 1863, [n.p.]

1177   C.D. Dally, Asst. Surg., 5th V.V.I., in charge to Col. Brown, 13th V.V.I. (manu. dated "Sept. 21, 1863"), RG 94 Box 4839 Misc. unbound regimental papers 13th W.Va., NARA.

1178   RG94 Morning Reports of Capt. Mathews' Co. H 13th Regiment Va. Vols. for September 1863, [n.p.]

1179   W.R. Brown, Col. Com'dg, by John S. Cunningham, Adj., Special Order No. 44 (manu. dated "Head Quarters 13th Regt V.V.I. Barboursville West Va. Sept 21/63"), RG94 Order Books, [n.p.]

1180   A.I. Boreman, Governor of West Virginia to John S. Cunningham, Adjutant, 13th Regt. Va. Vol. Inftry, "Order to arrest William and Charles Furguson as Hostages for Morgan Garret + to report them to Hd. Qtrs. State of West Va. at Wheeling" (manu. dated "Sept. 18, 1863"), RG94 Box 4838 Misc. unbound regimental papers 13th W.Va. Spec. Orders and Gen. Orders, NARA.

1181   The Ferguson families of Wayne County were largely loyal to the 'Old Dominion' Virginia, and the Confederate army had numerous volunteers surnamed Ferguson recruited from Wayne County. These served predominantly in the 8th and 16th Virginia Cavalry Regiments. See Dickinson, *Tattered Uniforms and Bright Bayonets. West Virginia's Confederate Soldiers*, pp. 161-62.

1182   William R. Brown, Col. Comdg, by John S. Cunningham, Adj., Special Order, No. 50 (manu. dated "Head Quarters 13th Regt. Va. Vol. Inftry, Barboursville West. Va., Sept. 25, [18]63"), RG94 Box 4838 Misc. unbound regimental papers 13th W. Va., Spec. Orders and Gen. Orders and W.R. Brown, Col. Com'dg, by J.S. Cunningham, Adj., Spec. Order No. 50 (manu. dated "Head Quarters 13th Regt V.V.I. Barboursville West Va. Sept. 25th /63"), RG94 Order Books 13th Regt. Inftry. Vols., [n.p.].

1183   Col. William R. Brown to Gov. A.I. Boreman (manu. dated "Head Quarters 13th Regt. V.V. Infty., Barboursville, October 13, 1863"), A.I. Boreman Papers, WVSA.

1184   Wm. R. Brown, Col. Comdg 13th Regt. V.V.I. [to Hd Qtrs. 3rd Div. Dept. W.Va., at Charleston] (manu. dated "Head Quarters 13th Regt. Va. Vol. Inftry, Barboursville, W. Va., Oct. 15, [18]63"), RG94 Box 4838 Misc. unbound regimental papers 13th W.Va., Spec. Orders and Gen. Orders.

1185   RG94 Morning Reports of Capt.Mathews' Co. H 13th Regiment Va. Vols. for September 1863, [n.p.]

1186   Field History 13th Regt. W.Va. Inftry, trans. p. 4.

1187   RG94 Morning Reports of Lieut. William Feazel's Co. I 13th Regiment Va. Vols. for September 1863, NARA, [n.p.]

1188   Lieut. Wm. E. Feazel Co. I. 13th Regt. Va. V.I to Hd. Qtrs. 1st Brig. 3d Div. 8th A.C. (dated "H'd Qt. 13 Regt. Va. V. Inf. Barboursville, Va. Sept. 14th 1863"), Letters Received Head Quarters 1st Brig. 3d Division 8th Army Corps, Letters Received Miscellaneous, Papers of R.B. Hayes (micro. Reel 272), HPL, trans. p. 28.

1189   Wm. E. Feazel, 1st Lt., 13th Va. Vol., Com. Post Guyandotte, W.Va. to Lt. John S. Cunningham Adjt. 13th Va Vol. Infty, Barboursville, W.Va., (manu. dated "Head Co. I 13th Reg Vol Infty, Guyandotte, W.Va., Sept. 27, 1863"), RG94 Box 4839 Misc. unbound regimental papers 13th W.Va., NARA.

1190   Henry Stump to Wm R Brown, Col. Comdg 13th Regt. V.V.I., Head Quarters Coals Mouth W.V. (manu. dated "Charleston, West Va, Sept. 3, 1863"), RG94 Box 4839 Misc. unbound regimental papers 13th W.Va.

1191   Henry Stump, 1st Lieutenant of the 13th Regt. V.V. Infty. and Recruiting Officer to Arthur I. Boreman, Governor of West Virginia (manu. dated "Charleston, Kanawha Co., W.Va., September 21, 1863"), Uncatalogued 13th W.Va. Inftry. Regt. Box, WVSA, [n.p.]

1192   Henry Stump, Recruiter 13th Regt. V.V. Infty. to Colonel Brown, HeadQuarters Chapmansville (manu. dated "Charleston W.V., Sept, 21, 1863"), RG94 Box 4839 Misc. unbound regimental papers 13th W.Va.

1193   Record of Events Field and Staff, Regimental Returns 13th W.Va. Vol. Infantry for October 1863 (micro. #594 Roll 196), NARA.

1194   Registry # 393 Part 2 Entry 1184 Dept. of W.Va. Press Copies of Letters Sent and Special Orders Vol. 139, NARA.

1195   "No Draft in Mason," Point Pleasant *Weekly Register*, April 7, 1864, [p. 2].

1196   Fragment of a telegraph message sent to "Hd. Qtrs. 13th Regt. W.V.I." (manu. dated probably 1864) RG94 Box 4838 Misc. unbound regimental papers 13th W.Va. Telegrams, NARA.

1197   RG94 Box 4839 Misc. unbound regimental papers 13th W.Va. Letters A.G.O., NARA and Annual Return of Casualties + c. 13th W. Va. Infty. to Dec. 31/63 (manu. "Return completed Jan. 13, 1864"), RG94 Box 4838 Misc. unbound regimental papers 13th W.Va., NARA.

1198   A "main spring swivel" was a metal piece that fastened on the butt end of a gun to which a sling for carrying the gun was attached.

1199   John M. Butler, P.M. + M.S.K. to Col. W.R. Brown, Comg 13th Regt. Va. Vol. Infy., Barboursville (manu. dated "Allegheny Arsenal, Oct. 3, 1863"), RG94 Box 4839 Misc. unbound regimental papers 13th W.Va. The Springfield rifled musket, manufactured at the United States Armory at Springfield, Massachusetts, was carried army-wide and used .58 caliber bullets. The Enfield rifled musket was one the best of the foreign made weapons purchased by the Federal government to supplement the supply of domestic weapons. It was made at the government armory at Enfield, England. The Enfield required .57 caliber bullets. Exception of the caliber of bullets required, the Enfield was in every respect like the 1861 Springfield rifle manufactured at Springfield. Both Enfields and Springfields were esteemed by volunteers. They did not require heavy charges, were relatively easy to load and handle (weighing only about fourteen pounds) and were accurate to about seven hundred yards.

1200   Brig. Gen. Scammon to Col. R.B. Hayes (manu. dated "Hd Qtrs 3rd Div. Dept. West Va., Charleston W. Va. Oct. 3rd, 1863"), RG94 Misc. Dept. of W.Va. Papers, NARA.

1201   W.R. Brown, Col. Com'dg, by J.S. Cunningham, Adj. (manu. dated "Head Quarters 13th Regt Va Vol Inftry Barboursville West Va Oct 5th 1863"), RG94 Order Books 13th Regiment Infantry W.Va. Vols., [n.p.]

1202   W.R. Brown, Col. Com'dg, by J.S. Cunningham, Adj., Special Order No. 56 (manu. dated "Head Quarters, 13th Reg't V.V.I. Barboursville Oct. 5/63"), RG94 Order Books, [n.p.]

1203   W.R. Brown, Col. Com'dg, by J.S. Cunningham, Adj. Spec. Order No. 57, (manu. dated "Head Quarters, 13th Regt V.V.I. Barboursville West Va. Oct. 7/63"), RG94 Order Books, [n.p.]

1204   Brig. Gen. E.P. Scammon, by James L. Botsford A.A. General, Spec. Order No. 112 (manu. dated "Head Quarters 3d Div. Dept. West Va., Charleston, W.Va., Oct. 7, 1863"), RG94 Box 4838 Misc. unbound regimental papers 13th W.Va., Spec. Orders and Gen. Orders, NARA.

1205   C[olumbus] Shrewsbury to his wife (manu. dated "Barboursville, Cabell Co., W.Va, October 10, 1863"), A & M — 3216 —Lt. Columbus Shrewsbury Co A, 4th WVI, Cook Coll., WVU.

1206   RG94 Misc. Records Descriptive Roll Co. D. 13th Regiment V.V.V.I., NARA, [n.p.]; Original Muster Roll of Company D 13th W. Va.V.I., manu. Records of Mason County Soldiers #1541, WVU and email to the me from John Dawson, who owns a copy of Blackburn's pension file and his service medallion. "John Blackburn needed money very badly and sold his service medallion to a Robert Dewees," son of Benjamin Dewees, Co. K 11th W. Va. Inf. (Correspondence to me from John Dawson, dated April 26, 1998.)

1207   Co. E Enlisted Men 13th Regiment Infantry, Roster of the Regiment, WVSA, p. [27].

1208   Ibid.

1209   RG94 Misc. Records Descriptive Roll Co. F. 13th Regt. V.V.V.I., NARA, [n.p.].

1210   Ibid.

1211   RG94 Misc. Records Descriptive Roll Co.B 13th Regt. V.V.V.I., [n.p.].

1212   Wm. R. Brown, Col. Comdg, by J.S. Cunningham, Adj. to Capt John S. Witcher 3d Va Cavalry, Special Order No. 64 (manu. dated "Head Quarters 13th Regt V.V.I. Barboursville Oct 15th /63" RG94 Order Books 13th Regiment Infantry W.Va. Vols., [n.p.]

1213   Wm. F. Dusenberry to Col. Brown (manu. dated "Guyandotte Cabell Co West Va., Oct. 16, 1863"), RG94 Box 4839 Misc. unbound regimental papers 13th W.Va., NARA.

1214   Capt. and Adjt. Gen. Avery to Col. W.R. Brown (manu. dated "Hd Qtrs. 1st Brig., 3 Div., Dept. W. Va., Camp White, W. Va., Oct. 20, 1863"), RG94 Box 4839 Misc. unbound regimental papers 13th W.Va. and Annual Return of Casualties + c. 13th W. Va. Infy. to Dec. 31/63, RG94 Box 4838 Misc. unbound regimental papers 13th W. Va.

1215   Col. W.R. Brown, 13th Va.V.I. to Hdqtrs. 1st Brig. 3rd Div. 8th A.C. (dated "Head Qrs 13th Regt. Va. V. I. Barboursville W.Va. 22 Oct. 1863"), Letters Received Head Quarters 1st Brig. 3d Division 8th Army Corps, Letters Received Miscellaneous, Papers of R.B. Hayes (micro. Reel 272), trans. p. 35.

1216   Col. R.B. Hayes to Col. Wm. R. Brown (manu. dated "Head Qtrs. 1st Brigade 3rd Division, Dept 8th Army Corps, Camp White, Va., Oct. 24, 1863"), I 24 Registry # 393 Part 2 Entry 1184 Dept. of W.Va. Press Copies of Letters Sent and Special Orders, NARA and Hayes to Brown, by Avery (manu. dated "Hd. Qtrs. 1st Brig. 3d Div., Dist. of W. Va., Camp White, W.Va., Oct. 24, 1863"), RG94 Box 4839 Misc. unbound regimental papers 13th W.Va.

1217   Col. R.B Hayes to Col. W.R. Brown (manu. dated "1st Brig. 3rd Div. 8th AC. Camp White, Oct. 24, 1863"), Registry #393 Part 2 Entry 1184 Dept. of W.Va. Press Copies of Letters Sent and Special Orders Vol. 139, NARA.

1218   I.H. Duval, Col. 9th Va., by C.B. Hayslip, Adjt to Col. Brown, comdg 13th Va. (manu. dated "October 25, 1863"), RG94 Box 4839 Misc. unbound regimental papers 13th W.Va.

1219   James R. Hall, Lieut. Col., 13th Va. Vols to James L Botsford, A.A.G. (manu. dated "Head Quarters 13th Va Vol Infty, Barboursville, W.Va., Oct. 27, 1863"), RG94 Box 4839 Misc. unbound regimental papers 13th W.Va. Requests for leave of absence Letters A.G.O., NARA. On the back of the manuscript, Col. Brown approved Hall's request and forwarded it. The request was approved all the way up the chain of command, i.e., by Col. Hayes at 1st Brig. Hd. Qtrs. at Camp White and by Brig. Gen. Scammon at 3d Division Hd. Qtrs. at Charleston.

1220   J.S. Cunningham to his wife, manu. dated "Head Quarters 13th Regt V.V.I. Barboursville West Va Oct. 28th 1863," Letters of John S. Cunningham, WVSA.

1221   W.R Brown, Col. Com'dg, by J.S. Cunningham, Adj., Special Order No. 69 (manu. dated "Head Quarters 13th Regt V.V.I. [Barboursville] Oct 29th /63"), RG94 Order Books 13th Regiment Infantry W.Va. Vols., [n.p.]

1222   Registry #393 Part 2 Entry 1184 Dept. of W.Va. Press Copies of Letters Sent and Special Orders Vol. 139, NARA.

1223   RG94 Morning Reports of Capt. Greenbury Slack's Co. A 13th Regiment Va. Vols. for October 1863, [n.p.]

1224   W.R. Brown, Col. Com'dg, by J.S. Cunningham, Adj., Special Order No. 62 (manu. dated "Head Quarters 13th Regt V.V.I. Barboursville Oct 14th /63"), RG94 Order Books, [n.p.]

1225   W.R. Brown, Col. Comdg, by J.S. Cunningham, Adj., Special Order No. 67 (manu. dated "Head Quarters 13th Regt V.V.I. Barboursville Oct 26th /63"), RG94 Order Books 13th Regt. Inftry. W.Va., [n.p.]

1226   RG94 Morning Reports of Capt. Milton Stewart's Co. B 13th Regiment Va. Vols. for October 1863, [n.p.]

1227   RG94 Morning Reports of Capt. Lemuel Harpold's Co. C 13th Regiment Va. Vols. for October 1863, [n.p.]

1228   Col. W.R. Brown, Spec. Order No. 64 (manu. dated "13th Regt. Headquarters, Barboursville, W.Va., Oct[ober] 11th, 1863") and W.R. Brown, Col. Com'dg, by J.S. Cunningham, Adj., Spec. Order No. 63 (manu. dated "Head Quarters 13th Regt V.V.I. Barboursville W.Va. Oct. 15th/63"), both orders from RG94 Order Book Co. C 13th W.Va. Regt. Inftry. Vols., NARA, [n.p.]

1229   William R. Brown, Col. Comdg, by J.S. Cunningham, Adj., Gen. Order No. 26 (manu. dated "Head Quarters 13th Regt Va Vol Infty Barboursville West Virginia October 31st 1863"), RG94 Order Book Co. C, [n.p.]

1230   Col. Wm. R. Brown, by J.S. Cunningham, Adj., Special Order No. 68 (manu. dated "Head Quarters 13th Regt V.V.I. Barboursville, West Va Oct 27th/63"), RG94 Order Book Co. C, [n.p.]

1231   RG94 Misc. Records Descriptive Roll Co.C. 13th Regiment V.V.V.I., NARA, [n.p.]

1232   RG94 Morning Reports of Capt. Lemuel Harpold's Co. C 13th Regiment Va. Vols. for October 1863, [n.p.]

1233   William R. Brown, Col. Comdg, by J.S. Cunningham, Adj., Gen. Order No. 24 (manu. dated "Head Quarters 13th Regt Va Vol Infty Barboursville West Virginia October 9th 1863"), RG94 Order Books 13th W.Va. Regiment Inftry., [n.p.]

1234   RG94 Carded Medical Records Vols. 1846-1865 Entry 534 Box 3849.

1235   W.R. Brown, Col Com'dg, by J.S. Cunningham, Adj., Spec. Order No. 60 (manu. dated "Head Quarters 13th Regt V.V.I. Barboursville West Va. Oct 10th/63"), RG94 Order Books, [n.p.]

1236   RG94 Regimental Books Misc. Records Descriptive Roll Co.D. 13th Regiment V.V.V.I., [n.p.] The word *ankylosis* describes a fused joint. Graham applied for Invalid Pension (#50609 filed August 29, 1864). He received pension pursuant to Certificate #71795.

1237   John M. Graham, "A Card of Thanks," Point Pleasant *Weekly Register*, Oct. 29, 1963, [p. 2].

1238   RG94 Morning Reports of Capt. Simon Williams' Co. D 13th Regiment Va. Vols. for October 1863, [n.p.]

1239   Wm. R. Brown, Col. Comdg, by J.S. Cunningham, Adj., Special Order No. 53 (manu. dated "Head Quarters 13th Regt Va Vol Inftry Barboursville West Va. Oct 3d 1863"), RG94 Order Books 13th W.Va. Regt. Inftry., [n.p.]

1240   Ibid.

1241   W.R. Brown, Col. Comdg, by J.S. Cunningham, Adj., Spec. Order No. 55 (manu. dated "Head Quarters 13th Regt V.V.I. Barboursville W.Va. Oct. 5/63"), RG94 Order Books, [n.p.]

1242   Robert Cobbs to his mother and father (manu. dated "October 9th, 1863, Camp at Barboursvill[e] Cabell Co[unty]," Letters of William Robert Cobbs, courtesy of Terry Lowery.

1243   West Virginia Clothing Books Co. E 13th Infantry, [n.p.]

1244   Mark E. Robinson to his father, Mr. John Robison (manu. dated "Camp Barbers Ville Cabelle Co West Va October the '14' 1863"), Robinson, Military Pension File, NARA.

1245   A.F. McCown, Capt. Commanding, by James R. Walkup, A.O.S., Spec. Order No. 4 (manu. dated "Head Quarters Company F 13th V.V.V.I., Barboursville, W.Va., Oct. 22, 1863"), RG94 Box 4838 Misc. unbound regimental papers 13th W.Va., Spec. Orders and Gen. Orders, NARA. Col. Brown's Gen. Order No. 25 confirming Edwards' reduction and Fry's promotion is as follows (the discrepancy in the dates between McCown's and Brown's orders were present in the original manuscript): "Order No 4 issued from Head Quarters Co F reducing Corporal James Edwards to the ranks at his own request is hereby approved — Private John W Fry is promoted to Corporal vice James Edwards reduced he will be obeyed and respected accordingly —"William R. Brown, Col. Comdg, by John S. Cunningham,

Adj. Gen., Order No. 25 (manu. dated "Head Quarters 13th Regt Va Vol Infantry Barboursville West Virginia October 18/63"), RG94 Order Books 13th W.Va. Regt. Inftry., [n.p.]

1246   RG94 Morning Reports of Capt. A.F. McCown's Co. F 13th Regiment Va. Vols. for October 1863, [n.p.]

1247   W.R. Brown, Col. Comdg, by J.S. Cunningham, Adj., Spec. Order No. 59 (manu. dated "Head Quarters 13th Regt V.V.I. Barboursville West Va. Oct 10th /63"), RG94 Order Books, [n.p.]

1248   W.R. Brown, Col. Com'dg, by J.S. Cunningham, Adj., Spec. Order No. 61 (manu. dated "Head Quarters 13th Regt V.V.I. Barboursville West Va. Oct. 12th /63"), RG94 Order Books, [n.p.] Barret and Andrews had familial ties to Ohio and were probably visiting their homes there when Brown ordered them to be apprehended. Records show that Private Andrews was mustered out with the regiment at Wheeling on June 22, 1865, with the charge of desertion still upon his record. Despite the fact that Andrews had deserted his company at Barboursville, on October 16, 1863, the charge was finally removed "by order of the President" with "S[pecial] O[rder] 582 A.G.O., dated "November 21, 1866 vide Ppr 399=66 ams-105-1868." Company muster rolls for September and October 1863 show that Andrews deserted on October 16, 1863, and that his pay was stopped "for 2 haversacks and 1 canteen in the amount of $1.40." Co. muster rolls for March through June reported Andrews again present with the regiment. See Andrew/ Andrews, Gustavus A., Private, Co. F 13th W.Va., Military History Cards (micro.), WVU.

1249   W.R. Brown, Col. Com'dg, by J.S. Cunningham, Adj., Spec. Order No. 58 (manu. dated "Head Quarters 13th Regt V.V.I. Barboursville West Va. Oct. 9th/63"), RG94 Order Books 13th W.Va. Regt. Inftry., [n.p.]

1250   Joseph Brumley, 2d Lt., Co. F 13th V.V.I. to Lieut. J.S. Cunningham, Adjt. (manu. dated "Head Quarters Co F 13th Regt Va Vol Infty Barboursville West Virginia Oct 14, 1863"), RG94 Box 4839 Misc. unbound regimental papers 13th W.Va., NARA.

1251   RG94 Morning Reports Capt. A.F. McCown's Co. F 13th Regiment Va. Vols. October 1863, [n.p.]

1252   Timothy Russell, 1st Lieut., Co. F to Lieut. John S. Cunningham, Adj. (manu. dated "Head Quarters Co. F 13th Va. Vol. Infty, Barboursville, W.Va., Oct. 12, 1863"), RG94 Box 4839 Misc. unbound regimental papers 13th W.Va., NARA.

1253   RG94 Morning Reports of Capt. A.F. McCown's Co. F 13th Regiment Va. Vols. for October 1863, [n.p.]

1254   Timothy Russell, 1st Lieut., Co. F to Lieut. J.S. Cunningham, Adj. (torn manu. dated "Head Quarters Co. F 13th Regt Va. Vol Infty, Barboursville W.Va., Oct. 30, 1863"), RG94 Box 4839 Misc. unbound regimental papers 13th W.Va.

1255   William R. Brown, Col. Comdg., by J.S. Cunningham, Adj., Spec. Order No. 66 (manu. dated "Head Quarters 13th Regt Va Vol Infantry Barboursville West Va Oct 26th /63"), RG94 Order Books, [n.p.]

1256   RG94 Morning Reports of Co. F 13th Regiment Va. Vols. for October 1863, [n.p.]

1257   Col. Daniel Frost commanded the 11th W.Va. Infantry Regiment. He was from Ravenswood, West Virginia. Prior to joining the army he had worked as editor for the *Virginia Chronicle*, a pro-Union newspaper. Frost, like George Tippett of Point Pleasant, was much hated by those in sympathy with the South. On September 4, 1862, when Albert G. Jenkins cavalry went raiding into the northwest, he briefly captured the town of Ravenswood and burned the building occupied by the *Chronicle*.

1258   Young to Paulina, dated "Barboursville Cabell Co., W.Va. Oct. 11, 1863," Young Civil War Papers (Vol. 31 Box 10), WVU, trans. p. 84.

1259   Young to Paulina, dated "Barboursville, Oct. 13, 1863, Young Civil War Papers Vol. 31 Box 10), trans. p. 85.

1260   Although the roll is faded and difficult to read, it appears as if Doliver returned to the company from desertion October 31, 1863. RG94 Regimental Books Misc. Records Descriptive Roll Co. H. 13th Regiment V.V.V.I., [n.p.]

1261   Wm. R. Brown, Col. Comdg, by J.S. Cunningham, Adj., Spec. Order No. 65 (manu. dated "Head Quarters 13th Regt Va Vol Inftry Barboursville Oct 20th /63"), RG94 Order Books, [n.p.]

1262   W.R. Brown, Col. Comdg, by J.S. Cunningham, Adj., Spec. Order No. 58 (manu. dated "Head Quarters 13th Regt V.V.I. Barboursville Oct. 10th /63"), RG94 Order Books, [n.p.]

1263   W.R. Brown, Col. Comdg, by J.S. Cunningham, Adj., Spec. Order No. 67 (manu. dated "Head Quarters 13th Regt V.V.I. Barboursville Oct 26th /63"), RG94 Order Books, [n.p.]

1264   RG94 Morning Reports of Capt. W.I. Mathews' Co. H 13th Regiment Va. Vols. for October 1863, [n.p.]

1265   RG94 Morning Reports of Lieut. William Feazel's Co. I 13th Regiment Va. Vols. for October 1863, [n.p.]

1266   The wounded soldier being maintained by McCullough may have been Lewis G. Pine, Private, Co. E 8th Virginia Cavalry, of Kanawha County. See Dickinson. *Tattered Uniforms and Bright Bayonets,* p. 300.

1267   P.H. McCullough to Lieut. E. Feizel (manu. dated "Mcullough Landing, Oct. 26, [18]63" "Received A.O. Oct. 26, 1863"), RG94 Box 4839 Misc. unbound regimental papers 13th W.Va. McCullough's letter was forwarded to Col. Brown with the following note from Lieut. Feazel appended: "Sir I think it best with your approval for me to arrest Pine immediately and if you will send an ambulance to me I will proceed at once."

1268   Lt. Wm. E. Feazel to Lt. John S. Cunningham, A[j]t. 13th V V I. (manu. dated "Head Quarters Co I 13th V.V.I., Guyandotte W. Va., Oct. 26, 1863"), RG94 Box 4839 Misc. unbound regimental papers 13th W. Va.

1269   West Virginia Clothing Books Co. C 13th Infantry, WVSA, [n.p.]

1270   See Wm. Gray, Co. A 13th West Va. Vol. Infantry, List of West Virginia Prisoners at Wheeling August 14, 1865, manu. Misc. Records of the Adjutant General of W.Va., Uncat. 13th W.Va. Inftry. Regt. Box, WVSA.

1271   Thota, "Letter from West Va. Correspondence Cincinnati Commercial," *Cincinnati Daily Commercial,* Nov. 26, 1863, p. 1.

1272   Ibid.

1273   Ibid

1274   Record of Events Field and Staff, Regimental Return 13th Va. Vol. Infantry Nov. and Dec. 1863 (micro. #594 Roll 196), NARA.

1275   Barboursville was located on well-travelled conduits by water and land. The town was located on the Guyandotte River, a tributary of the Ohio, and the James River turnpike ran right through the town. This was one of the better roads in the area. It had been maintained in good condition before the war with revenue collected at tollgates set up at four mile intervals. This contributed to Barboursville becoming, in the years before the war, a bustling town of manufacture and commerce. All manner of items: furniture, wagons, buggies and harnesses, barges, steamboat bottoms to name a few items, were made and sold there..

1276   Field History 13th Regiment W.Va. Infantry, trans. p. 4.

1277   Col. W.R. Brown to Adj. Gen. of the State of W.Va. Pierpont (manu. dated "Hd Qtrs. 13th Regt., Barboursville, W. Va., Nov. 4, 1863"), RG94 Box 4839 Misc. unbound regimental papers 13th W.Va.

1278   T. Melvin, A.A.G. to Lt. Col. J.B. Frothingham, Com. of Musters, Dept. W.Va. Copy to John S. Cunningham, Adjt., 13th Regt. V.V.I. (manu. dated "Head Quarters Dept of West Va, Clarksburg W. Va. Nov. 5, 1863"), RG94 Box 4839 Misc. unbound regimental papers 13th W.Va.

1279   Field History 13th Regiment W. Va. Infantry for the year 1864, p. 1.

1280   Brig. Gen. B.F. Kelley, Commanding the Department of West Virginia to Gov. of W.Va. A.I. Boreman (dated "Cumberland, Md. Nov. 28, 1863") *Calendar of Arthur I. Boreman Letters in the State Department of Archives and History* Prepared by The Historical Records Survey Division of Women's and Professional Projects Works Progress Administration (Charleston, W.Va.: The Historical Records Survey, 1939), p. 21.

1281   Col. R.B. Hayes to Col. Wm. R. Brown, by Avery (manu. dated "Hd. Qtrs. 1 Brig. 3d Div. D. W.Va., Vamp White, Va., Nov. 23, 1863"), RG94 Box 4839 Misc. unbound regimental papers 13ᵗʰ W.Va., NARA and Col. R.B. Hayes to Col. Wm. R. Brown (manu. dated "Head Qtrs. 1ˢᵗ Brigade 3ʳᵈ Division, Dept 8ᵗʰ Army Corps, Camp White, Va., Nov. 23, 1863"), Registry #393 Part 2 Entry 1184 Dept. of W.Va. Press Copies of Letters Sent and Spec. Orders Vol. 139, NARA.

1282   E.B. Long with Barbara Long, *The Civil War Day By Day. An Almanach 1861-1865* (Garden City, N. Y.: Doubleday & Co), 1971.

1283   "Summary of the Week. Virginia," *Frank Leslie's Illustrated Newspaper*, Nov. 28, 1863, p. 146.

1284   Brig. Genl. Scammon to Hdqtrs. 1ˢᵗ Brig. 3 Div. 8ᵗʰ A.C. (dated "Head Quarters 3ʳᵈ Division D. W.Va. Camp White West Va Nov. 1, 1863"), Letters Received Head Quarters 1ˢᵗ Brig. 3d Division 8ᵗʰ Army Corps, Letters Received Miscellaneous, Papers of R.B. Hayes (micro. Reel 272), HPL, trans. p. 35.

1285   Col. R.B. Hayes to Capt[ain] Botsford, [A.A.G.], 3ʳᵈ Div. Dept. of W.Va. Charleston, W.Va. (manu. dated "Head Qtrs. 1ˢᵗ Brigade 3ʳᵈ Division, Dept 8ᵗʰ Army Corps, Camp White, Va., Nov. 3, 1863"), Registry #393 Part 2 Entry 1184 Dept. of W.Va. Press Copies of Letters Sent and Spec. Orders Vol. 139, NARA.

1286   Brig. Genl. Scammon Hdqtrs. to 1ˢᵗ Brig. 3ʳᵈ Div. 8ᵗʰ A.C. (dated "Hdqtrs. 3 Div. D. W.Va. Camp White West Va. Nov. 5ᵗʰ, 1863") Letters Received Head Quarters 1ˢᵗ Brig. 3d Division 8ᵗʰ Army Corps, Letters Received Miscellaneous, Papers of R.B. Hayes, trans. p. 36.

1287   M.P. Avery, Capt. and A.A.A.G. to Col. Wm. R. Brown, Comdg 13ᵗʰ Va V.I. (manu. dated "Head Qrs 1ˢᵗ Brig 3ʳᵈ Divis Dep W. Va., Camp White, Va, Nov. 5, 1863"), RG94 Box 4839 Misc. unbound regimental papers 13ᵗʰ W. Va.

1288   Col. Wm. R. Brown to Col. R.B. Hayes (manu. dated "Head Quarters R 13 Regt Va Vol Infty Barboursville West Va Nov 8 1[8]6[3]"), RG94 Misc. Dept. of W.Va. Papers.

1289   W.R Brown, Col. Com'dg, by J.S. Cunningham, Adj., Special Order No. 75 and Spec. Order No. 77 (manuscripts dated "Head Quarters 13ᵗʰ Regt Va. Vol. Inf'try Barboursville West Va. Nov. 6ᵗʰ /63"), RG94 Order Books, [n.pp.]

1290   H.J. Samuels, Judge, 8ᵗʰ Circuit W.Va. to Col. Brown (manu. dated "Ironton, Ohio, Oct. 23, 1863"), RG94 Box 4839 Misc. unbound regimental papers 13ᵗʰ W.Va. and Annual Return of Casualties + c. 13ᵗʰ W.Va. Infy. to Dec. 31/63, RG94 Box 4838 Misc. unbound regimental papers 13ᵗʰ W.Va.

1291   Eli Chapman and William Chapman, both of Cabell County, are probably the privates named here. They belonged, respectively, to the 8ᵗʰ and 16ᵗʰ Virginia Cavalry Regiments, C.S.A..

1292   Isaac M. Rucker to Col. W.R. Brown, Com[mading] 13ᵗʰ Va. Vol. Inft. (manu. dated "Hd Qtrs. Co. A 18ᵗʰ Regt W.Va. Vol Militia, Mason Co., W.Va., Nov. 1, 1863"), RG94 Box 4839 Misc. unbound regimental papers 13ᵗʰ W.Va. and Annual Return of Casualties + c. 13ᵗʰ W. Va. Infy. to Dec. 31/63, RG 94 Box 4838 Misc. unbound regimental papers 13ᵗʰ W.Va.

1293   W[illia]m E. Feazel, Lt. Com. Post to [Col. Brown] (manu. dated "Head.Quarters Co. I, V.V.I., Guyandotte, W.Va., Nov. 1, 1863"), RG94 Box 4839 Misc. unbound regimental papers 13ᵗʰ W.Va.

1294   John Slack, Jr., to Col. Wm. R. Brown, Official Copy John S. Cunningham (manu. dated "Charleston Va, October 29, 1863"), RG94 Box 4839 Misc. unbound regimental papers 13ᵗʰ W.Va.

1295   Dixon R. King, Comg U.S. Forces at Ravenswood + Capt Co. I 11ᵗʰ Regt W. Va. Vols Comdr U.S. Forces (manu. dated "Head Quarters Ravenswood, W. Va, Nov 15, [18]63"), RG94 Box 4839 Misc. unbound regimental papers 13ᵗʰ W.Va.

1296   John. S. Cunningham, Adj., 13ᵗʰ W.V.V.I., notation at the bottom of John Slack, Jr., to Col. Wm. R. Brown, Official Copy John S. Cunningham (orig. letter from Slack dated "Charleston Va, October 29, 1863"), RG94 Box 4839 Misc. unbound regimental papers 13ᵗʰ W.Va.

1297 H.J. Samuels, Judge, 8th Circuit Court of West Virginia to John Alford, Sheriff of Cabell County, West Virgina, notation in upper left corner dated "December 4th, 1863," orig. from Slack to Brown dated "October 29, 1863," RG94 Box 4839.

1298 Notice of Recruitment for the 13th West Virginia Volunteer Infantry Regiment, "Attention!!," Point Pleasant *Weekly Register*, Dec. 3, 1863, Vol. 2 No. 37, [p. 4]. It is worth mentioning here, that pro-Union West Virginians came forward with a will whenever a call for more men was published. The *National Telegraph*, a periodical published out of Washington City, reported in June 1864 that West Virginia had furnished sixteen hundred men above her quota and that as of that date there had been no necessity to impose the draft in West Virginia. As for Hawkins, his occupation in the 1860 Kanawha County Census was given as: "Master Painter" with an apprentice. As the image differs somewhat from others published across the North, one wonders what input he had in its creation.

1299 Wm. R. Brown, Col. Comdg, by J.S. Cunningham, Adj., Special Order No. 82 (manu. dated "Head Quarters 13th Regt V.V.I Barboursville West Va. Nov 13th 1863"), RG94 Order Books, [n.p.]

1300 Cassander Spurlock and Samuel A.G. Mguire to Col. Wm. Brown., Attest Wm. A. McGinnis (manu. dated "November 15, 186[3]"), RG94 Box 4839.

1301 William R. Brown, Col. to F. Pierpont, A.G., State of W.Va. (dated "Head Quarters 13th Regiment V.V.I., Barboursville, Nov. 16, 1863"), Samuels-Pierpont Papers, WVSA.

1302 H.J. Samuels, Judge, 8[th] Cir., W.Va. to Col. Brown, comdg 13[th] Va Vols U. S. Service (manu. dated "Guyandotte West Va, Nov. 18, 1863"), RG94 Box 4839 Misc. unbound regimental papers 13[th] W.Va., NARA.

1303 William R. Brown, Col. Comdg to Ja[me]s L. Botsford, Capt + A.A.G. (manu. dated "Hd Qrts 13[th] Regt V. V. Inft, Barboursville, Nov. 24, 1863"), RG94 Box 4839.

1304 Ibid., Brig. Gen. A.N. Duffié, his endorsement on reverse side (no date) of William R. Brown, Col. Comdg to Jas. L. Botsford, Capt + A.A.G.

1305 John Holroyde, by J.K. Heath to Col. Brown (manu. dated "Barboursville, W.Va., Nov. 21, 1863"), RG94 Box 4839 Misc. unbound regimental papers 13th W.Va.

1306 Col. William R. Brown to James L. Botsford, Capt. and A.A.G. (dated "Hd Qtrs. 13th Regt. Barboursville, W.Va., Nov. 24, 1863"), M508 Compiled Service Records of Vols. W.Va. (micro. #M508 Roll 204), NARA.

1307 Brig. Gen'l. Duffié to Hdqtrs. 1st Brig., 3rd Div., 8th A.C., Letters Received Head Quarters 1st Brig. 3d Division 8th Army Corps, Letters Received Miscellaneous, Papers of R.B. Hayes (micro. Reel 272), trans. p. 37.

1308 William R. Brown, Col. Comdg to Th. Melvin, Capt. + A.A.G. (manu. dated "Head Quarters 13th Regt V.V.I, Barboursville, Nov. 27, 1863"), RG94 Box 4839 Misc. unbound regimental papers 13th W.Va.

1309 Ibid., Frothingham notation on reverse.

1310 Lt. James W. Johnson, A.A.I.C., 3d Brigade to Col. W.R. Brown, Commanding 13th Regt. Va Infty. Vols. (manu. dated "Morris Island I C, Nov. 28, 1863"), RG94 Box 4839 Misc. unbound regimental papers 13th W.Va.

1311 W. Va. Clothing Books Co. C 13th Infantry, [n.p.]

1312 John C. Bayler to Col. Brown, 13th Va., manu. dated "Ashland, Ky., Nov. 1863," RG94 Box 4839.

1313 RG94 Morning Reports of Capt. Greenbury Slack's Co. A 13th Regiment Va. Vols. Army of the U.S. for November 1863, [n.p.]

1314 Wm. R. Brown, Col. Comdg, by John S. Cunningham, Adj., Gen. Order No. 28 (manu. dated "Head Quarters 13th Regt Va Vol Infty Barboursville West Virginia Nov 26th 1863"), RG94 Order Book Co. C 13th W.Va. Regiment Inftry., [n.p.]

1315   Wm. R. Brown, Col. Comdg, by J.S. Cunningham, Adj., Spec. Order No. 72 (manu. dated "Head Quarters 13th Regt Va Vol Inftry Barboursville West Va Nov 2nd 1863"), RG94 Order Books, [n.p.]

1316   Milton Stewart, Capt. Comdg Post, by J.S. Cunningham, Adj., Spec. Order No. 88 (manu. dated "Head Quarters 13th Regt V. Vol. I. Barboursville West [Va] Nov. 29th 1863"), RG94 Order Books, [n.p.]

1317   RG94 Morning Reports of Capt. Milton Stewart's Co. B 13th Regiment Va. Vols. for November 1863, [n.p.]

1318   W.R. Brown, Col. Com'dg, by J.S. Cunningham, Adj., Spec. Order No. 78 (manu. dated "Head Quarters 13th Regt V.V.I. Barboursville West Va Nov. 9th /63"), RG94 Order Books, [n.p.]

1319   Wm. R. Brown, Col. Comdg, by J.S. Cunningham, Adj., Spec. Order No. 86 (manu. dated "Head Quarters 13th Regt V.V.I. Barboursville W. Va. Nov. 26th 1863"), RG94 Order Books, [n.p.] and for the foregoing paragraph see RG94 Morning Reports of Capt. Lemuel Harpold's Co. C for November 1863, [n.p.]

1320   RG94 Morning Reports of Capt. Simon Williams' Co. D 13th Regiment Va. Vols. for November 1863, [n.p.]

1321   W.W. Harper, dated "Barboursville, W.V., Nov. 28th, [1863]," "Married," Point Pleasant *Weekly Register*, Thursday, December 3, 1864, [p. 2]. Reverend, Methodist Episcopal Church and Sergeant Major, Co. K 13th West Virginia Infantry, William Wiley Harper performed civilian marriages during his term of service. Another marriage performed by him was that of Henry McCalister and Mrs. America Harber, both of Putnam County, West Virginia, on May 2, 1863. See "Married," Point Pleasant *Weekly Register*, May 7, 1863, [p. 3].

1322   Wm. R. Brown, Col. Comdg, by J.S. Cunningham, Adj., Spec. Order No. 71 (manu. dated "Head Quarters 13th Regt Va Vol Inftry Barboursville West Va Nov 1st /63;" W.R. Brown, Col. Com'dg, by Cunningham, Adj., Spec. Order No. 73 (manu. dated "Head Quarters 13th Regt V.V.I. Barboursville West Va Nov. 3d /63;" and W.R. Brown, Col. Com'dg, by Cunningham, Adj., Spec. Order No. 74. (manu. dated "Head Quarters 13th Regt V.V.I. Barboursville West Va Nov. 5th /63"), RG94 Order Books, [n.p.]

1323   W[illia]m N. Hawkins to Sergeant [Vincent. A.] Hays (manu. dated "Charleston Kanawha, W.Va., Nov. 9, 1863"), RG94 Box 4839 Misc. unbound regimental papers 13th W.Va.

1324   Wm. N. Hawkins to Col. Brown appended to Hawkins to Sergeant Hays, RG94 Box 4839.

1325   W.Va. Clothing Books Co. E 13th Infantry, WVSA, [n.p.]

1326   J.D. Carter, Capt., 13 Va. Vols to Wm. R. Brown, Col., 13 Va Vols. (manu. dated "Headquarters Co. E 13 Va Vols, Barboursville W.Va., 10th Nov. 1863") and Wm. N. Hawkins, Com. Det. 13 Reg, Va. Vol. Inf. to Col. W[illia]m R. Brown, Col. Comdg 13th Regt. Va. Vol. (manu. dated "Mouth [of] Sandy West Va., Nov. 20, 1863"), RG94 Box 4839 Misc. unbound regimental papers 13th W.Va.

1327   Wm. R. Brown, Col. Comdg, by J.S. Cunningham, Adj., Spec. Order No. 84 (manu. dated "Head Quarters 13th Regt Va Vol. Inft Barboursville West Va Nov. 15th 1863"), RG94 Order Books, [n.p.]

1328   Sylvester Keith, Military Pension File, NARA.

1329   RG94 Morning Reports of Capt. A.F. McCowns' Co. F 13th Regiment Va. Vols. for November 1863, [n.p.]

1330   Clark Elkins, 2d Lt., Comdg. Co. G 13th Regt. Va. Vol Inf., Special Order No. ___ (manu. dated "Head Quarters Co G 13th Regt Va Vols, Barboursville [Va.] Nov. 10th, 1863"), RG94 Box 4838 Misc. unbound regimental papers 13th W.Va., Spec. Orders and Gen. Orders. Colonel Brown approved Elkins' order on November 12 in Gen. Order No. 27: "Order No ___ issued from Head Quarters of Co 'G' reducing Corporal Jacob C May to the ranks for using unbecoming and disrespectful language to his superior officer, is hereby approved. Private David Stevenson is appointed corporal vice Jacob C May reduced he will be obeyed and respected accordingly." (William R. Brown, Col. Comdg, by John S. Cunningham, Adj., Gen. Order No. 27 (manu. dated "Head Quarters 13th Regt Va Vol Infty Barboursville West Virginia Nov 12th 1863"), RG94 Order Books, [n.p.]

1331   Brig. Genl. Scammon to 1st Brig., 3rd Div., 8th A.C. (dated "Charleston West Va. Nov. 10th 1863"), Letters Received Head Quarters 1st Brig. 3d Division 8th Army Corps, Letters Received Miscellaneous, Papers of R.B. Hayes (micro. Reel 272), trans. p. 36.

1332   Col. R.B. Hayes to Col. Brown (manu. dated "Hd. Qtrs. 1st Brig. 3rd Div. Dept. W.Va., Charleston, W. Va., Nov. 10, 1863"), RG94 Box 4839 Misc. unbound regimental papers.

1333   W.R Brown, Col. Com'dg, by J.S. Cunningham, Adj., Spec. Order No. 79 and No. 80 (manuscripts dated "Head Quarters 13 Reg't Va. Vol. In'ftry Barboursville, West Va. Nov. 11th /63"), RG94 Order Books 13th W.Va. Regt. Inftry., [n.p.]

1334   Wm. R. Brown, Col. Com'dg, by J.S. Cunningham, Adj., Spec. Order No. 81 (manu. dated "Head Quarters 13th Regt V.V.I Barboursville West Va Nov. 12th 1863"), RG94 Order Books, [n.p.]

1335   J.V. Young, Capt. Comdg Post to Wm. R. Brown, Col. Comdg 13th (manu. dated "Head Quarters Co. G Herricane Bridge, Nov. 14/63"), RG94 Box 4839 Misc. unbound regimental papers 13th W.Va.

1336   William R. Brown, Col. Comdg, by J.S. Cunningham, Adj., Spec. Order No. 83 (manu. dated "Head Quarters 13th Regt V.V.I. Barboursville West Va Nov. 14th 1863"), RG94 Order Books 13th W.Va. Regt., [n.p.]

1337   S. Benedict to Capt. A.I. Bottsford, manu. dated "Coals Mouth, November 15, 1863," Scammon's orders on the reverse, RG94 Misc. Dept. of W.Va. Papers, NARA.

1338   J.V. Young, Comdg. Post at Hurricane Bridge, Putnam Co., W.Va. to his wife, Paulina (dated "Hurricane Bridge Nov. 20, 1863"), Young Civil War Papers, trans. p. 86.

1339   Muster-In Roll of a Detachment of Recruits for Co. G 13th W.Va. Inf. Muster-In Jud'l 1862-1865 in RG94 Box 4838 Misc. unbound regimental papers 13th W.Va.

1340   Shumaker was enrolled at Kygerville by Taylor W. Hampton. See "Muster-In Roll of Detachment of Recruits for the 13th Regiment of Va Infty Volunteers commanded by Colonel William R. Brown called into the service of the United States by the President from the 11th day of December 1862, (date of this muster,) for the term of three years unless sooner discharged," Uncat. 13th W.Va. Inftry. Regt. Box, WVSA.

1341   W.R. Brown, Col. Com'dg, by J.S. Cunningham, Adj., Spec. Order No. 76 (manu. dated "Head Quarters 13th Regt V.V.I. Barboursville West Va. Nov. 7th/63"), RG94 Order Books 13th W.Va. Regt. Inftry., [n.p.]

1342   Wm. R. Brown, Col. Comdg, by J.S. Cunningham, Adj., Spec. Order No. 70 (manu. dated "Head Quarters 13th Regt Va Vol Inftry Barboursville West Va Nov 1st 1863"), RG94 Order Books, [n.p.]

1343   Possibly Corporal Jesse B. Dodson of Cabell County serving in Co. E 8th Virginia Cavalry Regiment, C.S.A.

1344   O.W. Griswold, Lt. Comdg. Detachmt Co H 13th V.V.I. to J.S. Cunningham, Adj., 13th V.V.I. (manu. dated "Head Quarters Co. H., 13th V.V.I., Barboursville, West Va., Nov. 6, 1863"), RG94 Box 4839 Misc. unbound regimental papers 13th W.Va.

1345   Col. R.B. Hayes to Col. Brown (manu. dated "Head Quarters 1st Brig. 3d Div. D. W. Va., Camp White, W. Va., Nov. 9, 1863"), RG94 Box 4839.

1346   Wm. R. Brown, Col. Comdg, by J.S. Cunningham, Adj., Spec. Order No. 87 (manu. dated "Head Quarters 13th Regt V. V. I. Barboursville West Va Nov. 27th 1863"), RG94 Order Books, [n.p.]

1347   Col. Wm. R. Brown, Cmdg. 13 Va. V.I. to Headqtrs. 1st Brig. 3rd Div. 8th A.C. (dated "Headquarters 13th Regt. Va. V.I. Barboursville, W. Va. Nov[ember] 27th 1863"), Letters Received Head Quarters 1st Brig., 3d Division, 8th Army Corps, Letters Received Miscellaneous, Papers of R.B. Hayes (micro. 272), trans. p. 38.

1348   RG94 Morning Reports of Capt. W.I. Mathews' Co. H 13th Regiment Va. Vols. for November 1863, [n.p.]

1349   RG94 Morning Reports of Lieut. William Feazel's Co. I 13ᵗʰ Regiment Va Vols. for November 1863, [n.p.]

1350   S. Bowden, Citizen, Affidavit of Allegiance to the United States, subscribed to and Sworn before Lt. William E. Feazel, in the presence of Ernest M. Ong, in Guyandotte, W.Va., Nov. 7, 1863 (Duplicate RG94 Box 4839 Misc. unbound regimental papers 13ᵗʰ W.Va. Letters A.G.O., NARA). Compare this oath of allegiance with the one Alexander McCausland swore. (See the 1862 volume of *In Letters of Fire,* Martinsville, Indiana: Fideli Publishing, 2023.) It is the second and additional paragraph that Bowden had to swear that contains more unusual terms as to violation of revenue and the stringent language that the oath was sworn to by affiant "without hesitation, mental reservation, or self evasion of mind." The new precise language in the Bowden oath and the serious penalty for breaking the oath (banishment from the State or death) indicates that the gloves had been "taken off." Whether the second paragraph was added by Feazel to address local problems plaguing the regiment or was handed down by military authority higher up the chain of command is at this point mere conjecture.

1351   Wm. E. Feazel, Lt. Com. Post to [Regimental Head Quarters] (manu. dated "Head Quarters Detacht. 13ᵗʰ V.V.I., Guyandotte, W.Va., Nov. 9, 1863"), RG94 Box 4839 Misc. unbound regimental papers 13ᵗʰ W.Va.

1352   One of the sons mentioned here was perhaps Roderick Noel originally from Cabell County, Private in Co. E 8ᵗʰ Virginia Cavalry. See Dickinson, *Tattered Uniforms and Bright Bayonets West Virginia's Confederate Soldiers,* p. 287.

1353   F.D. Beuhring to Col. Brown, manu. dated "Nov. 19, 1863," RG94 Box 4839 Misc. unbound regimental papers 13ᵗʰ W.Va.

1354   Wm. R. Brown, Col. Comdg, by J.S. Cunningham, Adj., Spec. Order No. 85 (manu. dated "Head Quarters 13ᵗʰ Regt V.V.I. Barboursville West Va Nov. 23ᵗʰ 1863"), RG94 Order Books 13ᵗʰ W.Va. Regt. Inftry., [n.p.]

1355   Co. K 13ᵗʰ Va. Vol. Inftry. Co. Muster Roll Nov. and Dec.1863 (micro. #594 Roll 196), NARA.

1356   Henry Stump to Wm. R. Brown, Col. Comdg 13ᵗʰ Regt. Va. Vol. Inftry, Barboursville, West Va. (manu. dated "Charleston, West Va., Nov. 28, [18]63"), RG94 Box 4839 Misc. unbound regimental papers 13ᵗʰ W.Va.

1357   Both quotations are from Jas. L. Botsford, A.A.G. to Lt. Col Comly (manu. dated "Meadow Bluff, Dec. 14, 1863"), Letter Book July 25, 1862 to 16 December 186[?], Headquarters 23ʳᵈ Regt Ohio Vols., James Comly Diaries (micro. Roll 1), OHC, p. 348.

1358   Gen. B.F. Kelly to Gov. A.I. Boreman (telegraph dated "December 19ᵗʰ, 1863"), reprinted from the Fairmont *National*, "A Fight at Lewisburg," Point Pleasant *Weekly Register*, Dec. 24, 1863, [n.p.]

1359   13ᵗʰ Regiment returns for December carry the notation that "Co G commanded by Capt. John V. Young, which has been reported on the Return of this Regiment since its organization, belongs to and is part of the 11ᵗʰ Regt. Va. Vols. Inf. It will hereafter be dropped from this Return or be reported separately." Regimental Return 13ᵗʰ Va. Vol. Inftry. for Nov. and Dec. 1863 (micro. #594 Roll 196), NARA.

1360   Ibid.

1361   See correspondent to the Cincinnati *Commercial* complaining of the myth of the loyal West Virginia population. He took aim at West Virginians but also conveyed the situation in West Virginia as seen from without. "I am disappointed in West Virginia — her loyalty is a very questionable article. It is decidedly of the Kentucky order, the furthest remove from the free-handed and open-hearted patriotism of glorious East Tennessee, or North Georgia and Alabama. The people are sour and sullen, and suspicious, and though there are many noble and honorable exceptions—though in the belt of territory which has been 'Abolitionized,' as the Richmond papers complain, by the Baltimore & Ohio Railroad, in the Panhandle, and along the Ohio Valley, these exceptions are the rule — still it remains the hard stern fact that about the whole of the interior of West Virginia has got to be flogged into obedience, or rather, to be kept there. In the fertile river valleys, where it was profitable to own slaves, the people inevitably imbibed Southern ideas, and as a consequence became rebellious; in the upland, interior regions, those of them who were not too ignorant to know anything of the issue involved whatever (and their name is legion), were pretty equally divided between loyalty and rebellion; while in the mountainous country they either have the goitre and are idiots entirely, or are plunged in such impenetrable darkness that they are the merest tools of the party which is strongest for the hour. This thing of the mountains and their inhabitants, is a matter which has been greatly obscured and befogged by much fustian rhetoric and a sort of romantic declamation about the sacred Goddess of Liberty perching eternally upon the huts of the sturdy mountaineers, keeping alive in their bosoms a virtuous patriotism and an undying hatred of despotism, all that; but it needs only that one should mix himself up, for a brief space, with the realities to be most effectually disenchanted of such school-boy notions, in respect to these particular mountains at least." (Q.P.F. "Letter from West Virginia, Correspondence Cincinnati Commercial," Cincinnati *Daily Commercial*, Dec. 11, 1863, p. 1.)

1362   General Orders No. 1 (dated "Head Quarters Post of Charleston, Charleston West Va, Dec. 9, 1863"), Headquarters Letter Book, 23rd Regt Ohio Vols., Comly Diaries (micro. Roll 1), p. 334.

1363   "Local and Miscellaneous. The Draft in Mason County," Point Pleasant *Weekly Register*, Dec. 3, 1863, [p. 2].

1364   "Volunteering," Point Pleasant *Weekly Register*, Dec. 24, 1863, [n.p.]

1365   Ibid.

1366   W[illiam] W. H[arper] to the Editor of the Point Pleasant *Weekly Register* (dated "Barboursville, Dec. 2, [1863]"), "From the Thirteenth Va.," Point Pleasant *Weekly Register*, Dec. 17, 1863, Vol. II No. 39, [n.p.] "So mote it be" is a phrase used in Masonic ritual meaning "so be it."

1367   Wm. McKinley, Jr., 1st Lt. + A.A.Q.M. to S. Comstock R.Q.M. 13th Regt. Va. Vols. at Barboursville (manu. dated "Hd. Qtrs. 1st Brig. 3rd Div. 8th A.C. Camp White, Va. Jan. 8, 1864"), Registry No. 393 Part 2 Entry 1194 Register of letters received by Quartermasters, Letter 85 NARA, [n.p.]

1368   Wm. E. Feazel, Lt. Comdg. Post to Lt. J.S. Cunningham, Adj. 13th Va Vols. Inft. (manu. dated "Head quarters Co. I 13th Va Vols Inft., Guyandotte West Va., Dec. 1st, 1863. Received A.G.O. Dec. 2, 1863"), RG94 Box 4839 Misc. unbound regimental papers 13th W.Va.

1369   Col. Wm. R. Brown to [A.I.] Boreman (manu. dated "Head Quarters 13th Regt V.V.I., Barboursville, December 7th, 1863"), Uncat. 13th W.Va. Inftry. Regt. Box, WVSA, [n.p.]

1370   William R. Brown, Col. Comdg to James L. Botsford, Capt. + A.A.G. (manu. dated "Head Quarters 13th Regt V. V. I., Barboursville Dec. 8, 1863"), RG94 Box 4839 Misc. unbound regimental papers 13th W.Va.

1371   William R. Brown, Col., 13 Regt. Va. to James L. Botsford, A.A. General (manu. dated "Head Quarters 13th Regt Va. Vol. Infty., Barboursville, W. Va., Dec. 10, 1863"), RG94 Box 4839.

1372   Col. [W].R. Brown to Headqrts. 1st Brig., 3rd Div,. 8th A.C. (dated "Hdqrts 13th Va. Vol. I. Barboursville, West Va. Dec. 12, 1863"), Letters Received Head Quarters, 1st Brig., 3d Division, 8th Army Corps, Letters Received Miscellaneous, Papers of R.B. Hayes, trans. p. 39.

1373   W.R. Brown, Col. Comdg, by J.S. Cunningham Adj., Gen. Order No. 29 (manu. dated "Head Quarters 13th Regt Va Vol Infty Barboursville, West Virginia Dec 13th, 1863"), RG94 Order Books 13th W.Va. Regiment Infantry, [n.p.] The 21st Article of War provided that "[a]ny con-commissioned officer or soldier who shall without leave from his commanding officer, absent himself from his troop, company or detachment, shall upon being convicted thereof, be punished according to the nature of his offense, at the discretion of a court-martial."

1374   Francis Lord, *They Fought For the Union* (Harrisonburg, Pennsylvania: The Stackpole Company, 1960), p. 367.

1375   Wm. F. Dusenberry, Dep[uty] Prov[ost] Marshal to Col. Brown (manu. dated "Guyandotte, West Va., Dec. 4th, 1863"), RG94 Box 4839 Misc. unbound regimental papers 13th W.Va.

1376   J.M. Comly, Lt. Col. Comdg to Capt. Jas L. Botsford, A.A.G. en route, (manu. dated "Charleston, Dec. 10, 1863"), Letter Book Headquarters 23rd Regt Ohio Vols., Comly Diaries (micro. Roll 1), p. 345.

1377   John S. Witcher, Captain, Co. G 3rd Virginia Cavalry to Lieutenant Harry Thompson, Post Adjutant (manu. dated "Charleston West Va., Dec. 10, 1863"), Letter Book Headquarters 23rd Regt Ohio Vols., Comly Diaries, p. 335.

1378   Ibid.

1379   Information given by Capt Witcher. The document from which this information was taken is without date but its relation to Witcher's report of December 10, 1863, seems certain. Ibid., p. 336.

1380   Ibid., information given by Capt Witcher, p.336.

1381   Comly to Botsford, manu. dated "Charleston, Dec. 10, 1863," Headquarters Letter Book 23rd Regt Ohio Vols., Diaries, p. 345.

1382   Information given by Capt Witcher, p. 336.

1383   Ibid.

1384   C[olumbus] Shrewsbury to his wife (manu. dated "Barbersville W Va December 8th, 1863"), A & M 3216 — Lt. Columbus Shrewsbury Co A 4th WVI, Roy Bird Cook Coll., WVU.

1385   Wm. R. Brown, Col. Comdg, by J.S. Cunningham, Adj. (manu. dated "Head Quarters 13th Regt V.V.I. Barboursville W Va Dec 15th 1863") RG94 Order Books, [n.p.]

1386   Brig. Gen. E.P. Scammon General Order No. 33 (manu. dated "Head Quarters 3rd Division, Department of West Virginia, Charleston, W. Va., Dec. 18, 1863"), RG94 Order Book Co. K, [n.p.]

1387   Col. Wm. R. Brown Comdg. 13 Va. Vol. I to Hd. Qres. 1st Brig. 3rd Div., 8th A.C. (dated "Head Quarters 1st Brig. Va. Vol. I. Barboursville West Va. Dec. 22, 1863"), Letters Received Head Quarters 1st Brig., 3d Division, 8th Army Corps, Letters Received Miscellaneous, Papers of R.B. Hayes, trans. p.39.

1388   Wm. R. Brown, Col. Comdg, by J.S. Cunningham, Adj., Spec. Order No. 104 (manu. dated "Head Quarters 13th Regt V.V.I. Barboursville W Va Dec 23rd 1863"), RG94 Order Books, [n.p.]

1389   Henry Stone, Assistant Adjutant General In Charge of Bureau of Deserters to the Commanding Officer 13th Regiment Virginia (manu. dated "War Department, Provost Marshal General's Office, Washington, D.C., Dec. 23, 1863"), RG94 Box 4839 Misc. unbound regimental papers 13th W.Va.

1390   "Christmas Dinner," Point Pleasant *Weekly Register*, Dec. 24, 1863 Vol. II No. 40, p. 4.

1391   Tho. [or Jno.] Heaton, Asst. Sp. Ag + M [?] Dept. to Col. Brown (manu. dated "Office of Supervising Special Agent, State Agency U.S. Treasury Department, Cincinnati, Dec. 26, 1863"), RG94 Box 4839 Misc. unbound regimental papers 13th W.Va.

1392   Joseph C. Wheeler, Capt. and Prov. Marshal 3d Dist., W.Va., (manu. dated "Charleston, Dec. 25, 1863"), RG94 Box 4839.

1393   Mrs. Mary Eggars to General Scammon (manu. dated "Barboursville, Cabell C.H., Va., Dec. 28, 1863"), RG94 Box 4839.

1394   Brig. Gen. E.P. Scammon, by Jas. L. Botsford, Assistant Adjutant General, Special Orders No. 158 (manu. dated "Head Quarters 3rd Division, Dept. of West Va., Charleston, West Va., Dec. 30, 1863"), RG94 Box 4838 Misc. unbound regimental papers 13th W.Va., Spec. Orders and Gen. Orders, NARA.

1395   R.B. Hayes, Col. Comdg to Col. W.R. Brown, 13th Va. Vols. (manu. dated "Hd Qtrs. 1st Brigade, 3d Div., 8th Army Corps, Dec. 24, 1863"), RG94 Box 4839 Misc. unbound regimental papers 13th W.Va.

1396   R.B. Hayes, Col. 23d to Col. W.R. Brown, 13th Va. Vols (manu. dated "Camp White, Dec. 30, 1863"), RG94 Box 4839.

1397   RG94 Misc. Records Descriptive Roll Co. A 13th Regt. W.V.V.I. First List, [pp 10-11 and pp. 12-13] and Second List, [pp. 31-32], NARA.

1398   G[reenbury] Slack, Capt., 13th W.V.I. Com. Co. A to Lt. J.S. Cunningham, Adj. 13th Va. Inf. Vols. (manu. dated "Camp Barboursville, Dec. 4, [18]63"), RG94 Box 4839 Misc. unbound regimental papers 13th W.Va.

1399   James R. Hall, Lt. Col. Comdg, by J.S. Cunningham, Adj. Spec. Order No. 92 (manu. dated "Head Quarters 13th Regt V.V.I. Barboursville W.Va. December 3rd 1863"), RG94 Order Books 13th W.Va. Regt. Inftry., [n.p.]

1400   Wm. R. Brown, Col. Comdg, by J.S. Cunningham, Adj., Spec. Order No. 94 (manu. dated "Head Quarters 13th Regt V. V. I. Barboursville W. Va December 5th 1863"), RG94 Order Books, [n.p.]

1401   Wm. R. Brown, Col. Comdg by J.S. Cunningham, Adj., Spec. Order No. 97 (manu. dated "Head Quarters 13th Regt V.V.I. Barboursville W.Va. Dec 8th 1863"), RG94 Order Books, [n.p.]

1402   See Special Orders issued by Lt. Col. James R. Hall on December 1 directing that "Sergeant Miletus Grinstead and Five Men of Co (A) are hereby ordered to proceed to Kanawha County for the purpose of recruiting for this Regiment — He will report to his Command within the next Twenty (20) Days." James R Hall, Lt. Col. Comdg, by J.S. Cunningham, Adj., Spec. Order No. 85 (manu. dated "Head Quarters 13th Regt V.V.I. Barboursville West Va Dec 1st 1863"), RG94 Order Books, [n.p.]

1403   RG94 Morning Reports of Capt. Greenbury Slack's Co. A 13th Regiment Va. Vols. for December 1863, [n.p.]

1404   Wm. R. Brown, Col. Comdg, by J.S. Cunningham, Adj., Spec. Order No. 101 (manu. dated "Head Quarters 13th Regt V.V.I. Barboursville W Va Dec 16th 1863"), RG94 Order Books, [n.p.]

1405   Wm. R. Brown, Col. Commanding, by J.S. Cunningham, Adj., Spec. Order No. 102 (manu. dated "Head Quarters 13th Regt V.V.I. Barboursville West Va Dec 20th 1863."), RG94 Order Books, [n.p.]

1406   Col. Wm. R. Brown to M.C. Avery, Capt. and A.A.A.G. (manu. dated "Hd. Qtrs. 13th Regt., Barboursville, Dec. 20, 1863"), RG 94 Box 4839 Misc. unbound regimental papers 13th W.Va.

1407   RG94 Morning Reports of Co. A 13th Regiment Va. Vols. for December 1863, [n.p.]

1408   John R. Gaskins was charged with being absent without leave on about Dec. 5, 1863. He remained absent "until brought back to his command at Barboursville, W.Va. about Dec. 15, 1863." He was "tried by Court Martial at Charleston, West Virginia and Found guilty. To forfeit 3 months pay and perform 3 months fatigue duty with his Regt. under guard." LL1439 #24 John R. Gaskins Co. B 13th Va.V.I. Court Martial convened at Charleston, W.Va. [no date], NARA. Gaskins whatever his failings, did not it seems, lack for martial spirit. The young man (about 22 in 1863 and a farmer) seems to have served in the early days of the war with Co. A 116th re-organized Mason County Militia. See "Solgers," "80 Gaskins, John," Private Co. A 116th Regiment Va. Militia Muster Roll (manu. dated "March 10, 1861," Capt. John A. Greer, cmdg., from an old "Ledger" owned by Greer and given in 1889 to Livia Simpson-Poffenbarger, ed., Point Pleasant *State Gazette*, a copy retained in Poffenbarger Papers, Vol. 5, Mason Co. Public Lib., Point Pleasant, p. 57.) Gaskins' name then appears on the rolls of Co. A 4th Virginia Regiment Infantry Vols. (Union); "enlisted June 17, 1861" and "discharged Feb. 23, 1862." Pierpont, Adj. Gen. State of W.Va., *Annual Report of the Adjutant General of the State of West Virginia for the Year Ending December 31, 1865*, Wheeling: John Frew, Public Printer, 1866, p. 109. Gaskins was thereafter enlisted by Capt. Milton M. Stewart for Co. B 13th Virginia Infantry at Point Pleasant, Va., on July 15, 1863, and mustered into service Aug. 3, 1863 at Charleston. He was mustered out May 25, 1865, at Gallipolis, Ohio—probably at the U.S. General Army Hospital there. This last discharge was likely due to disability caused by a gun shot wound to the thigh received in action Sept. 3, 1864, in the battle of Berryville, Virginia.

1409   James R. Hall, Lt. Col. Comdg, by J.S. Cunningham, Adj., Spec. Order No. 91 (manu. dated "Head Quarters 13th Regt V.V.I. Barboursville W. Va December 3rd 1863"), RG94 Order Books 13th W.Va., [n.p.]

1410   RG94 Carded Medical Records Vols. 1846-1865 Entry 534 Box 3849.

1411   RG94 Morning Reports of Capt. Milton Stewart's Co. B 13th Regiment Va Vols. for December 1863, [n.p]

1412   James R. Hall, Lt. Col. Comdg, by J.S. Cunningham, Adj., Spec. Order No. 90 (manu. dated Head Quarters 13th Regt V.V.I. Barboursville W Va December 1st 1863"), RG94 Order Books, [n.p.]

1413   James R. Hall, Lt. Col. Comdg, by J.S. Cunningham, Adj., Spec. Order No. 92 (manu. dated "Head Quarters 13th Regt V.V.I. Barboursville W. Va. December 3rd 1863"), RG94 Order Books, [n.p.]

1414   Wm. R. Brown, Col. Comdg, by J.S. Cunningham, Adj., Spec. Order No. 103 (manu. dated "Head Quarters 13th Regt V.V.I. Barboursville W Va Dec 21st 1863"), RG94 Order Books, [n.p.]

1415   Wm. R. Brown, Col. Comdg, by J.S. Cunningham, Adj., Spec. Order No. 106 (manu. dated "Head Quarters 13th Regt V.V.I. Barboursville W Va Dec 24th 1863"), RG94 Order Books, [n.p.]

1416   Dr. Abraham D. Williams, Assistant Surgeon, 13th W.Va. Vols.

1417   William Jackson "and [O]liver" to Col. W.R. Brown (manu. dated "Hartford City, December 17, 1863"), RG94 Box 4839 Misc. unbound regimental papers 13th W.Va.

1418   RG94 Morning Reports of Capt. Simon Williams' Co. D 13th Regiment Va Vols. for December 1863, [n.p.]

1419   Wm. R. Brown, Col. Comdg, by J.S. Cunningham, Adj., Spec. Order No. 107 (manu. dated "Head Quarters 13th Regt V.V.I. Barboursville W Va Dec 25th 1863"), RG94 Order Books 13th W.Va. Regt. Inftry., [n.p.]

1420   RG94 Morning Reports Co. D 13th Regiment Va Vols. for December 1863, [n.p.]

1421   Ibid.

1422   Brig. Gen. E.P. Scammon to Hdqrs. 1st Brig., 3rd Div., 8th A.C. (dated "Charleston Va. Dec. 2, 1863"), Letters Received Head Quarters 1st Brig., 3rd Division, 8th Army Corps, Letters Received Miscellaneous, Papers of R.B. Hayes (micro. Reel 272), trans. p. 38.

1423   W.Va. Clothing Books 13th Infantry, [n.p.]

1424   Wm. R. Brown, Col. Comdg, by J.S. Cunningham, Adj., Spec. Order No. 93 (manu. dated "Head Quarters 13th Regt V.V.I. Barboursville W.Va December 4th 1863"), RG94 Order Books, [n.p.]

1425   Brown, Col. Comdg, by Cunningham, Adj., Spec. Order No. 97 (manu. dated "Head Quarters 13th Regt V.V.I. Barboursville W.Va. Dec 8th 1863"), RG94 Order Books, [n.p.]

1426   Brown, Col. Comdg, by Cunningham, Adj., Spec. Order No. 99 (manu. dated "Head Quarters 13th Regt V.V.I. Barboursville W.Va. Dec 12th 1863"), RG94 Order Books, [n.p.]

1427   Wm. N. Hawkins, 1st Lt. Com[manding] Det[achment], 13 West Va. Volunteers to Col. Wm. R. Brown (manu. dated "Charleston Kan C[ourt] House, West Va., Dec. 23, 1863"), RG94 Box 4839 Misc. unbound regimental papers 13th W.Va.

1428   Roush, *If Thou Wilt Remember,* pp. 41-42.

1429   A.F. McCown, Capt Comdg. Co. F to J.S. Cunningham, Adj. (manu. dated "Head Quarters Co F 13th V.V.I., Barboursville, W.Va., Dec. 5, 1863"), RG94 Box 4839.

1430   RG94 Morning Reports of Capt. A.F. McCown's Co. F 13th Regiment Va Vols. for December 1863, [n.p.]

1431   Ibid.

1432   Wm. R. Brown, Col. Comdg, by J.S. Cunningham, Adj., Spec. Order No. 105 (manu. dated "Head Quarters 13th Regt V.V.I. Barboursville W Va Dec 24th 1863"), RG94 Order Books 13th W.Va. Regt., [n.p.]

1433   Wm. R. Brown, Col. Comdg, by J. S. Cunningham, Adj., Spec. Order No. 106 (manu. dated "Head Quarters 13th Regt V.V.I. Barboursville W Va Dec 24th 1863"), RG94 Order Books, [n.p.]

1434   RG94 Morning Reports Co. F 13th Regiment Va Vols. for December 1863, [n.p.]

1435   Brown, Col. Comdg, by Cunningham, Adj., Spec Order No. 98 (manu. dated "Head Quarters 13th Regt V.V.I. Barboursville W.Va. Dec 8th 1863"), RG94 Order Books, [n.p.]

1436   J.V. Young, Capt., Co. G., comdg post to William R. Brown, Col. Comdg. 13th Regt. V.V.I. at Barboursville (manu. dated "Hurricane Bridge Putnam Co., W.Va., Dec. 8, [18]63"), RG94 Box 4839 Misc. unbound regimental papers 13th W.Va.

1437   Brig. Gen. Scammon, by Jas. L. Botsford to Capt. Young, Comdg Detachment (manu. copy dated "Head Quarters 3d Division, Department West Virginia, Charleston West Va., Dec. 9, 1863"), RG94 Box 4839.

1438   Comly wrote to Capt. Young as follows: "I have a dispatch from Capt Botsford A.A.G. dated last night at Gauley in which he says: 'If you think it necessary to send a Lieut and 25 men (of Capt Youngs Company) to Coals Mouth do so.' From representations made to me by Mr Benedict, I do think it is necessary and therefore request that you will comply with the Adjutant Generals Order and notify your Regimental Commander thereof immediately." J.W. Comly, Lt. Col. Comdg. to Capt. Young, Comdg at Hurricane Bridge (manu. copy dated "Head Quarters Charleston W.Va., 10th Dec. 1863"), RG94 Box 4839.

1439   J.V. Young, Capt. comdg P[o]st to Wm. R. Brown, Col. Comdg 13 Regt. Va. Vol. Infy. (manu. dated "Hurricane Bridge Putnam Co., W.V., Dec. 11, [18]63"), RG94 Box 4839 Misc. unbound regimental papers 13th W.Va.

1440   James M. Comly, Lt. Col., Comdg. to Brig. Gen. Cullum, Chief of Staff (dated "Charleston, W.Va., Dec. 15, 1863 Received 9.10 p.m."), Spaulding, "Records of Official Correspondence of Union and Confederate Armies in the Kanawha, Valley, W.Va.," quoted in Geiger, *Civil War in Cabell County*, p. 87 and cited in Ch. 7 fn. 14, p. 139.

1441   I.M. Rucker, Capt., Com. Scouts to Col. Brown (manu. dated "Head Qrs Indpt Scouts, Mason Co. W.Va., Dec. 13, [18]63"), RG94 Box 4839.

1442   W.S. Rice, Capt., Comd'g Post to Lt. Col. J.M. Comly (dated "Camp Piatt, Dec. 13, 1863"), Letter Book Headquarters 23rd Regt. Ohio Vols., Comly Diaries (micro. Roll 1), OHC, p. 342.

1443   Col. W.R. Brown, by Wm. [E.] Feazel Capt. Comd'g Post to Telegraph Operator (manu. dated "Hd. Qrs. Guyandotte, Dec. 13, 1863"), Letter Book Hdqters. 23rd Regt. Ohio, p. 342.

1444   W.H. Zimmerman, Capt., Comdg Post at Gallipolis, Ohio to Lieut Col. Jas. M. Comly (dated "Gallipolis, [Ohio], Dec. 13, 1863"), Letter Book Hdqters. 23rd Regt. Ohio, p. 342.

1445   J.M. Comly, Lt. Col., Comdg to Capt. Jas. L. Botsford, A.A.G. (dated "Charleston, Dec. 13, 1863"), Letter Book Hdqters. 23rd Regt. Ohio, p. 345.

1446   J.M. Comly, Lt. Col., Comdg to Capt. Zimmerman (dated "Charleston, Dec. 13, 1863"), Letter Book Hdqters. 23rd Regt. Ohio, p. 346.

1447   W.H. Zimmerman, Capt., Comdg Post [Gallipolis, Ohio] to Lieut. Col. J.M. Comly (dated "Gallipolis, Dec. 14, 1863"), Letter Book, p. 343.

1448   J.M. Comly, Lt. Col. Comdg to Capt. Zimmerman, comdg at Gallipolis (dated "Charleston, Dec. 14, 1863"), Letter Book, p. 346.

1449 C.E. Elkins, Lieutenant, Commanding at Coals Mouth to Lieutenant H. Thompson, Post Adjutant, Charleston (dated "Coals Mouth, West Va., Dec. 15, 1863"), Letter Book, [n.p.]

1450 J.M. Comly, Lt. Col., Comdg to T. Melvin A.A.G., Cumberland (dated "Charleston, Dec. 15, 1863. 4-30 P.M."), Letter Book, pp. 347-48.

1451 Ja[me]s R. Hall, Lt. Col. to J.S. Cunningham Adjt. 13th Va. (manu. dated "Barboursville W. Virginia, Dec. 17, 1863"), RG94 Box 4839 Misc. unbound regimental papers 13th W.Va.

1452 Ibid.

1453 Col. Wm. R. Brown by J.S. Cunningham, Adj. to I.M. Rucker, Captain (manu. dated "Head Quarters 13th Regt V.V.I."), AR 373 Putnam County Militia Captain Isaac M. Rucker's Co., WVSA, courtesy of Terry Lowery.

1454 W.R. Brown, Col. 13th Regt. V.V.I. to Capt. M. Avery, A.A.A.G., 1st Brigade, 3rd Div., Dept. W.Va., "Report of the part taken by Detachment of the 13th Regt. under command of Lt. Col. Hall in trying to intercept the Rebel force which passed by Mud Bridge Dec. 13, 1863" (manu. dated "Hd Qrts. 13th Regt. V.V.I. Barboursville, W.Va. Dec. 18, 1863"), RG94 Box 4839 Misc. unbound regimental papers 13th W.Va.

1455 The enemy's strength was variously reported as was also what direction the enemy moved from Hurricane Bridge. Later, on December 14 Lieut. Elkins conveyed to Comly that the enemy that proceeded to Barboursville was reported at "700 strong." (C.E. Elkins, Lieut., Cmd. at Coals Mouth to Lieut. Col Comly, Comdg. at Charleston, dated "Coals Mouth, West Va., Dec. 14, 1863," Letter Book Headquarters 23rd Regt. Ohio Vols., Comly Diaries, p.338.) Comly reported to department headquarters that the rebels who attacked Young numbered "about 300." (Comly, Lt. Col., Comdg to T. Melvin A.A.G., Cumberland dated "Charleston, Dec. 15, 1863. 4-30 P.M." Letter Book, pp. 347-48.) Brig. Gen. George C. Bowyer, cmdg. Virginia militia, wrote to Comly on December 14 that "the Rebels who made their appearance at Hurricane Bridge, [were] said to be five hundred (500) strong." (Bowyer to Comly, dated "Red House, Dec. 13, 1863," Letter Book, p. 343.) Lieut. Hicks, upon his return to Gallipolis on December 14, reported that he had "received information at Buffalo that the Rebes numbering five hundred (500) strong were on Freaswell's Bottom [Frazier's Bottom?] seven (7) miles back from the river and advancing towards Hurricane Bridge." (W.H. Zimmerman, Capt. Comdg. Post [at Gallipolis, Ohio] to Comly dated "Gallipolis, [Ohio], Dec. 14, 1863," Letter Book, p. 343.) Capt. Rice, cmdg. post at Camp Piatt, reported to Comly on December 15 that "Lieut. Nessle returned from a scout late last night. He was within five (5) miles of Boone C.H. — saw nothing of any rebels but secured positive information that Col. Smith had gone down Guyandotte river with three hundred (300) men. Also a company of Infantry had gone down Coal River, expecting to go to its mouth." (W.S. Rice, Capt., Comdg Post to Lieut Col. J.M. Comly dated "From Camp Piatt to Charleston, Dec. 15, 1863," Headquarters Letter Book, p. 344.) As Comly rightly observed to Capt. Gilmore in correspondence dated December 15, while the rebels were not disturbing them much as yet, they were indeed "poking around down below." (Comly to Capt. G.W. Gilmore, Gauley, dated "Head Quarters Charleston, Dec. 15, 1863," Letter Book, p. 344.) On December 15, Comly issued orders to troops at Camp Piatt to determine whether the rebels who attacked at Hurricane Bridge "went in the direction of Boone or Logan C.H." (Comly to Capt. Rice, Comd'g Camp Piatt dated "Charleston, Dec. 15, 1863," Letter Book, p. 349.) Rice, who executed Comly's order to scout, reported that it was learned that "about one hundred (100) rebels" "nearly all infantry" had been at Boone Court House on Monday night but "part of them left the next morning stating they were going to Coal River." (W.I. Rice to Comly dated "Camp Piatt, Dec. 16, 1863," 23rd Ohio Letter Book, p. 349).

1456 C[lark] E. Elkins, Lieutenant, Commanding at Coals Mouth to Lieutenant Harry Thompson, Post Adjutant, Charleston (dated "Coals Mouth, West Va., Dec. 14, 1863"), Letter Book Headquarters 23rd Regt. Ohio Vols. Comly Diaries (micro. Roll 1), p. 337.

1457 Ibid.

1458 C[lark] E. Elkins, Lieut., Cmdg. at Coals Mouth to Lieut. Col. Comly, Comdg. at Charleston (dated "Coals Mouth, West Va., Dec. 14, 1863"), Letter Book 23rd Ohio Vols., p. 338.

1459   J.M. Comly, comdg at Charleston to C.E. Elkins, Lieut. Cmdg. at Coals Mouth (dated "Head Quarters Charleston, [West Va.], Dec. 14, 1863"), Letter Book, p. 338.

1460   James M. Comly, Lt. Col., Comdg. to Brig. Gen. Cullum, Chief of Staff (dated Charleston, W.Va., Dec. 15, 1863 Received 9.10 p.m.), Spaulding, "Records of Official Correspondence of Union and Confederate Armies in the Kanawha Valley, W.Va.," quoted in Geiger, *Civil War in Cabell County,* p. 87 and cited in Ch. 7 fn. 14, p. 139.

1461   C.E. Elkins, Lieut., Cmdg. at Coals Mouth to Lieut. H[arry] Thompson, Post Adj., Charleston (dated "Coals Mouth, West Va., Dec. 15, 1863,"), Letter Book Hdqtrs. 23ʳᵈ Ohio, [n.p.]

1462   On December 15, Comly ordered Capt. Young to "collect everything remaining at Hurricane Bridge, and remove it to Coals Mouth. Remain at Coals Mouth until further orders." J.M. Comly, Lt. Col., by Harry Thompson, Lieut. and Post Adjutant to Capt. Young, comdg at Coals Mouth, dated "Head Quarters Charleston, [West Va.], Dec. 15, 1863," "Headquarters Letter Book," [n.p.]

1463   James M. Comly, Lt. Col., Comdg. to Brig. Gen. Cullum, Chief of Staff. (dated Charleston, W. Va., Dec. 15, 1863 Received 9.10 p.m.), Spaulding, "Records of Official Correspondence of Union and Confederate Armies in the Kanawha, Valley, W. Va.," quoted in Geiger, *Civil War in Cabell County,* p. 87.

1464   By December 15, Comly could report that the telegraph between Charleston and Gallipolis was again operating and "[t]he rebels seem to have left the river entirely." Boats to Loup Creek were running. The go ahead for all other boats to return to their regular runs would be given on December 16. (J.M. Comly, Lt. Col. Comdg to Capt. J[ames] L. Botsford, A.A.G., by Capt. Gilmore, dated "Charleston, Dec. 15, 1863," Headquarters Letter Book 23ʳᵈ Ohio Vols.," p. 349.

1465   J.M. Comly, Lt. Col Comdg to Capt. Zimmerman, Comdg Post [at Gallipolis], dated "Charleston, Dec. 15, 1863," "Headquarters Letter Book", p. 349.

1466   James M. Comly, Lt. Col., Comdg. to Brig.Gen. Cullum, Chief of Staff (dated Charleston, W.Va., Dec. 15, 1863 Received 9.10 p.m.), Spaulding, "Records of Official Correspondence of Union and Confederate Armies in the Kanawha, Valley, W. Va.," quoted in Geiger, *Civil War in Cabell County,* p. 87.

1467   "False Alarm," Point Pleasant *Weekly Register*, Dec. 24, 1863, Vol. II No. 40, [p. 2].

1468   J.V. Young, comdg. Co. G 11/13 Regt V.V.I. to Col. Brown (manu. dated "Dec. 23, 1863 Rec'd Dec. 24"), RG94, courtesy of Darl Stephenson.

1469   Isaac M. Rucker, Capt., Com. Co. Independent Scouts to Col. Brown, 13 W.Va. Vol. Inft, Barboursville, W.Va. (manu. dated "Head Qrs Independt Scouts, Mason Co W [V]a Dec. 23, [18]63"), RG94 Box 4839 Misc. unbound regimental papers 13ᵗʰ W.Va.

1470   Mrs. A.R. Shaw to Cols. Hall and Brown (manu. dated "Jan. 12/64") and S.B. Thompson to Cols Brown + Hall (manu. dated "Winfield West Va, Jan 12ᵗʰ, 1864"), both documents in RG94 Box 4839.

1471   J.V. Young, Captain, Comdg. Post to W[illia]m R. Brown, Col. Comdg 13ᵗʰ Regt. Va. Vol. Inftry. (manu. dated "Hurricane Bridge Putnam Co. W. Va. Dec. 27, [18]63"), RG94 Box 4839.

1472   J.R. Hall, Lt. Col. 13ᵗʰ Va. to Col. W.R. Brown (manu. dated "Head Quarters 13ᵗʰ V. V. I. Barboursville, W. Va., December 30ᵗʰ,1863"), RG94 Box 4839.

1473   Stiarwalt, Diary 1861-64, p. 11 of manu.

1474   Charles A. Lattin, M.D., Hastings, Fla., "Captured A General," Confederate Veteran, Vol. XXX 1922, p. 344.

1475   Ibid.

1476   Col. W.R. Brown, 13th V.V.I., by J. Cunningham, Adjt. to Hd Qtrs. 1st Brig. 3rd Div. 8th A.C. (dated "Head Quarters 13 Regt. V.V.I. Barboursville W. Va. Dec. 28, 1863"), Letters Received Head Quarters 1st Brig., 3d Division, 8th Army Corps, Letters Received Miscellaneous, Papers of R.B. Hayes (micro. Reel 272), trans. p. 40.

1477   RG94 Misc. Records Descriptive Books 13th Regt. W.Va. Inf. Co. G, List of Non-Commissioned Officers, [n.p.]

1478   The original story of Saunders which appeared in the *Philadelphia Press* was reprinted widely. See, "A Man Who Has Not Slept For Fourteen Years" in the Wheeling *Daily Intelligencer*, Feb. 20, 1865, p. 3; the Morgantown *Weekly Post*, Saturday, March 11, 1865, p. 1; and the West Virginia *Journal* of March 15, 1865, p. 1.

1479   W[illia]m Jackson to Col. Brown (manu. dated "Hartford City West Va., Dec. 29, 1863"), RG94 Box 4839 Misc. unbound regimental papers 13th W.Va.

1480   John S. Cunningham, Adjutant for the 13th Regt. V.V.I. appended the following note from Isaac M. Rucker to Feazel's list: "A correct List of the above mentioned Recruits." Isaac M. Rucker to Col. W. R. Brown, Com[mading] 13th Va. Vol. Inft. (manu. dated "Hd Qtrs. Co. A 13th Regt W. Va. Vol Militia, Mason Co., W. Va., Nov. 1, 1863") RG94 Box 4839. Feazel's letter to "13th Head Quarters" was dated "December 30, 1863," RG 94 Box 4839.

1481   A notation in the margin of Feazel's letter to Brown indicates that the three enlistments were probably Thomas W. Davis, Franklin Elkins and Floyd Turley although only Elkins actually went into Company G. Davis and Turley went into Feazel's command, Co. I. Wm. E. Feazel, Capt. Commdg. Post to Col. W.R. Brown, 13 V.V.I. (manu. dated "Head Quarters Co. I 13th V.V.I., Guyandotte, West Va., Dec. 31, 1863"), RG94 Box 4839.

1482   Ibid.

1483   RG94 Morning Reports of Capt. W.I. Mathews' Co. H 13th Regiment Va Vols. for December 1863, [n.p.]

1484   Levett Perdue to his father, Lewis Perdue and mother (manu. dated "Dec. 27, 1863 Barboursville, W. Va."), Levett Perdue, Military Pension File, NARA.

1485   On December 12, Lieut. Feazel wrote to Lieut. J.S. Cunningham informing him that "Co. I is now full aggregate 101 enlisted men 98 and prospects are that I will recruit several more." See Wm. E. Feazel, Lt. Comdg Post to Lt. John S. Cunningham, A[d]jt. 13th Va. Vols. Inft. (manu. dated "Head Quarters Co. I, 13th Va. Vols. Infty, Guyandotte, West Va., Dec. 12, 1863"), RG94 Box 4839 Misc. unbound regimental papers 13th W.Va.

1486   RG94 Morning Reports of Lieut. William Feazel's Co. I 13th Regiment Va. Vols. for December 1863, [n.p.]

1487   Laban T. M[oor]e to Col. Brown (manu. dated "Cattlettsburg, Ky., Dec. 9, 1863"), RG94 Box 4839. Moore (b. Jan. 13, 1829 in Wayne Co., Va.), a lawyer by profession, had served in the U.S. House of Representatives from Kentucky's 9th District from March 4, 1859, to March 31, 1861. As he indicated to Brown, he commanded the 14th Kentucky Infantry Vols. (Union) as its colonel beginning Nov. 19, 1861. He resigned his commission Jan. 1, 1862, to return to the practice of law at Catlettsburg, Ky., where he resided.

1488   Wm. E. Feazel, Lt. Comdg Post to Lt. J.S. Cunningham, Adjt. 13th Va. Vols. Inft. (manu. dated "Head Quarters Co. I, 13th Va. Vols. Infty, Guyandotte, West Va., Dec. 12, 1863" RG94 Box 4839.

1489   I.H. Buffington to Col. Brown, manu. dated "Guyandotte, Dec. 16, 1863," RG94 Box 4839. One wonders if the barns referred to in this letter belonged to historic Buffington House, built 1816 at Guyandotte.

1490   RG94 Morning Reports of Capt. Henry Stump's Co. K 13th Regiment Va Vols. and "Remarks" for the month of December 1863, [n.p.]

1491   Col. Wm. R. Brown to F[rancis] Pierpont, Adj. Gen. State of Va. (dated "Head Quarters 13th Regt. Va. Vol. Infy., Barboursville, W.Va., Dec. 17, 1863"), Samuels-Pierpont Papers, WVSA.

1492   Field History 13th Regiment W.Va. Infantry, manu. addressed to F. Pierpont, Adj. Genl. W.Va. and dated "Head Quarters 13th Regt. W. Va. Vol. Inft., Barboursville, W.Va., Jan'y 13th, 1864," Uncat. 13th W.Va. Infty. Regt. Box, WVSA, [n.p.]

1493   Col. William R. Brown, Commanding the Regiment, "Remarks," Annual Return of Casualties + c. 13th W. Va. Infy. to Dec. 31/63" (manu. "Return completed Jan. 13, 1864"), RG94 Box 4838 Misc. unbound regimental papers 13th W.Va., NARA.

1494   Ibid.

1495   Annual Return of Casualties + c. 13th W.Va. Infy. Oct. 10/62 to Dec. 31/63 (manu. "Return completed Jan. 13, 1864"), RG94 Box 4838.

1496   Ibid.

1497   Annual Return of Casualties + c. 13th W.Va. Infy. Oct. 10/62 to Dec. 31/63 (manu. "Return completed Jan. 13, 1864"), RG 94 Box 4838.

1498   For the foregoing paraphrase and quotations above, see Virgil A. Lewis, *First Biennial Report of the Dept. of Archives and History of the State of West Virginia* (Charleston, W.Va.: Tribune Printing Co., 1906), pp.62-71.

1499   "Point Pleasant, West Va.," *Illustrated Industrial Edition. The State Gazette Supplement*, Thursday, Feb. 2, 1905 p. 9. Smith's appointment was in place at the latest by April of 1862.

1500   "Virginia House," Point Pleasant *Weekly Register*, Aug. 14, 1862, p. 2.

1501   "Looking Backward Eighty Years. Reminiscences As Gleaned from Mr. Hiram H. Swallow and others Who Remembers the Best of this Locality Article IV," *The State Gazette*, Point Pleasant, W. Va., August 19, 1909, in *History of Mason County, W.Va.*, Accession #203668 Call #RL 975.433H, Huntington Public Library, [p. 1].

1502   Francis P. Pierpont, Adjutant General State of W.Va., "Adjutant General's Report for 1863," dated State of West Va., Adjt. Generals Office, Wheeling, January 18, 1864.

1503   Moorfields Clinic was began in 1805 at Charterhouse Square as the London Dispensary for curing diseases of the eye and ear. The clinic came into being as a result of the epidemic trachoma, a type of tropical conjunctivitis. It was highly contagious and capable of causing blindness. It had been brought back by British troops returning from the Napoleonic Wars in Egypt. In 1822, the Charterhouse Dispensary was moved to a larger site on the corner of Lower Moorfields on Blomfield Street and renamed the London Ophthalmic Infirmary.

Index